SWITCH ENGINEERING HANDBOOK

SWITCH ENGINEERING HANDBOOK

John R. Mason

McGRAW-HILL, INC.

New York St. Louis San Francisco Auckland Bogotá
Caracas Lisbon London Madrid Mexico Milan
Montreal New Delhi Paris San Juan São Paulo
Singapore Sydney Tokyo Toronto

Library of Congress Cataloging-in-Publication Data

Mason, John R. (John Robert), date.
Switch engineering handbook / John R. Mason.
p. cm.
Includes bibliographical references and index.
ISBN 0-07-040769-X
1. Electric switchgear—Handbooks, manuals, etc. I. Title.
TK2831.M37 1993
621.31'7—dc20 92-23508
CIP

1 2 3 4 5 6 7 8 9 0 DOC/DOC 9 8 7 6 5 4 3 2

ISBN 0-07-040769-X

The sponsoring editor for this book was Robert Hauserman, the editing supervisor was Peggy Lamb, the production supervisor was Suzanne W. Babeuf, and the copy editorr was Susan Sexton. It was set in Times Roman by McGraw-Hill's Professional Book Group composition unit.

Printed and bound by R. R. Donnelley & Sons Company.

To
Mary
John & Mahtab
Mike & Donna & Kelley & Michelle
Dave & Susan & McCoy
Melissa
Daniel

About the Author

John R. Mason is marketing director for the XCEL Corporation, makers of switches, panels, keypads, keyboards, and displays for the aerospace, industrial, and medical marketplace. His broad experience in the field of switch and related electromechanical technologies spans 26 years. Mr. Mason is considered one of the foremost authorities on switch technology and is the author of numerous publications on the subject. He resides in Valencia, California.

CONTENTS

PREFACE

It is literally impossible to get through our daily routines without at least once, and probably dozens of times, coming into contact with a switch of some type or style. Our use of switches is so prevalent and so mundane that we hardly even think of their importance. Nothing electrical operates unless it is turned on, sometime, somehow, by a switch. The kitchen microwave gets started in the morning by the simple push of a finger...and this same push can start the launch sequence of a multibillion dollar space vehicle. Engineers tasked with designing control panels constantly evaluate switch devices to select the one most suited to the task, the one that is cost effective, the one that will withstand the environment, the one that will fit the design budget. Often, this evaluation is done with very little opportunity to really survey the switch marketplace and understand what is available, and just as importantly, what *isn't* available.

The engineer, always under time and budget constraints, can ill-afford the time it takes (and it can be considerable) to meticulously weigh the hundreds of possible switch configurations which meet the operational requirements of the system. Collecting catalogs is one way to try to keep up with the changing switch marketplace, reading trade journals and attending trade shows is another. But all this painstaking investigation means very little if the engineer doesn't have at least a fundamental understanding of switch technology which would provide a headstart on switch choice.

That is the purpose of this book: to give the engineer some insight into the functioning, construction, and, yes, even the physics of what makes switches work and why. Perhaps this book will give the engineer ammunition in the form of questions to ask the potential switch supplier, questions which can not only help the engineer specify precisely what is needed for the application, but also to verify that the manufacturer selected can, indeed, supply what the engineer needs and specifies, at a reasonable cost and on schedule. It is hoped this book will help the engineer develop an intuition about what will work, and what won't work, when confronted with having to choose from a headspinning variety of switch types, each with its own assets and liabilities. If this book gives the engineer even the slightest edge in the daily battle of time, schedule, and budget, then it has served its purpose.

When a particular switch technology is being considered, always consult with the switch manufacturer(s) about the technical details contained in this handbook; the switch business moves quickly, and what is valid today may be obsolete tomorrow.

John R. Mason

ACKNOWLEDGMENTS

In researching this book over the last year and a half, the author has had an opportunity to talk to people all over the world who are involved with switches. This group of "switch" people all shared one thing in common: an enduring interest in switches and switch technology. Whether it was the small, three-employee switch manufacturer in Germany, or the large, multiemployee circuit breaker manufacturer in the United States, I continually found an enthusiasm for the products they design and build, and an inquisitive nature about switch technology. It has been a pleasure meeting with and talking to these interesting people.

Unfortunately, it is impossible to single out each and every person who contributed to this handbook in at least some way. However, suffice it to say that this handbook would simply *not* exist without the assistance provided by the following people and companies:

Abraham Gohari and Rob Turner at Transparent Devices; Dick Melius, Gary McKittrick, Jim Lisi, and Jim Carpenter at Eaton Corporation; David Olson at Mechanical Products; Holly Pasciak and Dan Diedrick at W. H. Brady Co.; Pamela Hamilton and Dennis Kuzara at Design Technology; Julie Beach at Micro Switch; Bob Wersen of Panel Design Components; J. W. Jecklin of Chur; John Carlson, Mike Higgenbotham, Shelley Koss, and Debbie Maxie at Elographics; Herr Ringwald of Marquardt GmbH; Chris Carroll with Electroswitch; Mike Furczak of C&K Components; Floyd Schneider of Banner Engineering; Larry Kocon at Berquist; Sheryl Fischer at Memtron; Judy Borowski and Karen Sullivan of 3M; Erick Schumacher and Leon Iverson with IEE; Sheila Smith, John Crego, and Stan Ferrell of Glolite Sales; Kurt Kuhn of Kuhn GmbH; Carmine Oliva and Bill Miller at XCEL-Digitran; Keith Huebner of Cherry Electrical Products; Scott Birch, Dennise Seidel, and Kaaren Baekgaard of Conductive Rubber Technology; Rollin Ryan of Advanced Connector Technology; Alan Burk and Karen Keith of GM Nameplate (INTAQ); George O'Hanlon of Shin-Etsu; Tom Menzenberger of Grayhill; Cindy Shurtleff of Korry Electronics; Mike Hartman and Bob Marion of Luminescent Systems; Jay Dokter of Square D Company; Colman Daniel of C.Itoh; Mark Austin of Carroll Touch; Bruce Baunach of Lumitex; Carol Tinen and Theo Notaras of Weber US; Julia Lundquist of Aerospace Optics; Ron Sparks and Georgette Smith at Otto Controls; Homer Baustug of General Silicones, Bill Lang and Pat McMullen of Interface Products, and Stuart Siegel of SMTEK.

Special thanks to Bob Garwood of Banner Engineering who, having written his own book, provided encouragement when it was needed most; to the

McGlynn brothers, Ken and Bill, for continual motivation, and to John Deegan of Lockheed Corporation for his prodigious proofreading skills.

Special recognition also goes to James Satterfield of SCI Systems whose timely suggestion to add several switch technologies, not originally included, was right on the money.

To my original switch mentor, Don Tubbs... thanks for the start!

And, finally, to my wife, Mary, a heartfelt thanks for her support, always 110%.

John R. Mason

CHAPTER 1
INTRODUCTION TO SWITCH TECHNOLOGY*

Electricity is a form of energy. Energy is defined as the ability to do work. Energy can be stored before it is used to accomplish useful work. When energy is stored, it is called "potential energy." For example, a simple flashlight battery is a source of "potential electrical energy." There are other forms of energy, but for the purposes of this handbook, we concentrate on electrical energy.

The stored electrical energy can be released and put to work, but it must be controlled to accomplish the desired work.

Electrical energy is particularly useful because:

- It is readily converted into other forms of energy and other forms of energy are readily converted into electricity.
- It is easily transported from one place to another.
- It can be controlled very precisely.
- It is usually readily available.

Examples of electrical energy being converted into other forms of energy are electricity into sound in a stereo speaker, electricity into rotary motion in a motor, electricity into heat in an electric heater, and electricity into light in an incandescent lamp. Examples of these same forms of energy being converted into electricity are sound into electricity in a microphone, rotary motion into electricity in a generator, heat into electricity in a thermocouple, and light into electricity in a solar cell.

Electrical energy is easily transported by means of conductors such as wires or bus bars, and is readily controlled by relays, potentiometers, and switches. Electrical energy is converted, transported, and controlled in an electric circuit. An electric circuit can be simple or complex. An ordinary flashlight is an example of a simple electric circuit consisting of a battery, which provides the electrical energy; an incandescent lamp, which converts the electrical energy into light; connecting wires, which transport the energy between the battery and the lamp; and a switch, which controls the electrical energy.

This simple electric circuit, like its complex counterpart, consists of three principal parts:

- A *source* of electrical energy

*Numbers in parentheses indicate items in the References at the end of this chapter.

- A *load* (converting device)
- A *complete path for current*

If any one requirement is not fulfilled, current will not travel in the circuit. A switch is used by an operator to open and close the path for current. As such, a switch is a basic element used for control of current in a circuit.

The *source* in the circuit is that device which provides the necessary energy to cause an electrical action to take place.

The *load* in an electric circuit is that device which converts electrical energy into some other form of energy. Regardless of the purpose of the circuit, a load is necessary to produce the desired output (energy conversion or signal development).

Energy is transferred between the source and load by means of an electrical current. This current travels from one terminal of the source, through the load, and back to the other terminal of the source. The source provides the energy which causes the current to travel this path. Unless this path is present, current cannot travel through the load; electrical energy will not be converted, and no useful work will be accomplished. Thus the last requirement for an electric circuit is a complete path for current.

A switch performs its function by opening or closing the path for current in a circuit. When the path for current is open, the load is disconnected from the source, and there can be no current in the circuit. When the switch closes the circuit, the requirements are met and current travels through the load. This action by the switch is referred to as "making and breaking" the circuit. For our purposes, then, the basic definition for a switch is:

"A switch is a device for making or breaking an electric circuit."

The definition suggests the ultimate in simplicity—that a switch need be no more than the bare ends of two wires that can be touched to *make* a circuit or separated to *break* a circuit.

If we touch or separate the two bare wires of our "switch" while current is flowing in the wires, an arc appears between the two wires. Arcing is a natural phenomenon attendant to switching, the arcing being more intense on break (or separation) because of the induced current created by the collapsing magnetic field. Because arcing is damaging to switch contacts, much of switch design is devoted to taming the arc. As will be seen later in this chapter, the physics of arc extinction is a complex science, and many switch engineers have devoted entire careers to the understanding, and conquering, of this phenomenon.

Obviously, even though the two bare wires meet our fundamental definition for a switch, a device is needed to permit opening and closing the circuit in a more sensible way. A great improvement over our crude bare wire "switch" is the familiar knife switch with its hinged copper blade, break jaw, and insulated handle and base. All switches today can trace their lineage back to this crude, but effective, switch which, for all its simplicity, did exactly what it was designed to do: make a circuit and break a circuit in an efficient, straightforward manner.

All switches have a common denominator (Fig. 1.1) in basic components:

- The *actuator,* which initiates switch operation
- The *contacts* made of low-resistance metal that make or break the electric circuit
- The *switch mechanism,* linked to the actuator, which opens and closes the contacts

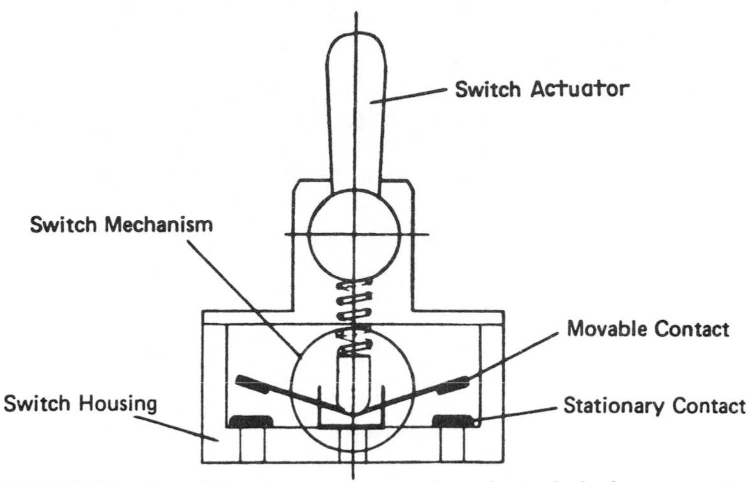

FIGURE 1.1 All switches have a common denominator in basic components. [*Adapted from (4).*]

If the original knife switch is compared with one of today's high-precision avionic-grade pushbutton switches, the same three basic components of a switch are in evidence. However, our advanced knowledge of such areas as vibration analysis, metallurgy, and polymer chemistry has pushed modern switches to high levels of performance and reliability.

The fact that a switch is nothing more than a device for making or breaking a circuit might suggest that the technology is equally simple. Nothing could be further from the truth! The discussions of switching technology in this and the following chapters emphasize to the reader that switch design is a complex interaction of a wide variety of disciplines, from chemistry to human factors, requiring careful attention to detail by switch manufacturers, regardless of whether the switch is destined to control an appliance in the innocuous ambience of a contemporary kitchen or to initiate the imaging system operation of an earth-launched probe in the hostile environment of deep space.

VOCABULARY OF SWITCHES

Three major terms designate a switch's function—*pole, throw,* and *break.* A single-pole, single-throw, single-break switch will be used to define these terms. This switch is also abbreviated SPST, for single-pole (SP), single-throw (ST).

Pole

The term pole refers to the number of conductors that can be controlled by the switch. In our example, the single-pole switch is capable of interrupting the current in a *single* conductor of the circuit (Fig. 1.2*a*). A double-pole switch, by comparison, is capable of simultaneously interrupting the current in two separate

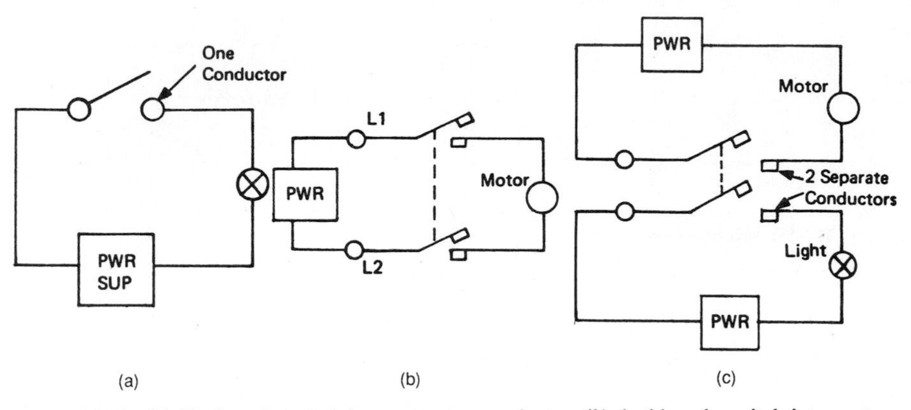

FIGURE 1.2 (*a*) Single-pole switch interrupts one conductor, (*b*) double-pole switch interrupts both sides of line, and (*c*) double-pole switch simultaneously interrupts two conductors in separate circuits. [*Adapted from (4).*]

conductors (Fig. 1.2*b* and *c*). It is standard switch convention to abbreviate single-pole SP, double-pole DP, triple-pole 3P, and four-pole 4P.

Throw

The term throw indicates the number of circuits the switch can control. In our example, the moving contact member of the single-throw switch completes only one circuit (Fig. 1.3*a*). However, a double-throw switch permits its moving contact element to alternately complete two extreme positions (Fig. 1.3*b*). Again using standard switch convention, single-throw is abbreviated ST and double-throw DT.

Putting together our abbreviated switch terminology for poles and throws produces an acronym of four characters; i.e., to describe a double-pole, double-throw switch, the abbreviation is DPDT; for a four-pole, double-throw switch, it is 4PDT. These are commonly used abbreviations in the switch industry.

Break

The term break is self-explanatory. It refers to the breaking or opening of a circuit. For example, *single*-break means that the contacts are separated at only *one*

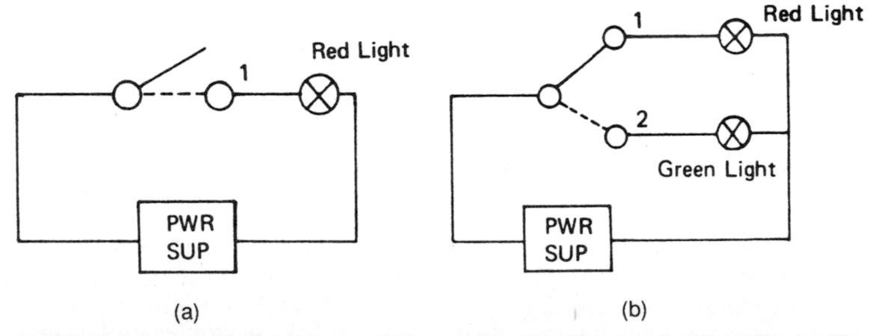

FIGURE 1.3 (*a*) Single-throw switch completes only one circuit while double-throw switch (*b*) alternately completes two circuits. [*Adapted from (4).*]

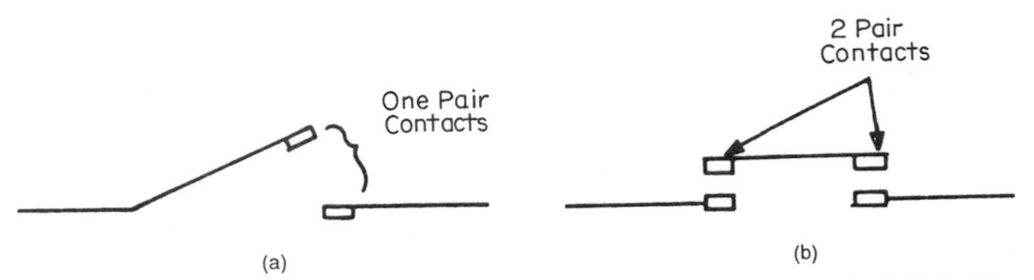

FIGURE 1.4 (*a*) Single-break means contact separation at only one place, and (*b*) double-break has two pairs of contacts which open circuit at two places. [*Adapted from (4).*]

place (Fig. 1.4*a*), while a *double*-break switch has two pairs of contacts that open the circuit at *two* places (Fig. 1.4*b*). Incidentally, the reason for having a double-break switch configuration is that the double-break switch provides greater volume of contact material, resulting in greater heat dissipation and, thereby, longer switch life. The double-break switch also has twice the voltage breaking capacity, a desirable feature for dc circuit applications, as discussed in more detail later in the chapter.

Slow Break vs. Quick Break. Should switch contacts be broken slowly or quickly? As will be seen later in this chapter, it all depends on whether alternating or direct current is being switched. At first, this may seem odd, since electricity is electricity, and one electron is no different from another. However, since alternating current varies in magnitude and direction while direct current maintains a steady unidirectional flow, an interesting phenomenon exhibits itself when ac and dc circuits are broken.

Consider an ac and a dc circuit, each carrying the same amperage. When we slowly break the ac circuit, the arc is extinguished relatively soon—a desirable condition, as will be seen later in the chapter. But when we slowly break the dc circuit, the arc can be drawn much longer before it is extinguished—an undesirable condition.

The reason for the smaller arc in the ac circuit is found by studying the ac sine wave (Fig. 1.5). It will be noted that no matter where the alternating current is broken—even at maximum current—it takes only a fraction of a second for the current to go through zero. This brings up the question: why not break the circuit slowly and give the alternating current time to go through zero and extinguish the arc? As it happens, this is exactly what switch engineers do—purposely design ac switches with *slow make–slow break* mechanisms.

Basic Break Mechanisms. Two basic switch break mechanisms are:

- Slow make–slow break
- Quick make–quick break

All other mechanisms are simply variations of these fundamental devices.

Slow Make–Slow Break. This type of mechanism is usually associated with ac applications for the reasons previously stated; i.e., its relative slowness of operation provides the slight delay in time which permits the ac wave to go through its zero energy level. An analogy of the slow make–slow break mechanism is a teeter-totter or seesaw. Figure 1.6 depicts a toggle switch (Chap. 5) with a slow make–slow break mechanism. The device has a center support member, a sta-

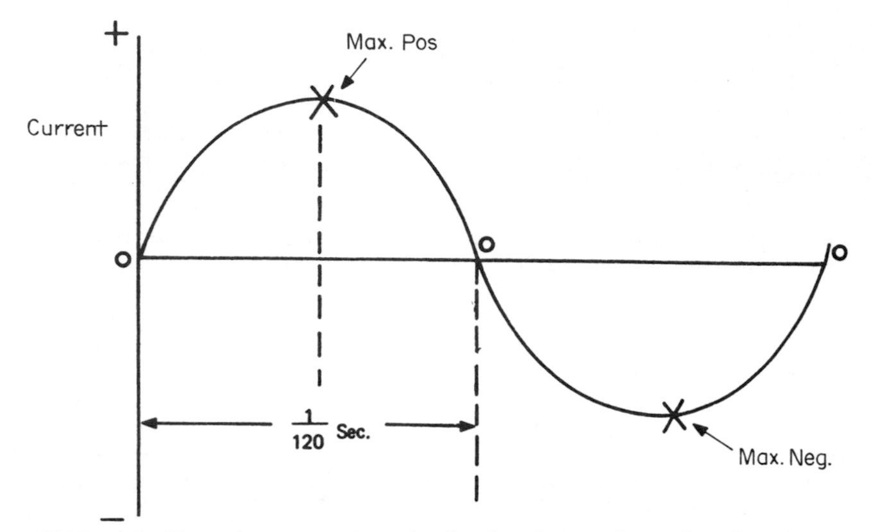

FIGURE 1.5 Alternating current takes only a fraction of a second to go through zero, even though switched off at maximum (*x*) points on waveform. [*Adapted from (4).*]

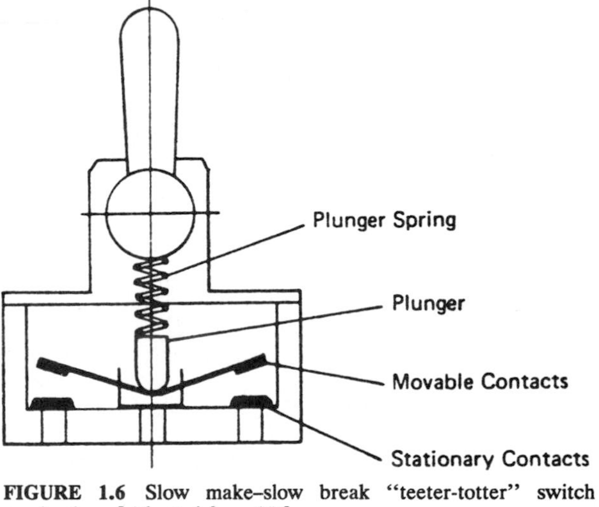

FIGURE 1.6 Slow make–slow break "teeter-totter" switch mechanism. [*Adapted from (4).*]

tionary contact on one end, and a stationary contact on the opposite end. A movable contactor pivots on the center support. Manipulating the toggle lever in either extreme throw position will either make or break the circuit. Since this mechanism is spring-loaded, a very positive force is necessary to close the contacts. But despite this force, the contacts can still be teased.

Teasing is the very slow manipulation of the teeter-totter mechanism by the operator. A characteristic of the teeter-totter mechanism is that it can be operated as slowly or as quickly as the user chooses.

Another look at Fig. 1.6 reveals that the movable contacts are positioned to meet face-to-face with the stationary contacts. This contact construction is referred to as a *butt contact*. Butt contacts are employed in slow make–slow break ac applications and perform well under normal applications. But what if the contacts are subject to oxidation or atmospheric contaminants? The resultant film on the contacts, being nonconductive, could seriously affect electrical continuity. While high-energy circuits would probably not be affected because of their breakthrough ability (discussed later in this chapter), the deleterious effects on low-energy circuits could be significant. In such cases, the self-wiping contacts associated with the quick make–quick break mechanism must be considered.

Quick Make–Quick Break. The quick make–quick break mechanism differs dramatically from the slow make–slow break mechanism. The device depicted in Fig. 1.7 shows a toggle switch with a quick make–quick break mechanism which employs a compression-type spring to provide the motive power to produce a nonteasable, snap-action response. The spring has one of two positions in its free state. Movement of the switch actuator compresses the spring, causing it to move from its end position to the trip position. It is at this point that the switch actuator, like the trigger on a gun, causes the contactor mechanism to snap irrevocably from one position to the next. During this change of position, the movable contact physically *wipes* across the stationary contact. The resultant abrasive action cleans the contact surface, thereby minimizing contact resistance.

The snap-acting fast speed of the mechanism lends itself to dc applications where, as previously noted, the more quickly the contacts are separated the sooner the arc is extinguished.

Normally Open and Normally Closed

Two other terms, *normally open* and *normally closed*, refer to the physical position of the contacts in reference to each other. In a *normally open* switch (Fig.

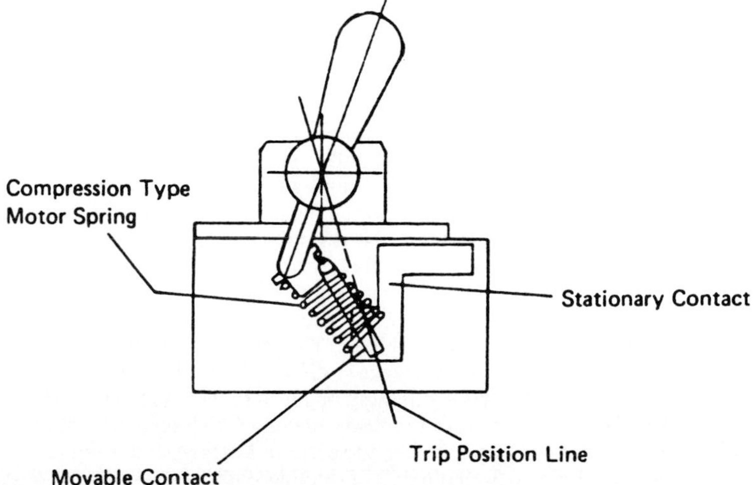

FIGURE 1.7 Quick make–quick break snap-action switch mechanism. [*Adapted from (4).*]

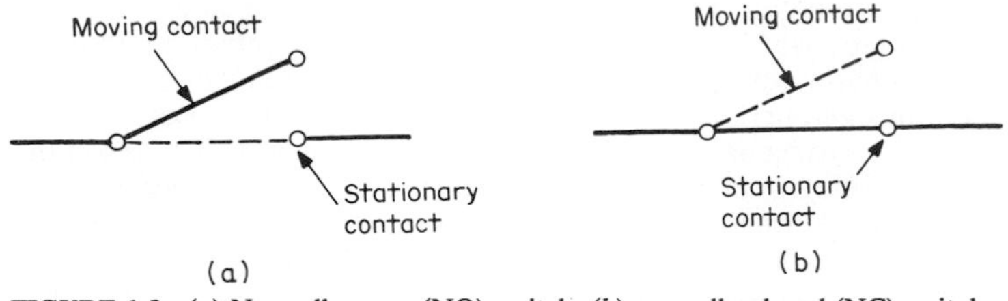

FIGURE 1.8 (*a*) Normally open (NO) switch, (*b*) normally closed (NC) switch. [*Adapted from (4).*]

1.8*a*), the contacts are separated or open before actuation; thus the circuit is open and no current can flow through the switch. A typical example of this is a pushbutton switch where depressing the pushbutton causes the contact element to move from its normally open condition (no current flow) to the other of its extreme positions and close the circuit (current flow).

In a *normally closed* switch (Fig. 1.8*b*), the contacts are closed before actuation, thereby making electrical contact and permitting current flow in the circuit. Operation of the switch causes the contact element to move to its opposite extreme position and open the circuit.

It is important to note that the term "normally" *always refers to the switch before it is actuated.* This is true even though in some applications the switch is actuated a greater percentage of the time.

Figure 1.9 depicts industry-standard contact arrangements. The terms "Form A, Form B," etc., are standard *relay* contact forms sometimes used to describe switch contact arrangements. Not all switch manufacturers use these form codes, but they are included here for reference. The dashed line shown in several of the contact arrangements indicates that the switch actuator moves both poles at the same time.

With an understanding of switch vocabulary and the phenomenon of switch arcing, we now investigate the physics of switch technology [adapted from (1)–(5)].

PHYSICS OF CIRCUIT INTERRUPTION

The voltage and current in a complete electric circuit obey Kirchhoff's voltage and current laws. These laws, simply stated, are: the rises and drops in voltage around any closed circuit (a circuit loop) must sum to zero; and the total current flow into any one junction (connection point) must also sum to zero. If we wish to interrupt the current in a circuit, we must do so in accordance with these laws.

Although it sounds simple—interrupt the circuit, break the conduction path, or open the switch—it is not. Forcing a conducting circuit to a steady-state condition of zero current interruption is obscured by the seeming triviality of the switching action—such as simply turning off a flashlight. But consider what actually happens when a flashlight is turned off. A steady-state direct current is flowing from the batteries to the light bulb as the switch contacts begin to move. At the last microscopic points of electrical contact, the current density becomes high enough that portions of the metallic surfaces actually melt owing to resistive heating; and a liquid metal vapor plasma state continues the electrical conducting path as the contacts physically part. As the contacts pull farther apart to distances of several microns (1

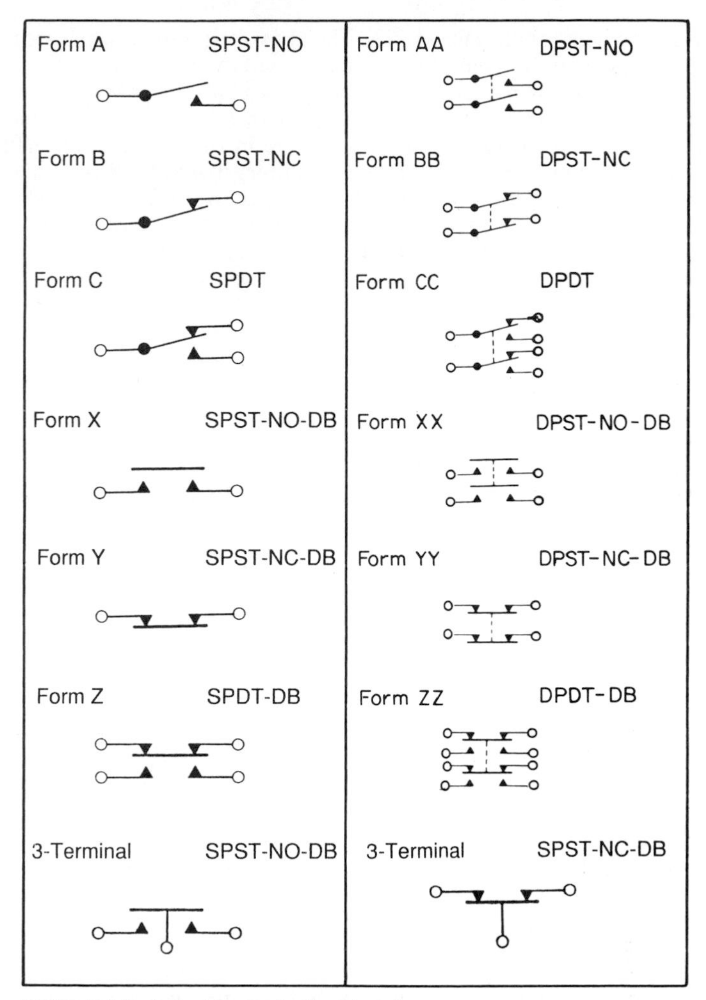

FIGURE 1.9 Industry-standard contact forms.

micron = 10^{-6} meter), electrons from the contact into which the current is flowing, the cathode contact, are emitted into the intercontact space region because of thermal emission (they boil off) and field emission (they are ripped from the cathode metal by electrostatic attraction forces).

Some of the electrons emitted from the cathode collide with air molecules within the contact gap and ionize the molecules. This frees still more electrons, which in turn ionize still more air molecules. This self-perpetuating action is an electrical breakdown phenomenon commonly referred to as an arc. It is the arc which enables the switch to open the circuit. The arc forms just as the contacts part, and continues to conduct the circuit current as the contacts move farther and farther apart.

The voltage drop across the arc—which is proportional to the arc length, and inversely proportional to the arc cross-sectional size—is in series with the voltages in the circuit loop which contains the switch. The arc voltage grows as the arc is lengthened by the physical movement of the contacts, and the arc cross section is diminished as the arc is cooled by contact with un-ionized air molecules.

The arc voltage in low-voltage dc circuits grows at such a rate that it soon exceeds, or at least matches, the source voltage in the circuit (in our flashlight example, the initial arc voltage exceeds the battery voltage). When this occurs, the circuit current is driven to zero in a short period of time. All circuits contain a small but finite inductance, so current cannot be driven to zero instantaneously. When the current does reach zero, no further arc ionization takes place, and the arc is cooled even more rapidly, since it has no energy input. If it is cooled momentarily to such a state that it is no longer a conducting medium, the interruption process is complete and the circuit has been opened. It is important to remember that it is the arc that forces the current to zero. The opening of the switch forms the arc, but it is the arc which enables the circuit to be interrupted.

A switch (or circuit interruption device) which is intended to open alternating current (ac) has a somewhat easier chore than its direct current (dc) counterpart. In ac circuits, there is no need to force a *current-zero* condition. Since the current alternates about zero already, there is a natural current-zero twice in each ac cycle. Any arc which forms in an ac switching device does not have to be stretched and cooled to the extent that the arc voltage exceeds the magnitude of the circuit source voltage. However, this can be done if one wishes to limit the magnitude of an overcurrent by driving it down to an unnatural current-zero.

Alternating currents can be interrupted at a natural current-zero, which is primarily determined by the circuit alone and practically unaffected by the presence of the interruption device. Alternating currents can also be interrupted at forced current-zeros, which are imposed by the action of the interruption device. Figure 1.10 illustrates these concepts of natural and forced current-zeros in an ac circuit.

All mechanical switches and mechanical circuit-interrupting devices depend on the rapid cooling of an arc medium to open an electric circuit. Solid-state switches do not need an arc to break a circuit, since they supply their own conducting medium, the semiconductor material itself. A semiconductor can conduct current only as long as mobile carriers (electrons and holes) are provided from supply or injection regions within the device. If the injection of mobile carriers in a semiconductor switch is turned off, the semiconductor material will revert to an insulating state and block the flow of current—that is, the semiconductor switch will turn off.

The allowable current density within a semiconductor switch is much lower than that which can safely flow in a metal contact–arc switch. Thus the cross-sectional size of a semiconductor switch, for equal rating devices, will always be larger than that of a mechanical switch. Even with this disadvantage, the ease with which a semiconductor switch can be controlled, and the reliability of a device with no mechanically moving parts, suggest a bright future for solid-state power switches and circuit breakers.

ARCS

In the previous section, we gave a short description of the arc that forms between two parting contacts when current is flowing at the time of initial contact separation. In this section, the contact arc is described in more detail. It must be recognized, however, that a contact arc in a switch or circuit breaker is an extremely

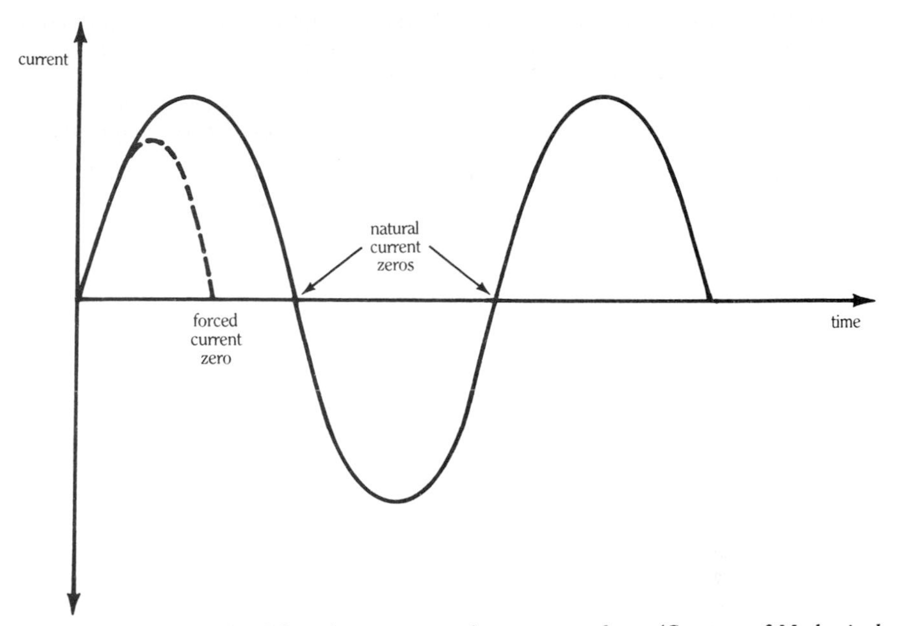

FIGURE 1.10 Natural and forced current-zeros in an ac waveform. (*Courtesy of Mechanical Products, Inc., Jackson, Mich.*)

complex process and that the detailed physics of an arc probably can never be fully described mathematically. The goal of this section is to develop an approximate arc model such that we can treat an arc as a circuit element, and analyze electric circuits containing arcs.

During normal switch operation, the arc, when present, is in a continual state of change. It is dynamically lengthened by parting contacts and by electromagnetic forces which push it away from its original trajectory. It is dynamically heated by its current. It is dynamically cooled by its environment and, perhaps, by other auxiliary means (forced gas flow, cool containment wall, etc.). And, dependent on the net rate of energy absorption (heating minus cooling), it dynamically grows in cross-sectional area.

As the arc changes physically and thermally, it also changes electrically. A change in the electrical characteristics of the arc, in turn, changes the amount of through current that the external electric circuit can supply. Therefore, an engineering description of a switch arc must include a dynamic description of the switch–electrical network interaction. This device circuit interaction is described by means of a simple circuit model, which is developed from the basic energy balance equations for the arc. First, however, the components of an electric arc are discussed (10).

Arc Cathodes, Anodes, and Plasma Columns

The two arc electrodes are referred to as the *cathode* and the *anode.* Electrons are injected into the arc by the cathode at a rate proportional to the arc current.

Arc electrons are collected by the anode at the same rate, since the current must be continuous. The region between the cathode and the anode is divided into three subregions: the *cathode fall region,* the *plasma column* (sometimes referred to as the positive column), and the *anode fall region.* A typical voltage profile along the path of a "short" arc is shown in Fig. 1.11. Short, in this case, means

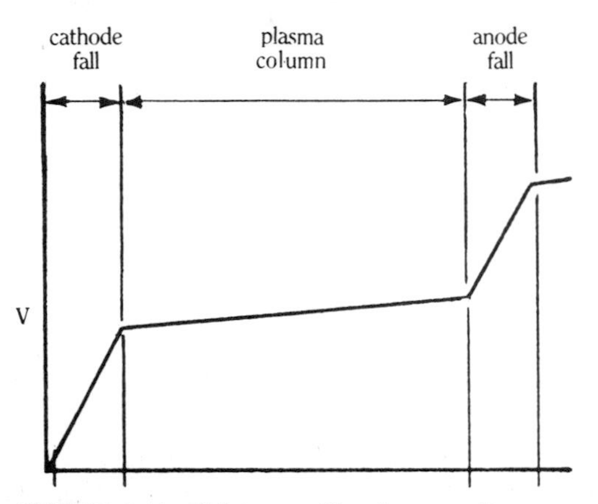

FIGURE 1.11 Voltage profile along a short arc. (*Courtesy of Mechanical Products, Inc., Jackson, Mich.*)

that the voltage drop across the plasma column is small in comparison with the combined voltage drops across the cathode and anode fall regions. Typically, this occurs when the physical length of the plasma column is small. The cathode and anode fall regions are the transition regions between the metallic cathode and anode electrodes and the gaseous plasma column. The magnitudes of the electric fields within the cathode and anode fall regions are much higher than the magnitudes of the fields within the metallic cathode and anode, and much higher than the magnitude of the field within the plasma region. Higher electric fields are, by definition, higher voltage drops per unit distance, and thus the use of the term "fall" as in "voltage fall," in the descriptions of the cathode and anode transition regions.

The voltage drops, or "falls," within the cathode and anode fall regions are strong functions of the materials used as cathode and anode electrodes, but relatively weak functions of the current level within the arc. The energy required to completely remove an electron from the surface of a material body is defined as the "work function" of that material. Expressed as an equivalent voltage (energy divided by the charge of one electron), the vacuum work functions of most metallic elements are approximately 4 to 5 V. The detailed physics of electron emission and collection in cathode and anode regions under arc conditions is of such complexity that only a limited number of low-current simplified cases have been theoretically analyzed by researchers [see (6) for a summary and further references]. For our purposes, it is sufficient to say that the cathode and anode voltage drops are "of the order" of the cathode and anode work functions.

The actual surface area of cathode electron emission and anode electron collection varies with the total arc current. The current densities within these active areas, however, are extremely large, particularly so for the cathode. Current densities exceeding 10^6 A/cm^2, and surface temperatures exceeding 4000 K have been postulated for cathode "spots." At these current densities and temperatures, electron emission is a combination of thermionic and field emission. Electrons with enough thermal energy can thermionically escape the surface of the cathode but, because of the large concentration of positive ions in front of the cathode, a high surface electric field is also present. It enables surface electrons to tunnel through a reduced surface work function energy barrier and be accelerated away or "emitted" by field emission.

Even higher surface "spot" temperatures can be present at the anode. When electrodes leave the cathode, they take energy with them. Therefore, the cathode is actually cooled by their exit (on a net basis, however, the cathode is heated by the I^2R heating within the cathode spot and the energy of incoming positive ions). When electrons arrive at the anode, they dump their energy into the anode surface and heat it up (in addition to the anode I^2R heating).

Dependent on the actual surface spot temperatures, anode and cathode evaporated surface material will transfer from hotter to cooler surfaces if the gap is sufficiently small (as in a switch at initial contact parting). Some investigations have shown that material transfer can be a function of peak arc current, where the peak surface temperature transfers from cathode to anode as the peak arc current increases beyond a certain threshold.

The plasma column in an arc is composed of a partially ionized gas. Gas molecules are "ionized" when neutral gas molecules separate into negatively charged free electrons and positively charged ions. This occurs by a number of different processes: high electric field electron and positive-ion collisions; absorption of radiation; and thermal ionization, ionization by means of collisions with high-temperature (i.e., high-energy) electrons, positive ions, and neutral molecules. All these processes occur in an arc; the relative importance of each is dependent on location within the plasma column and the strength of the arc. The energy input to the plasma column is the *Joule heating* due to mobile current carriers.

Since there is a large difference between the mass of an electron and the mass of a positive ion, there is a large difference between the response of an electron and a positive ion to an applied electric field. By far the majority of the current within the plasma column of an arc is carried by electrons. Therefore, the initial energy transfer to the plasma is to the electron gas within the plasma. But very rapidly, by means of collisions, this energy is shared with the plasma ions and the background neutral molecules. Thus, in time intervals of interest to the engineer, and to a very good degree of approximation, the plasma is in a state of thermal equilibrium. That is, all components (electrons, ions, and neutral molecules) within a small spatial region are at the same temperature.

At thermal equilibrium conditions, the rate of ionization within a particular differential region is balanced by an equal rate of ion-electron recombination. Also, the net concentrations or densities of electrons and positive ions are approximately equal and monotonically dependent on the plasma temperature.

The conductivity of the plasma region in an arc is a strong function of the plasma temperature. The higher the temperature, the higher the level of thermal ionization and carrier concentration. The more carriers, the less the value of electric field needed to support a given level of current density (i.e., the conductivity increases). This positive feedback effect—more current → higher heating → more carriers → more current for a given level of external excitation—partially accounts for the steady-state negative differential resistance of an arc.

Another contributor to the steady-state negative differential resistance of an arc is the cross-sectional spreading of the plasma column at higher current levels.

As the temperature of the active (ionized) plasma column increases, so too does the temperature of the gas surrounding the plasma column, owing to thermal conduction (and perhaps convection and radiation). At high enough temperatures above a threshold temperature, the immediate surrounding gas will also undergo thermal ionization. Additional carriers will then be present to carry the arc current, increasing further the net arc conductivity.

A typical static or steady-state voltage-current characteristic of an arc is given in Fig. 1.12. In general, for a given level of arc current, the arc voltage is pro-

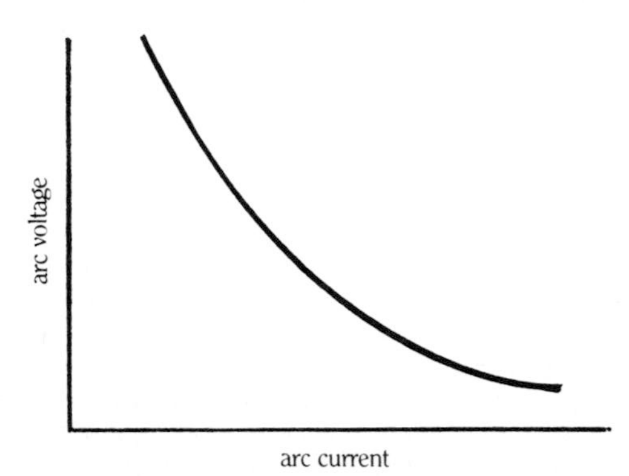

FIGURE 1.12 Static current-voltage characteristics of an arc. (*Courtesy of Mechanical Products, Inc., Jackson, Mich.*)

portional to the arc length. But for a given arc length, higher arc currents result in lower arc voltage drops because of the static negative differential resistance characteristic.

In switch and circuit breaker design and analysis, the static characteristics of the arc are certainly of interest, but it is the dynamic characteristics that are of prime concern. The arc carries the circuit current until an interruption current-zero. Whether or not this interruption will be successful, that is, whether or not the arc will reignite as the voltage across the breaker contacts rises, is a question that can only be answered by a study of the arc dynamic behavior.

The dynamic behavior of an arc is characterized by the arc thermal time constant. This thermal time constant must be much shorter than any circuit electrical time constant or period (for ac circuits) if the arc is to be interrupted at a current-zero crossing. McCleer has developed a simple dynamic arc model from the thermal behavior of an arc. The evolution of the final model is complex and beyond the scope of this book; however, a step-by-step development of the model itself is contained in (10). This model, in particular, can be used to simulate protection device–electric circuit interaction (10).

SWITCH RESISTANCE

The world's attention was centered on a steel tower, standing in a desert in the southwestern United States, as the moment approached for the test firing of a new atomic weapon. Years of painstaking and costly development had culminated in the sophisticated device that was now ready atop the tower. The area had been cleared; observers were at their posts in blockhouses; monitoring instruments were connected. A special radio broadcast was bringing the spectacular event to the world. The countdown to firing time was first announced in minutes, and then in seconds, reaching zero as everyone present tensed for the anticipated detonation...but the eerie silence lengthened into seconds, and then into minutes. No blinding flash had occurred; the tower could still be seen, silhouetted in the early morning light. No mushroom-shaped cloud enveloped the test site. E, it would seem after all, had *not* equaled mc^2!

Could the scientists and engineers, who had achieved such brilliant success with previous atomic devices, somehow have miscalculated this time? The question was of particular interest to the task force who had the unenviable chore of disarming a live nuclear device so the trouble could be traced. Eventually the fault was found in a control circuit and was traced to a miniature snap-acting switch (Chap. 2). During installation of the switch, solder flux had entered the switch case and coated the contacts. It wasn't much, but it was enough to abort the test. For practical purposes, the "closed" switch had infinite resistance.

This extreme example illustrates the fact that a little practical knowledge about switch resistance can save time, money, and an engineer's reputation. In the instance cited, careless installation disabled the switch. On the other hand, countless examples could be cited in which the user has heard of switch resistance, is apprehensive about it, and has tried to minimize it with stringent procurement specifications. These, too, can be costly, time-consuming, and at times, ludicrous. Consider the not unusual demand that switch resistance not exceed 0.025 Ω, even though the component that the switch is to control has a resistance of thousands of ohms. Such a sensitive circuit would demand a fantastically stable power supply!

Problems with switch resistance are often the result of carelessness or misdirected caution. These, in turn, derive from inadequate understanding of switch resistance. The solution lies in being informed about the nature and behavior of switch resistance so that sound technical judgment can prevail. The engineer can then recognize those relatively infrequent instances in which a switch resistance specification is needed and can write a meaningful document to define the necessary parameters.

The purpose of a switch is to close an electric circuit, carry current, open the circuit, and hold it open. Two important electrical characteristics of a switch are its resistance when open and its ability to conduct current when closed.

The Resistance and Capacitance of an Open Switch. When a voltage is applied between two electrodes in, or on, an insulating material, a feeble current flows from one electrode to the other. The applied voltage divided by the current is the electrical resistance of the insulation and is called the *insulation resistance.* When a voltage is applied between two nonconnected terminals of a switch, a feeble current flows through the insulation that separates and supports the terminals. The applied voltage divided by the current is the insulation resistance of the switch, measured between those terminals. It usually is expressed in megohms, i.e., millions of ohms. Insulation resistance is often confused with another property, the *dielectric strength.* The dielectric strength of an insulating material is the highest electrical potential gradient that the material can withstand without breaking down. As a material property it is calculated by dividing the breakdown voltage by the thickness of the insulating material between a pair of test electrodes. The dielectric strength of a switch, between two nonconnected terminals, is the maximum voltage that can be applied between the terminals without rupturing the insulation that separates them. Dielectric strength is expressed in volts.

Figure 1.13 illustrates the insulation resistance and dielectric strength of a switch as measured between a pair of nonconnected terminals. As voltage is increased from zero, a proportionate current flows through the insulation. The higher the voltage the higher the current. At several hundred volts the resulting current may be of the order of microamperes. A newly manufactured switch at normal voltages may have an insulation resistance well in excess of 100,000 MΩ between nonconnected terminals and between the terminals and switch mounting. The voltage puts an electrical stress on the material but does not damage it unless the potential exceeds the dielectric strength of the switch, usually well over 1000 V. When the voltage exceeds the dielectric strength, the insulating material is permanently damaged by the electrical stress and the insulation resistance decreases suddenly. Thus there is a sudden increase of current passing through the damaged insulator. The dielectric strength of a switch is somewhat lower if the voltage is applied suddenly rather than gradually. In other words, the rate at which the voltage is increased affects the dielectric strength.

During the life of a switch under normal conditions of application, both the insulation resistance and dielectric strength remain at satisfactorily high levels. Under some circumstances, which will be discussed later in this chapter, the insulating material deteriorates and its insulation resistance drops to such a low level that the "open" switch passes appreciable current. This sometimes is the result of a drop in dielectric strength of the insulation to a level at which the circuit voltage ruptures the material and ruins the switch.

When a switch is open, there is usually a voltage across the open contacts and the switch becomes a capacitor. Its capacitance is extremely low and is seldom of any practical concern, but occasionally it becomes a factor to consider in the design of a system. Because of the geometry and positioning of the internal parts, the capacitance of a switch is very difficult to calculate. Such computation involves assumptions about the composite dielectric constant of the insulating material which introduces a good deal of uncertainty. Then, too, the lead wires connected to the switch have capacitance, and their spacing affects its value. The

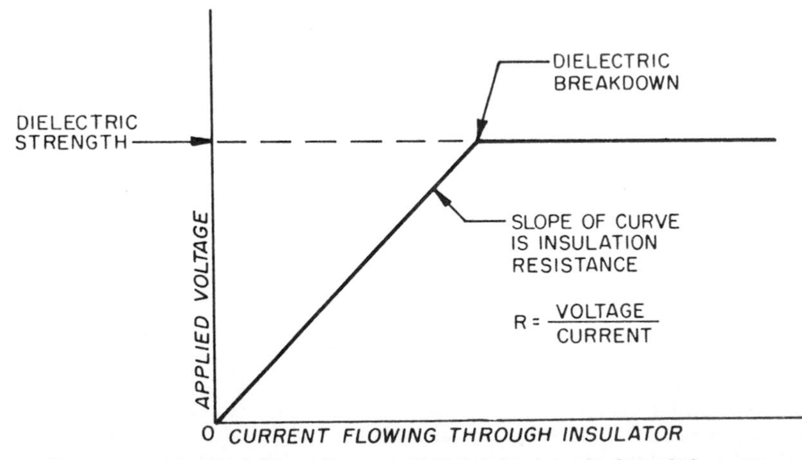

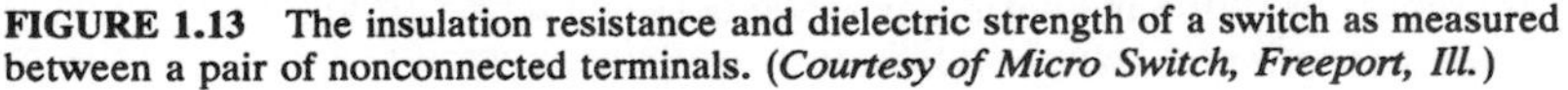

FIGURE 1.13 The insulation resistance and dielectric strength of a switch as measured between a pair of nonconnected terminals. (*Courtesy of Micro Switch, Freeport, Ill.*)

switch mounting and its environment also influence capacitance. If capacitance is important, it must be measured, but the measurement procedure involves additional variables. Even changing the length or position of a lead wire can alter the measured value. The practical fact is that, if capacitance is to be measured, the measurement must be made with the switch installed in the end-use equipment, with an environment similar to that of its application. Only then will a valid and useful capacitance figure be obtained.

Sources of Switch Resistance. *Switch resistance* is the total resistance of the conducting path between the wiring terminals of the switch. This is the resistance "seen" by the circuit and is the value usually specified and measured, although often it is erroneously called *contact* resistance. Switch resistance is the sum of the bulk resistance of all parts that make up the conducting path through the switch, plus the resistance of the joints or interfaces between these parts (Fig. 1.14). The joints may be staked, bolted, or welded. They may be bearings, such as knife-edge pivots through which current passes. The joint may be a contact interface, i.e., a connection that is joined and separated by the switch mechanism. The resistance of the pair of *closed* contacts is the *contact resistance,* about which more will be said later in this chapter. In a properly designed and manufactured switch the joint or connection between the contacts provides the only significantly variable resistance in the switch. Switch resistance can never be less than it would be if the contacts were soldered or welded together. This resistance typically is of the order of 0.005 to 0.050 Ω, and depends upon the materials and design of the switch. The design, in turn, is governed by the use for which the switch is intended. A switch designed for use at high temperature, for example, may require special spring material having higher than usual electrical resistivity.

Patterns and Magnitude of Switch Resistance. Because the resistance of the contact interface varies, so does the resistance of the switch. The particular pattern and magnitude of resistance depend upon several variables which will be discussed later in the chapter. To study the behavior of switch resistance during life, a manufacturer of precision snap-acting switches developed automatic recording equipment to plot a point on a resistance graph during each closure of the switch

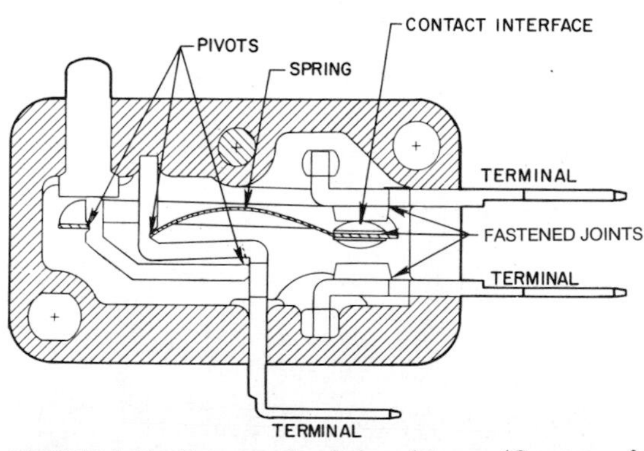

FIGURE 1.14 Sources of switch resistance. (*Courtesy of Micro Switch, Freeport, Ill.*)

under test. Switches from various manufacturers were tested on this equipment. Figure 1.15 shows typical resistance of various standard switches at 6 V dc, 0.1 A during 10,000 operations. The test circuit has a load resistance of 60 Ω. During these tests, switch resistance ranged, for the most part, from 0.005 to 0.020 Ω, with an occasional overshoot to as high as 0.038 Ω. In no instance did switch resistance exceed 0.063 percent of the 60-Ω load resistance. It is interesting and important to note that when the load resistance is lower, the switch resistance

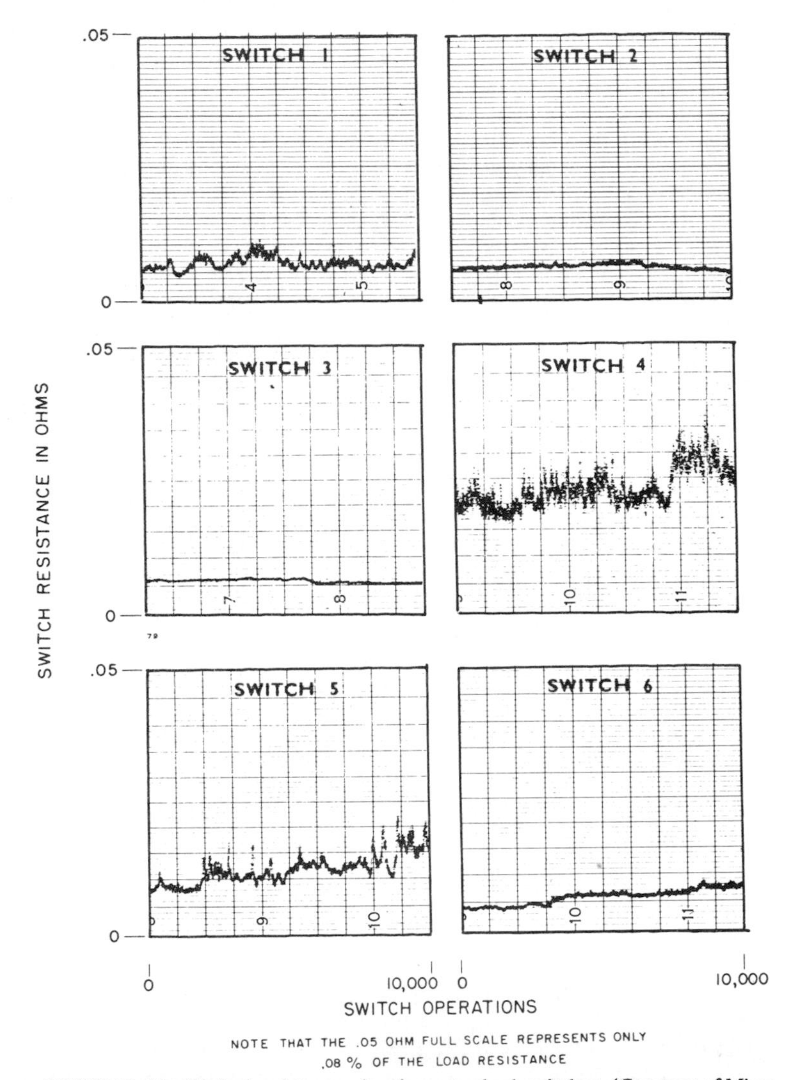

FIGURE 1.15 Typical resistance of various standard switches. (*Courtesy of Micro Switch, Freeport, Ill.*)

tends to be lower. When load resistance is higher, switch resistance tends to be higher, too. In general, if source voltage is 0.5 V or greater, switch resistance tends to be a small part of the load resistance.

Switch Resistance Behavior. The apparent dependence of switch resistance upon load resistance is not so mysterious if we consider events at the interface between the switch contacts. For practical purposes, it can be assumed that heat flows through a pair of contacts along the same path as that of the electric current. As a result, the voltage across a pair of current-carrying switch contacts is an approximate measure of the temperature of the metal at the interface between them. For this reason it is possible to speak of "softening and melting voltages" of contact material, just as one speaks of the softening and melting temperatures. Table 1.1 gives approximate values for the softening and melting voltages for silver and gold contacts.

TABLE 1.1 Approximate Softening and Melting Voltages of Gold and Silver

Material	Approximate softening voltage	Approximate melting voltage
Silver	0.09	0.37
Gold	0.08	0.43

Source: Lockwood (7).

Since the voltage measured across a pair of closed contacts is an index of temperature at the contact interface, it follows that the voltage impressed across a pair of closed contacts will determine the temperature. If this voltage exceeds the softening voltage of the contact material, the metal at the interface will soften. If the voltage reaches the melting voltage of the material, the metal at the interface will melt. The softening and melting are due to I^2R heating. They take place on a microscopic scale, but nonetheless they control contact resistance. If the metal at the contact interface softens or melts, the cross-sectional area of the conducting path between the contacts will depend upon the current. This area determines the resistance of the conducting path and therefore the contact resistance. If melting occurs, a metallic bridge will be established and its cross-sectional area will increase until it can carry the current in the solid state. The higher the current the greater the cross-sectional area of the bridge and the lower its resistance. If the softening voltage is not reached, the size of the conducting path is not affected by current, and contact resistance will be independent of current.

Voltage also serves to help establish electrical continuity during contact closure, by breaking through surface contaminants. The higher the voltage, the greater the assurance that the switch will close the circuit. In practice, if switch resistance is a *critical* factor in the operation of a system, it is best to consult with the switch manufacturer for specific advice (7).

Contact Resistance. A load converts electricity into a useful form of energy. A simple example is the conversion of electricity to visible light (luminance) by a light bulb. All loads are designed to operate at a specific voltage and current, which may be expressed as watts. If the load is operated at a higher voltage and current, it must dissipate more heat than its capability. As a result, the load "burns out" the same way a light bulb filament burns out. If the load is operated at a lower voltage and current, it does not operate efficiently. A 100-W light bulb would glow very dimly when operated below its rating. For maximum efficiency and protection, the load should be operated at the voltage and current for which it was designed.

In a circuit, the load is connected to a source which has a specified voltage. If the circuit is properly designed, the source voltage will be the required amount for operating the load. The opposition or resistance of the load allows the proper amount of current to travel through it. Thus the load operates most efficiently at its rated voltage and current.

The path for current through the load is completed by connecting wires and a pair of switch contacts. The switch is said to be in series with the load, and the same current flows through both. If the connecting wires or the switch contacts offer any appreciable opposition to the circuit current, the current in the circuit is reduced, and the operating voltage of the load is reduced. The ideal condition would be one where the connecting wires and the switch contacts would offer no opposition to current. Of course, in the real world this condition is not possible.

Although they have the same current flowing through them, series components such as the contacts and the load divide the voltage of the source depending upon their resistance. The larger the resistance of the component, the greater the amount of the source voltage across the component. In a practical situation, it is desirable to have as much as possible of the source voltage across the load and very little at the switch. Since the load has been designed to operate at a specific voltage and current, its resistance cannot be changed. As a result, the resistance of the switch contacts should be as low as possible for maximum voltage across the load (8).

When switch contacts are closed, current flows between the contacts only at very small physical contact points, or *asperities,* due to the surface roughness on the bulk contact faces. The actual area of electrical contact is only a small fraction of the apparent area of the bulk contact surface (Fig. 1.16). Current flowing in the contact bulk regions is constricted at these contact points, much like

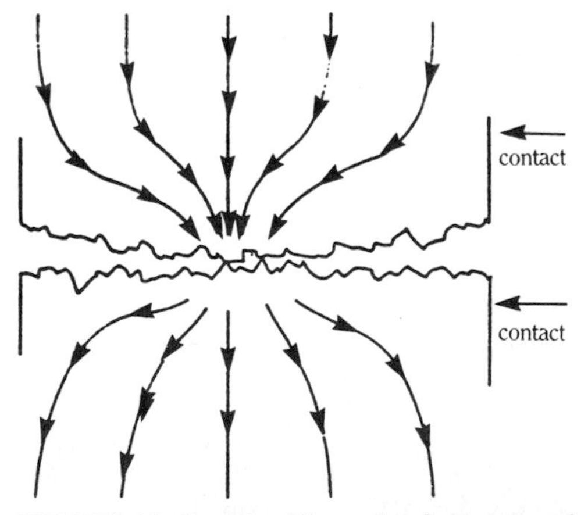

FIGURE 1.16 Constricted flow paths of current to and from surface contact spots in a set of contacts. (*Courtesy of Mechanical Products Inc., Jackson, Mich.*)

liquid flowing through a pipe with an insert containing very small holes. The extra electrical resistance due to this current constriction is referred to as the spreading or constriction resistance of the contact. As shown by Holm (9), the constriction resistance on each side of an individual contact "spot" is given by

$$R_c = \frac{\rho_r}{4a}$$

where ρ_r is the bulk resistivity of the contact material and a is the effective radius of the asperity or actual contact spot area.

If the contacts are constructed of two different materials, with respective bulk resistivities ρ_{r1} and ρ_{r2}, the total series spreading resistance due to current constriction in both contacts is

$$R_c = \frac{\rho_{r1} + \rho_{r2}}{4a}$$

Normally, contacts are fabricated with identical materials, and normally actual contact is made at N number of spots on the contact surfaces. The net constriction resistance for the contacts is then the parallel combination of all the individual contact values, or

$$R_c = \frac{\rho_r}{2\sum_{i=1}^{N} a_i}$$

The effective radius of each contact spot a_i is dependent on the preparation of the bulk contact surface, the normal forces applied to the contacts, the "hardness" of the contact material (i.e., will each contact asperity be under elastic or plastic deformation?), and the temperature at the contact interface.

In addition to constriction resistance at contact asperities, there may be a resistance due to a thin film or layer of material oxide between contacting asperities. Electrons either tunnel quantum-mechanically through this thin film or break through the film by a process Holm refers to as "fritting" (9). The film resistance is between the constriction resistances of individual asperities, so the net "contact" resistance would be a modification of the last equation:

$$R_{\text{contact}} = \frac{1}{\sum_{i=1}^{N} \frac{1}{R_{fi} + \rho_r/2_{ai}}}$$

where R_{fi} is the film resistance at asperity i.

In practice there is no attempt to determine contributions to R_{contact} due to individual contact spots. The net excess resistance of the contact system, beyond the bulk resistance of the two contacting bodies, is simply referred to as the *contact resistance.* The voltage drop across this resistance is commonly referred to as the *contact drop.* In most cases this contact drop does not exceed 0.1 to 0.2 V. Contact drops tend to saturate at these levels since, as the magnitude of the current rises, the asperity interface temperature rises, softening the asperity material. The softer material spreads out and increases the actual asperity contact area, thus lowering the contact resistance. When two bulk metallic contacts which are carrying an electrical current separate, the last point or points of physical and electrical contact will be at one or more (if more than one, a small number) constriction asperity spots. The current density at these points will be very large, easily enough to melt the asperity material and form molten bridges between the two contacts. These bridges are then heated and stretched to the point that they vaporize. The process initiates the arc between the two contacts. If the contacts are not metallic, such as carbon, the asperity points do not melt but rather arc immediately upon physical separation (10).

Contact resistance can be simply defined as the amount of *opposition* to current offered by the switch contacts. Opposition to current, or resistance, is measured in ohms. Although the resistance of a load is quite variable, a typical value is several thousands of ohms. By contrast, the value of contact resistance in most metal-to-metal contacts is thousandths of an ohm. This meets our previously noted requirement that the contact resistance be very low compared with the resistance of the load.

Typical values for the wide variety of switch styles available may range from 3 *milli*ohms for precision snap-acting switches and high-reliability rotary switches (Chaps. 2 and 3, respectively) to as much as 300 Ω, or higher, for some conductive ink membrane switches (Chap. 8). These values are generally listed in manufacturers' catalogs as the *initial values* of contact resistance, meaning these are the values *before* the switch has been subjected to any cycles of operation (8).

Catalog ratings can be misleading by the implication that the contact resistance will remain essentially the same value throughout the life of the switch. This, unfortunately, is not the case. Typically, contact resistance *increases* with the number of cycles of operation. If contact resistance over the life of the system being designed is important to the operation of the system, the design engineer should verify that the contact resistance values published by the manufacturer are indeed initial values and should request that the manufacturer provide data on contact resistance values over the life of the switch, usually indicated by the number of on/off cycles.

Initial contact resistance depends upon several factors. In rotary switches, for instance, contacts are held together by making one contact of spring material which then exerts pressure on the other contact. In a pushbutton switch, pressure is applied to the button to hold the contacts together. This pressure is necessary because, even under the most precise manufacturing techniques, it is not possible to produce two surfaces which will meet perfectly. Therefore, contact is made at several high points on the surfaces. Factors such as the shape of the material and the pressure exerted determine the number of actual points of contact. If there is a relatively *large* number of points of contact, the switch will have a *low* contact resistance. During the life of the switch, contact resistance is affected by the amount of arcing that occurs at the contacts and the film that is formed on the contact surfaces. As discussed earlier in this chapter, arcing is a natural phenomenon that occurs when a switch makes or breaks a circuit. The arc causes the contacts to carbonize and deteriorate. It is this carbonization and deterioration which causes the contact resistance to increase. Another factor that determines contact resistance is the formation of films on the contacting surfaces. One of the causes of a film formation is the heat generated within the contact material as the contacts are carrying current. This heat may be sufficient to create a film on the contacts, which usually causes the contact resistance to increase. Since the film is created from *within the contact material itself,* production of the film can be reduced by the proper selection of materials for the contact. Selection of contact materials and platings will be discussed in each chapter. Contact material and plating criteria specific to a particular switch technology are discussed in the chapter addressing individual switch styles. As might be expected, materials which do not readily create films at their surfaces are more expensive and, in some cases, not as long-wearing (8).

Switch manufacturers, then, strike a generally acceptable compromise in the selection of contact material for their products, while maintaining a database of special (even exotic) materials for unique switching applications where performance, not price or delivery, is of prime concern to the user.

Another cause of film formation on the switch contacts is contamination from *external* sources. Water, oil, dust, and dirt all have deleterious effects when deposited on the contacts. A large part of this problem is resolved by simply enclosing the switch.

Generally speaking, the published values of contact resistance are usually small enough so that the system designer should not anticipate a problem in the systems circuitry; but solid-state electronics (integrated circuits, microprocessors, etc.) have confused the contact resistance application problem. Some circuits require low contact resistance and others can tolerate very high contact resistance and still function. All solid-state electronics switch relatively low voltages (3 to 24 V) and currents, which may cause concern to some engineers. At *very* low voltage levels (0.03 V and below) and very low currents, the normal films that form on contacts are not broken down and contact is not made. This condition is known as a *dry circuit* or *dry circuitry.* Dry circuit conditions rarely occur, as standard solid-state electronics usually operate well above dry circuit voltage and current levels. If a dry circuit condition is unavoidable and a switch or switches must be used in this situation, it is critical that the system and design engineer discuss the application with the switch manufacturer as early in the design process as possible.

The effect that mechanical wear has on contact resistance depends upon the design of the switch. Typically, rotary switches are designed so the contacts wipe across each other. Pushbutton switches can be designed to incorporate this wiping feature. This wiping action may be helpful by partially removing the contamination due to arcing and the film formation. As already noted, the contact resistance of all switches increases during the life of the switch. In wiping contact switches, the contact resistance increases in a smooth, predictable manner. This can be contrasted to butt-type contacts in which the mating surfaces merely butt against one another. Butt contacts likewise exhibit an increase in contact resistance during life; however, the contact resistance is quite variable from operation to operation and can become quite high at times. If contact resistance is critical in an application, switch manufacturers recommend a wiping contact type switch [adapted from (8)].

Voltage Breakdown and Insulation Resistance

Voltage Breakdown. Voltage breakdown occurs in a switch when an arc takes place between two conducting or metal parts of a switch. The arcing that occurs at the contacts when a switch makes and breaks the circuit has been discussed earlier in this chapter. This arcing at the contacts is a natural occurrence that cannot be avoided. The voltage breakdown characteristic of a switch is not related to the arcing that occurs at the contacts when a switch makes and breaks the circuit. It is related to an unwanted arc which can permanently damage the switch. All this makes it necessary for the user to know and understand the circuit conditions at which this arc will take place so that a switch can be properly evaluated, selected, and specified.

An arc can occur between nontouching metal parts if the voltage is large enough. The voltage needed depends upon the distance and the type of insulation between the conducting members. The conducting members of a switch are all the metal parts, including the contacts, the mounting bushing, the shaft, rivets, etc. A large voltage can occur in a pushbutton switch when the contacts are open (separated). This is because the greatest voltage (potential) in a series circuit is between the points where the greatest resistance is. Open contacts have a much higher resistance than the load, so a greater voltage is evidenced. For all practical purposes, the voltage between the terminals (at the open contacts) is the voltage of the source. And if the source voltage is sufficiently large, an arc will occur between the terminals. An arc can also occur between a terminal and any other metal switch part, provided that it completes a circuit path. For example, the shaft and bushing in a rotary switch are conductors; the bushing is often mounted to a metal equipment panel; the panel is often connected to one terminal of the voltage source to provide a "ground." This is a partial circuit path. If there is a large voltage between a terminal and ground, an arc could occur between the terminal and the shaft—bushing—panel—ground path.

When an arc occurs in a switch, the terminals and the material between them deteriorate. The arc is maintained until the source voltage is reduced or removed. Each time the arc occurs, some of the contact material is burned away and some is carbonized. Eventually, the reduced contact area and the deposit of a form of carbon on the contact causes a high resistance, which generally makes the switch useless. The arcing also causes the insulating material between terminals to burn and carbonize. The carbonized area presents less opposition to current than the original material, so its insulating properties are reduced.

An arc can occur in any switch depending upon the amount of voltage between the terminals. The amount of voltage at which the arc will occur is idiosyncratic to the design of the switch. Generally, the amount of voltage required to cause an arc depends upon the distance between the terminals and upon the type of insulating material. This voltage is a switch characteristic called *voltage breakdown.*

The voltage breakdown parameter of a switch is usually listed in a manufacturer's technical literature as *x*—number V ac. This characteristic indicates the amount of ac voltage that will cause an arc to take place within the switch. Different values of voltage breakdown may be listed for the same switch. One value is the voltage required to cause an arc between terminals; the other value is between a terminal and ground. Often, the voltage breakdown is listed as the amount of voltage between mutually insulated parts. This covers situations both between terminals and between a terminal and ground. Typical values of voltage breakdown for some styles of switches are in excess of 1000 V ac. These values are usually determined by actual laboratory testing of switches and then assigning a more conservative rating.

For some miniature switches, the voltage breakdown characteristic is exceptionally high. This characteristic is stated for switch operation at room temperature and humidity. In a high-humidity atmosphere, an arc will occur at a lower voltage. To determine the exact value of voltage breakdown in such an atmosphere, the environmental conditions in which the switch will operate must be specified by the user. In turn, the manufacturer can then determine voltage breakdown value.

Voltage breakdown is often confused with the *dielectric strength* of an insulating material. Dielectric strength refers to a specific voltage placed between two conductors (between terminals or between a terminal and ground) to determine if an arc will occur or if a certain amount of current will travel between these points. The difference is that for dielectric strength, a certain amount of voltage is connected between conductors to determine *if* an arc (or a certain amount of current) will occur. For voltage breakdown, the voltage between conductors is increased until an arc *does* occur. If an arc or the specified current does not occur for a stated dielectric strength voltage, the switch is considered to have passed the test. Voltage breakdown is a destructive test.

Insulation Resistance. Insulation resistance is the amount of opposition to current that would be measured between terminals, or between a terminal and ground under normal conditions. Insulation resistance and voltage breakdown are measured between exactly the same points, the difference being that insulation resistance is the opposition to current measured under normal operating conditions, and voltage breakdown is a voltage between terminals which will cause an arc. The insulation resistance of a switch depends upon the type of insulating material used in the switch and the distance between contacts. Typical insulation resistance values are from 20,000 to 500,000 MΩ. Since 1 megohm equals 1 million ohms, these values are extremely large values of opposition which require measurement by special equipment.

Insulation resistance determines how well a switch will open a circuit. In a pushbutton switch a very high resistance between open terminals or contacts is necessary to prevent any current in the circuit. The same is true for rotary switches where they are used to open a circuit. A high insulation resistance is also necessary between the terminals of a rotary switch to prevent any unwanted current between circuits connected to adjacent terminals. A high insulation resistance is necessary between the terminals and ground to prevent any unwanted circuits back to the source. A high insulation resistance prevents all these undesirable effects.

When a switch is used to open a circuit, the switch contacts are in series with the load. Therefore, the insulation resistance which appears between the contacts is in series with the load. A value of load resistance comparable with the insulation resistance could cause problems because the circuit is never actually opened and there may be an appreciable amount of current in the circuit. This condition of a load resistance comparable with the insulation resistance of the switch is often referred to as a high-impedance circuit.

The value of insulation resistance of a switch is affected by humidity. If the insulating material will readily absorb moisture from the surrounding air, its insulation resistance is quickly reduced. What was once an insulator becomes a conductor of current. This deterioration of the insulating material usually occurs above 90 percent relative humidity. The insulating material of a well-designed switch will be a material which will not readily absorb moisture and which initially has a high insulation resistance [adapted from (11)].

Electrical Loads and Switch Ratings

A primary switch rating describes how long it is expected to function under a certain set of electrical conditions. The "how long" part of the rating is called the *switch life.* Pushbutton switches are life rated in number of operations. A pushbutton switch operation is an actuation and a release of the plunger. A pushbutton switch is rated by stating the amount of current it will make and break with a specific amount and type of voltage and type of voltage source and the number of operations it takes to reach failure. An example would be "1/4 A at 30 V dc, resistive for 35,000 operations." When the switch is used in such a circuit, 35,000 operations is the anticipated switch life. Rotary switches are life rated in cycles of operation, a cycle of operation being defined as the rotation of the movable contact through all its active positions and a return to the starting point.

"Failure" is based on specific criteria arbitrarily set by the switch manufacturer or defined by industry standards. Failure criteria often include one or more of the following:

- An increase in contact resistance to a certain level
- A decrease in insulation resistance to a specific amount
- A voltage breakdown value

The properties just mentioned are used with many switch products as a means of measuring failure. However, the value of the failure criteria can vary with the switch product. For example, for a switch used in applications where contact resistance is critical, the manufacturer may set a particularly low value of contact resistance as one of the failure criteria. In general, a manufacturer will test to in-house, arbitrary rating standards or to an industry standard such as a military specification.

A reduction in insulation resistance or voltage breakdown or an increase in contact resistance is partly caused by the natural arcing that occurs when a switch makes and breaks a circuit. In the foregoing discussion about voltage breakdown, it was noted that the distance and the insulating medium between the terminals determines how much voltage can appear between the terminals before an arc occurs and is maintained. Consider what happens as a switch makes and breaks a circuit. Just before contact is made or just as the contact is broken, there is a very small distance between, for example, the shorting bar and terminals in a pushbutton switch or the contact arm and the terminal in a rotary switch. Since this is an open circuit, the source voltage appears between these two points. Because the distance is considerably less than the distance between fully open contacts, an arc occurs between contacts just prior to making or just after breaking. Once contact is made, there is no longer the possibility of an arc's existing. When the contacts are broken, the distance increases owing to switch action (a spring in a pushbutton switch, a detent system in a rotary switch) and the arc is extinguished.

As discussed previously, the effect of this arcing is to cause the contact material to burn away and to create carbon deposits on the contacts. The result is an increase in contact resistance. Additionally, in rotary switches, as the contact arm slides across the insulating material between contacts, a track may be formed where the burned-away contact material can readily deposit. This will reduce the insulating properties and *lower* the voltage at which an arc will occur between contacts. Consequently, the natural arcing that occurs between contacts as the switch makes and breaks a circuit causes a deterioration of the switch's contact resistance, insulation resistance, and voltage breakdown properties.

How far these properties can deteriorate before the switch is considered a failure (failure criteria) is arbitrarily determined by the manufacturer or by industry standards. These limits are often more stringent than what the user requires. For example, a system engineer's application may be able to afford higher contact resistances than what have been established as failure criteria. At the same time, the application may require a larger number of cycles of operation than listed in the published ratings.

The amount of arcing that occurs as a switch makes and breaks a circuit depends upon the amount and type (ac or dc) of voltage source, the type of load, and the circuit current. The amount of voltage across the contacts will have an effect on when an arc will occur. Hence a larger voltage will cause an arc to occur sooner and burn longer as the circuit is being made, and to burn longer as the circuit is broken.

The type (ac or dc) of voltage source also affects arcing since it affects the amount of voltage at the terminals at the exact instant they are opened or closed. For direct current, this voltage is constant. For alternating current, the voltage is constantly changing as the direction of the current changes.

In alternating current the voltage level peaks and drops to zero over and over again. Rather than identifying the alternating current by its peak value, we label it by its effective value, which is somewhat smaller. This value for alternating current is equivalent to the amount of voltage it takes a direct current to produce the same amount of heat in a load. Accordingly, 115 V ac produces the same heat in a load as 115 V dc. Nevertheless, the alternating current has voltages higher, as well as lower, than 115 V, and it causes varying (usually smaller) amounts of arcing. As far as the switch contacts are concerned, 115 V ac and 115 V dc are *not* equal.

The amount of arcing that occurs at the contacts also depends upon the type of load in the circuit. Examples of common loads are light bulbs, heating elements, television receivers, and loudspeakers. Any electrical device or piece of electrical equipment is considered the load in a circuit. It is impossible to describe how a switch will operate in a circuit for each of these loads. Fortunately, loads can be classified into two major categories, which are applicable for the majority of the loads used in circuits. This classification describes the load by its effects in the circuit. The categories are *resistive loads* and *inductive loads.* The operation of the switch can then be described in terms of these two categories.

Inductive and Resistive Loads. An inductive load, because of its physical construction, has the ability to store electrical energy in a magnetic field as long as there is current in the circuit. When the circuit is broken, because a switch opens the current path, the stored energy must be returned to the circuit. The return of this electrical energy to the circuit causes a voltage that is larger than the source voltage across the switch contacts.

A resistive load is one which dissipates electrical energy in some form of energy (heat and/or light). A resistive load does not store any electrical energy while there is current in the circuit. As a result, more serious arcing can occur when an inductive circuit is broken than when a resistive circuit is broken. Therefore, a switch must be rated for an inductive load and for a resistive load even though both circuits may be supplied by the same amount of source voltage.

Switch contacts are deteriorated by excessive heat. The amount of heat dissipated by the switch contacts in the closed position depends upon the contact resistance and the amount of current through them. One common switch rating states that the switch is capable of carrying so many amperes continuously. This means that for an average value of contact resistance, this amount of current will cause the temperature of the contacts to increase not more than 68°F (20°C). This temperature rise is considered an industrial standard beyond which the contacts are overheated. Sufficient overheating causes the metal of the contacts to partially weld. Subsequently, when the contacts are opened, they are seriously deformed. The continuous current or carry rating does not apply to making or breaking a circuit carrying current of this amount. It merely states the amount of current that the contacts can safely handle after the circuit is completed.

The amount of current in the circuit is also partly responsible for the deterioration of the contacts when a switch must make and break a circuit. This is because the heat generated at the contacts is due to the current level. As a result, most switches must be rated for:

- Current
- Amount of voltage
- Type of voltage (ac or dc)
- Type of load (inductive or resistive)

Since arcing conditions are different for each type of source and for each type of load, a switch should be rated for several of these conditions. Once the voltages and loads have been established, it is a matter of the manufacturer's establishing the failure criteria and then testing the switches at various current levels. The number of operations or cycles of operation that a switch will perform at these levels, before the failure criteria are exceeded, can then be determined and the switches can be rated.

Often the stated conditions will not specifically fit the user's circuit conditions. Generally, the current rating at the stated voltage level can be increased if the number of operations or cycles of operation is decreased. This is not a linear relationship, however, and the switch manufacturer should be consulted for exact information relative to the actual relationship. Another method of rating switches is a life load curve, which is a plot of the number of cycles of operation against the current that a switch can make and break, at a particular voltage level and type of load. This curve can be used to predict the number of cycles of operation at a particular current level.

The user should be aware that published switch ratings can be misleading. A common practice is to list a contact resistance, an insulation resistance value, and a voltage breakdown value that apply to a new switch, and a switch life (cycles of operation) value that applies to the mechanical life of a switch. The mechanical life of the switch is the number of cycles of operation that a switch will perform with no voltage applied to the contacts. With no voltage applied to the switch there will be no current through the switch contacts. The number of cycles of operation at which the switch fails to meet the stated failure criteria is called mechanical life, since no voltage is applied and no current travels through the switch contacts. In normal circuit applications the required switch life probably will be less than the mechanical life. Hence catalog ratings that state the mechanical life seem to indicate that the switch will meet the life requirements; however, the switch may not meet the requirements when a source voltage is applied and there is current through the contacts.

Often in a particular application it is desirable to know the "useful" life of a switch. The *useful life* of a switch is the number of cycles of operation that a switch will perform without exceeding the failure criteria for the application. Useful life for a particular application can often be estimated by the switch manufacturer if specific information about the source voltage, the current level, the type of load, and the failure criteria for the application is provided by the user [adapted from (12)].

The Role of Contamination in Switch Resistance. In those infrequent instances in which the resistance of a switch is high enough to cause trouble in a circuit, the problem almost invariably can be traced to contamination of the contacts. Contact contaminants often are subtle, elusive, and nearly invisible even under a microscope. The high-resistance condition is intermittent as particles of dust or other foreign material move about on the contact surfaces. A particle 0.001 in (0,025 mm) in diameter falling from a height of 6 ft (1,82 m) would require more than 1½ min to reach the floor. And yet one particle much smaller than this can hold the contacts of a switch apart and cause an open circuit. Contact contaminants generally are classified according to their gross characteristics as films or particles. Contact films may be subdivided into those which are completely alien to the contact, such as an oil film, and those which are a chemical compound of contact material and contaminant. Usually the contaminant is readily identified and the problem is remedied at its source. There remain those instances in which the contaminant appears to have been generated spontaneously at the contact, and there is too little of the material for easy identification or analysis. In such cases the problem must be treated on an individual basis by the switch manufacturer, utilizing the best available knowledge and analytical techniques. The most common contaminants are the following.

Solder Flux. Minute amounts of flux can enter a switch as liquid or vapor during or after soldering of leads to the terminals. This flux can be trapped between the contacts. To avoid this, follow these practices:

1. Choose a nonactivated flux.
2. Use flux sparingly.
3. Use flux core solder.
4. *Never* allow liquid flux or flux vapor to enter the switch.
5. Use a soldering temperature of approximately 550°F (288°C).
6. *Never* use solvents on or near switches—they carry flux residue and other contaminants into the switch (unless there is a compelling reason to do so, it is not necessary to remove flux residue from switch materials).

Crushed Solids. Particles of foreign material can enter an unsealed switch and be crushed beyond recognition between the contacts. Most switches are designed to have a high contact force to assure good electrical continuity. This can pulverize brittle material and trap enough of the resulting powder between the contacts to cause problems.

Carbon. Organic materials, especially organic vapors, can decompose on the contact surface leaving an ash residue of carbon. This can occur whether or not there is an arc at the contacts. If there is no arc, energy comes from friction and I^2R heating of the material at the interface between the closed contacts. The vapor adsorbs on the contact surfaces and is decomposed by the energy dissipated at the point where the contacts touch. If an arc is present, it can decompose organic material directly without adsorption. Increasing the current will increase the yield of carbon, but if an arc is present it will also increase the tendency of the discharge to evaporate the carbon and to blast the contact surfaces clean. The most effective remedy, usually, is elimination of organic vapor contaminants.

Polymers. Under some conditions, simple organic contaminants can combine into long-chain compounds (polymers) in a reaction catalyzed by the contact material. For this to occur, the organic material must be in the form of vapor. The vapor adsorbs on the contact surface and the reaction proceeds from there. Silver does not form polymers because it does not adsorb vapor. Gold can form polymers. Polymers forming between contacts are of several types, having various characteristics. They form only during the sliding or wiping of contacts, but this may be on the microscopic scale as metals deform under the contact force.

Chemical Compounds of Contact Material and Foreign Material. When an unsealed switch is exposed to chemical vapors, the contact material may react with the vapor to produce sulfides, oxides, chlorides, nitrates, or other compounds. The most common occurrence of this kind is the tarnishing of silver contacts in the presence of H_2S and water vapor. The H_2S can come from many sources, such as decaying organic matter and vulcanized rubber insulation. This tarnishing seldom affects switch performance, but occasionally small amounts of silver sulfide collect between the contacts and increase the resistance of the switch. If, for some reason, the switch cannot be sealed and the sulfide cannot be eliminated from the environment of the switch, special contact materials may be used. Gold does not tarnish, but it is expensive. Various alloys of gold and silver, often with a ternary addition, preserve much of the tarnish-resistant quality of gold without the expense of gold. Another approach is electroplating. Its effectiveness depends upon the base metal, the nature and thickness of the plate and subplate, a number of variables in the cleaning and plating processes, and several factors in the design of the switch mechanism. Simply specifying a gold plate is useless. In fact, some kinds of gold plate can aggravate tarnishing of the base metal. In-depth discussion of contact plating is contained later in this chapter.

Condensed Metal Vapor. When an arc is present, evaporated metal may condense in the form of powder on the contacts and adjacent surfaces. Its color may be yellow, brown, black, or metallic. Condensed metal vapor seldom affects performance of the switch significantly.

Wear Particles. The generation of contact wear particles is affected by selection of contact material, contact fabrication (cold working and annealing), contact assembly (staking or welding), lubrication (intentional or otherwise), contact force, and contact sliding. The effect of wear particles on switch resistance is governed by these factors and by distribution of particles in the wear track, as well as by the number of cycles of switch operation.

The alleviation of contact contamination problems depends upon accurate identification of the type and source of the material between the contacts. To make this possible, it is best to return any switch suspected of having contact contamination, *unopened* (the switch case intact), to the switch manufacturer, in a plastic bag or other lint-free container. Prior to returning any switch, have the person in the user organization responsible for quality contact the manufacturer and discuss the reasons the switch is being returned. Send a copy of the receiving-inspection report that highlights obvious discrepancies in the switch operating parameters that point to contact contamination. The report will help the manufacturer's quality personnel quickly focus on the specific problem. The switch manufacturer has more than just a passing interest in determining what, if any, contact contaminants may be present in the manufacturer's environments or processes, or the environment and processes of one of the manufacturer's suppliers.

Generally speaking, the system or design engineer selecting a switch should not specify contact material; however, the user *is* entitled to know how the switch manufacturer decides what material to recommend. The use of gold contacts often is associated with low voltage and current. But just when should gold be used? The answer is not just a set of circuit parameters, and the decision often is difficult to make. The next section considers the various aspects of contact material selection.

Contact Materials

Before the discussion of contact materials, the problem of switch resistance should be put into perspective. As has been noted earlier in the chapter, practical problems with switch resistance are almost always due to careless soldering procedures, which contaminate the contacts with flux, or to particles of other contaminants settling between the contacts. Changing contact material has no effect upon such problems.

The most nearly universal contact material is silver. It combines the chemical, electrical, thermal, and mechanical properties that usually are needed for best contact performance in a wide range of applications. *If silver contacts are clean, there is no lower limit to the voltage and current that they will control reliably.* This applies, for example, to a switch that is sealed sufficiently to keep out contaminants. Silver has a definite drawback and that is its tendency to tarnish in the presence of H_2S and moisture. This characteristic encourages the use of gold contacts in some applications.

If silver contacts are exposed to sulfides and moisture for a long enough time and in sufficient concentration, the contacts will tarnish. This seldom affects performance of the switch. Nearly always, the combination of mechanical force and movement of the contacts, and the circuit voltage, rupture the tarnish film and reestablish good electrical continuity. Occasionally, however, small amounts of silver sulfide may collect at the contact interface and increase the resistance of the closed switch enough to constitute an open circuit. Generally such a malfunction clears up on the next switch closure, *but* it may not. The likelihood that silver contacts will experience this kind of problem depends upon the voltage, current, inductive characteristics of the circuit; the temperature, humidity, and purity of the environment; the degree of sealing of the switch enclosure; the mechanical forces and movement of the switch contacts; the exposure time and number of switch operations; and the amount of switch resistance that constitutes an effectively open circuit.

Gold is nearly inert chemically and does not form sulfides or oxides in normal switching environments. It has some important limitations as a contact material. It is expensive, it is soft and ductile, and its usefulness is very limited where an electric arc is present. It does, however, prevent sulfide tarnishing if properly applied. To reduce cost and make contacts more nearly universal, silver contacts are sometimes plated with gold. Accordingly, the theory goes, if the voltage and current are low enough to make gold contacts desirable, they are low enough that they will not disturb the gold plate. If voltage and current are too high for gold contacts, they will burn the plate off and expose the silver, which is suitable for higher loads. However, there are some practical limitations. If gold is plated directly on silver contacts, sulfur atoms in the presence of moisture can penetrate the pores in the gold plate and react with the silver base metal, forming silver sulfide. The sulfide then migrates rapidly over the surface of the gold plate as a spongy deposit that can cause more trouble than would a sulfide tarnish on an unprotected silver contact. The heavier the gold plate, the more slowly this will happen, but the more expensive will be the switch. The usual procedure is to use a nickel barrier plate between the silver base metal and the gold plate. This stops the sulfide problem but adds to the cost and sometimes is incompatible with some of the switch manufacturer's production processes.

Cost aside, the answer would seem to be a solid gold contact, but gold is very ductile and may experience plastic flow under the influence of contact force on closure. This can be remedied by alloying other elements with the gold to harden it. Sometimes it is possible to alloy a high enough percentage of other elements with the gold that the cost of the contact can be significantly reduced. But this introduces other considerations, for example, polymer formation. Gold and gold alloy contacts can generate polymers at the contact interface when organic contaminants are present in the atmosphere. Silver does not form polymers. The type of polymer usually formed on pure gold contacts does not increase switch resistance, but the same is not always true for gold alloys. If the atmosphere around gold or gold alloy contacts is clean, no polymer will be formed, but then under such conditions silver contacts will not form sulfides and are considerably less expensive. In short, the choice of contact material involves a number of considerations and often a trade-off decision.

Other contact materials sometimes are used for applications of this kind. The most common materials are platinum, palladium, and their alloys. Although these materials are sometimes used as pure metals, they tend to have poor wear properties and are very soft—hence the use of alloys, which preserve some of the desirable properties of the elemental materials and improve hardness and wear resistance. The principal alloying elements are other metals of the platinum group, such as iridium, ruthenium, osmium, and rhodium. Others are silver, copper, and nickel. Most of the platinum and palladium alloys can form polymers that increase switch resistance.

In summary, although no universal rules can be laid down, the following practices usually are followed:

- If the switch is sealed, sulfides cannot enter and silver contacts can be used.
- If the switch is not sealed, the electrical load should be considered. If there is an arc, silver contacts can be used.
- If there is no arc, the environment is a controlling factor. If *particle* contamination is likely to reach the contacts, gold is no help. Use a sealed switch or bifurcated contacts.
- If a completely alien film contaminant such as paint spray or oil mist can reach the contacts, gold does not help, so choose a sealed switch.
- If the environment of an unsealed switch contains significant amounts of moisture and H_2S (from sources such as decaying organic matter or vulcanized rubber), gold contacts can be a real help (7).

How Switch Design Affects Resistance

In designing switches, the logical way to prevent contamination of contacts by the environment is to enclose the mechanism in such a way that foreign material cannot enter the switch. For example, the molded plastic case of the common snap-acting switch does this adequately for most applications. Switches required to operate in particularly dirty environments are available with special enclosures which provide total sealing. Aside from a protective enclosure, however, there are four important factors which switch designers consider in reducing resistance problems due to contamination of the contacts:

1. Contact material
2. Contact geometry
3. Contact force
4. Contact movement after closure

These four factors must be considered *together,* and it is their combination that controls contact resistance. The important point to remember here is that the switch designer is the person responsible for the interaction of these factors, and generally speaking, the user gains nothing by specifying contact material, contact force, or other features of the *internal* design of the switch. In fact, by doing so, the user is likely to increase cost and may degrade switch performance. Instead, the discerning user specifies what the switch must *do,* and the switch manufacturer selects or designs the switch to perform, as required by the user specification, at as low a cost as possible.

How to Measure Switch Resistance

How should switch resistance be measured? If the resistance of a given switch is measured, in turn, with an ohmmeter, a bridge circuit, and the voltmeter-ammeter method (measuring voltage across the switch at a specified current), it will be found that no two methods yield the same result. This is understandable in view of the foregoing discussion. A switch does not have a fixed resistance, and the measured resistance of a switch depends to a large extent upon the way in which it is measured.

If this is the case, how can meaningful measurements be made? They can be made by keeping the variables as realistic as possible in view of the intended use of the switch (the discerning reader will realize that the following list of variables does not always apply to all switch technologies and is meant to be viewed generically):

Voltage. The source voltage of the measuring circuit should be the same as that of the circuit in which the switch is to be used. The higher the source voltage the higher the likelihood the switch will close the circuit with good electrical continuity. A measuring circuit having higher or lower source voltage than that of the end use can give unduly optimistic or pessimistic results.

Current. The current of the measuring circuit should be the same as that of the circuit in which the switch is to be used.

The Measuring Circuit. The above discussion has shown the importance of specifying and controlling the voltage and current at which switch resistance is measured. This is most easily accomplished by connecting the switch in a simple series circuit consisting of a power supply of the specified voltage, a variable resistor, an ammeter, and the test switch. The resistor is adjusted until the specified current is attained. With this current passing through the switch, the voltage drop across the switch terminals is measured with a voltmeter, using a separate set of test probes for the voltmeter connection. The measured voltage is divided by the current to calculate the switch resistance. It is essential that separate current and voltage connections be used, as described, to avoid including the terminal connection resistance as though it were part of the switch resistance.

Switch Plunger Position. In some switches, such as precision snap-acting types, the resistance varies with plunger position. When switch resistance is measured in such devices, the switch plunger should be held at the point of its travel where it will be held in actual use at the time when switch resistance is important (4). This is particularly true of membrane, metal dome, and conductive rubber switches in that relative position of the contacting materials may have an effect on switch resistance measurements. For example, pressing a membrane–metal dome switch until electrical contact is initially made will yield one measurement. Pressing the metal dome even farther could possibly yield another measurement entirely.

Sequence of Operations in Measuring Switch Resistance. If the switch closes the electric circuit, the voltage is present at the contacts at the time they close. The voltage assists the mechanical forces in breaking through any contaminant which may be present between contacts. If, on the other hand, the switch is closed before the voltage is applied, the mechanical, and electrical forces are applied to the contaminant in sequence rather than simultaneously. There is somewhat less likelihood, then, that continuity will be established. In measuring switch resistance, the sequence of contact opening and closure, and application of voltage to the switch should be the same as it will be in end use. Avoid the practice of measuring switch resistance at one combination of current and voltage, following an endurance test at a different current or voltage. Such a measurement has no practical value.

When to Give Switch Resistance Special Attention

The following circumstances should flag the user to pay particular attention to switch resistance and to discuss these aspects of the switch under consideration with the switch's manufacturer:

1. When the switch controls a circuit at less than 0.5 V
2. When several switches must be connected in series
3. When a switch is likely to be exposed to contaminating particles or fumes
4. When an occasional switch closure with switch resistance exceeding 1 percent of the load resistance will have *dire* consequences.

How to Be Constructively Cautious about Switch Resistance

The following checklist should assist the specifying engineer in applying a prudent amount of caution regarding switch resistance, resulting in a forceful specification governing performance and enhancing cost-effectiveness in the switch specified:

1. Design the circuit to be as insensitive to normal variations of switch resistance as possible.
2. Design the circuit so that the switch will see voltages well above the softening voltage of the contact material (see Table 1.1).
3. Design low-voltage circuits with a minimum number of switches in series.
4. Get advice from a switch manufacturer regarding choice of switch and suggestions about the specific application.
5. Store switches in a clean, reasonably dry environment that does not exceed the rated temperature of the switch.
6. Install the switch carefully. Do not modify the switch in any way without first checking with the switch manufacturer. Do not overtighten mounting screws

(if used). If the switch is mounted using an adhesive, use extreme care to avoid contaminating the interior of the switch. When soldering leads to terminals, use flux core solder and a soldering temperature of about 550°F (288°C). Do not use solvents on switches. Do not paint over switches after installation.

7. Protect unsealed switches from contamination by particles and fumes.
8. Actuate the switch as nearly as possible to the extremes of its travel without applying excessive force that might damage the switch. These force figures are generally available from manufacturers or can be calculated by the manufacturer.
9. Decide on *solid technical grounds* whether a switch resistance specification is needed and whether its additional cost is justified. If so, specify—*on the basis of end use*—the voltage, current, plunger position (if used), sequence of actuation and measurement, and maximum allowable switch resistance.
10. If additional assurance of low and stable resistance is desired, connect two or more switches in parallel and actuate them at the same time (obviously, this can't always be done, but it is a consideration in those situations where it is possible to do so).
11. Test the switch under conditions simulating end use to verify that it performs as required (7).

SWITCH LIFE

Switch life is a function of a number of variables in switch design and end use. A change of any one of these variables can alter the life of the switch. For this reason a switch life figure has meaning only if it is accompanied by a statement of the conditions under which it applies. The variables of end use that have the greatest effect on switch life are circuit parameters, actuation, the environment, and the criterion of failure. Under a given combination of these conditions, however, switch life is not constant. Ten switches of the same kind, tested as nearly identically as possible, are likely to have 10 different life figures with a fair amount of spread between highest and lowest. This is because subtle variations in the switch and test conditions have a significant effect on switch life.

When thinking of practical switch life at a given set of conditions, it is best to visualize it as a frequency distribution. If one figure is to represent the life of the switch at these conditions, the figure should be chosen with the distribution in mind. Many different switch types are manufactured with close control of materials, dimensions, and adjustment. Other switch types, by their very nature, may not be manufactured with the same amount of regulation (for example, once membrane switches are assembled, no further adjustment of the actuation stroke is usually possible). For satisfactory switch life, then, it remains for the end user to select the right switch and apply it with due regard for the electrical, mechanical and environmental factors that affect its life.

Just as other components do, switches can eventually wear out. However, "failure" of the switch may be subtle, such as a minor drift of operating point or reduction of insulation resistance that renders the switch unfit for further use in a specific application. Failures of this kind must be defined for the particular end use where certain characteristics of the switch are important. There are more obvious modes of switch failure that can affect a wider variety of application.

Contact Welding

If contacts weld together, the switch may fail to open the circuit reliably. As switch contacts close, they bounce for a few milliseconds. If the electrical load is high enough, the contacts draw an arc as they rebound. This melts metal at the contact surfaces, and they reclose on a pool of molten metal with considerable force. The higher the current the greater the volume of molten metal. As the molten metal hardens on both contacts, the interface between the contacts solidifies and the contacts remain closed, resisting any mechanical action by the switch mechanism as it endeavors to open the contacts.

Contact Material Migration

In some dc circuits, and occasionally in an ac circuit, a significant amount of contact material migrates gradually from one contact to the mating contact. This produces a mound or cone on one contact face and a corresponding crater in the other. The change of contact geometry may narrow the distance between the open contacts until they can no longer break the dc arc, and the switch is destroyed. In other cases, the cone hooks into the crater, and the switch fails to open the circuit. Typical temperature at the contact surface at the root of the arc equals that at the surface of the sun—several thousand degrees Celsius. Naturally, some of the contact material boils vigorously and some more is melted.

During contact bounce on closure, molten metal splashes. As the contacts separate to open the circuit, arcing occurs again. Meanwhile, things are happening on the atomic scale. As the contacts begin to separate, a bridge of molten metal is drawn between them. As it ruptures, it may leave more metal on one contact than on the other. This is called "bridge transfer," and opinions differ as to exactly how it occurs. A short arc is drawn as the bridge breaks. Electrons emitted from the cathode cross the gap without interference (the gap at this time is too short to contain many gas atoms) and bombard the anode. Their high energy causes ionization of some of the surface atoms of the anode. The resulting positive ions of anode material are repelled by the anode and attracted to the cathode. In this manner, metal is moved from anode to cathode. As the contacts continue to separate, a phenomenon occurs which moves material in the opposite direction in the form of vapor and continues until the contacts are about 4 μm apart (assuming silver contacts in air).

As the contacts continue to separate, a significant amount of electron avalanching begins and a plasma of ionized gas develops. Now the gap is wide enough to contain an appreciable amount of the ambient gas. Electrons emitted by the cathode strike gas atoms, losing some of their energy in the process, ionizing some of the gas atoms and thus releasing more electrons. The electrons reaching the anode have such low energy that ionization nearly stops. Pressure of the metal vapor in front of the cathode is higher than that in front of the anode and the pressure differential draws a jet of metal vapor from the cathode. The jet strikes the anode and the vapor condenses there. Called "plasma arc transfer," this causes migration of metal from cathode to anode.

Many other events take place at the contact interface before, during, and after contact separation, and the entire process may take place in a couple of milliseconds. On the other hand, if the circuit parameters so dictate, arc time may be 50 ms or longer. During contact separation, some processes act to transfer metal from anode to cathode, and others work in the opposite direction. The net effect often is removal of material from one contact and deposit of part of that material on the opposite contact. If electrical polarity is constant throughout switch life, as it is in most dc applications, the transfer of material during switch life can be considerable. In ac circuits, polarity reversals at 120 times per second (in a 60-Hz circuit) may result in general erosion of material from *both* contact surfaces with no buildup on either.

Contact Erosion

In many ac circuits and some dc circuits, there is gradual erosion of material from the faces of both contacts. If this proceeds rapidly enough, a contact may be entirely eroded away in the course of switch life. The result can be either failure of the switch to open or failure to close the circuit.

Dielectric Breakdown

Heating by the electric arc and deposit of eroded contact material can cause an insulator to lose its insulating properties. The result can be a gradual decline of insulation resistance and an increase of leakage current. If the leakage current becomes high enough, it can raise the temperature of the insulator and hasten deterioration. At some point, if the impressed voltage is high enough, the material can break down, destroying the switch. The result is conduction of current through an "open" switch or conduction from switch terminals to the grounded mounting.

Mechanical Deterioration

Wear can cause abrasive particles to accumulate on contacts. Fatigue can cause springs to fracture. Either can cause the switch to fail to open or fail to close the circuit.

Although failure modes have been described individually, two or more may combine. Considerable experience with switch evaluation is needed for accurate identification of the mode(s) of switch failure, but this is an important step in devising ways to increase switch life.

The terms "mechanical life" and "electrical life" sometimes are used in switch evaluation. *Mechanical life* is the life of the switch without an electrical load. It is determined by such factors as the wear life of bearings and the fatigue life of springs (if any) in the switch. *Electrical life* is the life of a switch controlling a specified electrical load. Electrical life may be limited either by mechanical factors or by the life of contacts or insulators. Neither of these terms is precise. It is much more useful to consider switch life under conditions of actual use, with meaningful criteria of failure (7).

Effects of Circuit Parameters on Switch Life

Current Rating of a Switch. The published current rating of a switch at a given voltage represents the maximum electrical load the switch is designed to control. As a rule it is based on connection of the circuit to either the normally open or normally closed throw of the switch, and does not necessarily apply where both throws of one pole are connected simultaneously. If the switch has more than one pole, the electrical rating usually applies with one throw of each pole connected. The current rating generally assumes that, in the case of pushbutton- and snap-action-type switches, the plunger of the switch is driven to full overtravel and full release during actuation.

The rated current is only one point on a curve of switch life versus electrical load. For example, a switch having a current rating of 15 A at 125 V ac at room conditions, based on a life of 100,000 operations with 95 percent survival, can be expected to control currents below 15 A satisfactorily. A life versus load curve for the switch at these conditions (see Fig. 1.17) shows the expected life at currents below the rated load. At each condition for which it is rated, the switch can close the circuit, carry the steady-state current indefinitely, and open the circuit, during each cycle of operation through life. The ability of the switch to close and

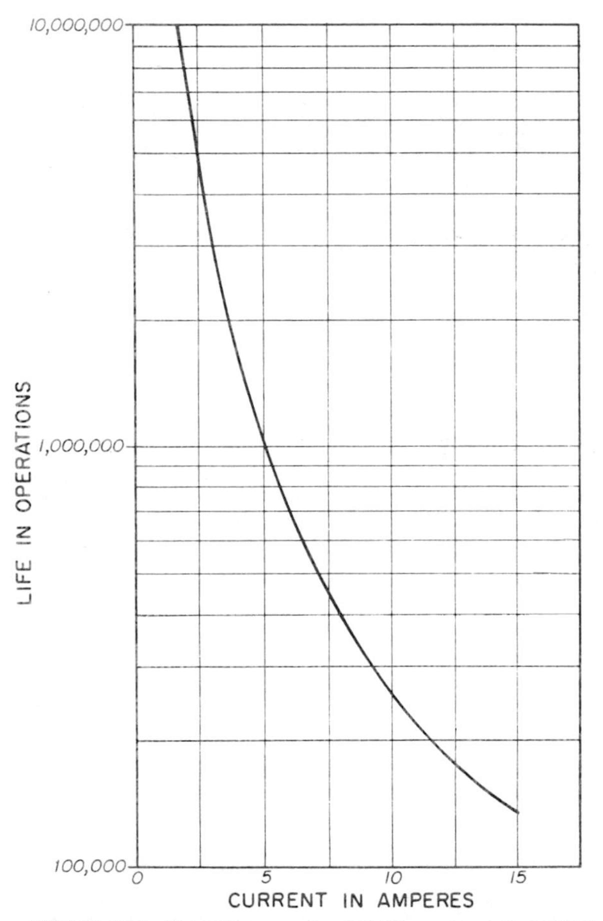

FIGURE 1.17 Typical curve of switch life vs. current at 120 V ac, 75 percent power factor, room conditions. (*Courtesy of Micro Switch, Freeport, Ill.*)

open the circuit reliably is affected by the current versus time characteristics of the circuit (Fig. 1.18).

Occasionally, the suggestion is made that switches be provided with minimum voltage and current ratings, i.e., values of voltage and current below which they should not be used. This derives from the erroneous impression that a given switch will develop performance problems below specific levels of voltage and current. In practice, this is not the case. A clean switch usually can control microvolt-microampere circuits without difficulty. There is no particular voltage or current level at which problems begin, and there is no technically valid way by which to set minimum electrical ratings. It is possible to establish minimum ratings on the basis of arbitrary resistance levels, but such figures are useless and meaningless for general application. Minimum ratings do not help prevent problems with switch resistance. Furthermore, they may mislead some users into

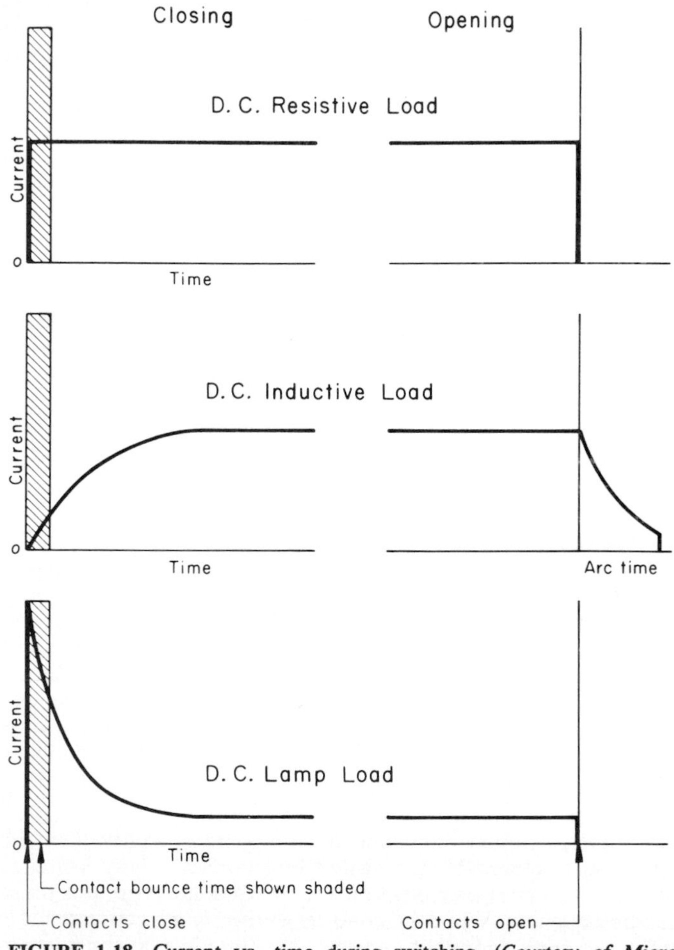

FIGURE 1.18 Current vs. time during switching. (*Courtesy of Micro Switch, Freeport, Ill.*)

thinking that the switch is satisfactory for any use above the specified current and voltage, without regard to the other variables which affect switch resistance. In short, minimum electrical ratings can increase the cost of switches without a corresponding increase in performance.

Closing the Circuit. In a dc resistive circuit such as an electric heater, the steady-state current is present at the instant of switch closure. As the switch contacts strike and then bounce apart, each rebound draws an electric arc. This melts metal on the surfaces of the contacts, and some of the metal is evaporated. There may be some general erosion of material from both contacts, and a net transfer of material from one to the other. The contacts then reclose on molten metal, sometimes forming a weld when the metal solidifies. The higher the current the stronger the weld is likely to be, and the higher the force that the switch mechanism will have to provide to open the circuit. The strongest welds occur when the load is characterized by a high inrush current, such as that of a motor or a tungsten-filament lamp. In the case of a lamp, the inrush current may be ten or more times as high as the steady-state current. If the inrush current persists during part or all of the contact bounce time, conditions are conducive to severe arcing and strong welds. Capacitive circuits often have high current at switch closure, encouraging contact welding. The highest current of all occurs when the switch is closed on a short circuit. Unless the switch is specially designed to withstand closure on a short circuit, life under this condition can be predicted as zero.

In a dc inductive circuit there is a time delay as the magnetic field builds up in the coil before current reaches its steady-state level. Since most of the contact bouncing occurs during the low-current part of the transient, there is little contact deterioration and almost no tendency to weld during closure. In ac circuits the current transients combine with the alternation of the current, and the switch closes at random points on the current wave, including current peaks and zero current. Ac inductive circuits, such as those containing solenoid coils, almost always involve moving iron and consequent inrush currents.

Switches are designed to resist contact deterioration and the effects of contact welding throughout life, but in all circuits, the lower the current on switch closure the longer will be the life of the switch.

Carrying the Steady-State Current. Once the contacts have closed and stabilized, the switch carries the steady-state current of the circuit. This is simply a matter of controlling the I^2R heating, and seldom presents a problem. One A ac has the same heating effect as 1 A dc, since the equivalence of the two is based on heating. Control bodies such as Underwriters' Laboratories may impose limitations on temperature rise at rated current, but the requirements usually are easily met. If a switch is used in calibrated thermostatic equipment, there may be practical limits on the heat the switch is permitted to dissipate. Except for very unusual overloads or short-circuit conditions, however, switch life is unaffected by the current the closed switch carries.

Opening the Circuit. Before the contacts can separate, any welds holding them together must be fractured by the switch mechanism. Many switches are designed to do this. In a dc resistive circuit the steady-state current is present at the instant of switch opening. When the contacts of the switch separate far enough to extinguish the arc, nothing further happens. Arc time in a dc resistive circuit usually is very short and arc energy is low.

There is some erosion and migration of contact material. In a dc inductive circuit arcing is more severe, because the energy stored in the magnetic field of the coil is partially dissipated in the arc as the field collapses. The arc often persists after the contacts are fully separated, and contact erosion and migration continue as long as the arc lasts. During the life of the switch, migration gradually narrows the space between the open contacts and eventually may draw and sustain an arc, destroying the switch.

Normal arcing melts and evaporates contact material, some of which may condense on surfaces of adjacent insulators. The intense heat of the arc itself may gradually deteriorate insulators that are near it. The general effect is to reduce their insulation resistance and dielectric strength. This is encouraged in a dc inductive circuit by the voltage transient that occurs just as the arc goes out. The current drops suddenly to zero, producing a voltage proportional to its rate of change. The duration of this high-voltage transient usually is so short that it has little effect on switch life. Switches are designed to withstand these conditions during life, but at the end of switch life, one possible mode of failure is electrical breakdown of an insulator. The higher the source voltage the more prevalent is this mode of switch failure.

Because dc inductive loads produce the most arcing, they are associated with another phenomenon. An unsealed switch exposes the electric arc to air, producing chemically unstable oxides of nitrogen, and nitric acid if the air is humid. Normally this is so diluted by the air that it causes no problem, but if the air is very humid, enough HNO_3 may be dissolved in the water vapor to cause internal corrosion of the switch. Dc motor circuits are no problem for the switch to open unless stalled rotor conditions prevail.

As a rule, ac circuits are easy for the switch to open because the arc is extinguished as the current passes through zero. At very high electrical frequencies (over 1000 Hz) the atmosphere between the contacts may not have time to deionize and the voltage may restrike the arc, but such conditions are rare. In general, when opening the circuit, anything that *reduces* arc energy will *increase* switch life.

Connection of opposite polarity between normally open and normally closed terminals of a switch can cause premature failure. The electric arc is a conductor that bridges the gap between the opening contacts of the switch. The arc may be extinguished while the contacts are separating or may persist for a few milliseconds after the contacts are completely open. This is normal, and most switch technologies are designed for it. A condition for which they are not designed occurs when the opposite sides of the line are connected respectively to the normally open and normally closed terminals of the same pole of the switch. When this is done, and when the arc persists after the contacts are fully apart, a short circuit is established through the arc and, hence, through the switch. This drastically reduces switch life and is a condition to be avoided. A connection of this kind can occur unintentionally. Figure 1.19*a* and *b* shows two circuits illustrating the principle. In the circuit of Fig. 1.19*a,* a double-throw switch is used in a conventional single-phase, three-wire installation, with the center wire grounded. This provides 240 V ac between outside conductors and 120 V ac between either of them and the center conductor. The switch is intended to change a heater from "high" heat (240 V) to "low" heat (120 V). Current values are 10 and 5 A, respectively. However, arc duration exceeds the transfer time of the switch, and an arc is drawn between normally open and normally closed contacts. This provides a short circuit through the arc with an impressed voltage of 120 V. The switch fails after a few operations. In the circuit of Fig. 1.19*b,* a 28-V dc permanent-

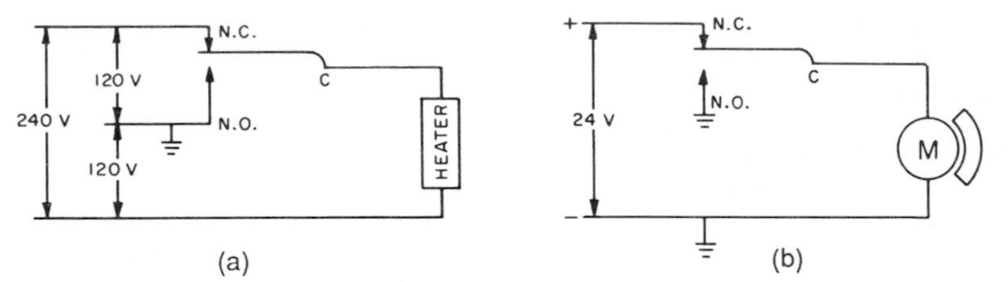

FIGURE 1.19 Connection of opposite polarity between normally open and normally closed terminals of a switch can cause premature failure. (*Courtesy of Micro Switch, Freeport, Ill.*)

magnet motor is connected to permit dynamic braking. Although the motor draws only 1.5 A, a short circuit occurs through the arc as soon as the contacts are fully apart, and the switch fails immediately (7).

Effects of Actuation on Switch Life

A number of factors related to the actuation of switches affect switch life. These are direction, distance, force, velocity, and frequency.

Direction of Actuation. If the particular switch style has a plunger actuator, longest life will be obtained by depressing the plunger with an actuating member that moves along the same axis as that of the plunger. The surface striking the plunger should be perpendicular to the plunger axis. The purpose is to avoid side thrust that can reduce switch life by causing unnecessary wear of bearings in the switch. It is wise to note that actuating force applied at an angle to the axis of the switch plunger will have a component perpendicular to the sliding bearing surface of the plunger. If the actuator movement is coaxial with the switch plunger, the extra bearing wear is eliminated. This is not to say that some side movement cannot be tolerated. In a wide variety of switch applications side thrust cannot be avoided conveniently, and in those types of switches where side thrust is most likely to occur owing to application-driven requirements (such as snap-acting switches), the switches are designed to withstand a certain amount of side thrust. But over many thousands of operations, the effects of a large side-thrust component are detrimental to switch life.

Distance of Actuator Travel. In many switch applications, the distance the actuating member moves is determined by the operating point of the switch. An example is a limit switch that stops movement of a sliding machine carriage at a certain point. In some applications, however, the amount of travel applied to the switch plunger is optional and controllable. In such cases, the plunger must receive enough travel to assure that the switch will operate properly, but additional travel may tend to reduce switch life. The farther the switch plunger (or switching member) is depressed or released, the higher will be the stresses in the internal mechanism of the switch and the shorter the life. All switches can tolerate a certain amount of overtravel (some more than others), and some are provided with special mechanisms that allow a very long overtravel without damage. But for longest switch life, unnecessarily long overtravel and release travel should be avoided.

Force Applied to Switch Plunger. With any switch, a certain force is necessary to depress the plunger to the end of its travel, i.e., the full overtravel point. Most switches are designed to tolerate additional force, and figures of maximum allowable actuating force are available from switch manufacturers. If this force is exceeded, the mechanical life of the switch can be reduced drastically. An example of misapplication of this kind is the use of the switch itself as a mechanical stop to limit the travel of a moving member. Dynamic loads can have disproportionately high static equivalents. It is essential that the force applied to the switch plunger (or switching member) not exceed the design limit of the switch. In the case of membrane switches (Chap. 8), however, the control of the applied actuating force is difficult since the operator usually initiates switch closure by pushing the appropriate membrane area. Every operator, and even the same operator on a repeat basis, will not push with the same force, and in fact, under duress or stress most will push with more vigor than under normal circumstances. The internal structure of the membrane does not incorporate force-limiting members, resulting in premature wear or failure of the switch.

Velocity of Actuating Member. As an example, in a precision snap-acting switch an actuating member striking or releasing the plunger of this style of switch with high velocity delivers a shock to the mechanism of the switch. This can cause high stresses, multiple flexures of the spring, and considerable extra bouncing of the contacts. Extreme forms of this, referred to as "impact" actuation or "drop-off" release, are noted for their tendency to reduce switch life. A logical question is: how high can actuating velocity be, without reducing switch life significantly? The answer depends largely upon the design of the switch itself and the nature of the electrical load. If long life is desired in an application involving high-velocity actuation, the switch manufacturer can help select a suitable switch, and some thought might be given to actuation through velocity-reducing linkage. In rare cases, the acceleration of the actuator must also be considered, but this requires individual treatment by the switch manufacturer.

Frequency of Switch Actuation. If the switch mechanism is still in motion from prior actuation when it is actuated a second time, dynamic effects can cause high stresses and reduce the life of the switch. High-frequency actuation can aggravate contact bouncing and hasten the deterioration of the contacts. For satisfactory switch life, the bulk temperature of the contacts must be controlled by the design of the switch. Most switches are designed with adequate heat sinks for normal requirements, but at heavy electrical loads, high-frequency actuation of the switch can overheat the contacts and reduce switch life. Again, limiting values of frequency of actuation depend upon the design of the switch, the electrical load, and the performance requirements of the circuit. Often, a switch manufacturer will have a standard frequency of operation for both ac and dc loads. These operation frequencies may be higher for ac loads because of its shorter inherent arc time. In either case, with actuations above these frequencies at full rated electrical load, it would be reasonable to expect some reduction of switch life. At lower current, higher frequencies of actuation can be tolerated. At the other end of the spectrum, mechanical endurance tests of switches with no (or negligible) electrical load are commonly performed with actuation frequency from 100 to 800 operations per minute, depending upon the design of the particular switch. Obviously, the switch manufacturer is the best source of advice on frequency of actuation and should be consulted as early in the application stage as possible (7).

Effects of Environment on Switch Life

Life figures and current ratings for general-purpose switches of all technologies are established at room conditions and are valid for the various combinations of temperature and barometric pressure associated with living areas. The most common deviations from room conditions that may tend to reduce switch life are reduction of atmospheric pressure and increase of ambient temperature. As atmospheric pressure is reduced (in high-flying aircraft, for example), an unsealed switch will experience an increase of arc energy and some reduction of dielectric properties. Under some conditions, it may be necessary to reduce the current rating to maintain the same switch life. Published current ratings often reflect this fact. A simple way to maintain life in a vacuum without reducing the current is to use a sealed switch.

Ambient temperature above that for which a switch is designed can reduce insulation resistance and dielectric properties of plastics, reduce the elastic modulus of any spring materials used, and interfere with efficient cooling of the contacts. The result can be reduced switch life, or the necessity to reduce the current to maintain the desired switch life. Specific questions about switch life at high temperature are best handled by consulting the switch manufacturer. To assure adequate switch life, however, it is always wise to test the switch under conditions simulating those of end use (7).

Evaluating and Estimating Switch Life

When Are Life Tests Needed? In most switch applications the electrical, mechanical, and environmental conditions and requirements are known to be well within the switch's capability (again, keeping in mind the parameters under which these factors were developed by the manufacturer), and no special testing of the switch is required. In some cases, however, evaluation of the switch is needed. A critical application, such as one in a manned space vehicle, may require that numerical life figures be established for reliability purposes. Similar data may be needed as a basis for preventive maintenance in electronic monitoring equipment for the surgical suite. An unusual condition or requirement of end use may dictate need for a life test.

Control of Life Test Conditions. Subtle differences between test conditions and those of end use can invalidate a test. The effect is to render the test not only useless but hazardous, since inadequate switches may pass the test but fail in end use. When the need for a test arises, time and money can be saved, and the tests rendered more useful, if the major variables are specified and controlled as they are in a switch testing laboratory. In order to be useful, a test must be both valid and reproducible. A valid test is one that measures what it is intended to measure. A reproducible test is one that gives essentially the same results on identical test specimens. The test must be carefully designed to reflect the electrical, mechanical, and environmental conditions of end use. The criteria of switch failure must be based on actual performance requirements of the application, and switch performance must be monitored accurately for this during the life test. While not necessarily applicable to every switch of each of the wide variety of switching technologies, the following test conditions are among the most important to consider:

- Mounting surface—electrically conductive or insulating
- Mounting means—panel mount, side mount, mounting hardware, adhesives, etc.
- Attitude of switch
- Mass and elastic properties of actuating member (plunger, finger, etc.)
- Direction of actuation with respect to the switch
- Overtravel and release travel applied to the switch
- Overtravel force applied to the switch
- Velocity of actuating member
- Frequency of actuation, and on-time vs. off-time
- Wire gauge and means of attachment of wires to switch
- Location of electrical ground points in the circuit, including grounding of power supply, switch mounting, and actuating member
- Polarity of connection to switch
- Connection of switch in hot or ground side of line
- Connection of poles and throws of switch
- Voltage magnitude and tolerance
- Current magnitude and tolerances on make, carry, and break
- Frequency of alternating current
- Inductive or capacitive characteristics of circuit (when testing, use actual load components of end use, if possible)
- Temperature
- Atmospheric pressure
- Relative humidity
- Definition of switch failure or end point of the test
- Instrumentation used to detect switch failure
- Unusual environmental conditions (heavy shock, vibration, etc.) expected to be encountered by the switch during operation

Criteria of Switch Failure. Most of the variables in the foregoing list are tangible and fairly obvious, although several are frequently overlooked. "Definition of switch failure" requires some clarification. If a switch is used to control a critical circuit in a surgical instrument, a single malfunction may have serious consequences. A switch in the operator control of an electronic game system may suffer repeated malfunctions and not be noticed, or cause nothing more than minor inconvenience. So the first consideration in deciding criteria for failure is the seriousness of a single malfunction. Bear in mind that the malfunction may take the form of either failure to open or failure to close the circuit. Next, how many malfunctions *can* be tolerated? Is the total number important, or is it the frequency of occurrence that should determine end of switch life? What if the switch experiences several malfunctions in succession, then reverts to correct operation? What if malfunctions occur singly and only occasionally, perhaps at the rate of 10 in 1000 operations?

The criteria of switch failure are set on the basis of answers to these questions. A very conservative criterion may be set by deciding that the first failure of the switch to open and close the circuit once, and only once, for each cycle of the actuating member constitutes end of switch life. In some circumstances, such a stringent criterion is justified. In many cases, however, a switch has considerable useful life remaining after the first malfunction, and a conservative definition of failure can be unnecessarily expensive. It may be more realistic and economical to set the end point of the life test in terms of X malfunctions per thousand operations, or a continuous sequence of Y malfunctions. The important thing is that the criteria of failure be significant, pertinent to end use, and clearly defined. The equipment used to monitor switch performance during the test must then be chosen with these criteria in mind.

Sample Size for Life Testing. How many switches should be tested? Unless the distribution of switch life is already known or the sample size is specifically governed by a military or industrial specification, this question cannot be answered before the test. Since the purpose of the test is usually to establish the shape and dimensions of the life distribution curve, the size of the initial sample is a matter of intuitive professional judgment. It is unsafe to assume that the life of a particular switch follows a normal, exponential, or other particular distribution, unless it has been demonstrated to do so by life tests. The sample must be large enough to be able to show the range of life which is characteristic of the switch under the conditions of test. In the absence of other guidelines, a good *initial* sample is 10 switches. Statistical treatment of the resulting life figures may show that this is sufficient, or may indicate a need for tests of additional samples to define the distribution more accurately.

Combining Tests. In the interests of efficiency and economy, it is not unusual for switch test specifications to call for several tests in sequence on one switch. If the tests preceding the life tests are simply measurements that do not damage the switch, this is a sound idea. If they degrade the switch, however, the subsequent life test may not be valid. Sometimes a test sequence is set up to "condition" the switch, i.e., intentionally render it more susceptible to failure during a subsequent life test. Unless the nature, severity, and sequence of such preparatory conditions accurately reflect those of end use, the life test will be difficult to interpret in realistic terms.

Postmortem Examination of the Switch. A switch may be examined after a failure has occurred either in end use or on a life test, and if the examination is properly performed, it can yield a great deal of useful information. An experienced eye is needed to perform this examination properly, and the best source of help is the switch manufacturer. In fact, an astute switch manufacturer will be eager to help determine the cause of any switch failure since it can aid the manufacturer in pinpointing an operation, material, or process in the production cycle that may be contributing to the failure(s).

It is important that the switch be opened without destroying useful information, and this task is best left to the manufacturer. The condition of the switch must be accurately observed and described, and the mode(s) of failure must be clearly determined. Any irregularity or unforeseen performance of the switch during its life must be accounted for during the examination. The x-ray and microscope are important tools for this purpose.

To get the most useful information, the user should not open the switch. Pack the failed switch in a plastic bag to keep out contaminants, and return it to the manufacturer with a full description of its history and performance. Include information on the environment, actuation, voltage, current, and frequency of actuation. Identify which throw of the switch controlled the load, and whether the switch failed to close or failed to open the circuit. Record any unusual environmental circumstances in effect at the time of switch failure, whether under test or in actual application. Complete information supplied with the returned switch will enable the manufacturer to make a rapid and conclusive analysis of the problem and to recommend a practical solution.

Interpretation of Life Test Results. Given the results of a life test of a sample group of switches, which of the life figures is to be considered the one that best represents the life of this kind of switch? It is a common practice to arrange the figures in order of magnitude, "eyeball" them, and make an educated guess. This is perfectly acceptable for many applications. Intuition is a fairly good guide, provided that the guess is truly an educated one and is based on an adequate and representative sample. Many statistical techniques have been applied to switch life figures, with varying degrees of success. Some of the unsuccessful attempts have assumed that the life figures were distributed in accordance with a gaussian or other standard curve, and have tried to force-fit the data to the curve. Such estimates have no practical value and may even be hazardous if trusted too far. It is a mistake to assume, without testing, that switch life tends to follow a normal, log-normal, or any other predetermined distribution (7).

How to Increase Switch Life in End Use

Given a switch designed for long life, what can the user do to take maximum advantage of this feature? The variables of end use can be classified under the headings of electrical, mechanical, and environmental factors. The following are some ways to prolong switch life.

Electrical Factors

Voltage. In some instances where a switch is used to control, as an example, a 480-V ac load, the circuit can be arranged so the switch performs the required function but controls only 120 V. In this instance, the contact life of a switch is about the same at 240 V as at 120 V ac, and is somewhat reduced at 480 V. The probability of dielectric breakdown, however, increases with the supply voltage, and switch life is appreciably longer at lower voltages.

Current. As a rule, the lower the current, the longer the switch life. Therefore, reducing the current that the switch must close and/or open usually increases the life of the switch. There are exceptions, such as cases where the switch utilizes a spring in its internal mechanism and spring life becomes the limiting factor. Circuits having high inrush current may limit switch life by contact erosion or welding. If the inrush is due to tungsten filament lamps, a resistor can be connected in parallel with the switch to keep the filaments warm and reduce the inrush current. If this is objectionable, a resistor or thermistor in series with the filament can reduce inrush current.

Arc Control. The less severe the arc, the longer the life of the switch contacts. The most severe arcing usually is encountered in dc inductive circuits. Here the energy stored in the field is partially dissipated in the arc. Several arc-suppression techniques are available. One of the simplest and most effective is a diode connected in a blocking mode across the dc coil. When the switch opens, the polarity of the voltage induced in the coil opposes that of the steady-state condition and the diode conducts, shorting the coil. This can greatly reduce arc energy at the switch contacts and give a corresponding increase of switch life. The diode can be of much lower current capacity than the steady-state current of the load, because the diode conducts only for an extremely short time. The diode should have a peak inverse voltage rating exceeding the source voltage. Use of a diode will delay drop-out time of the inductive device by a few milliseconds, but this is usually acceptable. The arc-suppression device should be tested to confirm its suitability. Reducing the arcing in this way can extend switch life greatly by reducing the rate of contact material migration and decreasing the heat dissipated by the arc.

Choice of Throw. On some switches, how a switch is designed internally may point to one throw as the preferred throw. For example, the normally closed throw of a snap-acting switch usually is better able to break contact welds than is the normally open, owing to the nature of the switch mechanism itself. If a circuit has a high inrush current on closure or requires the switch to open current near its rated load, switch life is likely to be longer if the normally closed throw is used to control the circuit.

Grounding. Switches are often mounted on a surface or bracket that is electrically grounded. In many cases this is desirable or necessary for safety. In other cases, it is optional. If the mounting is connected to a common electrical ground with the power supply, line voltage will be maintained between current-carrying parts of the switch and the mounting surfaces. This applies a continuous electrical stress and acts to encourage dielectric breakdown as switch life proceeds. The electrical ratings of switches are established with the mounting grounded, to provide "worst-case" conditions. However, if the switch is mounted on an insulator in end use and is actuated with an insulating member, the life of insulating parts of the switch will be prolonged.

Mechanical Factors

Coaxial Actuation. As much as possible, actuate the switch coaxially with the plunger or actuator. If this is not practical, install a roller-type plunger or similar device to prolong the life of the plunger bearings or actuator-bearing surfaces.

Overtravel. Do not apply excessive force to the plunger or actuator at full switch overtravel.

Frequency of Actuation. Consult with the switch manufacturer to obtain maximum frequency-of-actuation parameters and design in a reasonable safety factor as recommended by the manufacturer. For example, Micro Switch generally recommends that when their snap-acting switches are controlling heavy electrical loads, frequency of actuation should be no higher than 20 operations per minute on dc loads or 60 operations per minute on ac.

Mounting. Always follow the manufacturer's recommendations for mounting any switch, whether using mechanical fasteners or adhesives. If the user's requirements dictate an unusual installation method or atypical positioning of a switch, contact the switch manufacturer as early as possible in the design stage to avoid mounting or positioning the switch in such a way as to produce a less than optimum operating condition, greatly reduce switch life, or create a potentially hazardous condition (7).

TACTILE FEEL AND ACTUATION FORCE

Of all possible parameters of switch performance and specification, none creates more confusion, generates more misunderstanding between user and manufacturer, increases costs and delivery schedules, and yet is more arbitrary than *tactile feel.* Tactile feel is the *subjective* perception of switch actuation through the operator's finger or probe. Subjective is emphasized here because an optimum tactile feel suitable for one operator *may* not (and, more than likely, *will* not) be suitable for another person operating the same equipment. As the complexity of the equipment increases in scale and the sophistication and training of the equipment's operators increases concomitantly, the importance of tactile feel and actuation force also increases. An example of this is the immense amount of experimentation undertaken by human factors specialists at commercial and military aircraft manufacturers to establish acceptable ranges of actuation force along with relevant, useful, and reproducible parameters of tactile feel. The final arbiter in this decision process, of course, is the ultimate user: the pilot. It is not unusual for cockpit design and human factors personnel to devote countless hours to produce a seemingly ideal tactile feel and actuation force combination in a prototype device, only to see this hard work and effort obliterated by a simple statement from the pilot—"I don't like it." The point is, of course, that the pilot operates in a critical environment with many different types of stimuli. The safety of the aircraft may depend on the immediate and accurate actuation of a particular switch. The pilot needs reliable and precise feedback of successful operation of the switch, many times through protective gloves and in a noisy environment. Here is where tactile feel becomes so important. The pilot "feels" the switch actuate by the perception of closure provided by the tactile impression received through the finger. The maxim here is that in order to produce equipment employing switches which will be acceptable to the operator and enhance successful operation of the system, the designer must be or become familiar with the user population and must understand under what conditions the equipment is to be operated which may alter the type of tactile feedback or actuation force specified.

A common method of specifying actuation force is the force curve (also called a force-deflection curve) as shown in Fig. 1.20. The curve is a plot of the actuation force necessary to obtain a certain deflection, or displacement, of the switch actuating mechanism. The ideal way for a user to originate this curve is to secure samples of the type of switch desired in a variety of actuation forces. Next, determine empirically the device with the most suitable actuation force and tactile feel (this might mean getting a consensus among a number of operators). Have the switch manufacturer then measure the selected switch to develop the actual force curve, incorporating reasonable tolerances.

Note that in Chap. 11 distinct force curves are given for certain actuation force–tactile feel combinations. The reason conductive rubber force curves are so exact (and so common) is that the rubber molders can control the force curves with a great deal of precision in the mold design. The body of design knowledge possessed by these rubber molders allows them to exert such control over the final product.

APPLYING SWITCHES IN HOSTILE ENVIRONMENTS

In their simplest forms, electromechanical, electromagnetic, and solid-state control devices are designed for environments which may be called friendly, i.e., clean factory areas, offices, and other surroundings that are typical of room conditions. However, these devices often are needed for use in the presence of dust,

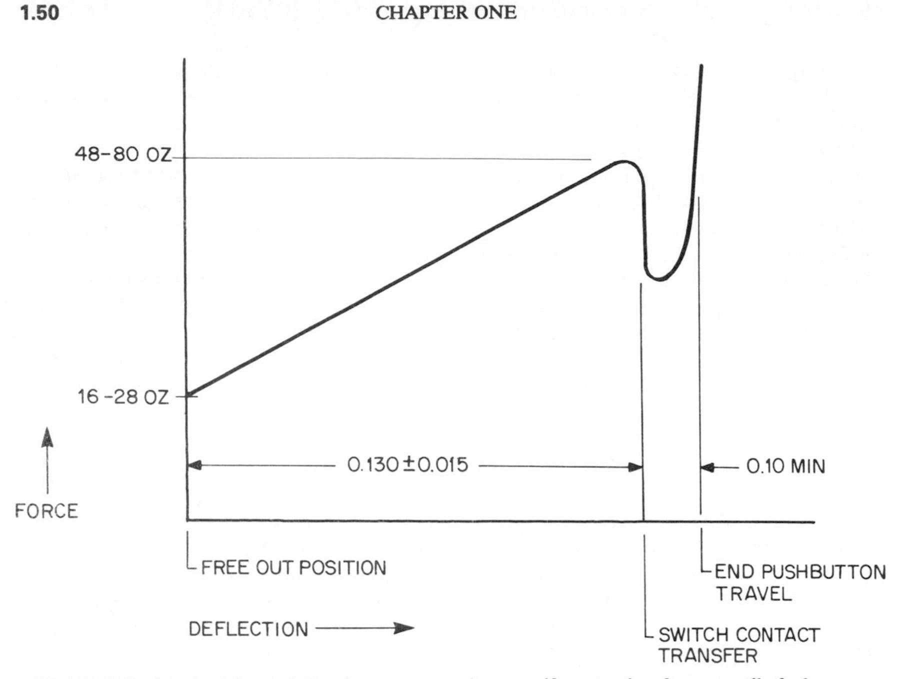

FIGURE 1.20 Typical force-deflection curve used to specify actuation force–tactile feel parameters. (*Courtesy of Eaton Corporation, Milwaukee, Wis.*)

dirt, metal particles, oil, corrosive agents, or very high or low temperatures. For extreme environments such components usually are made from special materials, provided with protective enclosures, or changed in other ways to alter the immediate area of control and enable the device to survive and perform satisfactorily.

Switches control circuits safely and reliably in millions of applications. As the link between the mechanical and electrical parts of a system, a switch must perform well, both mechanically and electrically. It usually does, despite the complication effects of temperature, humidity, and other environmental factors. Still, there are environments that reduce the reliability of switches and can even cause premature failure. High temperature can reduce contact life; a partial vacuum can encourage electrical breakdown to ground; oil can deteriorate plastics and disable the switch; ice can jam the actuating mechanism; the electric arc in an unsealed switch can detonate an explosive atmosphere.

The effects of environments on switches are not always obvious, and it is possible for a system design to be well advanced before a potential switch problem is recognized. Familiarity with the effects of environment on switch performance often can improve the design and save considerable time. This section discusses the factors to consider when applying switches in hostile environments.

Switch electrical ratings are shown in manufacturers' catalogs and often are printed on the side of the switch, but environments are much more difficult to classify. Any specific environment consists of a unique combination of temperature, pressure, humidity, contamination, and the like, and these conditions sometimes conspire to cause switch problems. The environment in which a switch must operate is determined not only by the geographical location but also by the equipment and circumstances in which it is to be used. A switch may be operating in an arctic oil drilling installation but in a heated cabin where it is never expected to function in extreme cold. On the other hand, a switch in the tropics, in the temperature control system of a refrigeration unit, may be exposed continuously to cold. The temperature climate of the central Atlantic Ocean imposes no undue stress on switches, aside from the humidity and corrosive effects of the marine environment. However, switches on aircraft flying above this same area may experience −65°F (−54°C) on the landing gear or 1000°F (538°C) near the afterburner, both in a partial vacuum. Below the ocean surface are switches on submarines and stationary undersea equipment where they are exposed to high hydrostatic pressure. On land, switches in sump pumps, flour mills, and ready-mix concrete trucks are exposed to contaminants, while those in paint spray equipment, printing presses, and surgical operating rooms may constitute an explosion hazard. Military vehicles and equipment are expected to survive a host of severe environmental conditions which can be, and often are, applied simultaneously. Table 1.2 illustrates some types of hostile environments to which switches may be exposed.

Selection of a switch for use in an adverse environment must be based on the required performance and the conditions of end use. The switch manufacturer is the best source of advice, and the earlier in the design process the manufacturer is contacted the better for all concerned. The switch should be tested by exposing it to simulated conditions of end use (electrical, mechanical, and environmental) and evaluating it to be sure it performs as required.

The following are some general guidelines to help the switch user understand and maneuver through the maze of environments often encountered in system and component specifications.

Acceleration, Shock, and Vibration

Before considering the effects of acceleration, shock, and vibration on switches, it is important to have the nature of these environments firmly in mind. Since shock and vibration are forms of acceleration, the first consideration is the nature of acceleration. When a vehicle travels along a straight road at constant speed, its velocity has two characteristics: magnitude and direction. Velocity always is characterized by magnitude and direction. The magnitude of velocity can be changed by speeding up or slowing down. The direction of velocity can be changed by changing the direction of motion. The direction of the velocity is always the same as that of the motion. Acceleration is any change of either the magnitude or the direction of velocity.

If the vehicle is coasting along a straight road (neglecting friction and the effects of wind), its velocity is constant and its acceleration is zero. If the vehicle speeds up, thereby increasing the magnitude of its velocity, it is accelerated in the same direction as its velocity. If the vehicle slows down, thereby decreasing the magnitude of its velocity, it is accelerated in the direction opposite to that of its velocity. If the magnitude of velocity (i.e., the speed) remains the same, but the direction of motion changes, the direction of velocity changes with it, and this also constitutes an acceleration. If the vehicle rounds a curve toward the right, while maintaining constant forward speed, its acceleration due to change of direction of velocity is toward the right, specifically toward the center of curvature of the curve itself. Similarly, a turn toward the left, at constant speed, involves acceleration toward the left. If both

TABLE 1.2 A Partial List of Typical Hostile Environments Encountered by Switches

Environment	Typical applications
High temperature:	Industrial and household furnaces, pasteurizing equipment, steam cleaning of food processing machinery, foundries, rolling mills, surfaces of high-performance aircraft, jet engine afterburners, missile launchers
Low temperature:	Commercial refrigeration, military and commercial equipment in arctic regions, aircraft flying above 35,000 ft, cryosurgical, liquid oxygen, and other cryogenic equipment
Temperature shock:	Transfer of equipment to and from heated shelters in arctic regions, airdrops of military supplies, spacecraft reentry
Vacuum:	Aircraft and spacecraft, aerial cameras and weather instruments, industrial vacuum processes
High pressure:	Undersea equipment, oil drilling instrumentation
Humidity:	Laundry machinery, dairy and meat packing equipment, textile plants, hothouses, carrier-based aircraft, pharmaceutical manufacture
Liquid splash or shallow immersion:	Sump pumps, aircraft landing gear, shipboard deck mounted equipment, gasoline pumps, hydraulic production machinery
Ice:	Snow removal machinery, ski lifts, refrigeration controls, aircraft, arctic installations
Corrosion:	Marine and seaboard applications, plating departments, battery manufacture
Sand or dust:	Earth moving machinery, desert vehicles, air conditioners, foundries, cement mills, concrete block manufacture, textile manufacture, flour mills
Fungus:	Tropical military gear, geological and meteorological instruments
Explosion:	Starch packaging, coal mines, petroleum refining, grain elevators, flour mills, coke manufacture, surgical operating rooms, machining operations producing aluminum or magnesium dust

Source: Lockwood (7).

speed and direction are changing at the same time, the total acceleration is the vector sum of the components of acceleration due to change of magnitude of velocity, and the change of direction of the velocity.

Now consider the effects on a passenger in the vehicle during acceleration. In all instances, the passenger feels a force in the direction opposite to that of the acceleration. If the vehicle accelerates forward, the passenger is forced back into the seat. If the automobile accelerates backward (by braking, for example) the passenger feels a forward force. If the vehicle accelerates toward the right, by rounding a curve toward the right, the passenger feels a force toward the left. If the vehicle is not accelerating (for example, when it is standing still or following a straight road at constant speed), the passenger feels no horizontal force.

The common contact of a switch can be considered a "passenger" in the switch. When the switch is accelerating in any direction, the common contact experiences an apparent force in the opposite direction. This may act to keep closed contacts closed and open contacts open, in which case there is no problem. The force may be directed perpendicular to the line of movement of the common contact, and have no significant effect. However, if the force acts to separate closed contacts or to close open contacts, there is the possibility that the switch may experience a malfunction. During the launching of a high-velocity missile, switches on the missile are subjected to high linear acceleration. Switches used in the hub of a propeller or in a spinning projectile have a component of acceleration toward the center of rotation. The movable contact, as a passenger in the switch, may be forced in an unfavorable direction.

Up to this point, we have considered acceleration that is fairly uniform. In practice, acceleration often is of a transient nature. When a device containing a switch is struck, dropped, or otherwise subjected to a disturbance, it undergoes a pulse of acceleration known as a shock. In its simplest form, this transient acceleration is all in one direction, but its magnitude varies with time. A graph of the acceleration versus time may be a simple half-sine wave, or it may have any of a wide variety of shapes and dimensions. Although ordinarily mechanical shock has little or no effect upon switch performance, a shock pulse having high acceleration and relatively long duration can cause a closed switch to open momentarily or an open switch to close momentarily. If acceleration is very high, in the thousands of gravity units, some switches may be permanently damaged by the shock.

To judge the effect of most mechanical shocks on switches, return to the vehicle analogy. If a standing vehicle is struck from the rear, the impact accelerates it forward. The passenger feels a force in the direction opposite to that of the acceleration and is forced back into the seat. If the standing vehicle is struck from the front, or if a forward-moving vehicle collides with an obstacle, it is accelerated toward the rear. The passenger feels a force in the opposite direction (forward). The common contact of a switch experiences similar forces when the switch is subjected to a mechanical shock, and the effect can be judged by considering the position of the contacts and the direction of the shock.

If a shock pulse has a fairly simple waveform (such as half-sine), it is usually specified in terms of acceleration versus time. If the shock wave is complex and cannot readily be expressed in this way, it is sometimes specified in terms of acceleration versus frequency, and the resulting graph is called a shock spectrum. If the duration of the shock pulse is of the same order of magnitude as the half natural period of some part of the switch mechanism, its effects on the switch may be amplified or attenuated. Thus a shock pulse having a $50g$ peak acceleration may separate the closed contacts of a switch, while a steady acceleration of $50g$ in the same direction would not.

Vibration is an oscillating movement which may have a consistent, repetitive pattern or may be irregular. Thus the acceleration may vary regularly or irregularly. Most laboratory vibration tests provide simple harmonic motion, which is a sine wave. The acceleration then follows a negative sine wave and is specified in terms of frequency and maximum acceleration. In applications where the vibration does not follow a simple waveform, conditions may be more difficult to specify. In some instances, the vibration is not periodic and the acceleration varies erratically. In the case of random vibration, a sampling of motion during a time interval shows that the instantaneous acceleration values are distributed in accordance with the gaussian curve. Such vibration usually is specified in terms of spectral density (g^2/Hz) versus frequency, and can be represented graphically on a loglog grid.

In vibration, as with the other forms of acceleration, the common contact is a passenger in the switch and experiences an apparent force in the direction opposite to that of the acceleration. With vibration, the acceleration is along an axis, first in one direction, then in the opposite direction. The magnitude and direction of acceleration are reversed rapidly, and the rate of change affects the response of the switch to the vibration. The closed contacts of a vibrating switch may remain closed at 10*g*, 50 Hz, but may separate momentarily at 10*g*, 500 Hz.

In summary, acceleration is any change of the velocity's magnitude or direction. Shock and vibration are forms of acceleration in which acceleration varies with time. The common contact, as a passenger in the switch, behaves as though it were forced in the direction opposite that of the acceleration. With this in mind, one can judge whether a given acceleration, shock, or vibration will act to cause closed contacts to open or open contacts to close. If it appears that there may be a problem of this kind, several steps can be taken to prevent occurrence of the problem:

1. Use a miniature switch. The mass of its moving parts is a major factor in the response of a switch to acceleration, vibration, or shock. Use of a subminiature version of a particular switch type is one of the most effective and least expensive solutions.
2. Orient the switch so the acceleration, shock, or vibration will not tend to separate the closed contacts or close the open contacts. It may even be possible to orient the switch so the acceleration helps to hold the contacts in the desired position.
3. On switches with these mechanisms, keep the switch plunger or actuation device fully released or fully depressed during acceleration, shock, or vibration. This takes advantage of the high contact force available at the extremes of plunger or actuator travel.
4. Be certain that the actuating device and switch mounting do not respond to acceleration, shock, or vibration in such a way as to cause movement of the switch plunger or actuator mechanism.
5. Where shock or vibration is a problem, install the switch on a shock mounted panel if the equipment itself is not shock mounted. The purpose is to attenuate the shock or vibration reaching the switch.
6. Use two or more switches oriented differently from each other. For example, use two switches with their plungers pointed in opposite directions but actuated by a common linkage. An acceleration, shock, or vibration tending to separate the closed contacts of one switch will tend to hold the contacts of the other switch closed. If the intent is to keep the circuit closed, connect the closed circuits of the two switches in parallel. If the intent is to keep the circuits open, connect the open circuits of the switches in series (Fig. 1.21).
7. Make the electric circuits less sensitive to momentary disturbance of the switch contacts. For example, if a switch is controlling a dc relay coil, a capacitor connected across the coil can increase the response time of the relay.

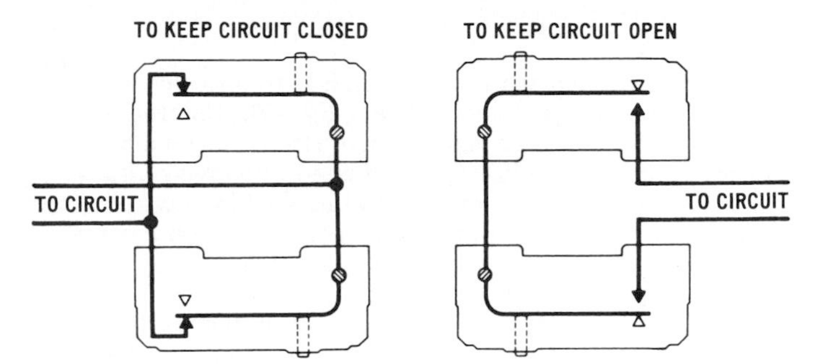

FIGURE 1.21 Use of redundant switches, actuated by a common linkage, to prevent disturbance of circuit by acceleration, shock, or vibration. (*Courtesy of Micro Switch, Freeport, Ill.*)

8. Test the switch in end use conditions. Laboratory acceleration tests performed with a centrifuge are quite valid if an application involves uniform unidirectional acceleration. Laboratory vibration tests are considerably less valid because it is difficult to duplicate conditions of end use accurately in the laboratory. Laboratory vibration tests are worthwhile if the results are applied with judgment. Incidentally, it is impossible to make a useful estimate of switch performance in random vibration, based on results of a sinusoidal test. Laboratory shock tests are used to demonstrate that switches conform to the shock requirements of applicable specifications. Beyond this, most laboratory shock tests have little practical use.

   With rare exceptions, every shock pulse encountered in the field is unique. Consequently, switches that pass a 50*g* shock test in the laboratory may experience contact chatter in a 20*g* shock of end use. Similarly, switches that have contact chatter during a 50*g* shock test in the laboratory may exhibit no chatter at 100*g* in the field. No laboratory test can determine the general shock resistance of a switch or even rank switches for general resistance to shock. If shock resistance is important, it warrants a test in end use equipment.

9. Specify performance requirements that reflect actual need. One of the best ways to save time and money is to put into the performance specification only those requirements that are needed for the switch to perform properly in end use. Acceleration, shock, and vibration are no exception. If a switch is required to hold a large solenoid energized during shock, and if a circuit opening of 1 ms would cause no problem, nothing is gained by requiring that contact separation time during shock not exceed 1 μs. A great deal of effort may be wasted in meeting unnecessarily stringent specifications. The inevitable result is higher cost with no corresponding benefit in end use.

10. If a problem arises, consult the switch manufacturer. The ability of a switch to resist the effects of acceleration, shock, and vibration is determined by a variety of factors such as the magnitude and direction of spring forces, the distribution of mass in the switch mechanism, and the elastic properties and physical strength of the various parts of the switch. These are a matter of switch design and are under the control of the switch manufacturer.

Corrosion

Unless it is specifically designed for the purpose, an unsealed switch in a corrosive environment is likely to have a short life. The internal mechanisms of switches are designed to meet a combination of electrical and mechanical requirements at normal operating conditions, and this may involve some highly stressed parts. A corrosive environment may lead to the combinations of stress and corrosion that bring early fracture. In addition, corrosion products may fall on the switch contacts, causing a short circuit. Unsealed switches can be built from materials selected to reduce galvanic effects. Many switches designed for use at high temperature make use of alloys which happen to be inherently resistant to corrosion. If a corrosion problem is confined to one area in an unsealed switch, such as the most highly stressed part of a leaf spring, it may be possible to protect the affected area with an organic coating so corrosive medium cannot reach it. The simplest solution is to use a sealed switch, chosen to withstand the corrosive environment in question.

Explosion

Explosive mixtures of gas, vapor, or dust with air can be detonated by an electric arc if the energy of the arc is at least as high as the ignition energy of the mixture. In general, gases produce the most violent explosions, with vapors ranking next and dusts the least violent. Obviously an unsealed switch is unsuitable for use in explosive environments. The use of switches and other electrical equipment in explosive atmospheres is closely governed by electrical standards which set minimum construction requirements, installation procedures, classifications of explosive atmospheres, and test criteria that must be met for approval. Among the control bodies having standards of this kind are National Electrical Code, Underwriters' Laboratories, Canadian Standards Association, U.S. Bureau of Mines, Department of Defense, and National Fire Protection Association. European standards are regulated by various agencies such as DKE, BSI, and SEMKO. Each of these organizations is concerned with preventing explosion, which may be ignited by arcing contacts in the case of flammable gas or vapor or by heating in the case of combustible dust.

In designing switches for explosive environments, switch manufacturers keep in mind the following major considerations:

- The tendency of the mixture to explode
- Explosion pressure of the mixture when detonated
- Ability of the switch enclosure to withstand an external explosion without damage
- Flame-arresting capabilities of the switch enclosure in the event of an internal explosion

Switches designed for use in explosive environments must be able to withstand an internal explosion without igniting the explosive mixture surrounding the switch enclosure. The enclosure (Fig. 1.22) is thus designed to withstand the maximum expected internal explosion pressures without damage or excessive distortion, and to provide venting for the pressure through channels of such dimensions that gases will be cooled below the ignition temperature before reaching the surrounding atmosphere. Accordingly, the design of an explosionproof switch enclosure involves careful consideration of thickness of housing, fit of cover, and fit of shaft joints.

The minimum ignition energy of an explosive atmosphere is affected by ambient temperature and humidity, as well as by pressure and other factors. If arc energy exceeds this ignition energy, increasing the electrical load has no effect on the tendency to detonate or on the resulting explosion pressure. Obviously, a switch that has passed a particular explosion test is not necessarily safe for use in all explosive or flammable environments. As always, it is best to test under the conditions expected in end use. When selecting an explosionproof switch, choose one with adequate mechanical life, as the fit of the actuator seal is critical and important and must be maintained throughout the useful life of the switch.

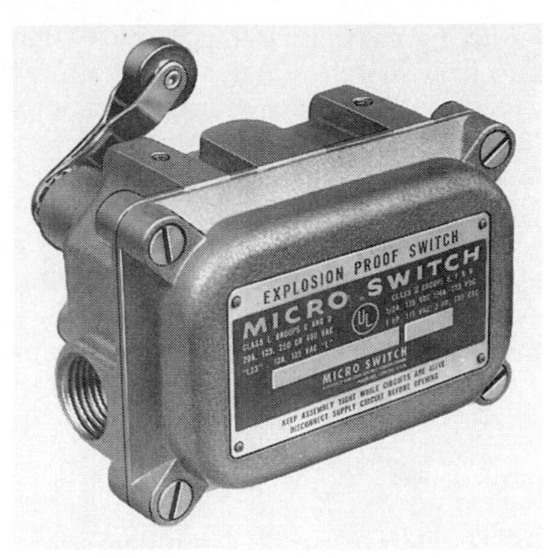

FIGURE 1.22 An explosionproof switch. (*Courtesy of Micro Switch, Freeport, Ill.*)

Fungus

Fungi grow fastest in warm, damp, and dark environments in the presence of certain organic salts. Fungus growths on or in a switch constitute contamination. On switch contacts in low-energy circuits, they may cause high resistance. On insulators they may set up a low-resistance path resulting in excessive leakage current to ground. Aside from such passive roles, fungi may attack switches actively. They can cause oxidation, reduction, or hydrolysis of organic materials. They may attack plastics containing cellulose fillers such as linen, cotton, or wood flour. They also may attack elastomers containing catalysts, plasticizers, or fillers that are susceptible to fungus. Such problems can be prevented by using sealed switches that have all exposed surfaces made from nonnutrient or fungicidal materials.

Humidity

Standard unsealed switches will perform satisfactorily in most environments, including those having high humidity. However, prolonged exposure to high humidity or to rapid changes of temperature and humidity can degrade switch performance, and it is wise to be aware of the possible effects. Humidity has both direct and indirect effects upon switch performance. It acts directly on plastics to cause dimensional changes, reduce physical strength, and degrade electrical properties. Humidity can also reduce fatigue life of springs. Humidity acts indirectly by combining with other environmental conditions to affect switch performance. For example, humidity makes it possible for sulfides in the air to tarnish silver contacts, causing occasional problems on low-energy circuits. An electric arc, in the presence of humidity, produces nitric acid and corrosive oxides of nitrogen, which can attack metallic parts of the switch mechanism. In applications involving fluctuating temperature, humidity may condense and even freeze on an unsealed switch. Carrier-based aircraft are exposed to this condition, in addition to the rapid increase of atmospheric pressure during a dive, which can force condensate into unsealed switches.

Minor effects of humidity, such as changes of operating characteristics of the switch, usually can be remedied by changes of plastics. Corrosion problems that sometimes occur with severe arcing at contacts in dc inductive circuits in moist air usually can be reduced or eliminated by adding an arc suppression circuit, such as a diode connected in the blocking mode across the coil. The more serious problems that may result from prolonged exposure to high humidity or rapidly cycling humidity are best avoided by use of sealed switches.

Ice

Ice formation can be a problem when it interferes with the normal actuation or release of a switch. An ice coating can withstand considerable force, its strength depending upon the conditions under which it is formed. Sealed switches are a necessity where ice formation is likely to occur. Unsealed switches may experience frosting at the contacts as well as external interference with actuation. The material, geometry, and finish of the external parts of a switch determine to some extent how tightly an ice coating will cling. The force of the plunger return spring or actuator return force of the switch is the most important factor in assuring that the switch will release despite an ice coating. Research has indicated that a surface to which ice cannot adhere is one coated with a liquid or semisolid lubricant which retains its lubricating properties at low temperature.

Some enclosed-style switches are designed with the plunger terminating in a clevis, connected to the actuating device with a pin. This allows the switch plunger to be driven positively in both the actuating and release directions and makes it unnecessary to rely upon the internal force of the switch to break the ice. Ice tests conducted in a laboratory generally rate "poor" in validity and reproducibility. In typical switch applications, ice almost never deposits as a uniform coating. Rather, it often tends to build up as a pile or prismatic ridge on one side of the switch, leaving the opposite side virtually bare. The quality, and thus the strength, of the ice varies widely. Ice formed on switches in the slipstream of flying aircraft has properties different from those of the ice buildup owing to freezing of condensed moisture in still air. Similarly, in the laboratory, ice applied to a switch using water from a spray gun is different from that formed by alternate immersion and freezing. The most nearly reproducible (but not necessarily valid) ice tests are those performed by alternately immersing the switch in ice water and freezing it in a low-temperature chamber. By controlling temperature, time, sample orientation, and the number of immersion and freezing cycles, one achieves the closest possible approach to a reproducible test. It is pointless to specify an ice test in terms of ice thickness. Ice is thin at outside corners and tends to form fillets at inside corners. Furthermore, ice thickness can be measured accurately only by breaking the coating, thus destroying the test. In short, actual trial of the switch in its end use environment is much more useful than a laboratory ice test.

Liquid

Unsealed switches sometimes are accidentally splashed when adjacent parts or equipment are being oiled, filled with hydraulic fluid or fuel, or washed down.

The effects of liquid on an unsealed switch can be devastating. The exposure of unsealed switches to liquid splash or immersion, whether accidentally or by design, is an extremely hazardous practice. Depending on several factors, it can cause the switch to fail to close the circuit, fail to open the circuit, start a fire, or cause an explosion. Because of the importance of precautions in this area, and because users sometimes feel that such precautions can be ignored "just this once," the following discussion illustrates how liquids affect unsealed switches.

Consider what happens when a typical plastic-cased unsealed switch is splashed with petroleum oil. The oil soaks into the switch case, dissolving the plastic to some extent. First the pigment dissolves, soon followed by some of the partially polymerized plastic. The effect is to erode the material and reduce its dielectric strength by releasing polar organic compounds into the solvent oil. Although phenol formaldehyde plastics resist oxidation, some of the compounds found in petroleum oils can oxidize the material when heated. Among these compounds are cresols, nitrates, and organic sulfates. The reaction produces compounds which are electrolytes and hence conductive. This provides a path for electric current to bypass the open switch. When the oil reaches the contacts of the switch, the electrical load determines what happens next. If the load is such that there is no arc at the contacts, the oil (and any debris carried by it into the switch) can deposit between the contacts and the switch may fail to close the circuit. If it does close the circuit, the resistance of the contamination between the contacts may be so high that the component being controlled by the switch will not have enough voltage for reliable operation. If the electrical load is high enough to cause an arc at the switch contacts, the electric arc burns some of the oil, forming water, CO_2, and carbon. The water contaminates the remaining oil and the phenolic case material, reducing the insulating properties of both. Carbon deposited on the surface of the phenolic provides a low-resistance path for electric current. At the same time, a carbon residue remains on the contact surface, increasing the energy of the arc itself (this assumes that the carbon residue has not prevented the contacts from closing the circuit). The result is that the insulating material begins to lose its insulating ability and leaks current, often to the grounded surface on which the switch is mounted. As leakage current increases, I^2R heating facilitates combustion of the oil and degradation of the case material. This, in turn, reduces the resistance of the current leakage path and increases the leakage current. In other words, the condition is self-aggravating and a conductive path is burned through the case of the switch. In its final stages, dielectric breakdown can be spectacular, with showers of sparks, tongues of flame, and a spray of molten metal. It is even more of a spectacle if breakdown occurs in a flammable or explosive environment.

What if an unsealed switch is immersed rather than splashed? The liquid may attack the plastics as described above. If the liquid is flammable, and if the switch is kept completely immersed so that oxygen cannot reach it, dielectric breakdown of the switch is not likely to ignite the liquid, unless hot metal ejected from the switch reaches the surface of the liquid while still above the ignition temperature of the vapor. Of course, if the liquid is a chemically unstable mixture that can burn without additional oxygen, dielectric breakdown of the switch may ignite it directly. A definite hazard exists if the switch is only partially immersed or is immersed only part of the time, as it might be in a fuel tank. This then reverts to a flammable vapor which can be ignited by the arc or dielectric breakdown of the switch.

What if the liquid is not flammable—water, for example? This sometimes happens when moisture condenses in a wiring conduit and runs into the attached switch enclosure. The moisture distorts the plastic parts of the switch and may change the operating characteristics. As the switch stands open, partially immersed in contaminated water and connected to the power line, leakage current passes through the water. This in itself is a hazard because the circuit is not completely open. Inside the switch, electrolysis begins to destroy the mechanism, and eventually the switch fails completely.

The lesson is clear—do not contaminate unsealed switches with oil, water, or other liquids, by either splash or immersion. Use sealed switches where such contamination is possible, and avoid the hazards mentioned.

High Pressure

Unsealed switches designed primarily for use at sea level will perform satisfactorily *in air* at high pressure, with no reduction of electrical rating. However, many high-pressure switch applications involve media other than air, such as seawater. In many of these instances, a sealed switch must be used. The case of a sealed switch will experience higher external than internal pressure, and the effects of the resulting pressure differential must be considered. If the switch is of a style that uses a plunger-type actuator, the ambient pressure will act upon it as a piston and provide a residual force (pressure times cross-sectional area of the plunger), tending to depress the plunger. At moderate pressures, this will appear only as a reduction of operating and release forces. If ambient pressure is high enough, however, it can cause spontaneous actuation or failure of the actuated plunger to release. At extreme high pressure, the switch case may suffer temporary distortion or permanent damage. Maximum allowable pressure figures are available from manufacturers of sealed switches, and special switches are available for use at very high ambient pressures.

Low Pressure (Vacuum)

Generally, a switch designed primarily for use at room conditions will give satisfactory performance at reduced atmospheric pressure, within the limits of the switch's electrical ratings for that pressure. Often the electrical ratings at reduced pressure are lower than those at sea level atmospheric conditions. If this reduction of ratings cannot be tolerated, remedies are available. They are discussed later in this section. Since switch applications in a vacuum usually involve aircraft or spacecraft, atmospheric pressures are expressed in terms of the corresponding altitudes. It is wise to remember that a switch intended for use at *X* thousand feet may also have to operate at altitudes *up to X* thousand feet.

As an unsealed switch is exposed to increasing altitudes, its dielectric strength and its ability to interrupt current first decrease, then increase until they exceed the capability at sea level. At the extreme altitudes of outer space where hard vacuum prevails, contacts or other metallic bearing surfaces may cold-weld, and sublimation of materials may present problems. The dielectric strength and current-interrupting ability of an unsealed switch vary continuously with altitude, but not all switches behave alike. The altitudes mentioned in the following discussion are the approximate levels or ranges at which the practical performance of most unsealed snap-acting-style switches is affected (see Table 1.3).

Reduction of Dielectric Strength. Vacuum can reduce the voltage an open switch will tolerate without breakdown. An unsealed switch at sea level may withstand

TABLE 1.3 Effects of Vacuum on Unsealed Switches

Absolute pressure, torr*	Approximate altitude, ft	Approximate altitude, m	Switch phenomena caused by reduced pressure
1.4×10^{-8} and below	1½ million and above	457,200 and above	Seizing and galling of metal surfaces in contact. Possible cold welding
3.5×10^{-6} and below	500,000 and above	152,400 and above	Aggravated electrical welding of contacts
8.5×10^{-5} and below	350,000 and above	106,680 and above	Enhanced current-interrupting ability but voltage transients may damage insulators. Dielectric strength of air so high that any breakdown is likely to be over-surface or through body of an insulator. Evaporation and sublimation of materials becomes significant and increases with altitude
0.00892	260,000	79,248	Dielectric strength and current-interrupting ability about equal to sea level values. Both increase above this altitude
2.27–13.21	130,000–90,000	39,624–27,432	Dielectric strength and current-interrupting ability lowest at some point in this range
33.66	70,000	21,336	Dielectric problems in unsealed switches on ac loads above this altitude
282.40	25,000	7620	Significant decrease of dielectric strength and current-interrupting ability above this altitude, up to about 260,000 ft
760.00	0	0	Basic performance figures established here

Source: Lockwood (7).
*U.S. Standard Atmosphere, 1962.

a potential of 1500 V impressed across the open contacts, without breakdown. Yet the same switch at 65,000 ft (19,800 m) may break down at 500 V. Although the dielectric strength of air decreases gradually with increasing altitude, no switch problems have been reported from this cause below 25,000 ft (7620 m). Above 25,000 ft, the reduction of dielectric strength is significant, and it reaches a minimum at some point between 90,000 ft (27,432 m) and 130,000 ft (39,624 m). Above this point it increases with altitude. At about 260,000 ft (79,248 m), dielectric strength again reaches its sea level value. At 350,000 ft (106,680 m), dielectric strength of air is so high that any breakdown of the switch is likely to be over the surface or through the insulating material rather than between the open contacts, regardless of their spacing. In the range from about 60,000 ft (18,288 m) to about 260,000 ft (79,248 m), breakdown may take the form of a momentary flashover or a partial breakdown in the form of corona discharge. Either of these may (or may not) be followed by a total breakdown, destroying the switch. Unsealed switches are seldom used to control ac circuits in this range because of the relatively high voltages involved.

Reduction of Current-Interrupting Ability. A partial vacuum can release the energy of the electric arc and thus reduce the current-interrupting ability of an unsealed switch. There is no significant change between sea level and 25,000 ft (7620 m). For safety, it should be assumed that arcing at the contacts of unsealed switches becomes increasingly severe as altitude is increased above 25,000 ft (7620 m), and reaches maximum severity at some point between 90,000 ft (27,432 m) and 130,000 ft (39,624 m). As altitude is increased beyond this point, arcing becomes less severe until, at about 260,000 ft (79,248 m), it is approximately the same as at sea level. At still higher altitudes, the arc becomes less severe. Above 350,000 ft (106,680 m), arc time becomes very short and current-interrupting ability of an unsealed switch is greatly enhanced. The extremely short arc time above 350,000 ft (106,680 m) means a very high rate of change of current with respect to time, resulting in high-voltage transients. This, in turn, can cause dielectric breakdown of insulation elsewhere in the circuit.

Increased Tendency of Contacts to Weld. Very high vacuum can aggravate contact welding. In the virtual absence of air, convection currents do not exist and heat is drawn away from the contacts of unsealed switches only by radiation or metallic conduction. At the same time, the vacuum helps to clean and degas the contact surfaces, making conditions more favorable for contact welding. Significant welding may occur above 500,000 ft (152,400 m), with heavy electrical loads, and at 1½ million feet (457,200 m) contacts may weld without an electrical load. This same cold-welding phenomenon can cause binding, seizing, galling, and accelerated wear of metal bearing surfaces exposed to vacuum.

Evaporation and Sublimation of Materials. As the prevailing pressure approaches the vapor pressure of the materials in an unsealed switch, the rate of evaporation and sublimation may become appreciable. The resulting vapors can leave deposits on surfaces of adjacent parts, forming conductive films on insulators and insulating films on contacts. Lubricants may be especially prone to evaporation in a vacuum unless they are designed for that environment.

Switches designed for use at reduced atmospheric pressures often have the contacts in a sealed, gas-filled chamber that maintains the gas around the contacts at sea-level pressure regardless of the ambient vacuum. In some cases, the switch terminals are exposed, while in others they are brought out of the switch as potted leads. The type of actuator seal used depends upon the intended use of the switch. If it is to be exposed to vacuum intermittently with appreciable recovery time between exposures, as in aircraft, the switch plunger can be sealed with an elastomer diaphragm or an O-ring. If the switch is to be exposed continuously to high vacuum for a matter of weeks or months, as in some space vehicles, elastomer-sealed switches will lose their fill gas gradually by diffusion. During any such leakage, the gas pressure in the switch will pass through pressures corresponding to all the altitudes below the one to which the switch is exposed, and the switch may experience any or all of the problems associated with an unsealed switch at each altitude. Switches intended for continuous use in a vacuum usually are hermetically sealed, with metal-to-metal and glass-to-metal joints.

When choosing a sealed switch, be sure to select one having the necessary mechanical life to assure that the seal will remain intact. Breakage of the seal results in gross leakage. If a switch with a gross leak is exposed to a vacuum, the gas in the switch leaks into the vacuum until pressure inside the switch equals that outside the switch. If the gross leak is a minor one, this process may take an hour or more to occur. If a switch on an aircraft landing gear develops a gross leakage, it will lose its internal pressure during flight. Then during landing, air will flow into the switch through the leak, perhaps carrying with it water and dirt splashed from the runway.

Normal variation of barometric pressure causes an unsealed switch at sea level to "breathe," but the velocity of air flowing into the switch is too low to carry contaminants in significant quantity. The operating characteristics (especially operating and release forces) of some sealed switches may change slightly when the switch is exposed to vacuum. This seldom is a problem, but if it is, switches are available whose operating characteristics are unaffected by vacuum. It is not unusual for a performance specification to call out a simultaneous exposure to vacuum and temperature. Since a partial vacuum contains very little air to have a temperature, it usually is best for test purposes to specify and control the temperature of the surface on which the switch is mounted.

Radiation

Switches are often used in the presence of nuclear radiation. As an example, precision snap-acting switches are used as limit switches to sense the position of control rods of nuclear reactors. Switches have numerous other applications in nuclear power generating plants and particle accelerators, where the levels of neutron or gamma radiation are high. Nuclear radiation can reduce the electrical performance of a switch by ionizing the gas between the contacts and by causing permanent damage to organic materials in the switch. Most material damage due to nuclear radiation occurs in one of three ways:

1. Elastic collision of radiation particles with atoms of the material, displacing the latter in the crystal lattice
2. Transmutation when an atom of the material captures a neutron and gains mass
3. Ionization caused by recoil atoms traversing the lattice, or by beta rays, x-rays, or gamma rays

Some styles of switch, such as the precision switches mentioned earlier in this section, are not damaged by gamma radiation exposures below 10^7 roentgens. Above that level, the effect depends largely upon the materials of the switch. The overall effect is approximately the same as that of exposing the switch to high temperature. Flexible rubber or elastomer parts and lubricants stiffen, and plastics gradually become brittle. At the same time, insulation resistance and dielectric withstanding voltage decline. On the basis of tests of a variety of switches in gamma radiation, the permanent effects of both gamma and neutron radiation on a given switch design can be estimated with reasonable accuracy.

The exact effect of ionizing radiation on electrical performance of a switch can be determined only by testing. A test is recommended if the switch is to control a circuit at or near its rated current or if an especially long life is required or if failure of the switch would have dire consequences. Ionizing radiation increases arc time, since the gas between the separating contacts requires longer to deionize. This is likely to reduce the life of the switch to some extent. However, if the switch is not required to close or open the circuit while exposed to radiation, only the permanent effects of radiation on the structure of the switch need be considered.

If radiation levels are high enough to affect performance of the switch, special materials may be employed. For example, in some instances ceramics can be used instead of plastics. Another measure is to shield the switch from the radiation, by using either a separate barrier or a metal-enclosed switch of suitable design. A third possibility is to install the switch at a location remote from the area of intense radiation, transmitting motion to the switch plunger or actuator by a mechanical linkage.

Switch manufacturers who have experience in providing switches for the nuclear industry can provide assistance in calculating radiation dosage, estimating effects of radiation, and recommending appropriate switches for specific applications.

Sand or Dust

Practical problems with switch resistance are rare, but those that do occur usually can be traced to particles of foreign matter on the contacts. An unsealed switch in an environment containing a high concentration of sand or dust is almost certain to have resistance problems and is likely to fail to close its circuit at all. If the particles are abrasive, they may, in addition to contaminating the contacts, reduce the life of bearing surfaces and even cause mechanical jamming of the switch mechanism. Such particles can cause leakage of sliding seals that are not designed to resist their effects. A sealed switch of suitable design is a necessity for reliable operation in sand and dust environments. When selecting a switch for this purpose, be sure to choose one having an adequate mechanical life rating, so that the seal will remain intact.

High Temperature

Many different styles of switches are designed to perform reasonably well at temperatures up to about 180°F (82°C). Above the maximum temperature for which a switch is rated, however, several effects may act to cause trouble. Some of these, such as differential expansion of parts, the increase of gas pressure in sealed switches, and the reduction of modulus of elasticity of spring material, take place immediately, while others are time-dependent. Among the latter are drying of lubricants and elastomer seals, deterioration of plastics (cracking, brittleness, and decreased ability to withstand moisture penetration), and accelerated degradation of contacts under electrical loading. These can cause changes of plunger (actuator) travel and force characteristics, binding of bearings, leakage of originally sealed switches, excessive leakage current through the open switch or from terminals to the grounded mounting surface, and reduction of switch life.

As a rule, if a switch is used above its rated temperature, the higher the temperature and the longer the exposure, the poorer will be the performance of the switch. A single, short-duration temperature overshoot with the switch not operating may have no significant effect on switch performance. A longer exposure to a less extreme temperature overshoot, with the switch operating, may render the switch unfit for further use. In practice, high temperature frequently occurs as part of a cyclic temperature condition to which the switch is exposed, as in the plenum control of a household furnace. The alternate expansion and contraction of air in the switch as a result of the cycling temperature can cause the switch to "breathe." This presents no problem unless the breathing draws contaminants into the switch.

The upper temperature limit of a switch can be raised by incorporating special materials (Fig. 1.23) and dimensions to reduce or avoid the problems mentioned above. Switches designed for use at high temperatures are also satisfactory for room temperature service, but they should not be used at low temperatures without consulting the manufacturer.

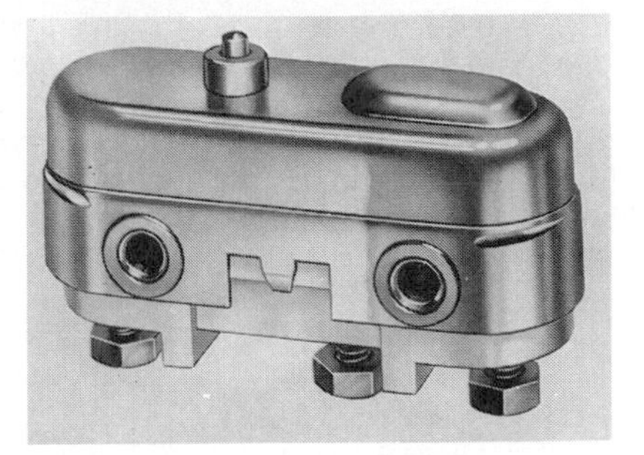

FIGURE 1.23 A switch designed for high-temperature use. (Courtesy of Micro Switch, Freeport, Ill.)

Avoid high current at high temperature, if possible, to reduce the I^2R heat supplied by the switch itself.

Low Temperature

Many switches designed for use at room temperature perform well at temperatures approaching absolute zero, but others are unsuited for use below −30°F (−34°C). Switches should not be used at low temperature without considering how they may be affected. Below the minimum temperature for which a switch is rated, lubricants may congeal, elastomer seals may stiffen and break, plastics may become brittle, springs stiffen, and at cryogenic temperatures, some metals shatter readily. Differential contraction can cause bearings to bind and sealed switches to leak. The result is that operating characteristics may change and the switch may be temporarily disabled or permanently damaged. If the plunger of a switch is sealed by an elastomer diaphragm or an O-ring, or if liquid or grease lubricants are employed, the return of the plunger from its actuated position may be delayed or even prevented by low temperature. Furthermore, a simple cold test may not show what will happen with long-term exposure to low temperature. For example, grease may congeal gradually over a period of days or weeks. Switches that are to be held in the actuated position at low temperature for long periods of time should be evaluated with this in mind.

Switches designed for low-temperature service perform satisfactorily at room temperature. The low-temperature feature is obtained by special materials and dimensions. Low-temperature switches should not be used at high temperature without consulting the manufacturer.

If a sealed switch containing dry nitrogen at atmospheric pressure and room temperature is cooled to a temperature approaching absolute zero, will the nitrogen liquefy and then freeze? For liquefaction and solidification to occur, three conditions must exist:

1. The gas must be at or below the critical temperature.
2. It must be at or above the critical pressure.
3. It must exceed the critical volume.

A sealed switch usually can be considered a constant-volume system. As temperature is reduced, the gas pressure decreases but the density remains constant. The reduction of pressure is caused by reduced molecular velocity. We leave it as an exercise for the reader to decide whether or not the gas remains gaseous. Meanwhile, just in case of a misplaced decimal point, it is best to test the switch before relying upon it at cryogenic temperatures.

Temperature Shock

Most switches are not damaged by sudden transitions from one temperature to another encountered in practice. An occasional problem may arise from accidental splash of a warm switch with liquid nitrogen or other cryogenic fluids, but damage from such causes is rare. The damaging effect of thermal shock is differential expansion or contraction between parts, or even within one part, when temperature changes rapidly. When damage does occur, it takes the form of slight cracking of plastics or separation of bonded joints, causing originally sealed switches to leak. Rapid fluctuations of temperature cause unsealed switches to "breathe" as the air within them expands and contracts. If moisture is present on the outside of the switch, it may be drawn into the interior of the switch as the switch is cooled. This may occur, for example, in the case of steam cleaning of switches on food-processing machinery if the cover of a switch enclosure is not properly tightened. As the enclosure cools following steam cleaning, moisture condenses and is drawn into the enclosure past the loosely fitting cover.

Radio-Frequency Interference (RFI)

Electrical discharges between switch contacts can "broadcast" a signal that interferes with sensitive equipment nearby. This can also happen with a switch controlling microampere circuits, although there is no electrical discharge and the broadcast signal is very weak. Such a weak signal can affect only extremely sensitive equipment near the switch. A telemetry system, for example, may be disturbed by spurious signals emanating from a nearby switch. Problems of this kind are rare, and when they occur they can be better treated individually than by wholesale preventive measures. In fact, misguided prevention can cause delays, increase costs, and accomplish nothing useful. If, as a *routine,* procurement specifications carry the requirement that "this product shall conform to the requirements of specification MIL-I______," the costs of negotiation, designing, testing, and correspondence can be considerable, and in the end may not benefit anyone.

The cause and cure of RFI problems lies more in the application of the switch than in the switch design. The design of the switch governs the shielding effectiveness afforded by the switch. It also affects the amplitude and duration of RFI to some extent, but the circuit parameters are primary determinants. In contrast to the switch design, the application governs the following:

- Susceptibility of the system to conducted and radiated interference
- Number, location, and physical arrangement of switches
- Shielding and wiring
- Circuit parameters
- Frequency of switch operation

- Number and type of noise generators other than switches

Thus the application, far more than the switch, determines the incidence and seriousness of RFI problems.

Given what appears to be an RFI problem, the first step is to ascertain that the problem actually is due to RFI. If careful tests of the equipment show that the problem definitely is caused by RFI, the next step is to locate the source of the RFI. If the problem is traced to the switch, and if the source of the interference is an arc at the switch contacts, an arc-suppression circuit will reduce the problem and may suffice as a remedy. If this is not feasible, or if an arc is not present at the switch contacts, determine whether the unwanted signal is being transmitted by conduction, radiation, or both.

For conducted interference from a switch, the best remedy is a filter attached to the switch wiring terminals. The filter should be designed to be compatible with the particular system. So-called universal filters tend to be large and expensive.

Radiated interference from a switch is remedied by shielding. If the switch could be completely covered by a conductive shell, all radiated RFI would be prevented. This is not always practical because terminals must be electrically isolated, and mechanical force and motion must be transmitted to the switch mechanism. The actuation problem can be remedied by operating the switch through a metal diaphragm or bellows. Hermetically sealed switches are available which are completely housed in metal except for glass terminal insulators. These switches are quite effective in reducing interference levels. Radiation from the lead wires can be controlled by the use of shielded wire (7).

A Final Word about Switches in Hostile Environments

If actual environments had only one significant component, such as temperature or pressure, switch application in hostile environments would be relatively easy. But in practice the numerous components of the environment combine in many ways to influence the kind and rate of electrical and mechanical deterioration of the switch. For help in choosing the right switch and applying it correctly, always consult the switch manufacturer as early in the design stage as possible. Many times, switch manufacturers produce special switches, often not shown in their product catalog, which may already solve unusual environmental problems faced by the design engineer. It doesn't hurt to ask, and more often than not the design engineer will find at least a partial answer to a particularly perplexing application problem or will be steered in the right direction toward an answer. The better switch manufacturers will have a wealth of information in their engineering files and should not be overlooked as a substantial resource for innovative and practical design solutions.

GLOSSARY

Actuator. The component of a switch which is the interface between the operator and the switch mechanism.

Anode. The electrode in an arc circuit which collects electrons during current flow. See *Cathode.*

Anode fall region. The high electric field region between the *anode* and the *plasma column* in an electric arc.

Arc. A self-sustaining discharge of a highly conductive ionized gas which conducts electric current between two electrodes.

Asperity. Very small physical contact points, on the switch contacts, caused by surface roughness on the bulk contact faces.

Bounce. The rebound effect that occurs when hard switch contacts close on each other. Relative mass of the contacts, forces, and frequency of supporting members are all components that determine the extent of bounce.

Break. Denotes the number of pairs of separated contacts a switch introduces into each circuit it opens.

Butt contact. A type of contact style in which the opposing contacts meet head-on. Opposite of *wiping contact.*

Cathode. The electrode in an arc circuit which supplies electrons during current flow. See *Anode.*

Cathode fall region. The high electric field region between the cathode and the plasma column in an electric arc.

Contact. The points of electrical interface in a switch.

Contact drop. The voltage drop across the contact resistance in a two-contact system.

Contact resistance. The total excess resistance in a two-contact system, beyond the bulk of body resistance of the contacts.

Current-zero. A condition (or time) at which a current waveform crosses the time axis. A current-zero is "forced" if it occurs because of a forcing voltage within a circuit such as a large arc voltage.

Dielectric strength. The maximum potential gradient that a material can withstand without rupture. The term, as applied to switches, means the maximum voltage a switch can withstand between specified terminals or between terminals and ground without leakage current exceeding a specified value.

Dry circuit. A low-energy circuit.

Electrical life. Life of a switch under a specified combination of electrical load, actuation, environment, and criterion of failure. See *Switch life.*

Inductive loads. An inductive load has the ability to store electrical energy in a magnetic field as long as there is current in the circuit. When the circuit is broken because a switch opens the current path, the stored energy must be returned to the circuit. The return of this electrical energy to the circuit causes a voltage that is larger than the source voltage across the switch contacts.

Insulation resistance. When a voltage is applied between two electrodes in or on an insulating material, a feeble current flows from one electrode to the other. The applied voltage divided by the current is the insulation resistance.

Insulation resistance of a switch. When a voltage is applied between two nonconnected terminals of a switch, a feeble current flows through the insulation that separates and supports the terminals. The applied voltage divided by the current is the insulation resistance of the switch, measured between those terminals.

Joule heating (I^2R). The process of transferring electrical kinetic energy in conduction electrons to the host lattice atoms in a conducting material via collisions between the electrons and the lattice atoms. In the process, the temperature of the lattice atoms is raised.

Load. The device in an electric circuit which converts electrical energy into some other form of energy.

Mechanical life. Life of a switch with no (or negligible) electrical load, and a specified combination of actuation, environment, and criterion of failure. Mechanical life usually is limited by the life of the switch's flexing parts and bearing surfaces.

Normally closed. The normally closed circuit of a switch is the electrically continuous path through the switch when the switch is in the normal contact position, i.e., when no force is applied to the *actuator.*

Normally open. The normally open circuit of a switch is the electrically continuous path through the switch when the switch is in the normal contact position, i.e., when no force is applied to the *actuator.*

Plasma column. The low electric field, intensely bright, ionized gas region of an electric arc between the cathode and anode fall regions.

Pole. The number of completely separate circuits that can pass through a switch at one time. The number of poles is completely independent of the number of *throws* and number of *breaks.*

Quick make–quick break. A nonteasable, snap-action type of switch mechanism designed to provide fast contact separation. Primarily designed for dc operation.

Resistive loads. A resistive load does not store any electrical energy while there is current in the circuit; thus, when the circuit is broken, only the source voltage appears across the opening contacts. Less arcing occurs when switching resistive loads. See *Inductive loads.*

Slow make–slow break. A switch mechanism with relatively slow operation to provide a slight time delay, permitting the ac wave to go through its zero energy level.

Source. The device in a circuit that provides the necessary energy to cause an electrical action to take place.

Switch life. See *Electrical life.*

Switch mechanism. That portion of a switch which opens and closes the contacts.

Switch resistance. The total resistance of the conducting members between the wiring terminals of the switch.

Tactile feel. The subjective perception of switch contact.

Tease (teasing). Slow manipulation of the switch actuator mechanism by an operator, causing electrical contact on an intermittent basis.

Throw. The number of circuits that each individual pole of a switch can control. The number of throws is completely independent of the number of *poles* and number of *breaks.*

Useful life. The number of cycles of operation that a switch will perform without exceeding the failure criteria for the application.

Wiping contact (wiping action). Lateral travel of movable contact over fixed contact while pressure between the two contacts exist. This action helps clean the contacts of contamination. See *Butt contact.*

REFERENCES

1. "The Electrical Circuit," Grayhill, Inc., La Grange, Ill., 1989.
2. "The Switch in an Electrical Circuit," Grayhill, Inc., La Grange, Ill., 1991.
3. "Pushbutton Switch Terminology," Grayhill, Inc., La Grange, Ill., 1989.
4. "Switch Fundamentals," Eaton Corporation, Cutler-Hammer Products, Aerospace and Commercial Controls Division, Milwaukee, Wis., 1991.
5. "Pushbutton Switch Terminology," Grayhill, Inc., La Grange, Ill., 1989.

6. Lee, T. H., "Physics and Engineering of High Power Switching Devices," MIT Press, Cambridge, Mass., 1975.
7. Lockwood, J. P., "Applying Precision Switches—A Practical Guide," Micro Switch, Freeport, Ill., 1973.
8. "The Meaning of Contact Resistance," Grayhill, Inc., La Grange, Ill., 1989.
9. Holm, R., "Electric Contacts, Theory and Application," 4th ed., Springer-Verlag, New York, 1967.
10. McCleer, Patrick J., "The Theory and Practice of Overcurrent Protection," Mechanical Products, Inc., Jackson, Mich., 1987.
11. "The Meaning of Voltage Breakdown and Insulation Resistance," Grayhill, Inc., La Grange, Ill., 1990.
12. "Electrical Loads and Switch Ratings," Grayhill, Inc., La Grange, Ill., 1991.

CHAPTER 2
PRECISION SNAP-ACTING SWITCHES*

Precision snap-acting switches are designed to have and maintain uniform mechanical and electrical characteristics while controlling a wide variety of circuits in many different environments. Because of their versatility and small size, precision switches are used in an endless variety of equipment.

In the 1930s, in response to the growing need for mechanical precision in electrical switching, a new switch mechanism was perfected and proved so successful that its principle is still used in most precision snap-acting switches today. The original patent drawing for this device is shown in Fig. 2.1. The device was designed in such a way that a very short travel of the switch plunger stored energy in a spring and used it to transfer the movable contact with a positive snap. Originally designed to provide thermostatic accuracy in a poultry incubator, today's precision snap-acting switch, represented by thousands of variations (a few of which are shown in Fig. 2.2), performs important functions in such diverse applications as production machinery, submarines, computers, space vehicles, and medical instruments. Precision snap-acting switches are so fundamental to the switch industry that they are even used as the contact mechanism for other types of switches (see Chap. 6).

HOW PRECISION SNAP-ACTING SWITCHES FUNCTION

The National Electrical Manufacturers Association (NEMA) defines a precision snap-acting switch as "a mechanically operated electric switch having predetermined and accurately controlled characteristics, and having contacts other than the blade and jaw, or mercury type, where the maximum separation between any butting contacts is 1/8 in (3,17 mm). A precision snap-acting switch consists of a basic switch alone, a basic switch used with actuator(s), or a basic switch used with actuator(s) and an enclosure."

In strict terms, *snap action* is a property of a switch such that the moving contact accelerates *without* added travel of the plunger beyond the amount of plunger travel required to separate the contacts.

*Numbers in parentheses indicate items in the References at the end of this chapter.

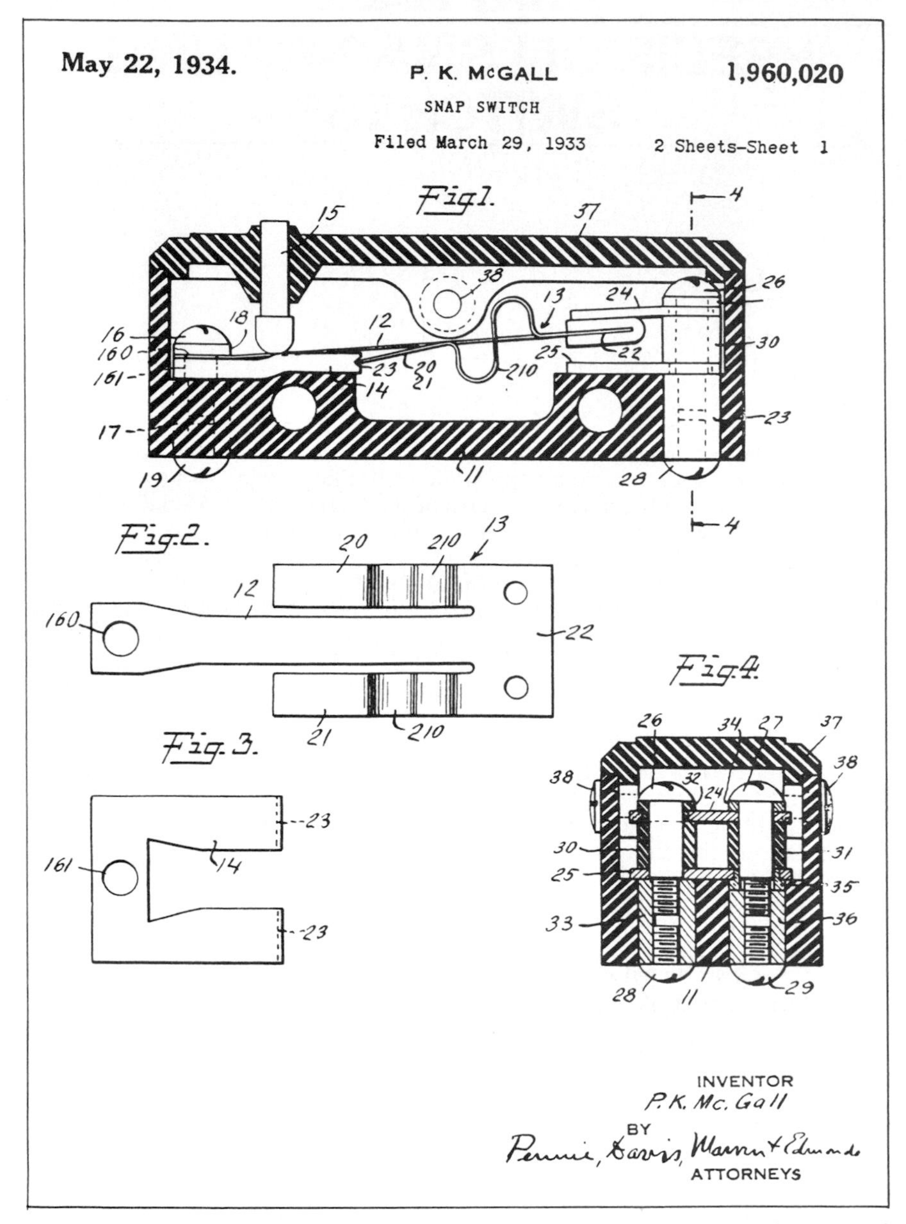

FIGURE 2.1 Patent drawing of the first practical precision snap-acting switch. (*Courtesy of Micro Switch, Freeport, Ill.*)

FIGURE 2.2 A few of the many varieties of precision snap-acting switches. (*Courtesy of Cherry Electrical Products, Waukegan, Ill.*)

The association further defines snap action as "a rapid motion of the contacts from one position to another position, or their return. This action is relatively independent of the rate of travel of the actuator." The term "relatively" is important. In fact, the acceleration of the moving contact is partially dependent upon the velocity of the plunger. The important point is that, once the plunger reaches the operating or release point, the movable contact immediately transfers to its opposite position without further travel of the plunger. A *non*-snap-acting switch lacks this feature.

Figure 2.3 illustrates a typical precision snap-acting switch. The plunger is at its *free position,* being completely released, and the common contact is against

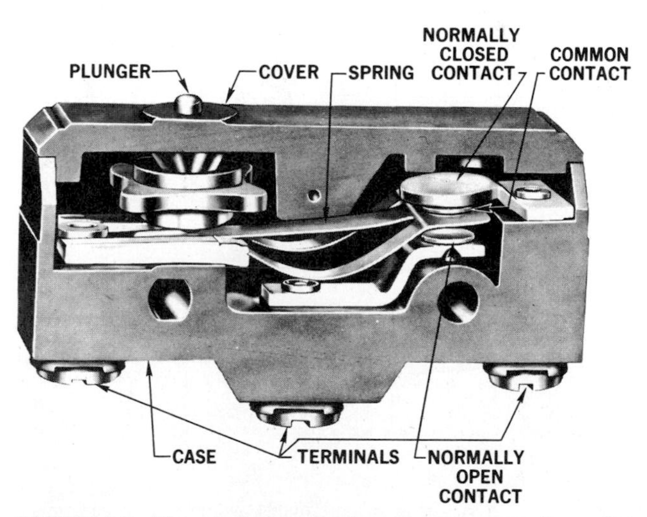

FIGURE 2.3 The principal parts of a precision snap-acting switch. (*Courtesy of Micro Switch, Freeport, Ill.*)

the normally closed contact. In this condition the normally closed circuit of the switch can carry current; there is electrical continuity between the common terminal and the normally closed terminal of the switch. The common terminal is electrically isolated from the normally open terminal. If the plunger is now depressed, it reaches the *operating point* (Figure 2.4). The distance from the free position to the operating point is called the *pretravel*. At the operating point, without further movement of the plunger, the common contact accelerates away from the normally closed contact. Within a few milliseconds the common contact strikes, bounces, and comes to rest against the normally open contact. Because the mechanism is designed for "snap action," the common contact cannot stop partway between the normally closed and normally open contacts. The normally closed circuit is now open and the normally open circuit is closed. As the plunger is depressed farther, past the operating point, the normally open circuit remains closed and the normally closed circuit remains open.

The distance the plunger travels past the operating point, in this direction, is called *overtravel*. When the plunger is fully depressed (i.e., at full overtravel), further depression of the plunger is prevented by the switch mechanism. The distance from the free position to the point of full overtravel is called the *total travel*. Total travel is the sum of the pretravel and overtravel. As the plunger is released from the point of full overtravel, it again passes the point at which the operating point has occurred, but this does not cause the position of the common contact to change. The normally open circuit remains closed and the normally

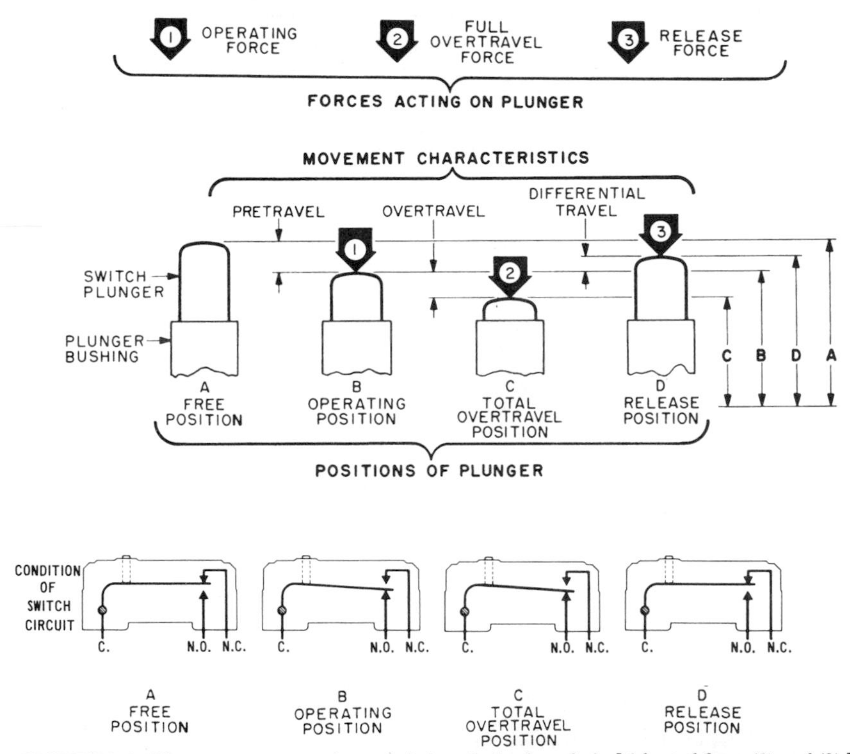

FIGURE 2.4 Plunger movement characteristics of a basic switch. [*Adapted from (1) and (2).*]

closed circuit remains open until the plunger reaches the *release point.* At the release point, without further movement of the plunger, the common contact accelerates away from the normally open contact. Once more, within a few milliseconds the common contact strikes, bounces, and comes to rest, this time against the normally closed contact. The normally open circuit is now open and the normally closed circuit is closed. This condition continues as the plunger is returned to its fully released (free) position. The distance between the operating and release points is called the *differential travel.* The distance from the release point to the free position is called the *release travel.* As a rule, the essential points of plunger travel (free position, operating point, release point, and full overtravel point) are measured and specified as dimensions from the centerline of the mounting holes of the switch.

The force required to depress the plunger to the operating point is called the *operating force.* As the plunger is released from the operating point or from the overtravel region, the force required to hold the plunger at the release point is called the *release force.* The difference between the operating force and release force is called the *force differential* or *differential force.* These characteristics vary with the design of the switch and, to some extent, with its application. These are discussed later in the chapter.

In rotary-actuated precision switches, the operating characteristics are expressed in terms of the analogous angles of rotation, and torques.

Relationship between Plunger Force and Plunger Position

The force and travel characteristics of a precision snap-action switch can be most clearly represented on a graph of plunger force vs. plunger position. The relation between these two variables is of particular interest to the designer when the switch is to be actuated by a force-sensitive device such as a dead weight, a thermostatic bimetal, or a gas-filled bellows. The shape and dimensions of the curve depend upon the design of the switch.

Figure 2.5 shows successive stages in the development of a curve of plunger force vs. plunger position for a typical precision snap-acting switch. With no force on the plunger, the plunger is at its free position, designated point *A* in Fig. 2.5. First, observe the behavior of the force as the plunger is driven through its travel by means of a rigid device, such as a micrometer head. As the plunger is depressed, the force increases from zero. In this instance, the design of the switch is such that the force is a linear function of the travel and, when the two are plotted, the locus of points lies along line *AB.* The vertical protection of point *B* on the horizontal axis represents the operating point of the switch, while the horizontal projection of point *B* on the vertical axis represents the operating force.

During travel of the plunger from *A* to *B,* mechanical energy is stored in the switch mechanism. When the plunger reaches the operating point, some of the stored energy is used by the switch mechanism to cause "snapover" of the common contact from the normally closed to the normally open position. During snapover, the plunger does not move from its position at point *B,* because the actuating device is rigid, but the force applied to the actuating device by the plunger drops to the level represented by point *C.* If the plunger is now moved into the overtravel range, the force will again increase, as shown by Line *CD.* Reversing the direction of plunger travel, and gradually releasing the plunger, the same line is retraced in the direction *DC.* However, the path continues in the same direction beyond point *C* to point *E,* the release point. At position *E,* with

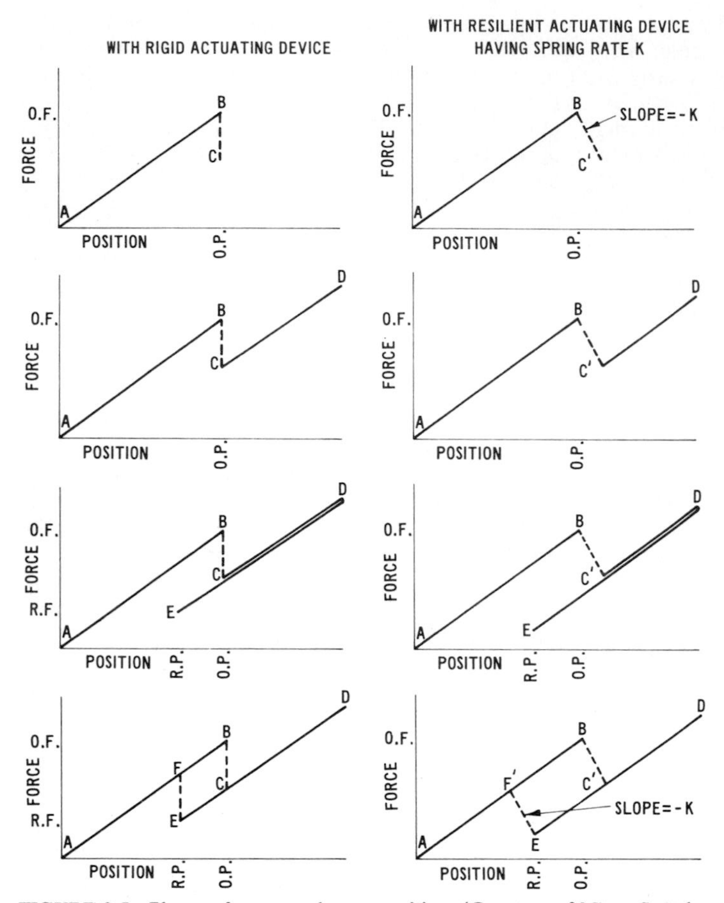

FIGURE 2.5 Plunger force vs. plunger position. (*Courtesy of Micro Switch, Freeport, Ill.*)

no further movement of the plunger, the switch mechanism utilizes stored energy to snap back to its original position, and the force applied by the plunger rises to point *F* on the graph.

Since the area under the curve *ABCD* represents the mechanical energy put into the switch, and the area under the curve *DCEFA* represents the energy returned by the switch to the actuating device, area *BCEF* represents the energy utilized by the switch mechanism in snapping over and back, as the plunger moves through differential travel.

When a resilient actuating device is used, it merely changes the location of points *C* and *F.* They are affected because the position of the actuating device is now determined by the varying force applied to it by the switch plunger. In this case, starting with the plunger at point *A,* the resilient actuating device depresses the plunger with force varying linearly with travel, as shown by line *AB.* At point *B,* the operating point, the mechanism snaps over, but this time the plunger moves because it is held by a nonrigid device while the force is changing. Instead of arriving at point *C,* as it did with a rigid actuating device, the plunger arrives at point C', the location of which is determined by the spring properties of the driving member.

Although path BC' is shown in Figure 2.5 as a straight line, the path is not necessarily linear. As before, it is assumed to be linear for simplicity in this case and is shown as a dashed line. Its actual shape usually is unimportant, but the location of C' can be of considerable interest to the designer. If the spring rate (force per unit of travel) of the resilient actuating device is K, draw a straight line with slope $-K$ through point B. Point C' then is located where this line intersects line CD. In similar fashion, the release point E remains unchanged; but after snapback, the plunger is at point F' instead of F. To locate point F', draw a straight line with slope $-K$ through point E until it intersects line AB.

Relationship between Contact Force and Plunger Position

Contact force is the force holding closed contacts together. It affects such diverse characteristics as electrical resistance and ability of the switch to maintain electrical continuity during acceleration, vibration, and shock. Snap-acting switches are usually designed to have as high a contact force as possible, consistent with other requirements, and to maintain high contact force during as much of the plunger travel as possible. As a rule, contact force is not held at specified values during switch production, and the user gains nothing but unnecessary cost by specifying it. As is emphasized throughout this book, it is much wiser to specify the performance of the switch as a whole rather than details of its internal design and adjustment.

Still, it is important for the switch user to know what is happening inside the switch, so that the switch can be utilized in a manner providing the maximum performance possible. It is relatively easy for a switch manufacturer's engineering department to design a switch having very high and constant contact force, regardless of plunger position, but this necessitates a wide differential travel and force differential, neither of which is desirable. In practical terms a trade-off is arranged in which contact force is maintained sufficiently high for good switch performance, without inordinately wide differentials.

Since contact force (as the term is used here) has meaning only when the contacts are against each other, the usual method of plotting the variation of contact force with plunger position is shown in Fig. 2.6. The horizontal axis represents plunger travel, from free position to full overtravel, and shows the location of the operating and release points. The upper half of the vertical axis represents the force of the common contact against the normally closed contact of the switch, while the lower half represents force of the common contact against the normally open contact. The graph shown in Fig. 2.6 is an instance in which the relation of contact force to plunger position is linear. In some switch mechanisms the contact force curve is nearly horizontal, turning steeply downward as the plunger approaches the operating point, or upward as the plunger approaches the release point. The important fact is that, as the plunger approaches the operating or release point, contact force decreases and reaches zero at the instant the contacts separate.

With this in mind, it is easier to visualize and understand snap-acting switch behavior under several conditions which are discussed later in this chapter.

Contact Bounce and Transit Time

Occasionally, a switch application requires knowledge of contact transit time or bounce time. These are best determined by measurement under the electrical and

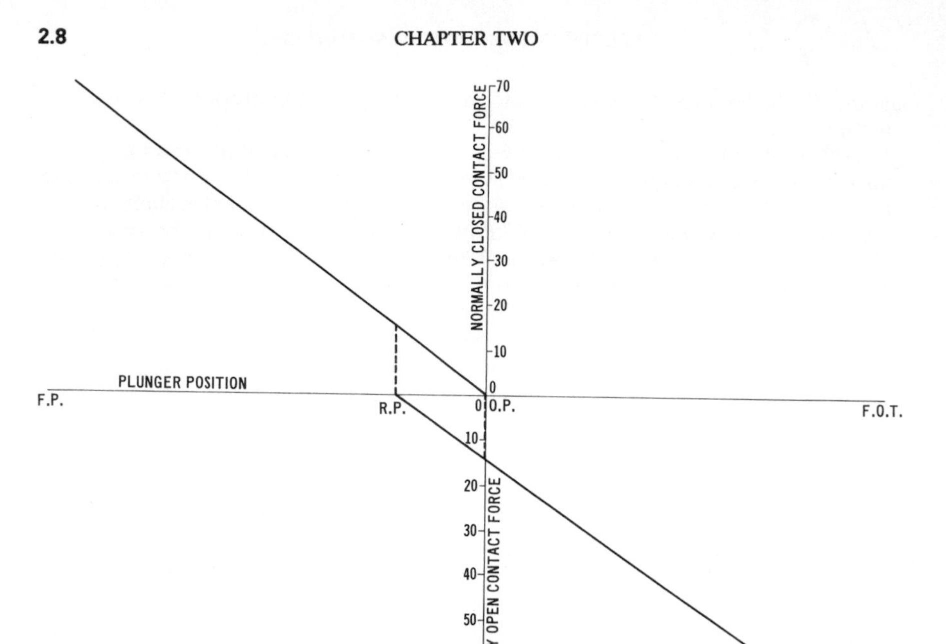

FIGURE 2.6 Contact force vs. plunger position for typical snap-acting switch. (*Courtesy of Micro Switch, Freeport, Ill.*)

mechanical conditions of end use, but some background information may be helpful. The transit time is the time required for the moving contact to leave one stationary contact and strike the opposite stationary contact. When the moving contact strikes the stationary contact, the kinetic energy is converted to potential energy in the form of heat and deformation of the contacts. As a result of elastic deformation the moving contact rebounds from the stationary contact, the contact pair being reclosed by the contact force. This can occur one or more times until bouncing ceases and the contact system reaches static equilibrium.

If the electrical load is insignificant, contact transit time and bounce pattern are affected primarily by plunger position vs. time, as well as by several variables in the design of the switch. A switch user seldom is concerned about, or even aware of, contact bounce unless the switch is controlling a device having a very fast response. Some circuitry may interpret each bounce as a separate signal. If this is objectionable, the remedy is a buffering circuit to remove the unwanted effects of contact bounce.

Reducing ambient vibration or increasing or decreasing plunger velocity may reduce contact bounce time. Increasing plunger velocity usually reduces transit time. No attempt is made to control contact transit time or bounce characteristics during switch manufacture; indeed, they are only partially controllable anyway. These characteristics are as much a function of the conditions of end use as they are of switch design.

Furthermore, contact transit and bounce time are inherently variable. In practice, two successive actuations of a switch seldom produce the same transit time, bounce pattern, or bounce duration, no matter how nearly constant the known variables are held. Given a switch installed in its end use equipment, measurement of an adequate number of contact closures enables one to make statistical estimates of contact behavior, assuming no new, unexpected influences act to change it. Some switch mechanisms have inherently low contact bounce time, and the switch manufacturer can provide specific information about them.

If the electrical load introduces significant thermal or arcing effects, contact transit and bounce time in themselves should seldom be of concern to the user. In addition to the effects mentioned above, separating contacts are affected by magnetic forces of repulsion associated with the arc. The pressure of metallic vapor at the interface also tends to drive the contacts apart. When separation of the contacts is due to bouncing, the contacts reclose with partially liquefied surfaces. From there on, things get quite complicated. If contact bounce or transit time of a snap-acting switch is a matter of special concern, it is prudent to enlist the aid of a switch manufacturer as early in the design process as is feasible.

Terminology of Precision Snap-Acting Switches

Although Chap. 1 discusses in detail the general terminology of switches, an examination of poles, throws, and breaks as related specifically to snap-acting switches may be helpful. Referring to Fig. 2.7, the term *pole* denotes the number of completely separate circuits that can pass through the switch at one time. It is independent of the number of *throws* and number of *breaks*. A double-pole switch can carry current through two circuits at the same time, since the circuits are completely insulated from each other. The circuits through the switch are mechanically connected (but electrically insulated) so they open simultaneously or close simultaneously.

Throw denotes the number of different circuits that each individual *pole* can control. It is independent of the number of *poles* and number of *breaks*. For example, a single-pole double-throw (single-break) snap-acting switch connects the "common" terminal of the switch to the "normally closed" terminal when the plunger is free but connects the "common" terminal to the "normally open" terminal when the plunger is depressed.

Break denotes the number of pairs of separated contacts the switch introduces into each circuit it opens. If actuating the switch breaks the circuit in two places, the switch is a double-break switch.

Electrical Characteristics of Precision Snap-Acting Switches

The purposes of a switch are to close an electric circuit, carry current, open the circuit, and hold it open. An ideal switch would be one whose electrical resistance could be readily changed at will, from zero to infinity. The design of such a switch would, however, first require the invention of a perfect conductor and a perfect insulator. It is interesting to note, however, that in most applications, an ideal switch of this kind would be no better than a precision snap-acting switch is today. An open snap-acting switch usually presents a circuit resistance of well over 100,000 MΩ, and closing the switch reduces its resistance to a few milliohms. This represents a resistance change of 13 orders of magnitude in a couple of milliseconds. Two important electrical characteristics of the switch, then, are its resistance when open and its ability to conduct current when closed. The resistance of both open and closed switches is discussed in Chap. 1.

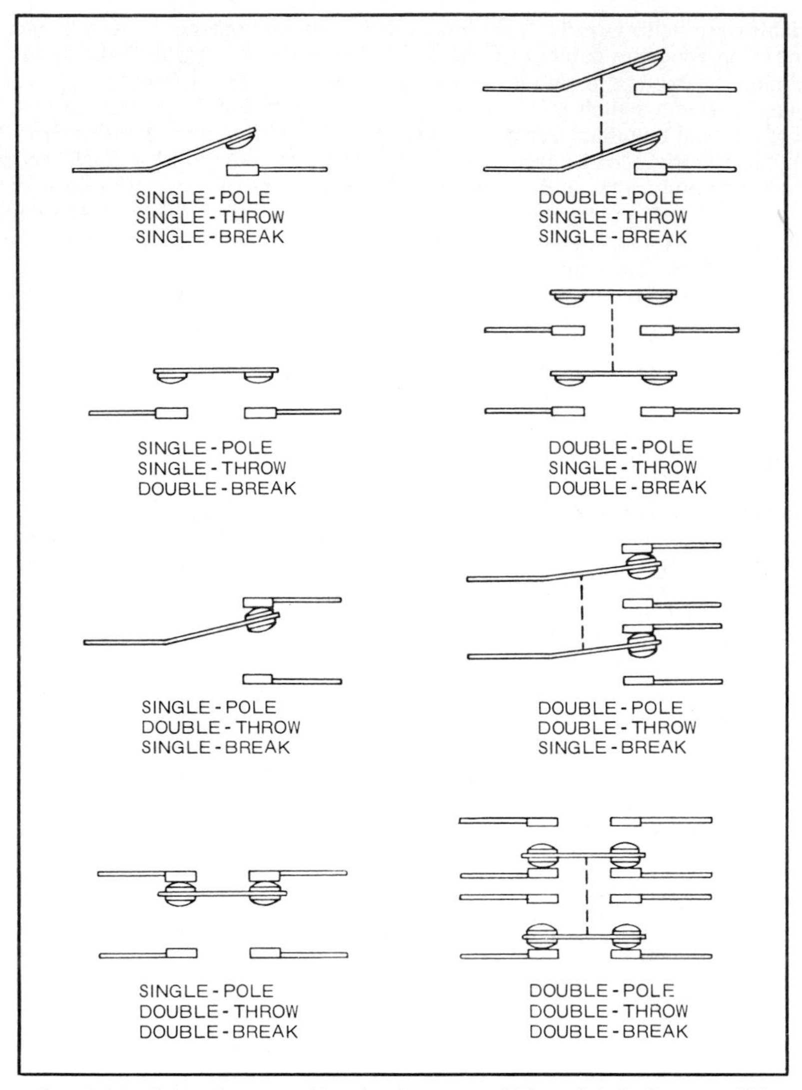

FIGURE 2.7 Poles, throws, and breaks. (*Courtesy of Micro Switch, Freeport, Ill.*)

Mechanical Characteristics of Precision Snap-Acting Switches

Normal Variation of Operating and Release Point. One of the advantages of precision switches is the extreme stability of their operating characteristics. Many control devices such as thermostats and pressure controls utilize this feature to simplify design. Currency-counting machines in large banks and department stores employ precision switches to detect bills passing between a pair of feed rollers and to stop the machine if two bills are fed through a pair of rollers at the same time. The operating characteristics of primary interest in this regard are the operating and release point (see Fig. 2.4). But just how stable they are depends upon the type and quality of the switch and the way in which it is applied.

A precision snap-acting switch, when actuated just through its differential travel (while controlling a negligible electrical load at room conditions), is likely to undergo a "settling in" period during the first several hundred operations, until the system (including the switch, its actuating mechanism, and associated operating cam or linkage) stabilizes. During this period, the operating and release point are likely to shift together, a distance of perhaps 0.00005 in (0,00127 mm), maintaining a fairly constant differential travel. If the system is then not disturbed mechanically, the operating and release point will stabilize, perhaps varying within a band no wider than about 0.0002 in (0,005 mm) per 100,000 operations. Needless to say, precision of this order is seldom required. Any significant disturbance, such as a mechanical shock, will result in another stabilization period, similar to the first one. The "settling in" is a stabilizing of the *entire system,* including the switch, so a preliminary run-in of the switch alone, before installing it in the system, is useless. In fact, it degrades the switch by accumulating bearing wear and spring fatigue.

How Application Factors Affect Stability. Many variables in end use can affect the stability of operating and release point of the switch during its life. The most important ones are described below:

1. *Selection of the Switch.* Some switches are designed especially for stability of operating and release points, while others are designed to optimize other features. If especially stable operating and release points are needed, the switch should be chosen with this in mind, with the help of the switch manufacturer.

2. *Mounting the Switch.* If the surface against which the switch is mounted is warped or rough, or if the mounting screws are overtightened, stability of operating and release point may be reduced. Precision switches are ruggedly constructed and will take considerable abuse. At the same time, the switch is a precise component and its precision can suffer if undue mounting stresses are applied.

3. *Soldering Lead Wires to the Terminals.* Overheating the terminals of a switch with a soldering iron can partially anneal the internal spring and distort the plastic insulators, either or both of which reduce the precision of the switch. Miniature and subminiature switches are especially sensitive to overheating from this source.

4. *Actuation.* Besides a standard pin actuator (Fig. 2.8), many basic switches are provided with leaf springs, levers, or other linkages (Figure 2.9) between the switch plunger and the actuating device in order to provide additional overtravel, reduce operating force, or for other reasons. The differential travel measured at the free end of the leaf or lever (Fig. 2.10) often is considerably greater than that measured at the switch plunger. A ratio as high as 20 to 1 is not unusual. This fea-

FIGURE 2.8 Precision snap-acting switch with standard pin plunger. (*Courtesy of Micro Switch, Freeport, Ill.*)

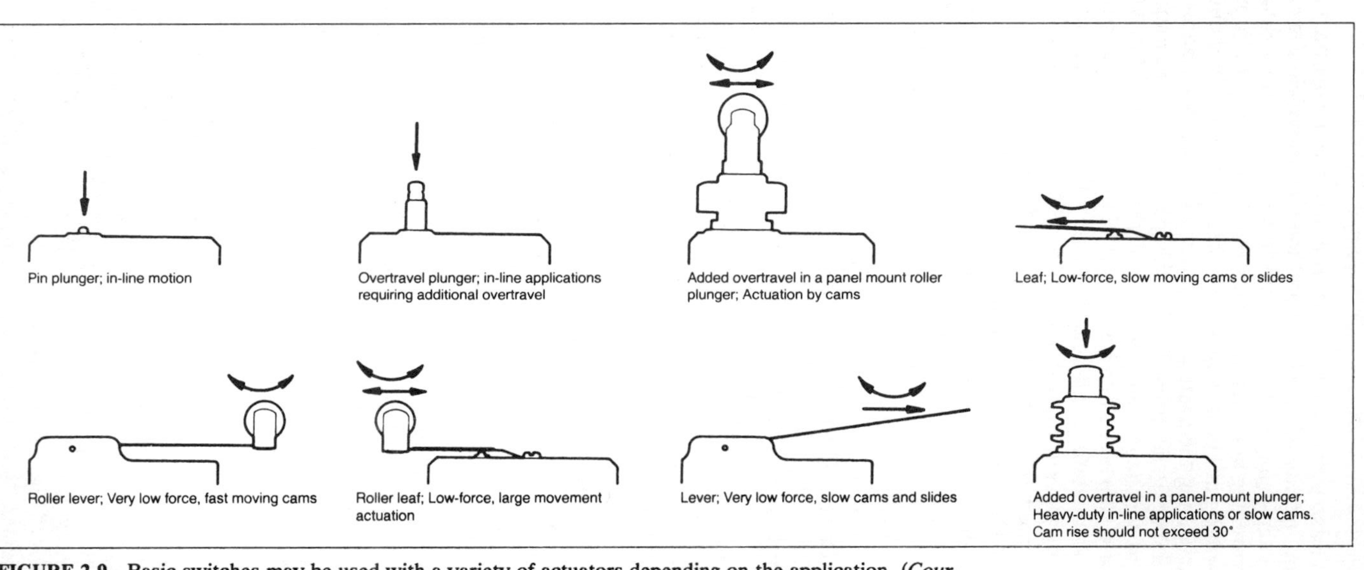

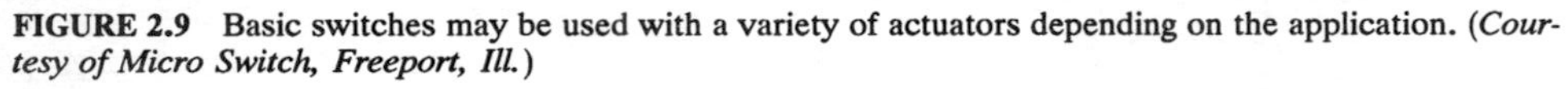

FIGURE 2.9 Basic switches may be used with a variety of actuators depending on the application. (*Courtesy of Micro Switch, Freeport, Ill.*)

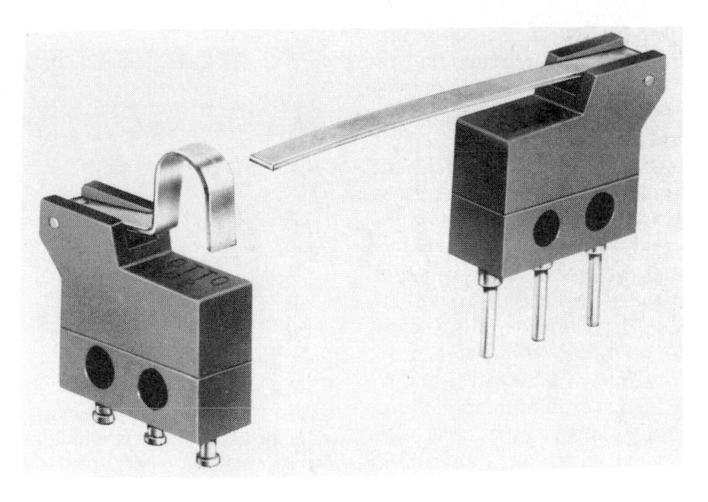

(a)

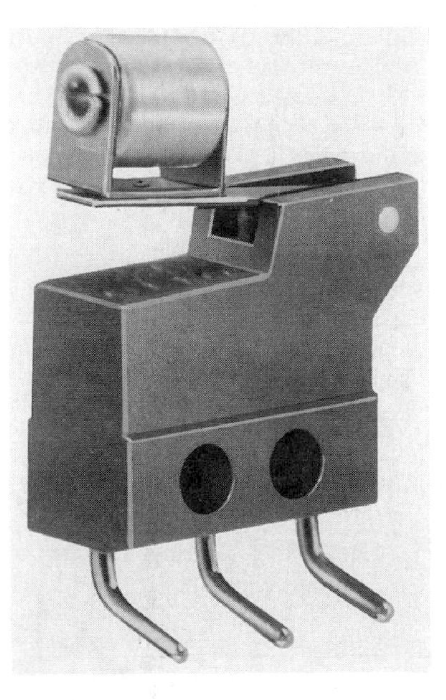

(b)

FIGURE 2.10 Detail of (*a*) leaf actuator and (*b*) roller lever actuator. (*Courtesy of Otto Controls, Inc., Carpentersville, Ill.*)

ture provides the operating characteristics needed in many applications. However, it may also magnify any instability of the operating and release point.

A typical application for a roller-leaf-actuated precision switch is depicted in Fig. 2.11. The device shown is a door safety interlock which turns power off when an access door to electrical equipment is opened for servicing. The cylindrical spring-loaded cam moves forward when the door is opened and the roller leaf actuator is depressed downward by the action of the cam. The actuator opens the normally closed circuit, and power is safely turned off. The wheel-like roller permits ease of movement by the cam.

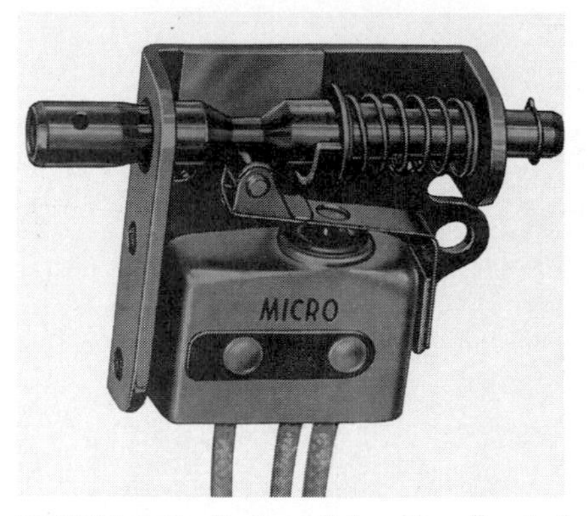

FIGURE 2.11 Basic switch with roller leaf actuator used in door interlock switch. (*Courtesy of Micro Switch, Freeport, Ill.*)

The greatest precision is achieved by actuating the switch at the plunger, parallel to the plunger axis. An actuating device that strikes the switch plunger at an angle will not achieve maximum stability of operating and release point. Such actuation applies side thrust to the plunger, promoting wear of plunger bearing surfaces and reducing the precision of the switch. Furthermore, when the actuating device strikes the plunger at an angle to the plunger axis, any shift of operating or release point will be magnified. If the switch plunger is driven directly by a cam, rubbing can cause excessive wear and changes of operating characteristics unless the plunger is designed to serve as a cam follower (Fig. 2.12). Switches can tolerate considerable force on the plunger without damage, but excessive force can shift the operating characteristics and cause permanent damage to the mechanism. Switch manufacturers can supply information on the maximum forces recommended.

5. *Electrical Load.* In the precision switch industry, it is standard to establish current ratings with life tests in which the switch plunger is fully depressed and fully released during each cycle of operation. This represents the actuation the switch will receive in many end use applications, but it gives no indication of the effect of current on operation and release points. Precision switches provide their most precise control at low current and are used at rated current in applications where maximum precision is not required. To learn more about the behavior of operating and release points during switch life, Micro Switch, Freeport, Ill., developed automatic recording equipment to plot these characteristics as points on graph paper. The completed graph presents a concise picture of operating and release points during switch life, and the effects of design variables and electrical load can be studied. Plunger travel is magnified 1000 times by the recording equipment, so 1 in (25,4 mm) on the vertical scale of the graph represents 0.001 in (0,0254 mm) of plunger travel. The time scale is so arranged that many thousands of operations are plotted in a short distance. If the points are closely clustered, they may appear as a solid line or a dark mass on the paper, but any "maverick" points will appear as individual dots. Figure 2.13 shows typical results.

Switches from various manufacturers have been tested on this equipment. With a switch controlling a negligible electrical load such as 100 mA, the first few hundred operations usually cause a slight shift of operating and release points in the same direction, as the system (consisting of switch and measuring equipment) settles and stabilizes. The differential travel remains virtually unchanged and stabilizes as shown in Fig. 2.13*a*. When a 5-A 120-V ac resistive circuit is connected to the normally open circuit of the switch, the trace changes as shown in Fig. 2.13*b*. The operating points are unaffected, but phenomena at the switch contacts (contact welding, erosion, etc.) cause some dispersion of release points. Most of the release points still lie along the line they occupied under essentially no-load conditions, but some of them shift in the direction that expands the differential travel.

If we connect the 5-A load to the normally closed circuit instead of to the normally open circuit of the switch, the pattern of Fig. 2.13*c* emerges. Here the release points are dispersed as shown. Most operating points lie along the no-load line, but some are shifted in the direction of expanding differential travel.

Most (but not all) switches tend to be less affected by an electrical load on the normally closed circuit than by the same load on the normally open circuit. The difference is more apparent at higher current. If the 5-A load through the normally closed circuit is increased to 10 A, the switch responds as shown in Fig. 2.13*d.* A glance at the vertical scale and the diameter of the human hair on the same scale (Fig. 2.13*e*) shows that the scatter of points is quite small.

FIGURE 2.12 Basic switches shown with cam-follower-style actuators. (*Courtesy of Micro Switch, Freeport, Ill.*)

6. *Summary of Facts about Operating and Release Point vs. Electrical Load.* Loading the normally closed circuit of the switch affects some of the operating points but none of the release points. Conversely, loading the normally open circuit affects some of the release points but none of the operating points. At a given current, differential travel will usually be more stable if the electrical load is controlled by the normally closed circuit of the switch rather than the normally open. Electrical loading tends to increase the differential travel during some but not all cycles of switch operation. The lower the current, the narrower the dispersion of differential travel and the fewer cycles of operation will experience widening of the differential travel. The amount of change of differential travel, of course, depends upon the design of the switch and conditions of end use.

7. *Acceleration, Vibration, and Shock.* As the switch plunger approaches the operating or release point, the contact force decreases, and the switch becomes more sensitive to acceleration. If the plunger is very near the operating or release point, a jar or other acceleration may cause the switch mechanism to snap over prematurely. The effect, then, is that of a shift of operating or release point, narrowing the differential travel.

An example of this involved the famous Norden bombsight of World War II. The design of this highly accurate device was a closely guarded secret throughout

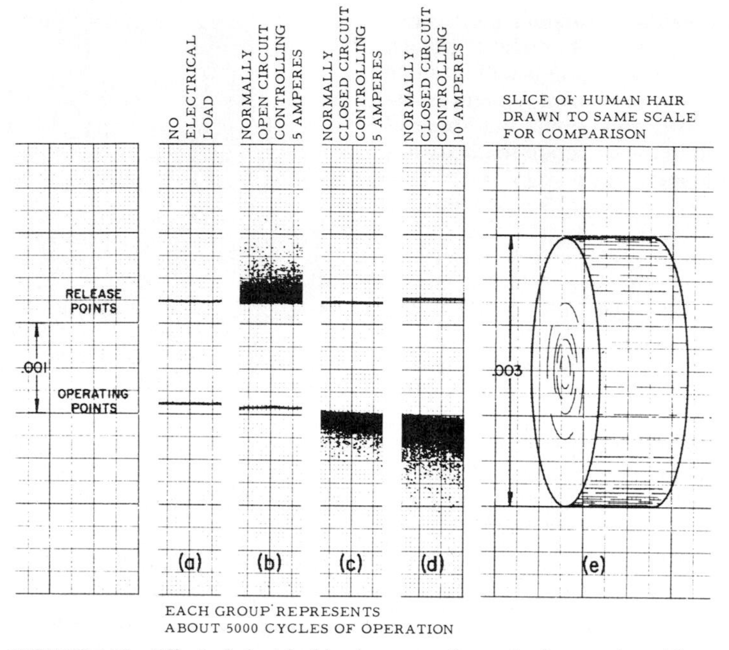

FIGURE 2.13 Effect of electrical load on operating and release points. (*Courtesy of Micro Switch, Freeport, Ill.*)

the war. To keep it from falling into enemy hands, the top-secret bombsight was rigged with an explosive charge to be detonated automatically by a shock-actuated switch assembly if the bomber should crash. The assembly had been carefully and logically designed. The actuating mechanism for the snap-acting switch was carefully adjusted to respond to the abrupt deceleration of a crash, but not to the buffeting of rough weather or the impact of a hard landing. Experience had shown that the snap-acting switch was unaffected by the vibration and shock usually experienced by military aircraft. When the assembly was ready, a technician installed it in the bombsight of an operational bomber, climbed out of the plane, and slammed the door. With ears still ringing, the technician turned to survey the scattered debris of what had once been a bombsight.

Study showed that neither the switch nor the actuating mechanism would have responded to the shock of the slamming door, but the combination did, since the shock-sensing mechanism held the plunger of the switch depressed almost to the operating point. This reduced the contact force, and the jar of the slamming door actuated the switch. The remedy was simple: Keep the plunger fully released instead of partially depressed.

8. *Temperature and Humidity.* Fluctuations of temperature and humidity influence operating and release points of a switch by affecting the dimensions of the working parts. Prolonged exposure to high temperatures may shift permanently the operating and release points of a switch (see Chap. 1).

9. *Contamination.* Splashing an unsealed switch with liquid or exposing it to steam or other vapors can cause drastic changes in operating and release points and have other undesirable effects (see Chap. 1).

How to Stabilize Operating and Release Point in End Use. The reasoning outlined in the previous paragraphs leads to the following precautions:

1. Early in the design, get advice about choice of switch from the manufacturer, especially if the application is unusual or performance requirements are critical. It is much easier to build the right switch into the equipment the first time than to change switches after problems appear.
2. Mount the switch against a smooth flat surface and do not overtighten the mounting screws. If in doubt, consult the switch manufacturer for recommended torque values. When attaching lead wires to switch terminals, avoid excessive soldering temperature or time.
3. Design the actuating device to strike the switch plunger in line with the plunger axis, unless using cam actuation. If a cam is used, choose a switch with the plunger designed as a cam follower. Do not apply excessive force to the plunger.
4. To stabilize the operating point, control the load with the normally *open* circuit of the switch. To stabilize the release point, control the load with the normally *closed* circuit of the switch. To stabilize both, minimize current during closure and opening of the switch.
5. If the switch is to be exposed to acceleration, vibration, or shock, orient the switch so that acceleration will not actuate the switch prematurely.
6. Minimize changes of temperature and humidity, and avoid exposing the switch to high temperatures for prolonged periods.
7. Avoid contaminating unprotected switches with water, oil, cleaning compounds, or other materials that can affect dimensional stability of plastics.

Thus far the discussion has dealt with ways to keep the operating and release points of switches as stable as possible. Another step would be to design the end use equipment to be as insensitive as possible to normal variation of switch operating characteristics. Finally, test the switch in the end use equipment under conditions simulating those expected in the field. Laboratory measurements of the switch itself, such as those described in this chapter, tell only what the switch will do under the particular conditions that prevailed during the test. A test in end use equipment brings into effect the combinations of mechanical, electrical, and environmental conditions to which the switch will be exposed in use. Then, if everything works satisfactorily, the equipment can be expected to perform as required in the field.

DESIGNING CAMS FOR SNAP-ACTING SWITCH APPLICATIONS

Many switch styles utilizing internal snap-acting switches are designed for actuation by rotating or reciprocating cams. Cam actuation of switches usually is easily arranged and seldom presents problems. To get the longest possible switch life with cam actuation, the following factors should be considered.

Mechanically, the switch and its roller plunger or roller lever constitute a spring-loaded cam follower. The common principles of cam design and follower positioning apply. Cams for this purpose should be designed to minimize wear, force, and impact. When a switch roller bears against a cam face, wear is governed by such factors as position and motion of switch plunger or lever with respect to the cam, applied force, cam surface speed, cam profile, cam material, finish, and lubrication. High-velocity cam movement against the switch roller can accelerate wear of both cam and roller. Misalignment of the roller with the cam increases wear (Fig. 2.14). They should lie in the same plane. An ideal cam profile would be designed so that the roller contacts the cam at all times. To reduce wear at low cam speeds, constant acceleration or simple harmonic curves can be used. At higher cam speeds, changes of acceleration with respect to time can be reduced with cycloidal curves.

As a rule, with roller-actuated switches, cam rise angle should not exceed 30° and drop-off release should be avoided (see Fig. 2.15). Impact between cam and roller tends to promote wear and loosen the follower on its bearing. Depressing the switch actuator with very high velocity may reduce the spring life, as may high-velocity release. In an extreme case of the latter, the cam can pull away from the roller before the switch plunger is fully released. Plunger travel then stops abruptly when the roller again strikes the cam or the mechanism of the plunger strikes its internal stops. The effect is to reduce switch life. In some lever-type switches designed for actuation in either direction, the lever may oscillate after drop-off release, giving a series of false operations with added wear and spring fatigue.

In studying the effect of cam profile on switch actuation, a stroboscope is often helpful. Excessive force applied by the cam at full plunger overtravel can reduce switch life (Fig. 2.16). Questions regarding cam material, finish, and lubrication generally are resolved by trial. Lubrication of the cam may require that the switch be sealed to avoid contamination. In testing switches designed for cam actuation, one switch manufacturer effectively uses test cams which are eccentric circles of 1 ft (304,8 mm) circumference. The cams are made from tool steel and hardened to 60 to 62 Rockwell C. The amount of eccentricity is specified to give the switch actuator the desired travel. A common rate of cam rotation for test purposes, where electrical load is negligible, is 200 rpm. [Adapted from (1) and (2).]

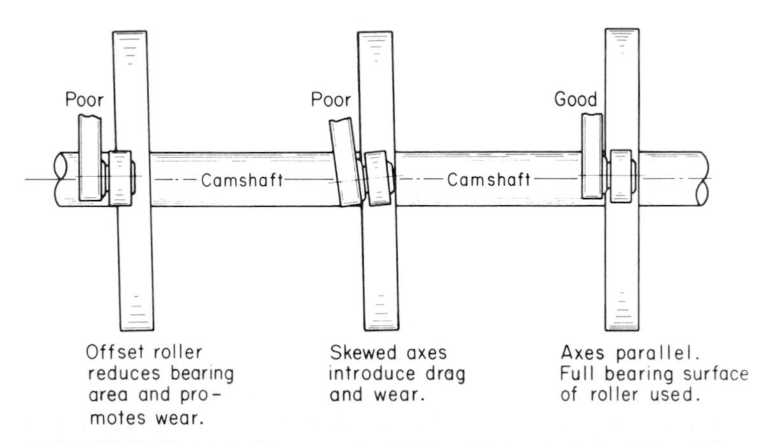

FIGURE 2.14 Proper and improper installation of cam-actuated switches. (*Courtesy of Micro Switch, Freeport, Ill.*)

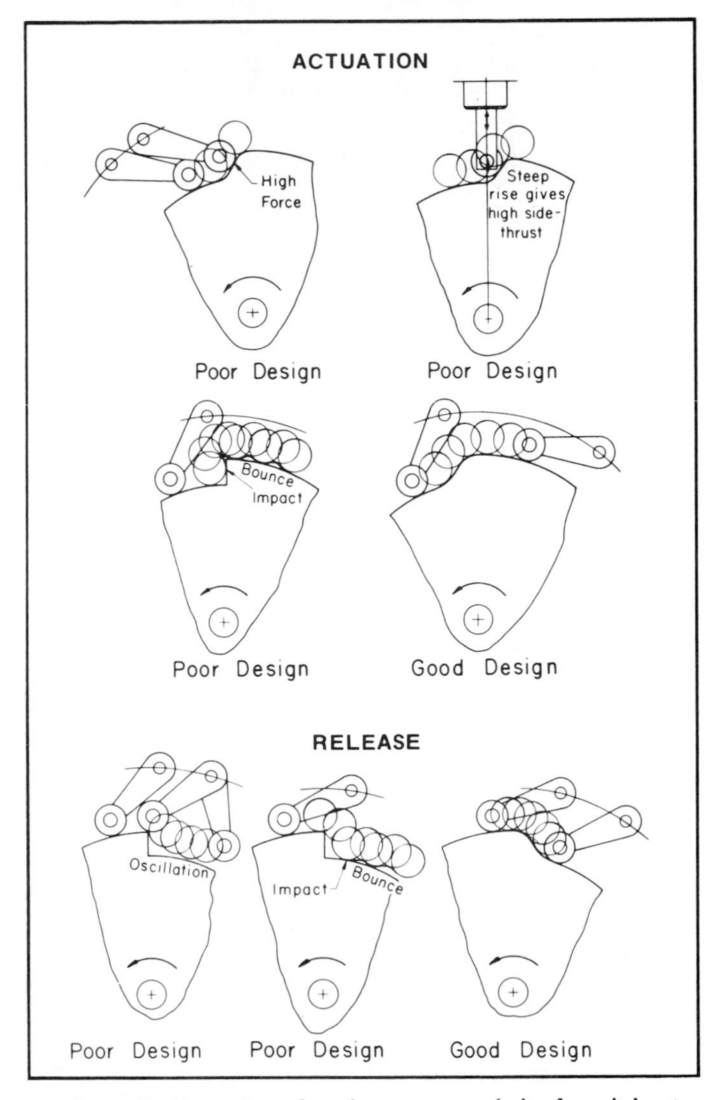

FIGURE 2.15 Comparison of good vs. poor cam design for switch actuation. (*Courtesy of Micro Switch, Freeport, Ill.*)

OTHER APPLICATIONS

The compact size and accuracy of precision snap-acting switches make them ideal for next-level applications. Basic switches can be assembled together to form multiple-switch modules. A switch module containing one, two, three, or even four basic switch(es) can be used interchangeably on the same control as shown in Fig. 2.17, which depicts the individual components which make up a standard oiltight control (3).

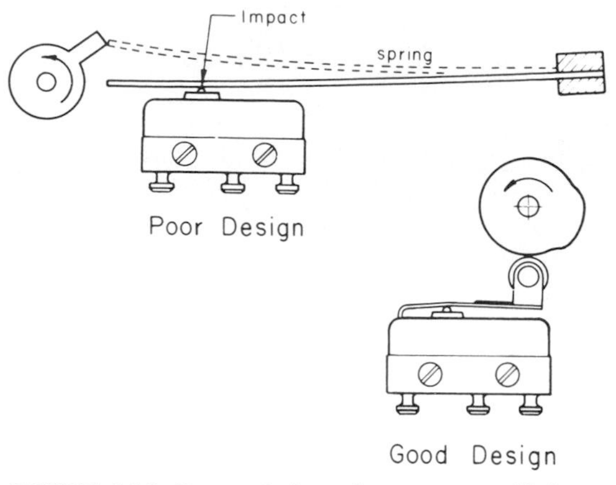

FIGURE 2.16 Proper design of cams to avoid impact actuation extends switch life. (*Courtesy of Micro Switch, Freeport, Ill.*)

Additionally, basic precision snap-acting switches are also used, either singly or in multiples, in special custom-designed switch arrays such as that depicted in Fig. 2.18. The device shown is part of a wide-body jet engine reverser system. The switch is subjected to continuous pressure and temperature changes during aircraft operation. Two basic switches, actuated by a simple mechanical mechanism, are shown mounted inside a hermetically sealed, inert-gas-filled chamber. This compact package, designed by Micro Switch, Freeport, Ill., replaced a previous device utilizing complex mechanical linkage which produced unacceptable tolerance buildups in the assembly. Incorporating this new compact switch package in the reverser assembly enhanced overall reliability and provided a weight saving and an overall 30 percent cost reduction. [Adapted from (4).]

Figure 2.19 portrays a snap-acting switch packaged to withstand IP67 environments. One of a number of environmentally resistant switches manufactured by Marquardt GmbH, Rietheim-Weilheim, Germany, the standard versions are available with different types of levers and a variety of mounting means for applications ranging from use in automobiles to electrical equipment operated in wet areas (5).

PREVENTIVE MAINTENANCE

Although snap-acting switches are designed for long life, and prudent application takes advantage of this fact, no switch is eternal. If actuated through enough cycles, any switch will wear out eventually. This may result in failure to open a circuit or failure to close it. The malfunction may be intermittent or permanent, and its onset may be gradual or sudden.

The way the switch is applied determines how serious a matter its failure can be. Shutdown of a production line, for example, can be expensive and can justify considerable effort to avoid it. If failure of a switch can have serious consequences, it is especially important that the right switch be chosen for the application, and it should be replaced at appropriate intervals to avoid failure due to

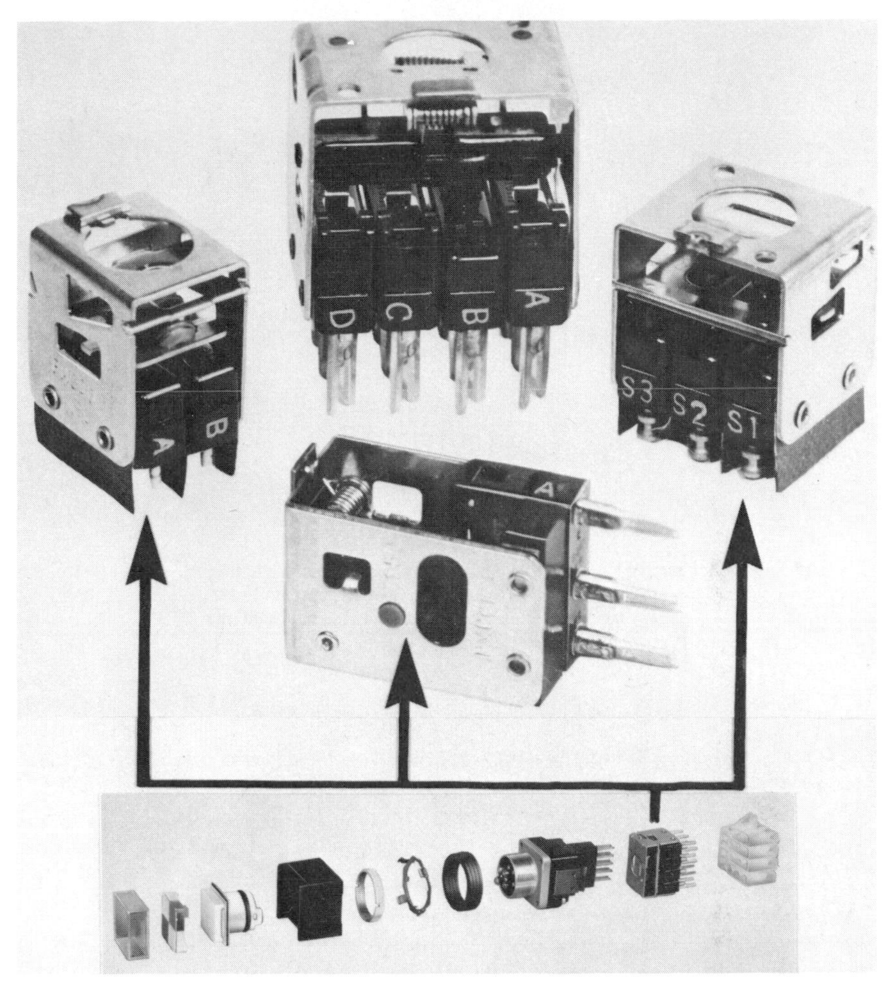

FIGURE 2.17 Interchangeable basic switch modules provide versatility for oiltight control component. (*Courtesy of Micro Switch, Freeport, Ill.*)

wear-out. A switch subjected to contamination in end use should be of sealed construction. If a switch is unsealed, it should be kept as clean and dry as possible. Oil spills or dust accumulation should be avoided. Watch for deterioration of the drive linkage that may apply excessive side thrust to the switch plunger. Switches which form a part of a safety system, such as are associated with safety gates on plastic molding presses, should not be used routinely to close and open a live circuit. Such use tends to wear out such switches electrically and reduces the ability of the system to prevent accidents.

If a switch malfunction does occur in a critical application, replace the switch *immediately*. Do not assume that all is well because the malfunction occurred only once. Return the switch *unopened* in a plastic bag to the manufacturer for expert advice and recommendations. Opening the switch yourself can destroy im-

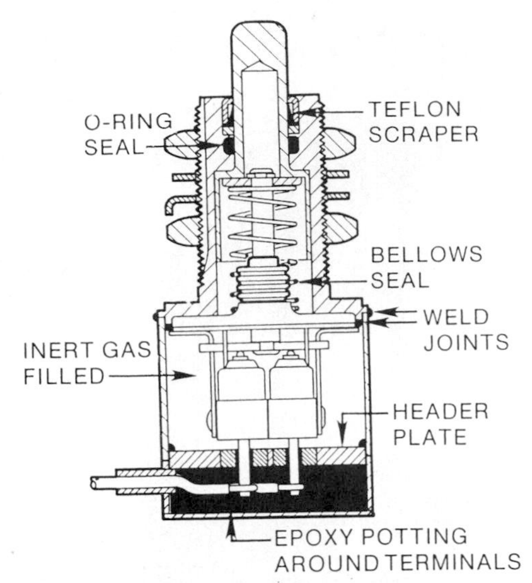

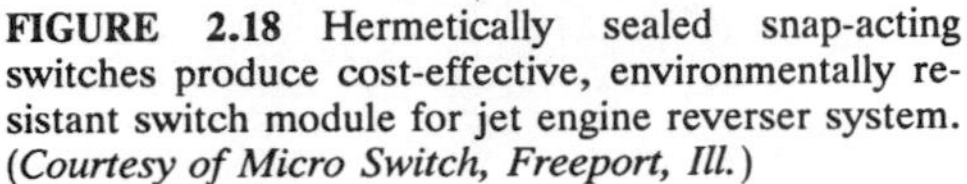

FIGURE 2.18 Hermetically sealed snap-acting switches produce cost-effective, environmentally resistant switch module for jet engine reverser system. (*Courtesy of Micro Switch, Freeport, Ill.*)

FIGURE 2.19 Specially sealed precision snap-acting switch meets IP67 environmental specification. (*Courtesy of Marquardt GmbH, Rietheim-Weilheim, Germany.*)

portant evidence needed for an accurate analysis of the switch failure. In discussing switch performance with the manufacturer, the following variables are of major importance:

Mechanical

1. Mounting means—side mount, panel mount, adhesive, bolts, etc.
2. Attitude of switch
3. Direction of actuation with respect to the switch
4. Mass and elastic properties of actuating member

5. Overtravel force applied to switch
6. Overtravel and release travel applied to the switch
7. Velocity of actuating member

Electrical

1. Mounting surface—electrically conductive or insulating
2. Wire gauge and means of attachment of wires to switch
3. Inductive or capacitive characteristics of circuit (when testing, use actual load components of end use, if possible)
4. Connection of poles and throws of the switch
5. Connection of switch in hot or ground side of line
6. Polarity of connection to the switch
7. Voltage magnitude and tolerances
8. Current magnitude and tolerances on make, carry, and break

Environmental

1. Temperature
2. Relative humidity
3. Atmospheric pressure

General

1. Definition of switch failure or end point of the test
2. Instrumentation used to detect switch failure

Being prepared to provide a full description of the end use conditions enables the manufacturer to recommend ways to increase switch life. Also, the experience of the switch manufacturer is helpful in setting up preventive maintenance programs. Often, the manufacturer can supply statistical estimates of switch life, and other valuable advice to reduce machine downtime. [Adapted from (1).]

The foregoing is, of course, not strictly limited to snap-acting switches, and switch manufacturers should be brought into the picture as early as possible regardless of the switch technology under consideration.

GLOSSARY

Actuator. The mechanical link that drives the plunger of a basic snap-acting switch.

Auxiliary actuator. A mechanism which may be attached to a switch to modify its operating characteristics.

Basic switch. A complete and self-contained switching unit of which there are many shapes and sizes. Basic switches may be used alone, gang-mounted, built into assemblies, or enclosed in metal housings.

Bifurcated contact. A movable or stationary contact which is forked or divided to provide two pairs of mating contact surfaces connected in parallel, instead of a single pair of mating surfaces.

Break distance. The minimum distance between separated mating contacts in their fully open position.

Dead break. Imperfect snap action in which the normally closed circuit of the switch opens before the plunger reaches the operating point, or the normally open circuit opens before the plunger reaches the release point.

Dead make. Imperfect snap action in which a switch fails to close its circuit when the plunger reaches the operating or release point.

Differential force. See *Force differential.*

Differential travel. The distance from the operating point to the release point.

Drift (of an operating characteristic). An inexact term referring in a general way to the degree of instability of a plunger force or travel characteristic under specified conditions and during a specified number of cycles of switch operation.

Enclosed switch. One or more basic snap-acting switches enclosed in a protective housing.

Force differential. The difference between the operating force and the release force.

Free position (of the plunger). The position of the plunger when no external force other than gravity is applied to it.

Full overtravel force. The force required to depress the plunger of a switch to the full overtravel point.

Full overtravel point. That position of the plunger beyond which further overtravel would cause damage to the switch or actuator.

Operated contact position. The position to which the contacts move when the plunger is traveled to the operating point or into the overtravel range.

Operating force. That force which must be applied to the plunger to cause the moving contact to snap from the normal contact position to the operated contact position.

Operating point. That position of the plunger at which the contacts snap from the normal contact position to the operated contact position.

Overtravel. As an operating characteristic of a switch, overtravel is the distance through which the plunger moves when traveling from the operating point to the full overtravel point. As a characteristic of the actuation applied to the switch, overtravel is the distance the plunger is driven past the operating point.

Precision snap-acting switch. NEMA defines a precision snap-acting switch as "a mechanically operated electric switch having predetermined and accurately controlled characteristics, and having contacts other than the blade and jaw, or mercury type, where the maximum separation between any butting contacts is 1/8 in (3,17 mm). A precision snap-acting switch consists of a basic switch used alone, a basic switch used with actuator(s), or a basic switch used with actuator(s) and an enclosure."

Pretravel. The distance through which the plunger moves when traveling from the free position to the operating point.

Release force. The level to which force on the plunger must be reduced to allow the contacts to snap from the operated contact position to the normal contact position.

Release point. That position of the plunger at which the contacts snap from the operated contact position to the normal contact position.

Release travel. As an operating characteristic of a switch, release travel is the distance through which the plunger moves when traveling from the release point to the free position. As a characteristic of the actuation applied to the switch, release travel is the distance the plunger is released past the release point.

Snap action. In strict terms, snap action is a property of a switch such that the moving contact accelerates without added travel of the plunger beyond that travel which was required to separate the contacts. NEMA defines snap action as "a rapid motion of the contacts from one position to another position, or their return. This action is relatively independent of the rate of travel of the actuator."

Total travel. The distance from the plunger-free position to the full overtravel point.

REFERENCES

1. Lockwood, J. P., "Applying Precision Switches, A Practical Guide," Micro Switch, Freeport, Ill., 1973.
2. "Specifier's Guide for Basic Switches, Catalog 10, Issue 12," Micro Switch, Freeport, Ill., 1991.
3. "Specifier's Guide for Oiltight Manual Controls, Catalog 70, Issue 10," Micro Switch, Freeport, Ill., 1990.
4. "Application Note 1, Value-added Solutions," Micro Switch, Freeport, Ill., 1988.
5. "Best for Use under Extreme Conditions: Snap Action Switches 1022," Marquardt GmbH, Rietheim-Weilheim, Germany, 1991.

CHAPTER 3
ROTARY SWITCHES*

A rotary switch is an electromechanical device which is capable of selecting, making, or breaking an electric circuit and which is activated by a rotational force applied to the shaft. Rotary switches have applications in electronic equipment that require a change in the mode of operation. Examples where changes in mode of operation are required are test equipment, communications equipment, data-processing input and output equipment, and industrial control.

STANDARD ROTARY SWITCH TECHNOLOGY

How Rotary Switches Function

The principal parts of a rotary switch are (*a*) a shaft and bushing, (*b*) a deck section, (*c*) a stop mechanism, and (*d*) a detent mechanism (Fig. 3.1). The *shaft* is the element to which rotational force is applied to cause a change in the switching elements. The *bushing* is an integral part of a rotary switch and is the usual means by which the switch can be mounted to a panel or a bracket of a piece of

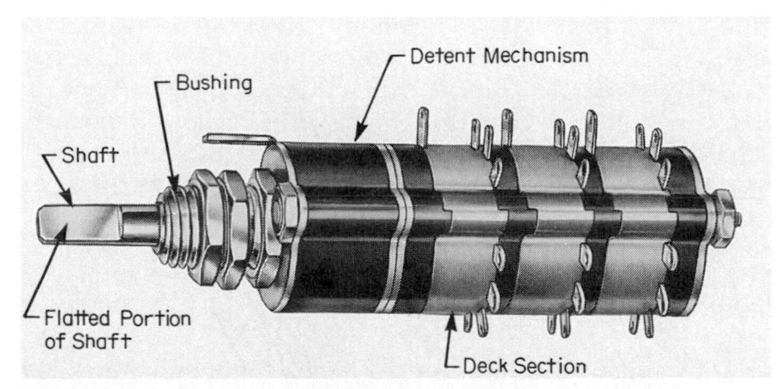

FIGURE 3.1 The principal parts of a rotary switch are the shaft and bushing, the deck section, the stop mechanism (not shown), and a detent mechanism. (*Courtesy of Grayhill, Inc., La Grange, Ill.*)

*Numbers in parentheses indicate items in the References at the end of this chapter.

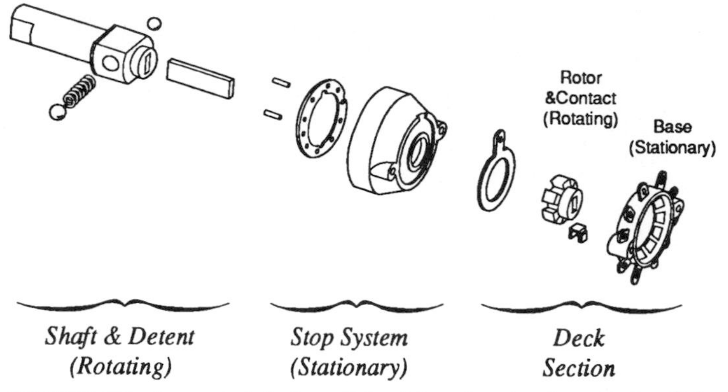

FIGURE 3.2 An exploded view of the internal parts of a rotary switch. (*Courtesy of Grayhill, Inc., La Grange, Ill.*)

equipment. The shaft projects from the bushing and internally drives the detent system and the switching elements. A *detent* system allows the operator of the shaft to positively feel the location of the switch contacts. The *stop* mechanism limits the rotation of the shaft to the active switch positions. The *deck section* contains the switching circuitry (Fig. 3.2).

Shaft and Bushing. The shaft is the rod-shaped device which is rotated by the operator to move the rotary switch from position to position. The rotor, with its U-shaped contact, is permanently attached to the shaft and is sandwiched between each common plate and terminal base. The shaft end nearest the operator has a flatted (milled) portion (Fig. 3.1) which permits positive mounting of a knob. The shaft flat orientation is the angular relationship of this shaft flat to a fixed point on the body of the switch with the switch in a specified position. Generally speaking, most rotary switch manufacturers specify this position as the first position of the first pole and the shaft flat is oriented 180° (or is opposite) from the point of electrical contact on the first pole. For example, if the switch were in position 1 of a 30° switch, the shaft flat would be opposite position 1, or facing position 7. With the switch in position 2, the shaft flat would be facing position 8, and so on. This particular shaft flat orientation is simply stated by saying "the shaft flat is opposite the point of contact of pole." (1) The engineer should pay particular attention to the shaft flat orientation because a mismatch between the switch shaft flat and the flat and/or setscrew of the knob specified for the switch could cause the index mark on the knob to be off as much as 180°. The index mark is the line or arrow on the knob that points to the legend defining the switch setting. This legend is usually marked on the panel to which the switch is mounted. You do not need the embarrassment of having the knob pointing to the wrong legend!

Deck Section. The deck section of a rotary switch contains the switch-position terminals along with the common terminal and provides some method of connection between a selected switch-position terminal and the common terminal. These terminals are located around the circumference of a structure which surrounds the shaft. Contact is made between a common terminal of the deck section and the selected switch-position terminal. As the shaft is rotated, contact is changed

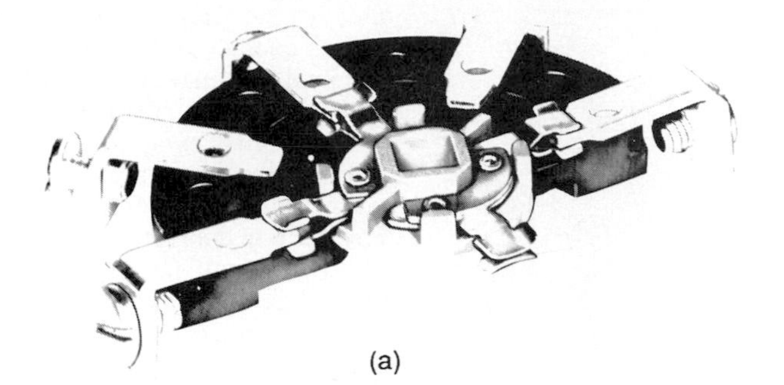
(a)

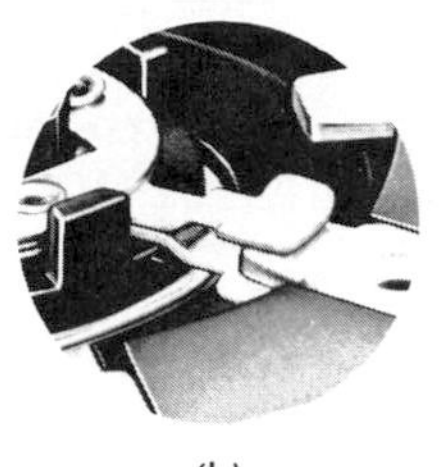
(b)

FIGURE 3.3 (*a*) Typical contact arrangement in a rotary switch. (*b*) Rotary switch contacts are usually wiping-action type. (*Courtesy of Electroswitch, Weymouth, Mass.*)

from the common terminal and one switch position to the common terminal and the next switch-position terminal (Fig. 3.3*a*).

Rotary switch contacts are usually wiping contacts since the rotating contact slides across the base terminals (Fig. 3.3*b*). Hence they have the characteristics of wiping contacts in that contact resistance remains relatively constant during switch life. However, the wear caused by the wiping action tends to shorten switch life.

Rotary switch contacts may also be either *shorting* or *nonshorting*. When switching from one position to the next, a *shorting* rotary switch contact (Fig. 3.4*a*) is one which makes contact with the adjacent base terminal *before* it breaks the existing contact with the original terminal. For example, a shorting contact between positions 1 and 2 would momentarily touch both base terminals as the switch is rotated from position 1 to position 2. When switching from one position to the next, a *nonshorting* contact (Fig. 3.4*b*) is one that completely breaks the con-

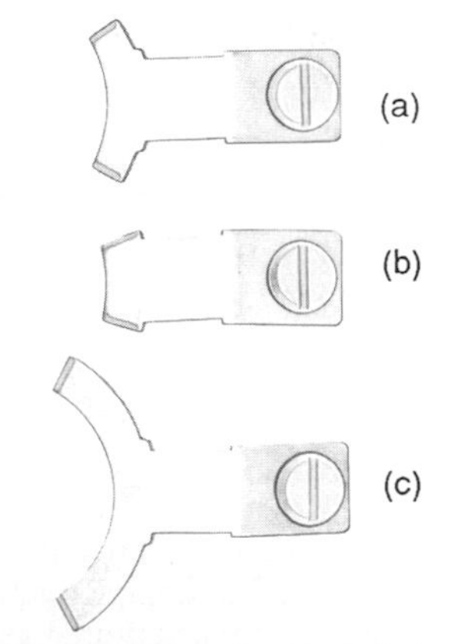

FIGURE 3.4 Types of rotary switch contacts: (*a*) shorting contact, (*b*) nonshorting contact, and (*c*) "sweep" contact. (*Courtesy of Electroswitch, Weymouth, Mass.*)

FIGURE 3.5 All decks in a rotary switch are driven by the same shaft. These multideck switches are commonly used in a variety of applications. (*Courtesy of Grayhill, Inc., La Grange, Ill.*)

tact at the original terminal *before* it makes contact at the next adjacent base terminal. When a nonshorting contact is rotated between position 1 and position 2, it will break away from terminal 1 before it touches terminal 2. There is a point between terminals where a nonshorting contact will not be touching *either* terminal. The term "make before break" is sometimes used to describe shorting contacts, and the term "break before make" is sometimes used to describe nonshorting contacts. (1)

A variation of the standard shorting contact is the so-called "sweep" contact depicted in Figure 3.4*c*. This type of contact maintains the connection with the rotor through consecutive positions. (2)

It is possible to provide rotary switches with more than one deck section. A three-deck and six-deck rotary switch are depicted in Fig. 3.5. All decks in an individual switch are driven by the same shaft. These deck sections are electrically independent but can be externally wired together.

In one popular style of rotary switch, a deck section consists of a *common plate* and its housing which surrounds the shaft, a U-shaped contact which touches or contacts the selected switch-position terminal and the common plate, a *rotor mounting plate* which keys to the shaft and holds the U-shaped contact in position, and a *base* to which the switch-position terminals are attached. An exploded version of a typical deck section is shown in Fig. 3.6. As the shaft is rotated, the U-shaped contact is moved between the switch-position terminals. However, contact is *always* maintained between the U-shaped contact and the common plate with its integral terminal.

It is general industry convention to number the switch-position terminals consecutively in a clockwise direction *when viewed from the shaft end of the switch.* When contact is made between the common terminal and the terminal at position 1, the switch is said to be in position 1. Rotating the shaft one position *clockwise,* the switch would then be in position 2. Each succeeding turn clockwise places the switch at the next higher position. A complete 360° rotation returns the switch to position 1. (1)

The number of terminals that are located on a base of a deck section can vary. The switch-position terminals are usually referred to as base terminals. Electrical contact is made between a selected base terminal and the common plate by a mechanical means. The operator opens the path between this base terminal and the common terminal and completes the path between the next successive base terminal and the common terminal as the shaft is rotated.

The base terminals are located around the circumference of the switch. The angular distance between these terminals is called the *angle of throw* of the switch. For example, if there are 12 evenly spaced terminals on the base of a

FIGURE 3.6 Details of a single deck section. (*Courtesy of Grayhill, Inc., La Grange, Ill.*)

deck section, the angle of throw is 30° (12 divided into 360° is 30°); 10 evenly spaced terminals yields an angle of throw of 36° (Fig. 3.7). A switch manufacturer may provide a number of rotary switches with different angles of throw, with industry standards being 15°, 18°, 22.5°, 30°, 36°, 45°, 60°, and 90°.

Once the angle of throw of a switch is known, the maximum number of switch positions can be determined. This is accomplished by dividing the angle of throw into 360° (12 positions for 30°, 10 positions for 36°, etc.).

Base terminals are numbered consecutively in a clockwise direction as viewed from the shaft end of the switch. If contact is being made between terminal at position 1 and the common terminal, rotating the shaft one position in a clockwise direction causes contact between the common terminal and position 2. This switching action then takes place for the remaining switch positions.

Since the common plate terminal performs a similar function to the pole in a pushbutton switch (Chap. 6), it is called a "pole." Many poles may be contained

FIGURE 3.7 The angular distance between terminals is called the angle of throw of a rotary switch. (*Courtesy of Grayhill, Inc., La Grange, Ill.*)

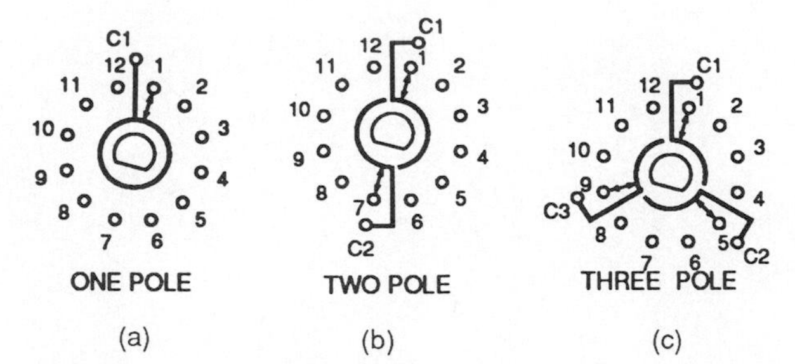

FIGURE 3.8 Circuit diagrams for (*a*) a one-pole switch with a 30° angle of throw, (*b*) a two-pole switch with a 30° angle of throw, and (*c*) a three-pole switch with a 30° angle of throw. (*Courtesy of Grayhill, Inc., La Grange, Ill.*)

in a single deck section of a rotary switch. In order to show which base terminals are associated with which poles, a circuit diagram is used.

The circuit diagram depicted in Fig. 3.8*a* is for a one-pole switch with a 30° angle of throw. This circuit diagram shows that there is presently a connection between position 1 and the common plate; the switch is said to be in position 1. The inner D shape represents the rotating shaft, and the line is the flat side; the shaft flat is opposite the rotating contact (arrows).

The circuit diagram also shows the location of the pole terminal. The pole (also called "common pole" or "common") is identified as C1 and is shown to be in line with a spot halfway between terminal positions 12 and 1.

The circuit diagrams depicted in Fig. 3.8*b* and *c* show a two-pole and three-pole switch, respectively, each with a 30° angle of throw. The two-pole circuit diagram shows that base terminals 1 through 6 are associated with pole 1, and base terminals 7 through 12 are associated with pole 2. As shown, contact is being made with pole 1 and terminal 1 and pole 2 and terminal 7. Moving the shaft one position clockwise would cause contact between pole 1 and terminal 2 and pole 2 and terminal 8.

The three-pole circuit diagram in Fig. 3.8*c* shows terminals 1 through 4 are associated with pole 1, terminals 5 through 8 with pole 2, and terminals 9 through 12 with pole 3. As shown, contact is made between common terminal 1 and terminal 1, common terminal 2 and terminal 5, and common terminal 3 and terminal 9. The circuit diagrams shown in Figs. 3.9*a* and *b* show a four- and a five- and six-pole switch, also both with a 30° angle of throw. The contact arrangements are similar to those discussed for the one-, two-, and three-pole arrangements. The five-pole switch is the same as the six-pole switch with one pole and its associated base terminals eliminated. (3)

The base of a deck section does not have to include the maximum number of terminals. If a switch has a 30° angle of throw, 12 terminals, or positions (30° × 12 = 360°), in a deck section are possible. The application may indicate that the switch requires only 4 terminals on a base. Figure 3.10 depicts both a 12-terminal and a 4-terminal deck. Since the angle of throw is 30°, less than one-third of the rotation is used in the 4-terminal deck. A switch can have from two to the maximum number of terminals on a deck, depending upon a customer's requirements. A minimum of two terminals, of course, is necessary if any switching is to be performed.

Up to this point, this chapter has described switches with deck sections that

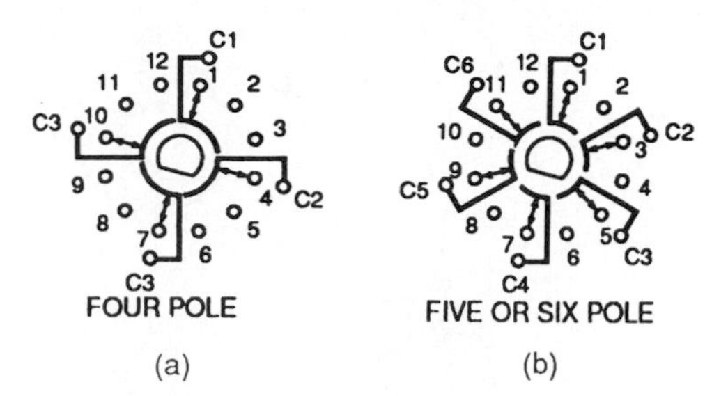

FIGURE 3.9 Circuit diagrams for (*a*) a four-pole switch with a 30° angle of throw, and (*b*) a five- or six-pole switch with a 30° angle of throw. (*Courtesy of Grayhill, Inc., La Grange, Ill.*)

have one common plate. Since this common plate may be connected to one of several base terminals or positions, it is often referred to as the *pole* terminal (this is in keeping with switch terminology). Switches with one common plate per deck section are referred to as "one pole per deck" switches. It is possible to have several commons in one deck section as shown in Fig. 3.11. Each common has its own U-shaped contact and base terminals associated with it. Thus, it is possible to have a switch with two, three, four (or more) poles per deck.

When less than the maximum number of terminals are located on the base of a deck section, some provision must be made to prevent rotation into the unused area of the deck section. This is accomplished by the incorporation of a stop mechanism into the rotary switch. (1)

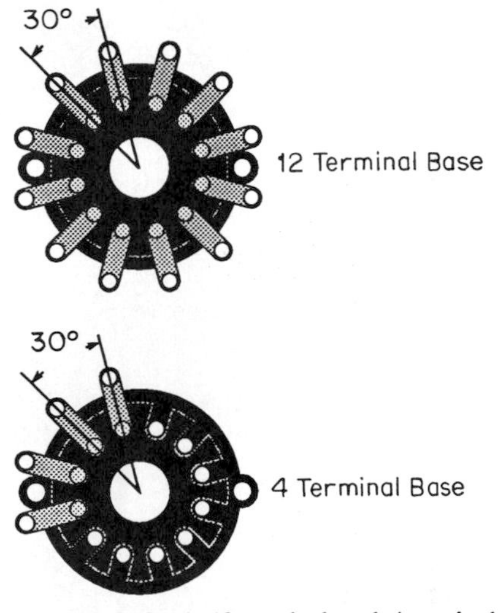

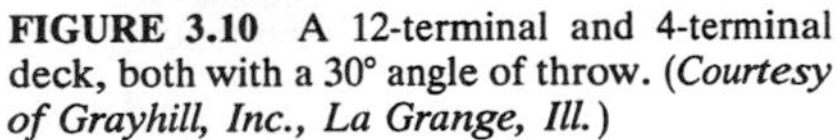

FIGURE 3.10 A 12-terminal and 4-terminal deck, both with a 30° angle of throw. (*Courtesy of Grayhill, Inc., La Grange, Ill.*)

Stop Mechanism. The stop mechanism limits the rotation of the contact between the common plate and its associated base terminals. One method of providing a stop function consists of a stop arm and stop washers (Fig. 3.12). A *stop arm* is a washer with an *outward* projection. It is keyed to the shaft and located in the detent cover. A pair of *stop washers* with *inward* projections are located at the inside perimeter of the detent cover. The stop washers can be arranged to provide any location of inward projection at the time the switch is being built. However, the stop washers are keyed to the cover so that they do not move once they are fixed in position. The outward projection of the stop arm, when it strikes the inward projection of the stop washer, prevents any further rotation of the shaft. Rotation of the detent system and the U-shaped contact is thus limited.

The stop mechanism provides the same function in a switch that has more

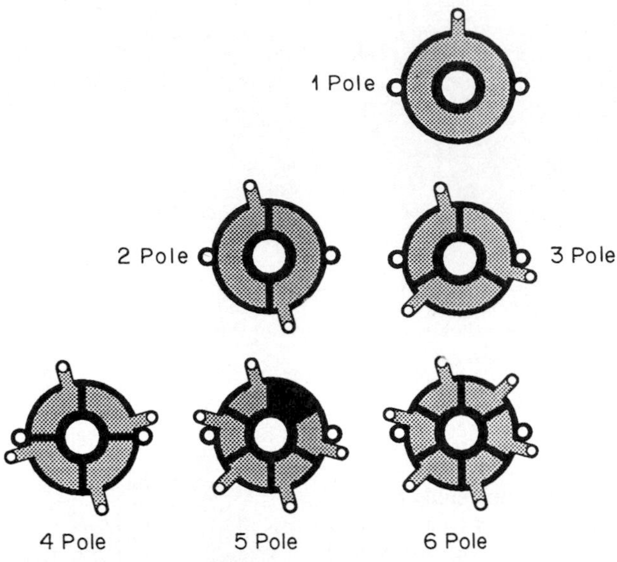

FIGURE 3.11 It is possible to have several commons in one deck section. (*Courtesy of Grayhill, Inc., La Grange, Ill.*)

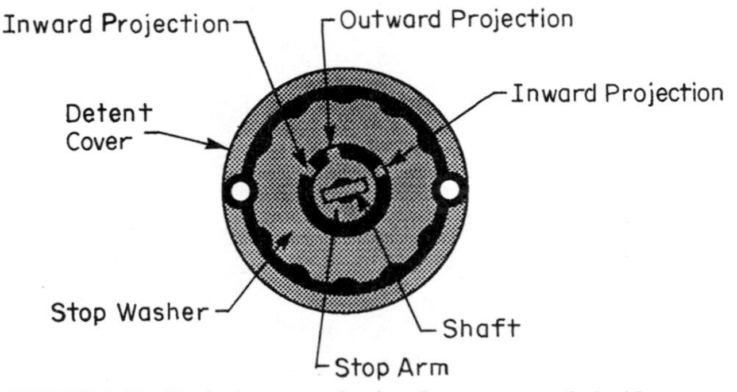

FIGURE 3.12 Typical stop mechanism for a rotary switch. (*Courtesy of Grayhill, Inc., La Grange, Ill.*)

than one pole per deck. Consider a switch with a 30° angle of throw. The maximum number of base terminals in a deck section is 12. If there were two poles in this deck section, there could be from two to six positions associated with each pole. The stop system would limit the rotation of the contact of the first pole between positions 1 and 6. This would automatically limit the rotation of the contact of the second pole to between positions 7 and 12. It is not necessary for the maximum number of positions to be associated with each pole. For example, our 30°, two pole per deck switch could have its rotation limited to two positions per pole. The stop system would limit the rotation of the contact of the first pole to terminals 1 and 2, which would limit the rotation of the contact of the second pole to positions 7 and 8. (1)

Referring back to the four-pole circuit diagram of Fig. 3.9*a*, the stop system prevents the contact between terminal 1 and base terminal 1 from rotating to base terminal 12 or base terminal 4. Rotation of the contact for common 1 is limited to positions 1 through 3. The remaining terminals will, of course, be respectively connected to their associated poles. (3)

A switch parameter that is associated with the stop mechanism is called the *stop strength.* This parameter indicates the amount of force that a stop mechanism can withstand before mechanical failure.

Detent Mechanism. The detent mechanism of a rotary switch is designed to provide the operator with a positive indication of rotation from one switch position to the next. Without a detent mechanism, there would be a tendency for the operator to turn the shaft through switch positions, for example, moving from position 1 through position 2 to position 3 when position 2 is the desired location.

One method of providing the detent is a "hill and valley" camlike raceway incorporating a spring and ball arrangement (Fig. 3.13*a*). The ball bearings are spring-loaded against the hill and valley race which is located in the detent cover. The pressure on the ball bearings is relaxed when they fall into a valley position in the detent cover. When the ball bearings are in a valley, the U-shaped contact is touching a base terminal. Turning the shaft forces the ball bearing up a "hill" and then down into the next "valley" as the next switch position is reached. In this manner, a positive action and feel is provided, preventing the operator from inadvertently turning through a desired switch position. (1)

Another method of providing detent is shown in Fig. 3.13*b*. This design uses a star wheel, rigidly attached to the rotary switch shaft, and up to four spring-loaded ball bearings. (2)

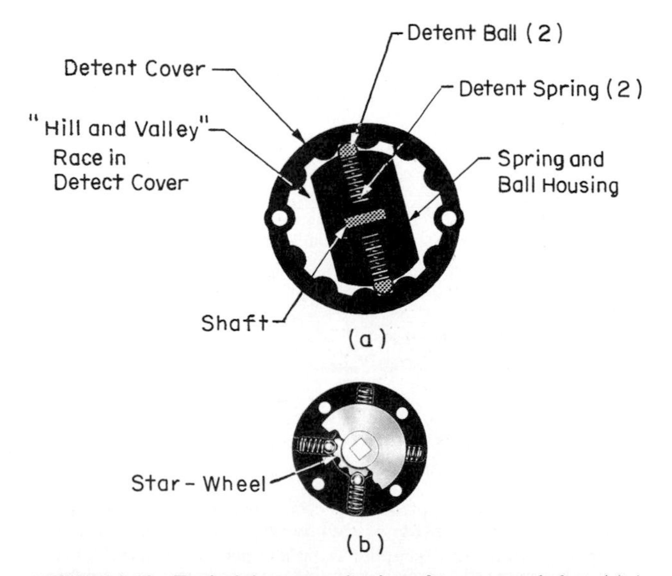

FIGURE 3.13 Typical detent mechanisms for rotary switches: (*a*) A "hill and valley" race design, and (*b*) a star-wheel design. Both styles use spring-loaded ball bearings. [(a) *Courtesy of Grayhill, Inc., La Grange, Ill. and* (b) *courtesy of Electroswitch, Weymouth, Mass.*]

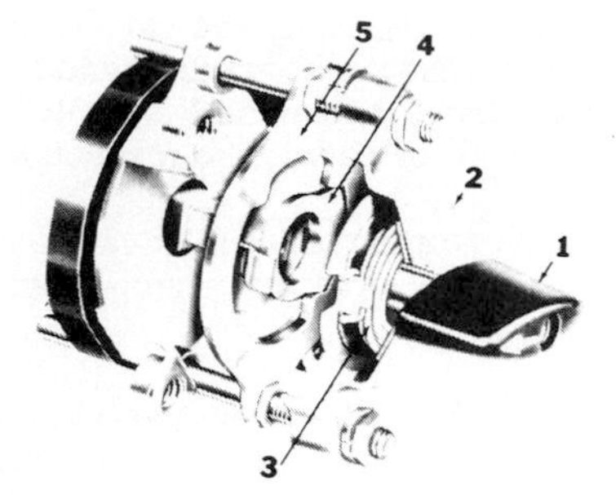

FIGURE 3.14 Details of a snap-action rotary switch mechanism: (1) operating handle, (2) mechanism cover, (3) coil spring, (4) indexing plate, and (5) locking plate. (*Courtesy of Electroswitch, Weymouth, Mass.*)

Snap-Action and Cam-Action Detent Styles. Two other mechanisms for providing an indication of positive rotation are the snap-action and cam-action devices.

The mechanical system of a *snap-action* rotary switch is designed to provide uniform high-speed make and break, regardless of whether the operating handle is turned rapidly or slowly (Fig. 3.14). Turning the handle through about 120° in either direction winds a powerful coil spring. When the spring is fully wound, the indexing plate is momentarily withdrawn from the locking plate by an eccentric cam. The shaft and movable contacts then snap rapidly to the next position, where the indexing plate holds them until the spring-drive mechanism is again operated. Transit time is about 10 ms.

In the *cam-action* rotary switch mechanism, the contacting consists simply of shunting two isolated contacts to make a circuit as depicted in Fig. 3.15*a*. Using this simple principle, two independent sets of contacts are placed in each deck. The moving portion is spring-loaded to close the contact. A notch on the cam that is affixed to the operating shaft allows the moving contact to spring closed, bridg-

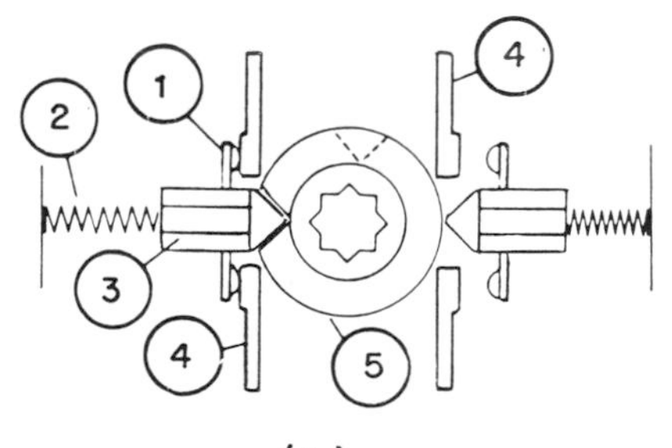

FIGURE 3.15 Details of a cam-action rotary switch mechanism: (*a*) contact design consists of shunting two isolated contacts to make a circuit, and (*b*) component parts of cam-action device: (1) movable contact, (2) spring loading, (3) cam follower, (4) stationary contact pair, and (5) notched cam. (*Courtesy of Electroswitch, Weymouth, Mass.*)

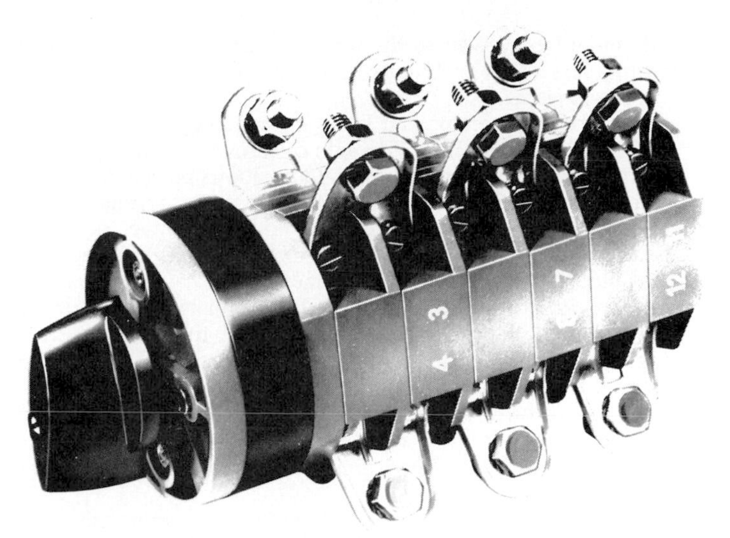

FIGURE 3.16 A cam-action rotary switch. (*Courtesy of Electroswitch, Weymouth, Mass.*)

ing the stationary contacts. This action is illustrated in Fig. 3.15*b*. The left side of the figure shows a cam follower detented in the cam notch, closing the spring-loaded movable contacts onto the two stationary contacts. The right side of the figure shows the circuit held open by the cam, which will close when the notch on the second independent cam is rotated around and comes in proximity to its cam follower (the second cam notch is illustrated by the dotted lines—the second cam is behind the first cam). (2)

A rugged cam-action rotary switch is shown in Fig. 3.16.

A parameter principally determined by the type of detent mechanism (and by the number of poles in the switch) is the *rotational torque*, which describes the amount of force required for the operator to change from one switch position to the next.

ADJUSTABLE STOP ROTARY SWITCH TECHNOLOGY

The initial part of this chapter described the function and construction of standard "plain vanilla" rotary switches. In the past, one aspect of standard rotary switch construction was responsible for annoying problems in the manufacturing and distribution of the switches, as well as for the user community. The basis of that problem was the stop mechanism which, up to that time, had to be built in at the factory. Since there were literally hundreds of different switch configurations and many different stop possibilities, each switch would have to be custom-built to match the specification produced by the customer's engineering department. This custom-built requirement meant more paperwork flowing through the manufacturer's plant, less opportunity for local distributors to stock the particular switch needed, and a longer wait in the queue for the engineer who wanted the switch in the first place. This problem was solved by the introduction of the adjustable stop rotary switch.

The purpose of the stop mechanism in rotary switches is to limit its rotation. To review a previous example, in a 30° angle of throw, one pole per deck rotary switch, it is possible to have a maximum of 12 positions per pole. An application may require only four of these to be active. The stop mechanism within the switch is set to limit the rotation of the shaft and the contact system between positions 1 and 4. Such a switch is unlikely to be in a distributor's (or even a manufacturer's) stock. Prior to the development of the adjustable stop mechanism, this type of switch had to be built to order. The introduction of the adjustable stop rotary switch allowed the stop mechanism to be *externally* adjusted without having to disassemble the switch. [Adapted from (4).] With the adjustable stop system, the user-engineer could now adjust the stops after receiving the switch from the manufacturer or distributor. The distributor could stock a larger variety of rotary switches and offer a faster turnaround to the end user. The development of this adjustable feature greatly enhanced the efficiency of the factory-to-user distribution system.

Figure 3.17*a* and *b* depicts how different manufacturers approach the design of adjustable stops. The adjustment of the stop mechanism shown in Figure 3.17*a* is accomplished by stop washers inserted into slots in the front cover plate. When the washers are inserted, the stop arm (internally keyed to the shaft) is prevented from rotating beyond the stop washers. The slots in the front plate are labeled FIRST STOP (indicating the location between positions 12 and 1) and the remaining slots are numbered 2, 3, 4, etc. The washers can be positioned to limit the rotation to the required number of positions. Stop washers are held in place by a knurled nut. If required, this nut can be removed when the switch is attached to a panel. (4)

In the stop system design depicted in Fig. 3.17*b,* adjustable stop pins are inserted in holes on the *rear* of the switch. The protrusion on a stop washer attached to the end of the rotary shaft contacts the stop pins to provide rotational limits. (5)

Because of its inherent flexibility, the adjustable stop feature is used primarily on switches destined for prototype or preproduction systems, while rotary switches with factory-installed stops are used for production systems. The latter eliminates the need for production-line personnel to handle the adjustment of the stops, thereby reducing labor costs.

ISOLATED-POSITION ROTARY SWITCH TECHNOLOGY

An isolated-position rotary switch is one in which one or more positions cannot be reached by normal rotation of the shaft. To reach these "locked-out" positions, an additional action is required by the operator such as pushing in on or pulling out on the shaft. This type of rotary switch is often referred to as a *push-to-turn* or *pull-to-turn* switch. There are various applications for this type of switch. For example, an aircraft navigation instrument may have several modes of operation and a "calibrate" function to be selected by a rotary switch. If switching to the "calibrate" position while the aircraft is in flight is undesirable,

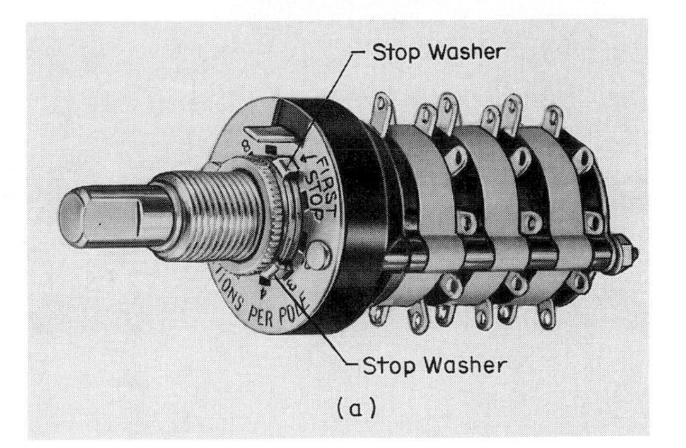

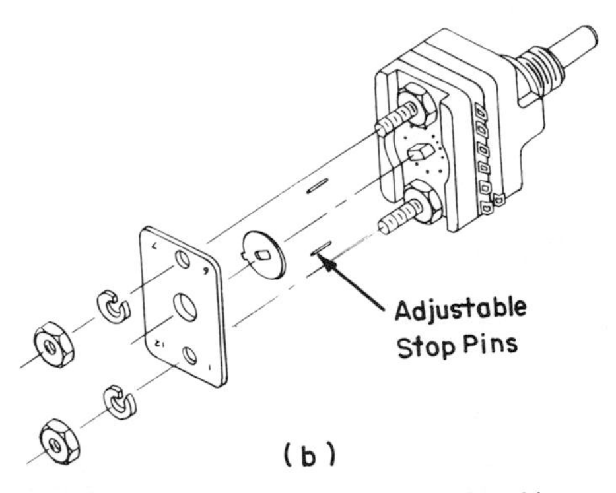

FIGURE 3.17 Two styles of adjustable stop mechanism: (*a*) design using stop arm and stop washer, and (*b*) design using adjustable stop pins and stop washer. [(a) *Courtesy of Grayhill, Inc., La Grange, Ill., and* (b) *courtesy of Electroswitch, Weymouth, Mass.*)

this switch position could be isolated to prevent accidental rotation to the "calibrate" position.

One such mechanism for isolating positions consists of a special shaft, two lock plates, a lock arm, lock posts, and a compression spring (Fig. 3.18). The shaft is specially configured to permit movement in or out within the switch. The lock arm is similar in appearance to a stop arm (Fig. 3.12). It has an outward projection and is keyed to the shaft. The lock plates have a circle of holes located between the switch positions. The compression spring maintains the shaft in its normal position until the shaft is pushed or pulled and this spring is compressed. The lock posts are staked into the proper holes in the lock plate at the factory. (6)

Figure 3.19*a* shows the assembly of the isolating mechanism within a switch. In the normal shaft position, there is conventional detent action between switch positions until the lock arm strikes the lock post. To turn to the next (the "locked

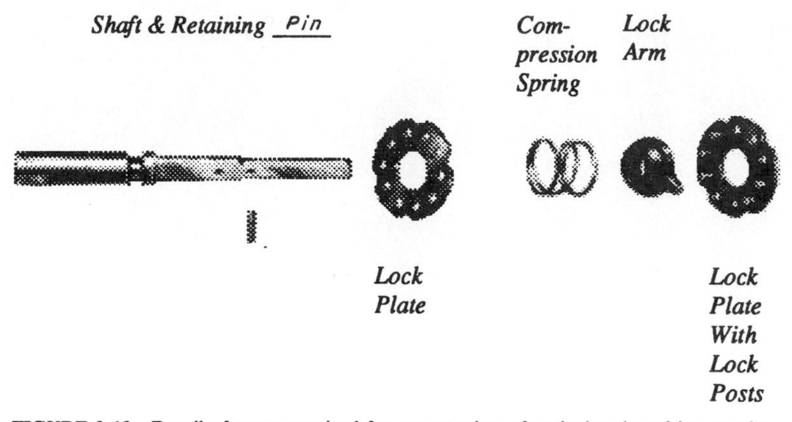

FIGURE 3.18 Detail of parts required for one version of an isolated-position mechanism. (*Courtesy of Grayhill, Inc., La Grange, Ill.*)

out") position requires the operator to pull the shaft to compress the spring and lift the lock arm over the lock post; the shaft can then be rotated to the next position. If there are lock posts on either side of this position, the shaft must be pulled again to rotate from this position.

Figure 3.19*b* illustrates the positioning of the lock arm and lock post combination in a pull-to-turn switch and a push-to-turn switch.

One potential disadvantage of this particular position-isolating mechanism is that the switch can be rotated to all active positions (up to the limits imposed by the stop mechanism) when the shaft is continually pushed or pulled and the lock arm is in a position over the lock posts. Some manufacturers incorporate special devices such as small cam-operated precision switches (Chap. 2) on the rotary switch to automatically "lock out" the nonisolated positions if the aforementioned situation occurs. The cost of these special items is generally significantly higher than switches utilizing the standard isolated-position mechanism. [Adapted from (6).]

Generally, either the pull-to-turn or push-to-turn function can be incorporated into rotary switches by manufacturers who offer the isolated-position option. During assembly the proper shaft must be chosen for push or pull to turn, and the compression spring must be positioned so that the lock arm rides on the lock plate either closest to the detent (push-to-turn) or closest to the deck section (pull-to-turn).

Since the lock posts can be staked into any or all of the holes in the lock plate, any combination of positions can be isolated. For example, a lock post in every hole will isolate every position. Stop posts between positions 1 and 2 and positions 2 and 3 will isolate position 2. This staking operation is performed at the manufacturer's factory; the isolation combination cannot be changed once the switches have been built. It is not currently possible to provide, on the same switch, a push-to-turn function in some positions and a pull-to-turn function in other positions. It is possible, however, to arrange the lock posts in such a way as to isolate any *combination* of positions.

When the switch is assembled, the entire mechanism is enclosed in a housing located immediately behind the detent system. The housing adds the length of one deck to the switch. In fact, the isolated-position mechanism has the appear-

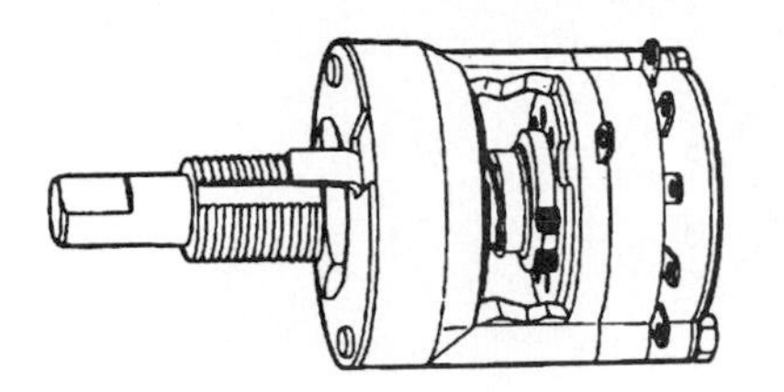

Lock Arm
Striking the Lock Post

Shaft Pulled, Lock Arm
Rises Above Lock Post

(a)

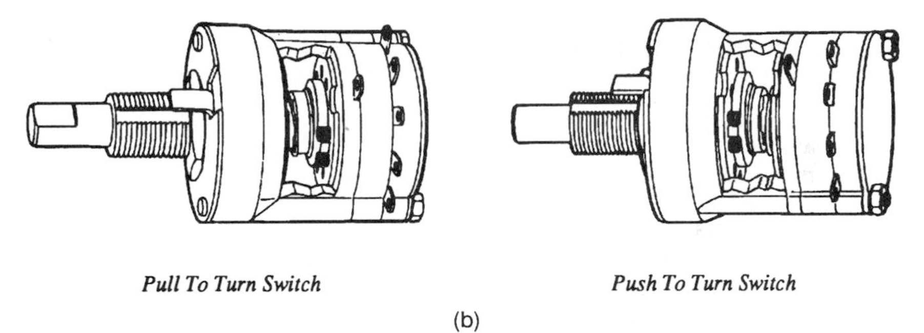

Pull To Turn Switch

Push To Turn Switch

(b)

FIGURE 3.19 Isolated-position mechanisms: (*a*) Assembly of the isolating mechanism within a switch, and (*b*) positioning of the lock arm and lock post in a pull-to-turn and push-to-turn switch. (*Courtesy of Grayhill, Inc., La Grange, Ill.*)

ance of an additional deck section, without terminals, located directly behind the detent system.

It is important to note that the isolated-position mechanism operates much like a stop mechanism of a rotary switch, but it operates *independently* of the stop mechanism. It is the stop mechanism which determines the active positions (isolated or not) of the switch. The isolated-position mechanism determines which of the active positions are to be isolated. If the stop mechanism, for example, limits rotation to six positions, it would be useless to isolate position 7, since it would never be reached.

In a multipole switch, the stop mechanism determines the active positions not only on the first pole but also on all other poles. Consider the example of a two-pole switch with a 30° angle of throw. If the stop mechanism is installed to limit the first pole to the first four positions, the second pole would likewise be limited to its first four positions. Positions 1 through 4 would be active on the first pole and positions 7 through 10 would be active on the second pole. Figure 3.20 illustrates this condition. If the last position of the first pole (position 4) is isolated, the last position of the second pole (position 10) is *also* isolated.

Electrical ratings for isolated-position switches are the same as for conventional rotary switches, but as noted previously, electrical ratings of standards within a series may vary, for example, because of distance between terminals. (6)

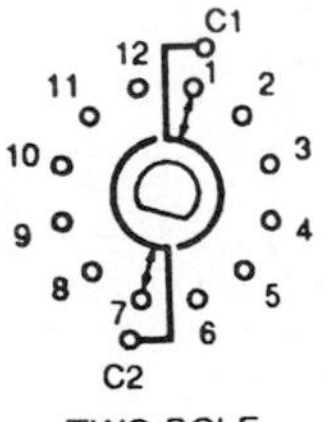

FIGURE 3.20 If positions 1 through 4 are active on the first pole and positions 7 through 10 are active on the second pole, isolating position 4 of the first pole would also isolate position 10 on the second pole. (*Courtesy of Grayhill, Inc., La Grange, Ill.*)

USEFUL ROTARY SWITCH VARIANTS

The popularity and usefulness of rotary switches has, over the years, made them subject to incorporation of a large number of design variations. A discussion of a limited number of these versions follows; the physical limitations of the chapter preclude discussion of the thousands of variations available in the marketplace. Suffice it to say, the engineer will find a wide variety of options available for rotary switches, as well as technically astute switch manufacturers ready and able to design unique solutions to special rotary switch application problems.

Miniature Rotary Switches

Miniature rotary switches are available from most rotary switch manufacturers. Figure 3.21 depicts several 0.500-in (12,7-mm)-diameter switches and some of the

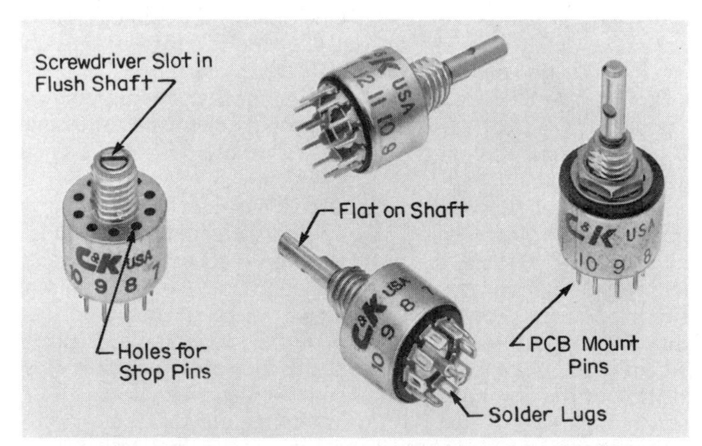

FIGURE 3.21 Miniature rotary switches. (*Courtesy of C&K Components, Inc., Newton, Mass.*)

options available within this small package. Note the holes in the front surface of one of the switches. These holes are used for the installation of stop pins for rotation limits. (7) Even smaller switches are available from some manufacturers.

PCB Mount Rotary Switches

Figure 3.22*a* and *b* illustrates two styles of PCB mount rotary switch. One style (Fig. 3.22*a*) features pins parallel to the axis of the rotary shaft to permit the device to be soldered into a motherboard which is at right angles to the shaft axis. The second style (Figure 3.22*b*) utilizes a pin arrangement which is at right angles to the shaft axis, allowing the rotary switch to mount parallel to a motherboard surface. The right-angle device enhances accessibility to the rotary switch from the side of densely packaged PCBs. (8)

Concentric-Shaft Rotary Switches

A rotary switch designed to save panel space features a concentric-shaft arrangement wherein one shaft operates a number of decks in a multideck device, and the other shaft operates the other decks, all within the same switch (Fig. 3.23*a*). Typically available with up to three decks per shaft, the switch appears to be a standard rotary switch, except for the smaller shaft, usually 0.125 in (3,17 mm) in diameter, exiting from the end of the larger shaft, typically 0.250 in (6,35 mm) in diameter. Some manufacturers even provide a double shaft seal, one for the small shaft and another for the large shaft. (8)

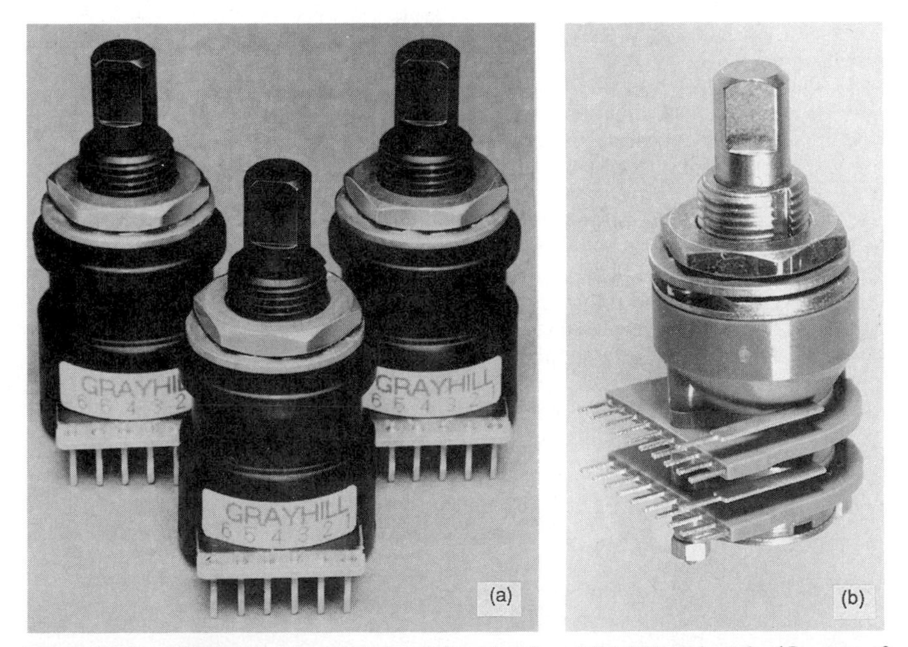

FIGURE 3.22 PCB mount rotary switches: (*a*) axial style and (*b*) right-angle style. (*Courtesy of Grayhill, Inc., La Grange, Ill.*)

Add-a-Pot Rotary Switches

A variation of the concentric-shaft rotary switch is the add-a-pot switch. A potentiometer mounting plate is furnished on which a standard potentiometer can be installed (Fig. 3.23*b*). An extension on the small shaft protrudes through the rear of the switch and is configured as a tapered tongue to provide coupling to the screwdriver slot in the potentiometer shaft. The large shaft operates the decks of the rotary switch in a normal manner.(8)

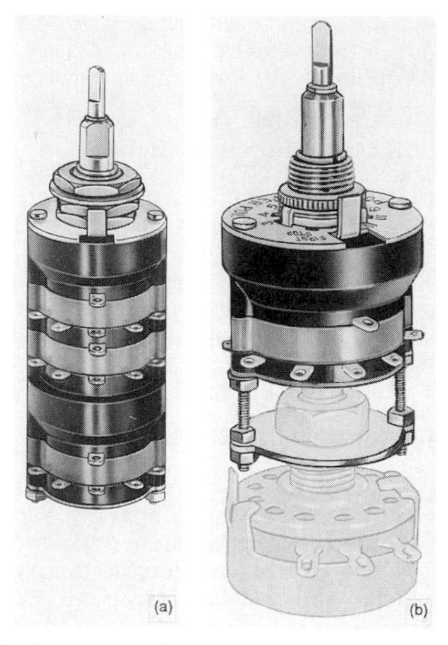

FIGURE 3.23 Two variations of concentric shaft rotary switches: (*a*) each shaft operates one or more decks of the switch, and (*b*) an added potentiometer is operated by one shaft and the switch by the other. (*Courtesy of Grayhill, Inc., La Grange, Ill.*)

Switchlock Rotary Switches

Two styles of switchlocks are depicted in Fig. 3.24*a* and *b*. These devices are available in a wide variety of configurations. A unique variation, essentially a hybrid, is shown in Figure 3.24*b*. This rotary device utilizes one or two precision snap-acting switch modules (Chap. 2) actuated by cams on the shaft of the rotary device, which is an extension of the key and tumbler mechanism. (9)

The engineer should be aware that not all rotary switch manufacturers offer adjustable stops and isolated-position features on their switches, and of those that do, all models in the manufacturer's product line may not incorporate these options. As stated a number of times in this book, it is to the engineer's advantage to discuss switch requirements as early as possible in the design stage with the switch manufacturer or switch manufacturer's representative. It is not unusual for delays of 6 months to 1 year to ensue between collection of technical data for a manufacturer's product catalog and the actual publication of the catalog itself, so that new technical developments or, worse, the discontinuation of a model the engineer wants to design into the system he or she is developing may not be noted in the technical catalog the engineer is referencing. It is always prudent for the engineer to confer with the manufacturer concerning any special requirements the engineer may need for the switches under consideration.

GLOSSARY

Angle of throw. The angular distance between switch terminals on a rotary switch.

Base. The portion of a rotary switch to which the switch position terminals are attached.

Bushing. The means by which a switch can be mounted to a panel. The threaded portion of the rotary switch through which the shaft protrudes.

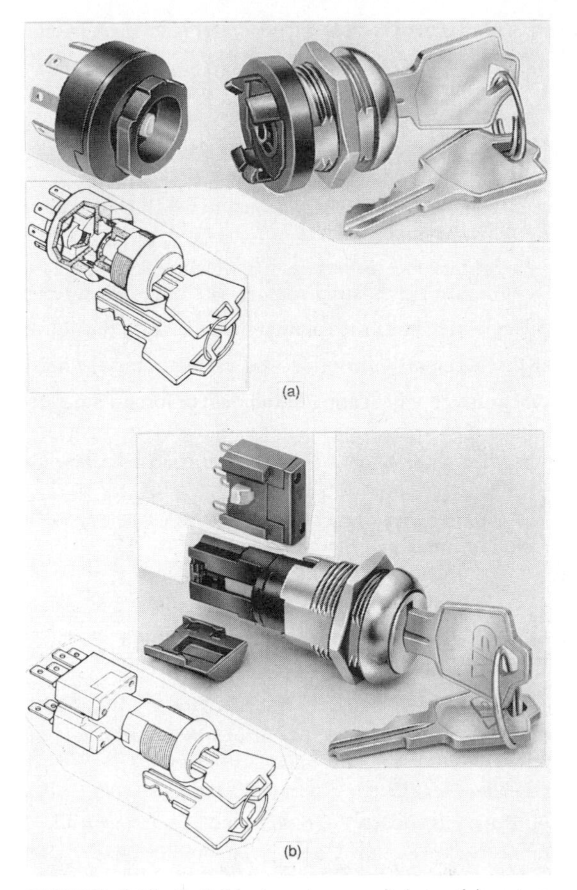

FIGURE 3.24 Switchlock rotary switches: (*a*) rotary switch is contained in modular housing, and (*b*) precision snap-acting switches are actuated by a cam on the rotary shaft. (*Courtesy of C&K Components, Clayton, N.C.*)

Common plate. The portion of a deck section which provides the common terminal for the switch.

Deck section. The portion of a rotary switch which contains the switching circuitry.

Detent. The mechanism within the rotary switch which allows the operator to positively feel the location of the switch contacts.

Nonshorting contact. A contact which completely breaks the electrical contact at the original terminal *before* it makes contact at the next adjacent terminal.

Pole terminal. The common terminal of a rotary switch.

Pull-to-turn. One of two types of additional action required to reach "locked-out" positions in an isolated-position rotary switch. The operator must pull out on the shaft to bypass the isolation mechanism. See *Push-to-turn.*

Push-to-turn. One of two types of additional action required to reach "locked-out" positions in an isolated-position rotary switch. The operator must push in on the shaft to bypass the isolation mechanism. See *Pull-to-turn.*

Rotational torque. The amount of force required for an operator to change from one switch position to the next.

Rotor mounting plate. A device which keys to the shaft and holds the U-shaped contact in position.

Shaft. The rodlike mechanism of the rotary switch to which is applied the rotational force to cause a change in the switching elements.

Shorting contact. A contact in a rotary switch which makes contact with the adjacent base terminal *before* it breaks the existing contact with the original terminal.

Stop. The mechanism which limits the rotation of the shaft to the active switch positions.

Stop arm. A washer with an *outward* projection used in rotary switch stop mechanisms.

Stop strength. A measurement indicating the amount of force a stop mechanism can withstand before mechanical failure.

Stop washers. A washer with an *inward* projection used in rotary switch stop mechanisms.

U-shaped contact. In some styles of rotary switch, this is the contact which touches the selected switch position terminal and the common plate.

REFERENCES

1. "Rotary Switch Terminology," Grayhill, Inc., La Grange, Ill., 1981.
2. "Rotary Switches for Industrial Control Applications—Catalog EMC—2," Electroswitch, Weymouth, Mass., 1990.
3. "The Switch in an Electrical Circuit," Grayhill, Inc., La Grange, Ill., 1991.
4. "Adjustable Stop Rotary Switches," Grayhill, Inc., La Grange, Ill., 1986.
5. "Electronic Rotary Switches for Commercial Applications," Bulletin COM-0001, Electroswitch Southern Operations, Raleigh, N.C., 1990.
6. "Isolated Position Rotary Switches," Grayhill, Inc., La Grange, Ill., 1991.
7. "Catalog No. 9105—Switches, Newton Division," C&K Components, Newton, Mass., 1991.
8. "Engineering Catalog," Grayhill, Inc., La Grange, Ill., 1990.
9. "Catalog No. C8909—Clayton Division Products," C&K Components, Inc., Clayton, N.C., 1989.

CHAPTER 4
THUMBWHEEL SWITCHES*

Thumbwheel switches, including the extended family of lever and pushwheel switches, are a useful variant of the rotary switch. Essentially a rotary switch which has been rotated 90° to the operator, the thumbwheel functions as a finger-actuated commutator capable of performing dozens of simple to complex coded switching functions. Thumbwheels, also called digital switches, have a smaller footprint than an equivalent rotary switch and can save a significant amount of front panel space. Connections are made to the device through hardwiring to terminals or feed-throughs on the rear of the PCB. Connectors can also be provided, although this is generally not an off-the-shelf feature and will cost extra.

CONSTRUCTION TECHNIQUES

As shown in Fig. 4.1, a basic thumbwheel switch is constructed of four parts: a case, a detent spring, a dial and commutator (brush) subassembly, and a printed circuit board. The case is usually produced as a plastic injection molding but may also be of die-cast metal for special applications. The case provides the housing into which the rest of the components are assembled. The detent spring is a flat spring, formed to conform to the detent requirements of the switch, which attaches to the housing and is the device which provides the distinct "snap-in-place" feel of the thumbwheel dial as it is revolved by the operator (usually with the thumb, hence the name "thumbwheel"). The dial subassembly does double duty as the brush holder *and* the carrier for the legend information, which is usually engraved, hot-stamped, or printed onto the injection-molded dial. The PCB is attached to the case with the dial commutators in contact with etched circuitry on the PCB. This etched circuitry provides the ability of the thumbwheel to produce a wide variety of encoded outputs, examples of which are shown in Fig. 4.2. The completed assembly, referred to as a module, can be combined with one or more other modules (with the same or different output coding) to produce a thumbwheel assembly (Fig. 4.3). Typically, a module will have single digits on the dial which are displayed through a clear window in the case. However, special-message unit modules, wider than the standard single module, allow multiple character legends to be displayed (Fig. 4.4). Further, special characters such as decimal points or degree signs can be marked on the case. Module spacers can also be incorporated to separate switch modules by function or to identify differ-

*Numbers in parentheses indicate items in the References at the end of this chapter.

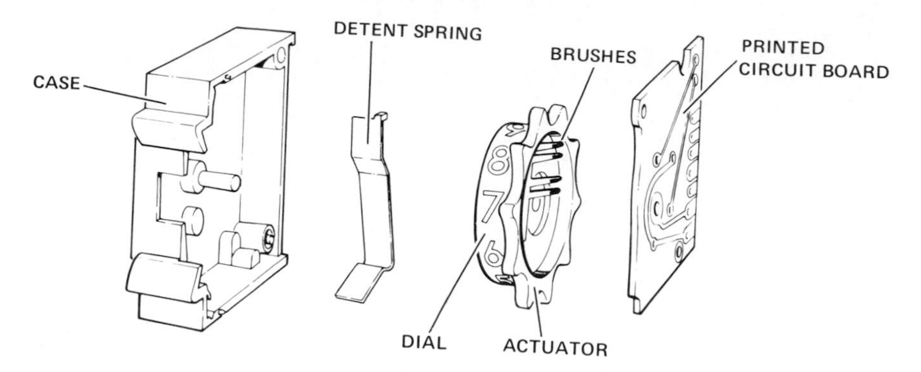

FIGURE 4.1 The component parts of a basic thumbwheel switch. (*Courtesy of XCEL-Digitran, Ontario, Calif.*)

ent digital parameters or define input information. These spacers can also be marked with special characters (Fig. 4.5).

SELECTION PARAMETERS

All digital switches are basically similar. Variations arise from method of actuation, environmental characteristics, mounting configuration, size, number of available positions, character height, character or word width, color, switching code capability, terminations, and price. Each of these factors individually, or any combination thereof, should be considered when selecting the ideal digital switch for the specific application.

Methods of Actuation

Finger-Detented Wheel. The finger-detented wheel, commonly referred to as the "thumbwheel," operates bidirectionally with the end of the operator's finger (Fig. 4.6). This style is recommended for applications where very positive positioning of the dial is desired or where protrusions from the front of the panel would be hazardous or undesirable and where more rapidly setting devices are not required.

Lever. The lever-style digital switch is operated bidirectionally by grasping the lever between thumb and finger (Fig. 4.7). Moving the lever through an arc of 90° will rotate the dial, by means of a rack and pinion gear, through full dial rotation. Another style of lever switch is also bidirectionally operated, but in this case, the lever changes one position at a time by pushing the lever either up or down (Fig. 4.8). Dial rotation is that of a standard thumbwheel. If the original setting of the switch is known, the operator need not watch the switch dial while setting the position, because the audible clicks can be counted by the operator (one click for each dial position). The operator can also be wearing gloves with no effect on accurate switch setting. Such switches are for use in applications where rapid setting is required or where the operator may be wearing gloves or where it is de-

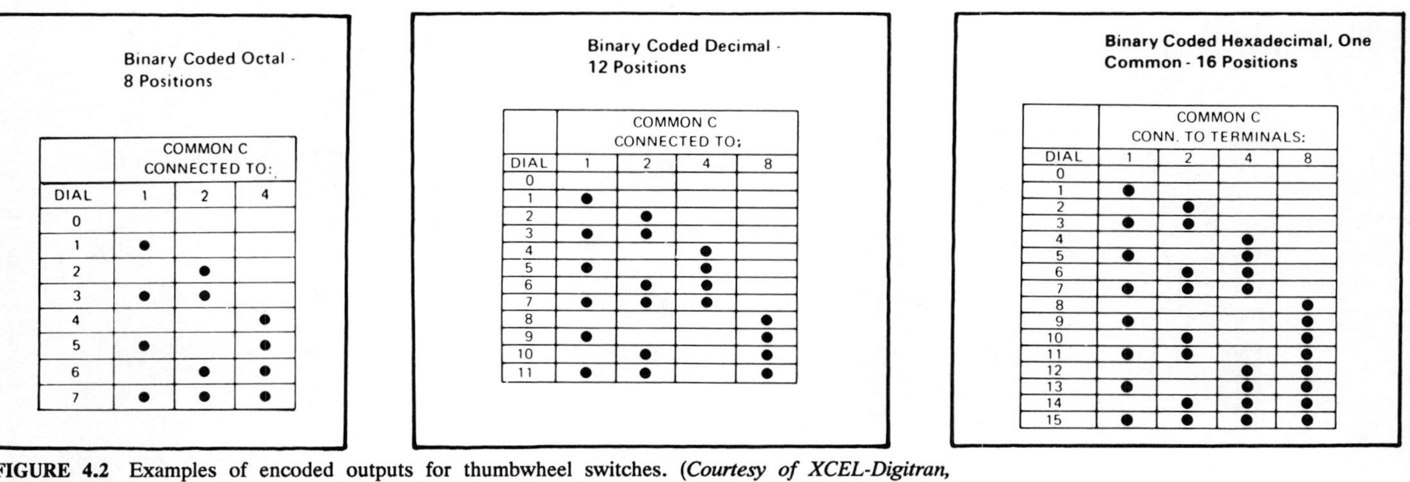

Binary Coded Octal - 8 Positions

	COMMON C CONNECTED TO:		
DIAL	1	2	4
0			
1	●		
2		●	
3	●	●	
4			●
5	●		●
6		●	●
7	●	●	●

Binary Coded Decimal - 12 Positions

	COMMON C CONNECTED TO:			
DIAL	1	2	4	8
0				
1	●			
2		●		
3	●	●		
4			●	
5	●		●	
6		●	●	
7	●	●	●	
8				●
9	●			●
10		●		●
11	●	●		●

Binary Coded Hexadecimal, One Common - 16 Positions

	COMMON C CONN. TO TERMINALS:			
DIAL	1	2	4	8
0				
1	●			
2		●		
3	●	●		
4			●	
5	●		●	
6		●	●	
7	●	●	●	
8				●
9	●			●
10		●		●
11	●	●		●
12			●	●
13	●		●	●
14		●	●	●
15	●	●	●	●

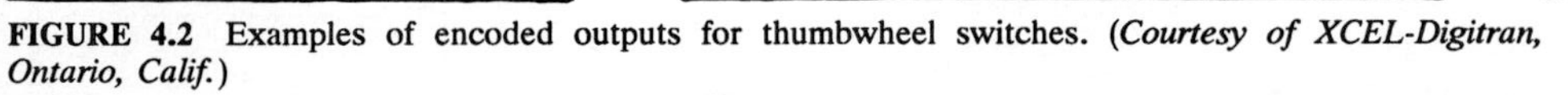

FIGURE 4.2 Examples of encoded outputs for thumbwheel switches. (*Courtesy of XCEL-Digitran, Ontario, Calif.*)

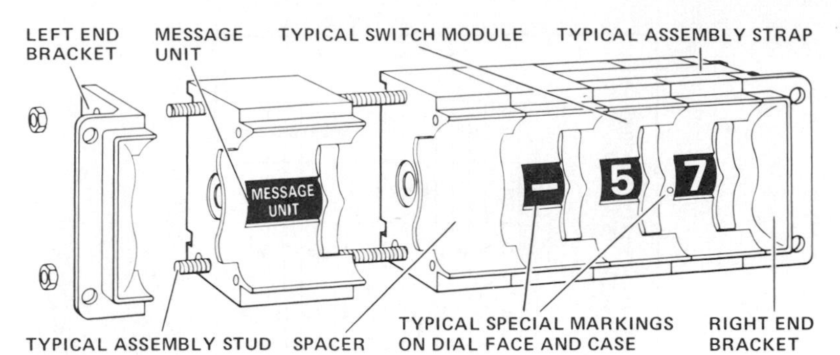

FIGURE 4.3 One or more thumbwheel modules may be combined to create a thumbwheel assembly. (*Courtesy of XCEL-Digitran, Ontario, Calif.*)

FIGURE 4.4 Multiple character legends are displayed in extra-wide message unit modules. (*Courtesy of Cherry Electrical Products, Waukegan, Ill.*)

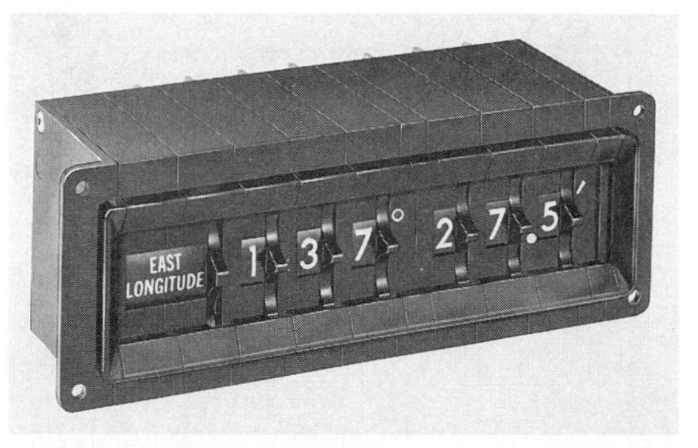

FIGURE 4.5 Special symbols and characters can be marked on the thumbwheel case and on spacers used to separate switch modules by function or parameter. (*Courtesy of XCEL-Digitran, Ontario, Calif.*)

FIGURE 4.6 The end of the operator's finger is used to actuate the thumbwheel, providing positive positioning of the dial. (*Courtesy of XCEL-Digitran, Ontario, Calif.*)

FIGURE 4.7 The operator grasps the actuating lever of the lever-style digital switch; moving the lever through 90° provides full dial rotation. (*Courtesy of XCEL-Digitran, Ontario, Calif.*)

sirable to have the convenience of resetting one or more switches in an assembly back to zero or some other mechanical reference point, with a sweep of the operator's hand.

Pushbutton. Both unidirectional and bidirectional pushbuttons are available, and one style may be more suited to an application than the other. The unidirectional pushbutton digital switch is operated by the thumb or finger moving in one step, in one direction, through a single-function ratchet mechanism (Fig. 4.9). The bidirectional pushbutton (Fig. 4.10*a* and *b*) is operated in the same manner with the thumb or finger through a dual ratchet mechanism; thus the operator advances or decreases the setting with greater speed.

Pushbutton switches are typically used in applications where the operator is wearing gloves (Fig. 4.11) or where precise knowledge of the switch position is critical. If the original position of the switch is known, the operator need not watch the dial while operating the switch. The operator can count the audible clicks and feel the button during operation.

FIGURE 4.8 Pushing the lever up or down on this lever-style switch operates the dial bidirectionally. (*Courtesy of XCEL-Digitran, Ontario, Calif.*)

FIGURE 4.9 A unidirectional pushbutton switch moves the dial in one step, in one direction. (*Courtesy of XCEL-Digitran, Ontario, Calif.*)

(a)

(b)

FIGURE 4.10 Although different in appearance, these two pushbutton switches operate similarly; the operator increases or decreases the setting by pushing the bidirectional pushbuttons. [*Courtesy of XCEL-Digitran, Ontario, Calif. (top) and Cherry Electrical Products, Inc., Waukegan, Ill. (bottom).*]

Environmental Sealing Methods. Several digital switch manufacturers, such as XCEL-Digitran, Ontario, Calif., produce a switch for almost every conceivable environmental requirement. Sealing is typically available in most sizes of switch under the following categories:

1. *Fully Sealed.* The switching elements are enclosed in a sealed chamber with an O-ring around the actuator shaft; a window seals the front of the case, and potting around the PCB or wire leads seals the back of the switch case. This type of construction results in a switch impervious to sand, dust, salt spray, high humidity, and temperatures.
2. *Dust Restrictive Semisealed.* An isolated switching chamber protects against dust and debris. No window or potting is used on the switch case as found on the fully sealed switches; however, the isolated switching chamber does aid in protecting the switch in applications of less severe environments.
3. *Panel Sealed.* There are typically two types of panel sealing: "gasket" and "gasket plus sealing can." Sealing cans are enclosures which are fitted over

FIGURE 4.11 Pushbutton-style thumbwheel switches are ideal for applications where the operator is gloved. (*Courtesy of XCEL-Digitran, Ontario, Calif.*)

the entire digital switch assembly. The sealing can is fabricated of either metal or plastic. Only certain types of digital switch can utilize the sealing can, so contact the switch manufacturer for assistance in selecting the proper type. A panel sealing gasket comes as part of the mounting package for all standard panel sealed switches. Panel sealing seals the switch into the panel so that no contaminants enter between the switches or around the switch mounting.

Always contact the switch manufacturer to review the environmental specifications for the digital switch style under consideration, and by all means, ask for clarification in the form of test reports for any environmental characteristic(s) of particular concern to the user.

Mounting Configuration. Mounting of any standard digital switch assembly is a simple, straightforward operation (Fig. 4.12*a, b,* and *c*). Whether back of the panel, fixed mounting, or front of the panel snap-in mounting is selected, only one panel cutout is required for mounting an entire assembly regardless of size (Fig. 4.13). Most "hard" mounted assemblies require only four mounting screws or studs. Snap-in units usually do not require any additional hardware.

Dial Positions. Most digital switches are available with 8, 10, 12, or 16 dial positions. Dial position capabilities are always indicated in switch manufacturer's published technical literature.

Dial Stops. Dial *stops* which limit the rotation of the digital switch dial to any number of positions less than normal are available for most styles of switch selected.

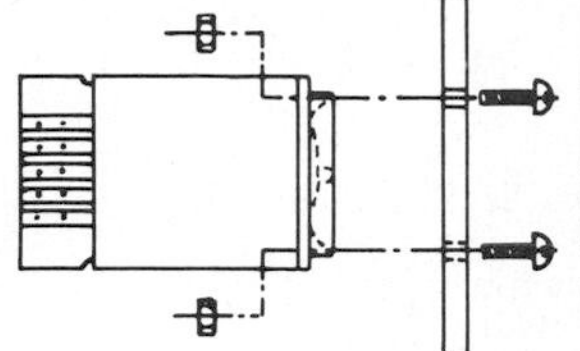

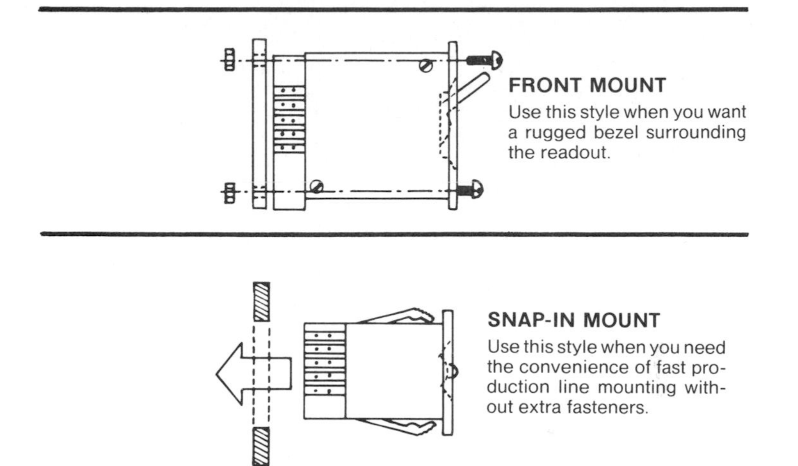

FIGURE 4.12 Three different mounting configurations provide simple installation in control panels. (*Courtesy of Cherry Electrical Products, Waukegan, Ill.*)

Standard and Optional Markings. Standard and optional markings are generally available for switch cases, switch dials, and assembly spacers (refer back to Fig. 4.3). Markings are applied directly to the surface by hot stamping, engraving, or pad printing. Standard off-the-shelf switch dials are typically hot stamped to ensure the permanency of the characters. Most manufacturers can apply any form of symbol, numeral, or message (such as degree, second, and minute signs), vertical lines to separate functions, slashes, arrows, and mathematical symbols to the dial, case, or spacer. Generally, the only limitation for either symbols or message length is the width of the dial window, switch case, or spacer. Special markings involve special tooling, and while not expensive, it is an extra cost. Digital switch manufacturers who do special markings retain the special tooling. If the user has a need for nonstandard markings, check with the switch manufacturer to determine if a particular special marking has been previously tooled. If so, it is important to determine if that special tooling is available to the user. If the user has some flexibility in the final form or appearance of the special marking, a cost-effective approach might be to tailor the special marking to fit tooling which is already available.

Message Units. Message units are designed for special applications that require more symbols or character space than the dial window width of a standard digital switch can support (refer back to Fig. 4.4). Message units are usually available to match the standard digital switch configurations of most manufacturers, and as is

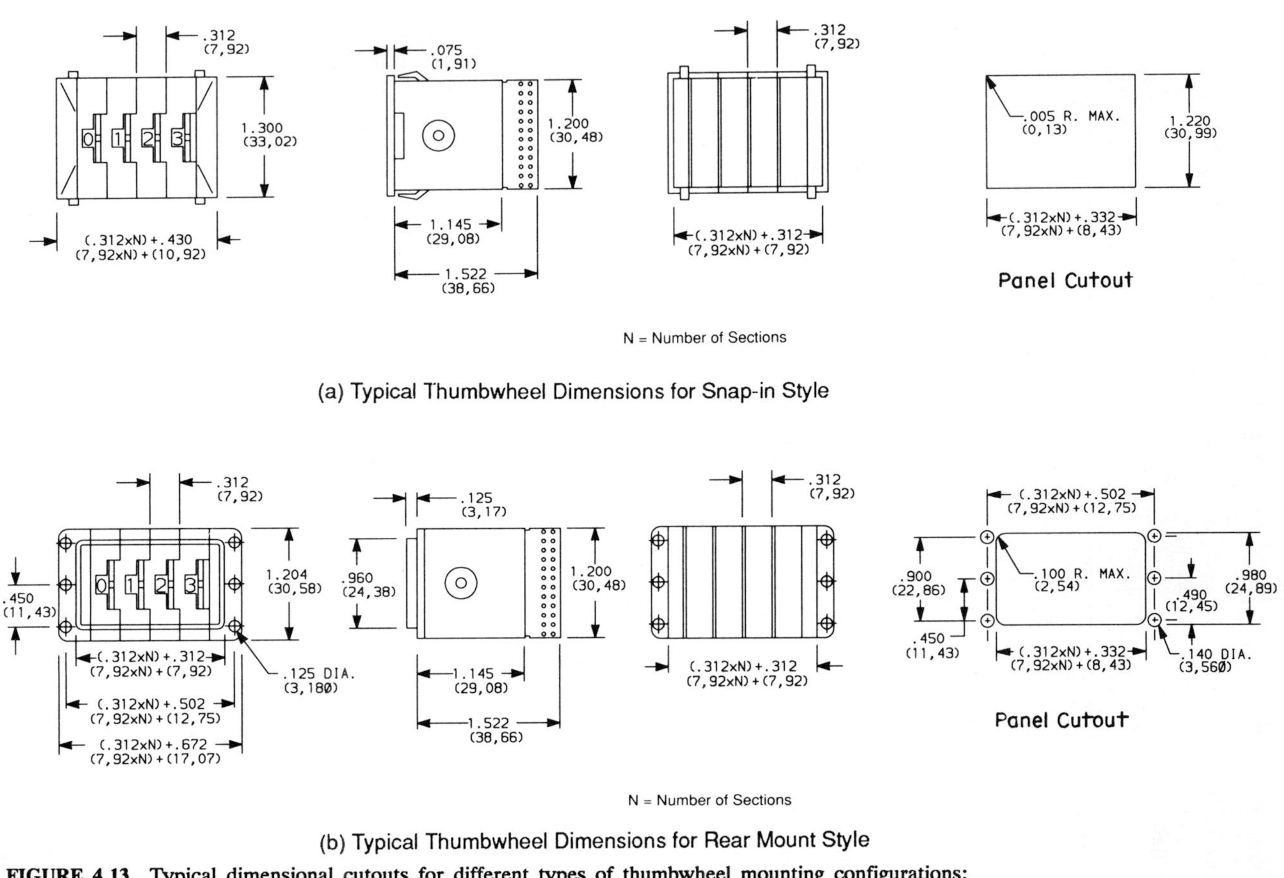

FIGURE 4.13 Typical dimensional cutouts for different types of thumbwheel mounting configurations: (*a*) snap-in style and (*b*) rear mount style. (*Courtesy of C&K Components, Inc., Newton, Mass.*)

the case with the standard and optional marking described in the previous paragraph, many special symbols and characters are already tooled.

Color and Finish. Matte or glossy black is the standard finish on digital switches offered by most manufacturers; however, special colors and finishes are usually available.

Switch Terminations and Connectors. A variety of switch terminations for standard, as well as special, applications is offered by most digital switch manufacturers (Fig. 4.14). Items such as standard PCB connectors, pin terminals, wrap post, and direct solder terminals are commonplace. Eyelet solder terminals, turret terminals, and wire loops are nonstandard options that are typically available.

It is common design practice to specify a multistation assembly with F-pins or J-pins that allow the assembly to be plugged directly into a motherboard. Cherry Electrical Products, Waukegan, Ill., recommends the following procedure for use of thumbwheel switches with a motherboard:

1. Specify a back-mounted assembly.
2. Follow this production sequence:
   a. Insert individual modules into the motherboard one at a time.
   b. Install tape (or ganging hardware).

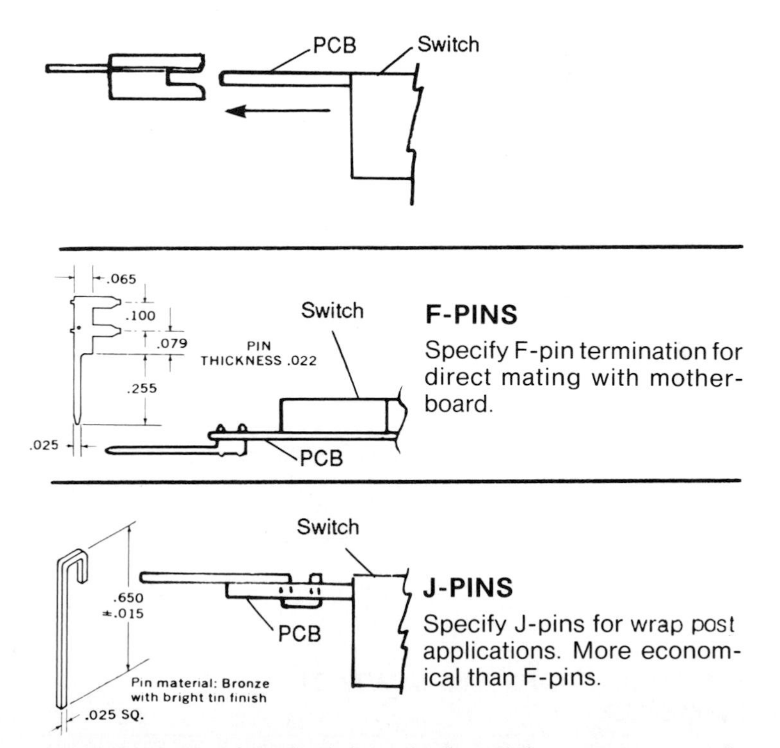

FIGURE 4.14 A variety of switch termination techniques cover most applications. (*Courtesy of Cherry Electrical Products, Waukegan, Ill.*)

c. With suitable fixturing, clamp assembly into its ultimate position relative to the panel cutout through which it will protrude.
d. Solder.
e. Install entire assembly into housing. (3)

In many applications, such as one-of-a-kind test instruments, it is desirable to fully wire and debug an instrument before installing it into its final cabinet or housing. In such situations, use back-mounted thumbwheel switches to avoid having to disconnect, make the final installation, and then reconnect. Another alternative is to use card-edge connectors if the thumbwheel PCB configuration will permit. [Adapted from (3).]

When soldering on, or near, a thumbwheel switch, *it is essential that flux or cleaning agents DO NOT enter the switch. Use only 40 percent isopropyl alcohol in distilled water for cleaning agents.* Always contact the switch manufacturer concerning other acceptable cleaning agents or cleaning methods.

Lighting Standards and Options. Top-, side- or back-mounted replaceable lighting (Fig. 4.15*a* and *b*) is available from several digital switch manufacturers. The purpose of this lighting is to illuminate the switch dial in applications where low ambient light conditions exist. In addition to producing lighting compatible with a number of commercial and military specifications, NVIS (Night Vision Imaging System) lighting is also offered by at least one digital switch manufacturer, XCEL-Digitran, Ontario, Calif. For an overview of NVIS lighting technology, the reader is referred to Chap. 6.

Double Modules. A double module is intended for use anywhere it is necessary to operate two switches at one time, with either similar or dissimilar electrical output codes. The double module consists of two internally interlocked switch modules with the same number of dial positions operated by a single actuator with one dial. This feature also doubles the coding capability of the switch, making it possible to derive such functions as four-pole, 10-position or four totally isolated binaries in one integral unit.

Component Mounting. The modular design of most digital switches makes it possible to include as part of the switch module all or a portion of the switch-associated circuitry such as diodes, resistors, integrated circuits, and complete encoding circuits. By having the necessary components mounted on the PCB of the digital switch during factory assembly operations (Fig. 4.16), the user receives a complete electromechanical assembly with the potential for significant savings in behind-the-panel space, as well as opportunities for cost reduction resulting from simplifying the user's assembly processes. (1) Check with the selected manufacturer(s) to determine what standard PCB configurations are available for this purpose. These same manufacturers can also produce specially configured PCBs to suit specific applications.

SWITCH OUTPUT CODES AND CONTACT ARRANGEMENTS

To determine the most suitable output code or contact arrangement for a digital switch application, it is convenient to study and understand the *truth tables,* a

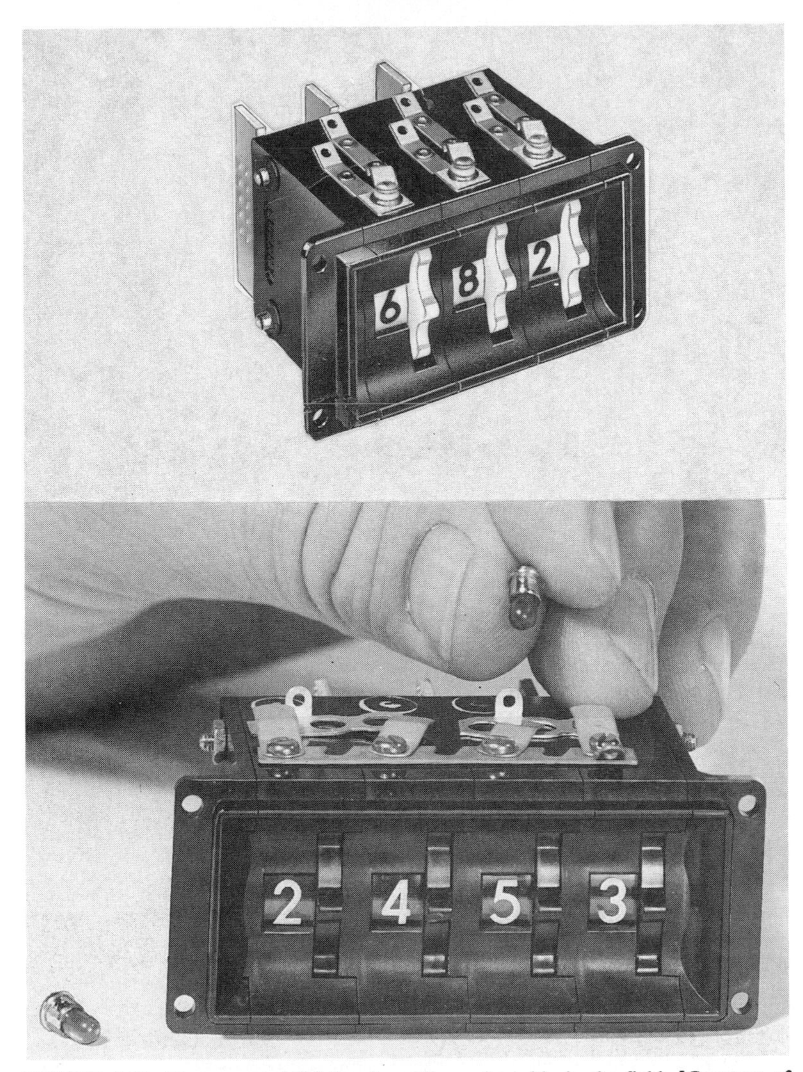

FIGURE 4.15 Top-mounted lighting is easily replaceable in the field. [*Courtesy of Cherry Electrical Products, Waukegan, Ill. (top) and XCEL-Digitran, Ontario, Calif. (bottom).*]

graphical method of describing the interrelationships of output line(s) to common line(s) in order to develop encodable output signals, from each device, which are compatible with the user's equipment. The truth table simply defines, for each dial setting of the switch, which output line or combination of output lines are connected to one or more common terminals.

Active terminals are those that are connected to a common (input) at a particular dial position. Figure 4.17 shows the relationship of a 10-position binary coded decimal output with a dial that reads zero to nine (0–9). There is one com-

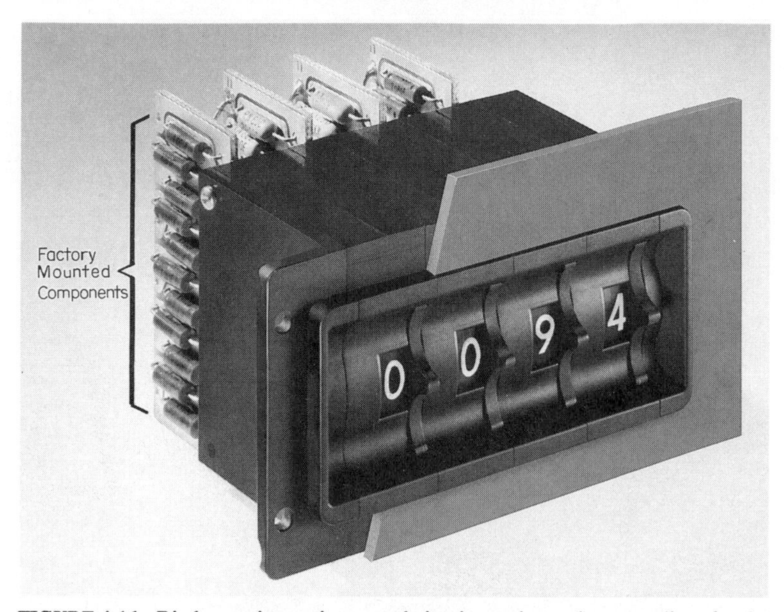

FIGURE 4.16 Diodes, resistors, integrated circuits, and complete encoding circuits can be factory-installed to simplify next-assembly operations. (*Courtesy of XCEL-Digitran, Ontario, Calif.*)

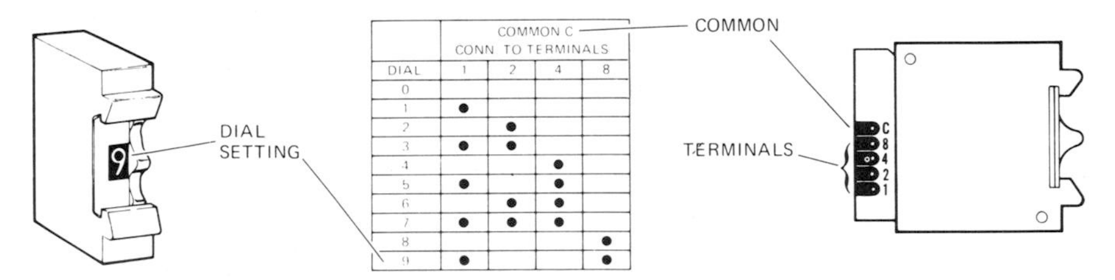

DIAL	COMMON C CONN. TO TERMINALS			
	1	2	4	8
0				
1	●			
2		●		
3	●	●		
4			●	
5	●		●	
6		●	●	
7	●	●	●	
8				●
9	●			●

FIGURE 4.17 Depicts the relationship of a 10-position BCD output with a zero to nine (0–9) dial. (*Courtesy of XCEL-Digitran, Ontario, Calif.*)

mon (C) terminal that can be connected to any one or combination of the four discrete terminals, numbered 1, 2, 4, and 8.

In this code, the common is connected to the terminals at the corresponding dial positions as depicted in Fig. 4.17. Terminal numbers on the printed circuit board correspond to the numbers in the truth table. The setting at number 9 indicates that the common (C) is connected to lines (terminals) 1 and 8. Lines (terminals) 2 and 4 are inactive, or "open."(1)

Figure 4.18 illustrates the 25 most commonly used codes. In actual practice, digital switch manufacturers are continually called upon to create special codes for a wide variety of applications. If the user requires a special code, always contact the switch manufacturer since there is a strong possibility the code design has already been completed.

Voltage Dividers, Resistance Decades, and Other Thumbwheel-Based Components

A number of unique applications utilizing standard thumbwheel technology which merit further discussion are:

- Voltage dividers
- Resistance decades
- Digital comparators
- Seven-segment drivers
- Counter and timer decades
- Capacitance decades
- Binary coded DIP switches

Voltage Dividers. A Kelvin-Varley circuit uses a cascade arrangement of resistors to subdivide voltages. Figure 4.19 depicts the four decades required to divide 100 V into increments of 10 mV. Input impedance of Kelvin-Varley circuits remains constant throughout the full range of incremental voltage setting. Input impedance of the circuit shown in Fig. 4.19 is 10,000 Ω. The dial setting is 38.55. The total voltage output is the sum of voltages contributed by all decks.

Electrical Stability. The maximum contact resistance change throughout the rated switch life will typically not exceed 5 mΩ. For the circuit depicted in Fig. 4.19, maximum output variation would be 0.0005 percent E_{IN} to the most significant digit.

Temperature Coefficient. The resistors in any matched group of each deck typically track closely to each other throughout the operating temperature range. Some voltage dividers have less than 0.02 percent maximum change in ohms per 212°F (100°C) change in temperature.

Accuracy. The accuracy of a voltage divider is expressed as a percent of input voltage, not as a percent of setting. A prime advantage of the Kelvin-Varley voltage divider is that, in order to achieve 0.025 percent linearity, the resistors need not have an absolute accuracy of 0.025 percent. Resistor matching during manufacture ensures the accuracy of the voltage divider. Linearity of the voltage divider is adversely affected by R_L. As R_L approaches Z_{IN}, the linearity error increases. Therefore, it is desirable to operate the Kelvin-Varley voltage divider into as high an impedance as possible. Typically, R_L should be at least 200 times greater than Z_{IN} for a full-scale accuracy of 0.1 percent of E_{IN} (Fig. 4.20). (1)

Resistance Decades. A resistance decade utilizes a weighted code of 1-2-2-2-2, using five resistors to achieve nine discrete steps of resistance from 0 to 9 or multiples thereof. The typical resistance decade has a string of five resistors mounted on its PCB, the appropriate resistors being shorted out to provide a linear progression yielding the desired total resistance (Fig. 4.21). Resistance values do not necessarily have to be in increments of 1, 10, 100, or 1000 Ω, but can be any desired value. As an example, if a resistance range of 0 to 2495 Ω with 5-Ω increments is desired, a three-digit resistance decade (based on thumbwheel technology) with dial readings 000 to 499 is selected. The proper resistors are then selected (i.e., one 5-Ω resistor and four 10-Ω resistors for the least significant digit, one 50-Ω resistor, four 100-Ω resistors, and two 1000-Ω resistors for the most significant digit) for an overall result of 499 × 5 Ω, which yields the 2495-Ω total desired.

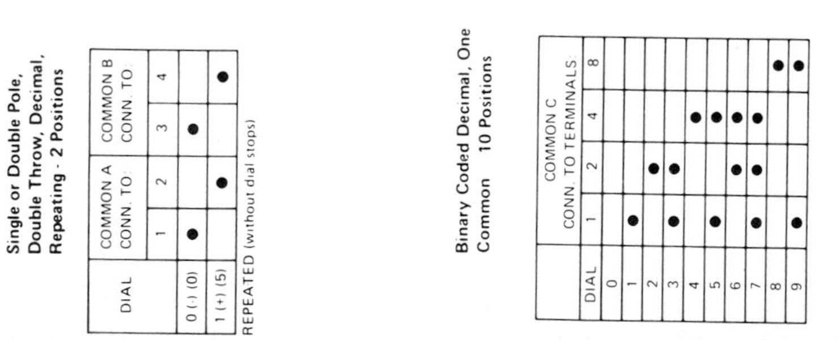

Single or Double Pole, Double Throw, Decimal, Repeating - 2 Positions

DIAL	COMMON A CONN. TO:		COMMON B CONN. TO:	
	1	2	3	4
0 (-) (0)	●		●	
1 (+) (5)		●		●

REPEATED (without dial stops)

Single Pole, Single Throw, Octal 8 Positions

DIAL	COMMON C CONNECTED TO:
0	0
1	1
2	2
3	3
4	4
5	5
6	6
7	7

Double Pole, Single Throw, Octal - 8 Positions

DIAL	COMM. A CONN. TO:	COMM. B CONN. TO:
0	A0	B0
1	A1	B1
2	A2	B2
3	A3	B3
4	A4	B4
5	A5	B5
6	A6	B6
7	A7	B7

Binary Coded Octal - 8 Positions

	COMMON C CONNECTED TO:		
DIAL	1	2	4
0			
1	●		
2		●	
3	●	●	
4			●
5	●		●
6		●	●
7	●	●	●

Binary Coded Decimal, One Common 10 Positions

	COMMON C CONN. TO TERMINALS:			
DIAL	1	2	4	8
0				
1	●			
2		●		
3	●	●		
4			●	
5	●		●	
6		●	●	
7	●	●	●	
8				●
9	●			●

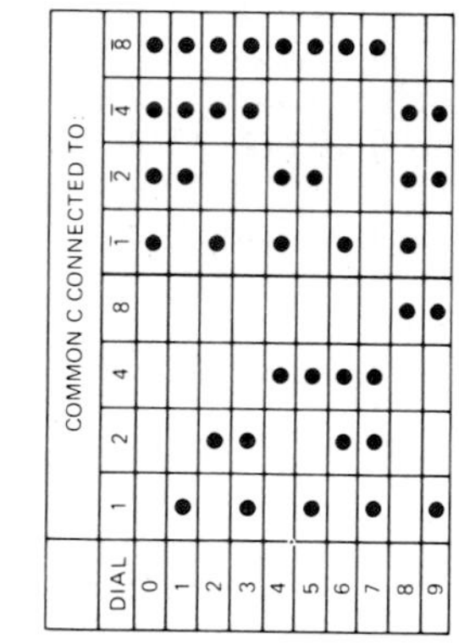

Binary Coded Decimal, Plus Complement, One Common - 10 Positions

	COMMON C CONNECTED TO:							
DIAL	1	2	4	8	$\bar{1}$	$\bar{2}$	$\bar{4}$	$\bar{8}$
0					●	●	●	●
1	●					●	●	●
2		●			●		●	●
3	●	●					●	●
4			●		●	●		●
5	●		●			●		●
6		●	●		●			●
7	●	●	●					●
8				●	●	●	●	
9	●			●		●	●	

Complement of Binary Coded Decimal, One Common - 10 Positions

	COMMON C CONNECTED TO:			
DIAL	$\bar{1}$	$\bar{2}$	$\bar{4}$	$\bar{8}$
0	●	●	●	●
1		●	●	●
2	●		●	●
3			●	●
4	●	●		●
5		●		●
6	●			●
7				●
8	●	●	●	
9		●	●	

FIGURE 4.18 The 25 most commonly used codes are illustrated, but digital switch manufacturers also have designed numerous special codes. (*Courtesy of XCEL-Digitran, Ontario, Calif.*)

9's Complement of Binary Coded Decimal With Separate Common To Not-True Bits - 10 Positions

DIAL READS	DIAL POS	COMMON "X" CONN TO ● COMMON "Y" CONN TO ○			
		1	2	4	8
0	0	●	○	○	●
1	1	○	○	○	●
2	2	●	●	●	○
3	3	○	●	●	○
4	4	●	○	●	○
5	5	○	○	●	○
6	6	●	●	○	○
7	7	○	●	○	○
8	8	●	○	○	○
9	9	○	○	○	○

9's Complement of Binary Coded Decimal Plus Complement, One Common - 10 Positions

	COMMON C CONNECTED TO							
DIAL	1	2	4	8	$\bar{1}$	$\bar{2}$	$\bar{4}$	$\bar{8}$
0	●			●		●	●	
1				●	●	●	●	
2	●	●	●					●
3		●	●		●			●
4	●		●			●		●
5			●		●	●		●
6	●	●					●	●
7		●			●		●	●
8	●					●	●	●
9					●	●	●	●

Complement of 9's Complement - 10 Positions

DIAL READS	DIAL POS	COMMON "C" CONNECTED TO			
		1	2	4	8
0	0		●	●	
1	1	●	●	●	
2	2				●
3	3	●			●
4	4		●		●
5	5	●	●		●
6	6			●	●
7	7	●		●	●
8	8		●	●	●
9	9	●	●	●	●

Single Pole Decimal, - 12 Positions

DIAL	COMMON C CONN. TO TERMINAL:
0	0
1	1
2	2
3	3
4	4
5	5
6	6
7	7
8	8
9	9
10	10
11	11

Binary Coded Decimal - 12 Positions

	COMMON C CONNECTED TO:			
DIAL	1	2	4	8
0				
1	●			
2		●		
3	●	●		
4			●	
5	●		●	
6		●	●	
7	●	●	●	
8				●
9	●			●
10		●		●
11	●	●		●

Single Pole Decimal - 16 Positions

DIAL	COMMON C CONN. TO TERMINAL:
0	0
1	1
2	2
3	3
4	4
5	5
6	6
7	7
8	8
9	9
10	10
11	11
12	12
13	13
14	14
15	15

FIGURE 4.18 *(Continued)*

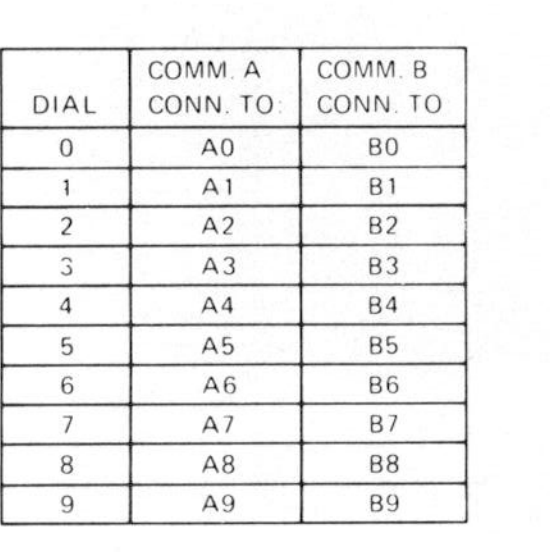

Binary Coded Octal, Plus Complement - 8 Positions

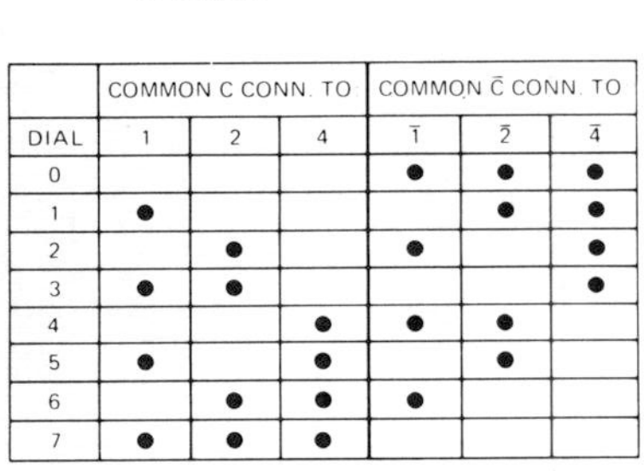

	COMMON C CONN. TO			COMMON $\bar{C}$ CONN. TO		
DIAL	1	2	4	$\bar{1}$	$\bar{2}$	$\bar{4}$
0				●	●	●
1	●				●	●
2		●		●		●
3	●	●				●
4			●	●	●	
5	●		●		●	
6		●	●	●		
7	●	●	●			

Binary Coded Octal With Separate Common To Not-True Bits - 8 Positions

	COMMON X (●) & COMMON Y (○) CONNECTED TO:		
DIAL	1	2	4
0	○	○	○
1	●	○	○
2	○	●	○
3	●	●	○
4	○	○	●
5	●	○	●
6	○	●	●
7	●	●	●

Single Pole, Single Throw, Decimal - 10 Positions

DIAL	COMMON C CONN. TO:
0	0
1	1
2	2
3	3
4	4
5	5
6	6
7	7
8	8
9	9

Double Pole, Single Throw, Decimal - 10 Positions

DIAL	COMM. A CONN. TO:	COMM. B CONN. TO
0	A0	B0
1	A1	B1
2	A2	B2
3	A3	B3
4	A4	B4
5	A5	B5
6	A6	B6
7	A7	B7
8	A8	B8
9	A9	B9

Binary Coded Decimal With Separate Common To Not-True Bits - 10 Positions

	COMMS. X (●) & Y (○) CONN. TO TERMINALS			
DIAL	1	2	4	8
0	○	○	○	○
1	●	○	○	○
2	○	●	○	○
3	●	●	○	○
4	○	○	●	○
5	●	○	●	○
6	○	●	●	○
7	●	●	●	○
8	○	○	○	●
9	●	○	○	●

Dual Binary Coded Decimal, Two Separate Isolated Outputs - 10 Positions

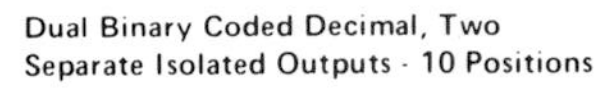

	COMMON X TO:				COMMON Y TO:			
DIAL	1	2	4	8	1	2	4	8
0								
1	●				●			
2		●				●		
3	●	●			●	●		
4			●				●	
5	●		●		●		●	
6		●	●			●	●	
7	●	●	●		●	●	●	
8				●				●
9	●			●	●			●

9's Complement of Binary Coded Decimal, One Common - 10 Positions

	COMMON C CONNECTED TO:			
DIAL	1	2	4	8
0	●			●
1				●
2	●	●	●	
3		●	●	
4	●		●	
5			●	
6	●	●		
7		●		
8	●			
9				

FIGURE 4.18 *(Continued)*

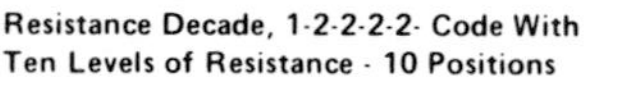
Resistance Decade, 1-2-2-2-2- Code With Ten Levels of Resistance - 10 Positions

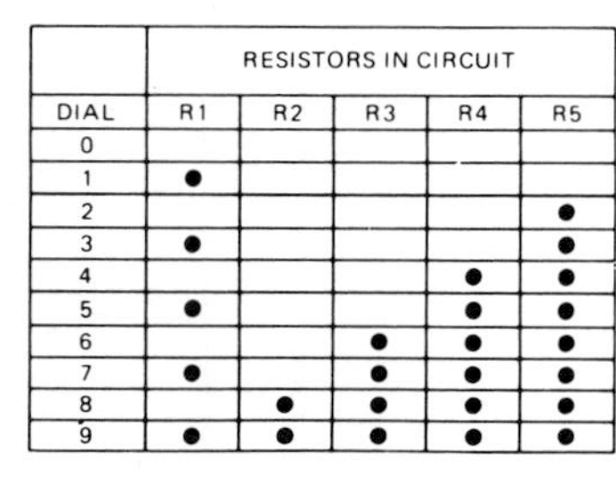

DIAL	RESISTORS IN CIRCUIT				
	R1	R2	R3	R4	R5
0					
1	●				
2					●
3	●				●
4				●	●
5	●			●	●
6			●	●	●
7	●		●	●	●
8		●	●	●	●
9	●	●	●	●	●

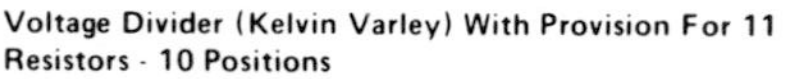
Voltage Divider (Kelvin Varley) With Provision For 11 Resistors - 10 Positions

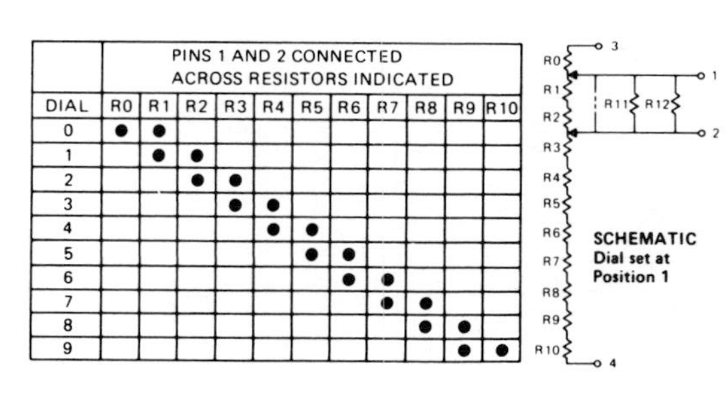

DIAL	PINS 1 AND 2 CONNECTED ACROSS RESISTORS INDICATED										
	R0	R1	R2	R3	R4	R5	R6	R7	R8	R9	R10
0	●	●									
1		●	●								
2			●	●							
3				●	●						
4					●	●					
5						●	●				
6							●	●			
7								●	●		
8									●	●	
9										●	●

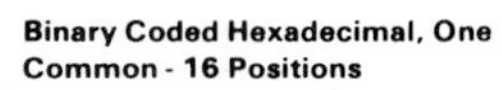
Binary Coded Hexadecimal, One Common - 16 Positions

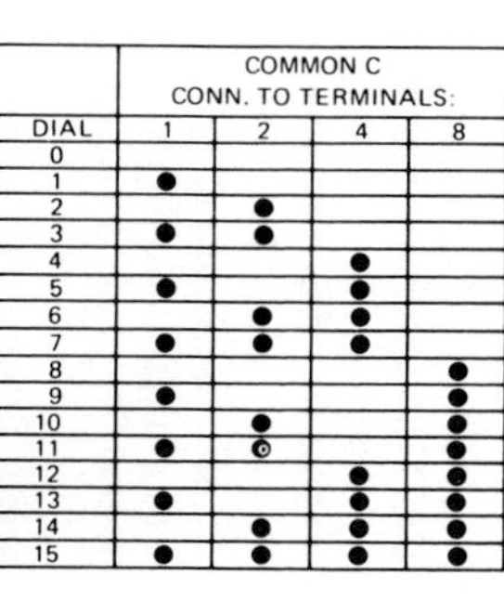

DIAL	COMMON C CONN. TO TERMINALS:			
	1	2	4	8
0				
1	●			
2		●		
3	●	●		
4			●	
5	●		●	
6		●	●	
7	●	●	●	
8				●
9	●			●
10		●		●
11	●	●		●
12			●	●
13	●		●	●
14		●	●	●
15	●	●	●	●

Binary Coded Hexadecimal, Plus Complement, Separate Commons - 16 Positions

DIAL	COMM. C CONN. TO:				COMM. $\bar{C}$ CONN. TO:			
	1	2	4	8	$\bar{1}$	$\bar{2}$	$\bar{4}$	$\bar{8}$
0					●	●	●	●
1	●					●	●	●
2		●			●		●	●
3	●	●					●	●
4			●		●	●		●
5	●		●			●		●
6		●	●		●			●
7	●	●	●					●
8				●	●	●	●	
9	●			●		●	●	
10		●		●	●		●	
11	●	●		●			●	
12			●	●	●	●		
13	●		●	●		●		
14		●	●	●	●			
15	●	●	●	●				

Binary Coded Hexadecimal, Non-Shorting With Separate Common To Not-True Bits - 16 Positions

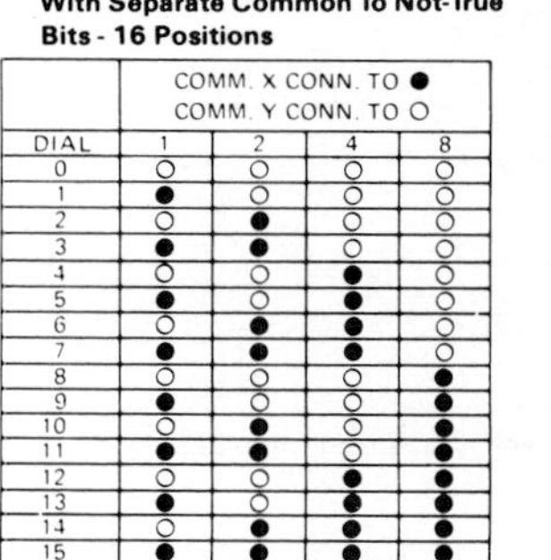

DIAL	COMM. X CONN. TO ● COMM. Y CONN. TO ○			
	1	2	4	8
0	○	○	○	○
1	●	○	○	○
2	○	●	○	○
3	●	●	○	○
4	○	○	●	○
5	●	○	●	○
6	○	●	●	○
7	●	●	●	○
8	○	○	○	●
9	●	○	○	●
10	○	●	○	●
11	●	●	○	●
12	○	○	●	●
13	●	○	●	●
14	○	●	●	●
15	●	●	●	●

FIGURE 4.18 *(Continued)*

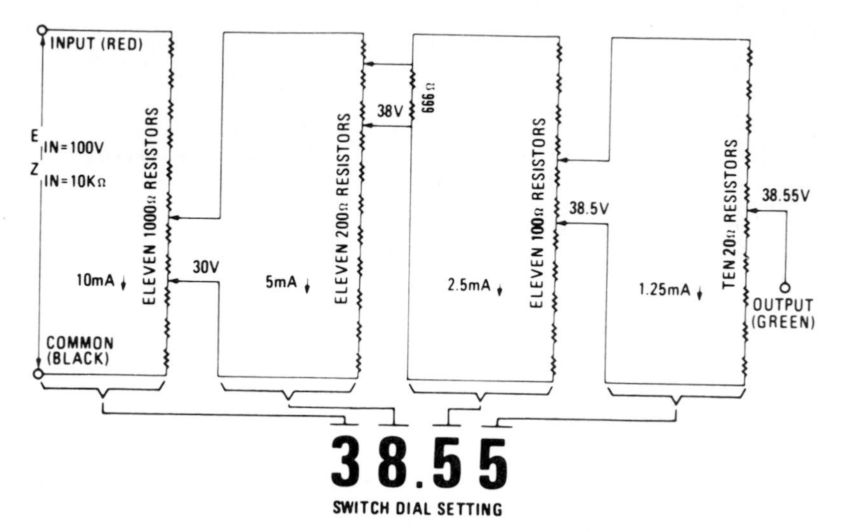

FIGURE 4.19 A standard schematic for a four-Digit Kelvin-Varley voltage divider. (*Courtesy of XCEL-Digitran, Ontario, Calif.*)

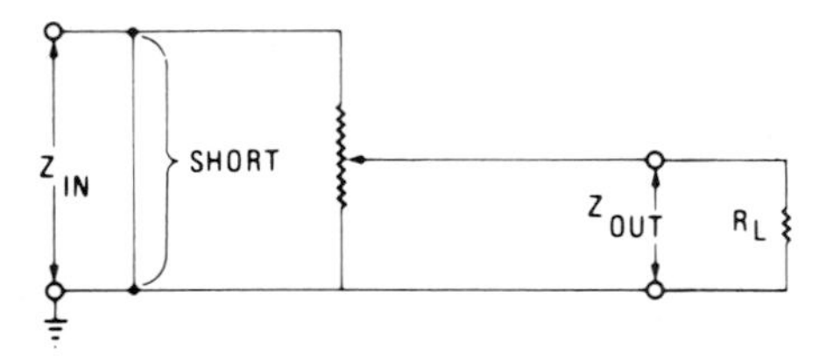

FIGURE 4.20 Kelvin-Varley voltage divider output impedance. (*Courtesy of XCEL-Digitran, Ontario, Calif.*)

Note that the accuracies of resistance decades are described in accuracy of setting, not of full scale as is the case with voltage dividers. (1)

Digital Comparators. Figure 4.22 depicts a thumbwheel switch which has an integrated BCD decade comparator. The switch converts the decimal number displayed by the switch into a BCD (1,2,4,8) format which is presented as word B to the comparator IC. Word A is presented at the PCB terminals in addition to the "high," "equal," and "low" carry inputs. "High," "equal," and "low" carry outputs are also available at the PCB to enable cascading of more than one decade to form multidecade digital comparators. "Pull-up" resistors are provided at the switch end for greater noise immunity. (2)

Seven-Segment Driver. The seven-segment driver shown in Fig. 4.23 has a built-in decoder driver. The decimal number displayed is converted to BCD and supplied to the decoder's BCD inputs. The seven segment (A through G) lines are available at the PCB terminal in addition to the V_{cc}, ground, ripple blanking input, and lamp test. This switch, available from C&K Components, Newton,

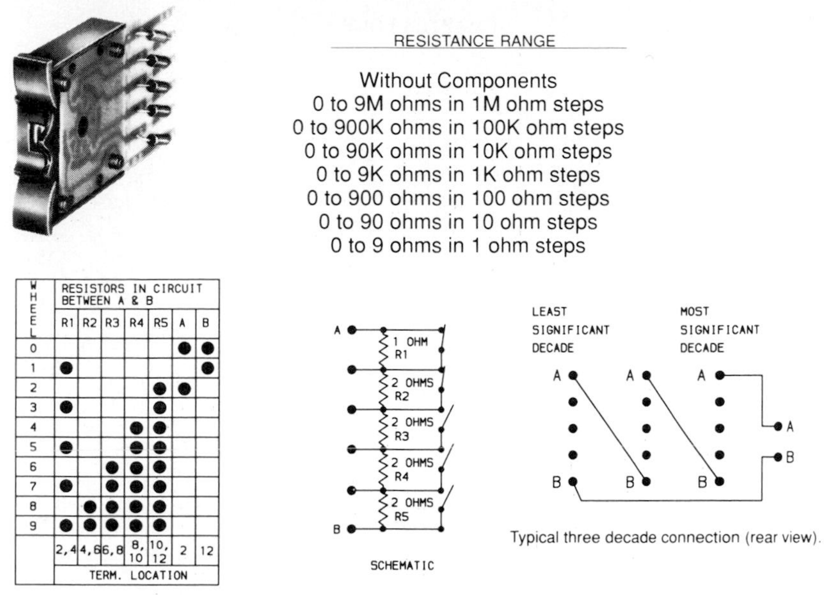

WHEEL	RESISTORS IN CIRCUIT BETWEEN A & B						
	R1	R2	R3	R4	R5	A	B
0						●	●
1	●						●
2					●	●	
3	●				●		
4				●	●		
5	●			●	●		
6			●	●	●		
7	●		●	●	●		
8		●	●	●	●		
9	●	●	●	●	●		
TERM. LOCATION	2,4	4,6	6,8	8,10	10,12	2	12

FIGURE 4.21 Typical resistance decade. (*Courtesy of C&K Components, Inc., Newton, Mass.*)

LOGIC TYPE	I.C. TYPE	NO. OF POSITIONS	WHEEL MARKING	OPERATING VOLTAGE
TTL	7485	10	0 – 9	5 V dc ± 5%
CMOS	4063B	10	0 – 9	6 to 15 V dc
TTL	7485	2	0/1 stopped	5 V dc ± 5%
CMOS	4063B	2	0/1 stopped	6 to 15 V dc
TTL	7485	2	+/– stopped	5 V dc ± 5%
CMOS	4063B	2	+/– stopped	6 to 15 V dc
Without Components				

Other I.C.'s available, consult factory.

TERM NO.	FUNCTION	
1	BCD 0	IN
2	BCD 1	IN
3	BCD 2	IN
4	GROUND	
5	LESS THAN	OUT
6	EQUAL TO	OUT
7	GREATER THAN	OUT
8	GREATER THAN	IN
9	EQUAL TO	IN
10	LESS THAN	IN
11	BCD 4	IN
12	Vcc	

LEAST SIGNIFICANT DECADE
MOST SIGNIFICANT DECADE
1 BCD 1 · 10 · 100
BCD 2 · 20 · 200
BCD 4 · 40 · 400
BCD INPUTS
< = > OUTPUTS
BCD 8 · 80 · 800
12 · Vcc

Typical three decade connection (rear view).

FIGURE 4.22 A typical digital comparator contains an integrated BCD decade comparator. (*Courtesy of C&K Components, Inc., Newton, Mass.*)

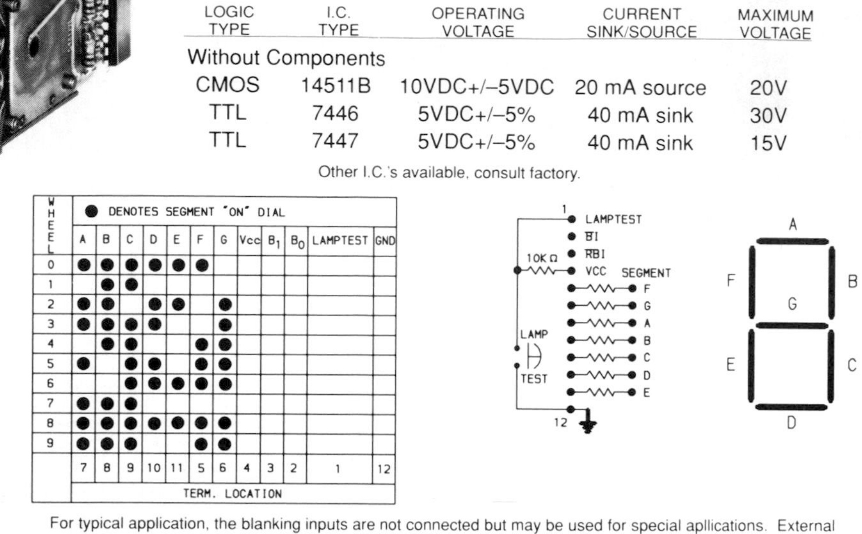

LOGIC TYPE	I.C. TYPE	OPERATING VOLTAGE	CURRENT SINK/SOURCE	MAXIMUM VOLTAGE
Without Components				
CMOS	14511B	10VDC+/–5VDC	20 mA source	20V
TTL	7446	5VDC+/–5%	40 mA sink	30V
TTL	7447	5VDC+/–5%	40 mA sink	15V

Other I.C.'s available, consult factory.

● DENOTES SEGMENT "ON" DIAL

WHEEL	A	B	C	D	E	F	G	Vcc	B_1	B_0	LAMPTEST	GND
0	●	●	●	●	●	●						
1		●	●									
2	●	●		●	●		●					
3	●	●	●	●			●					
4		●	●			●	●					
5	●		●	●		●	●					
6			●	●	●	●	●					
7	●	●	●									
8	●	●	●	●	●	●	●					
9	●	●	●			●	●					
TERM. LOCATION	7	8	9	10	11	5	6	4	3	2	1	12

For typical application, the blanking inputs are not connected but may be used for special apllications. External resistors must be used to limit the LED current at outputs A thru G.

FIGURE 4.23 This seven-segment driver contains a built-in decoder driver. (*Courtesy of C&K Components, Inc., Newton, Mass.*)

Mass., features active low outputs designed for driving common-anode LEDs or incandescent indicators directly. (2)

Counter and Timer Decade. When the BCD of the counter is equal to the BCD equivalent of the decimal number displayed by the switch (Fig. 4.24), an "equal" signal at the common of the switch is generated. Blocking diodes are connected between the counter's BCD output and the switch's coded input to assure proper decoding. The "equal" output has provisions for a "pull-up" (*Rp*) resistor for cascading purposes. Carry out, count input, reset, 9's preset, equal output, in addition to power inputs are available at the PCB terminals. (2)

Capacitance Decade. The thumbwheel switch depicted in Fig. 4.25 uses a 1-2-3-4 code specifically for low component count and high reliability. For best results, paralleling the commons of all the switches will obtain a high accuracy and resolution as shown in the schematic of Fig. 4.25. (2)

Binary Coded DIP Switches. While not a thumbwheel in the true sense, this switch provides an encoded output in a DIP (dual in-line package) configuration. Details on this type of switch can be found in Chap. 12.

Minithumbwheel Switches

So-called minithumbwheel switches less than 1.00 in (25,4 mm) square are available for mounting directly on PCBs. Figure 4.26 depicts one style of minithumbwheel switch with available function codes. Also of interest is the variety of wheel mark-

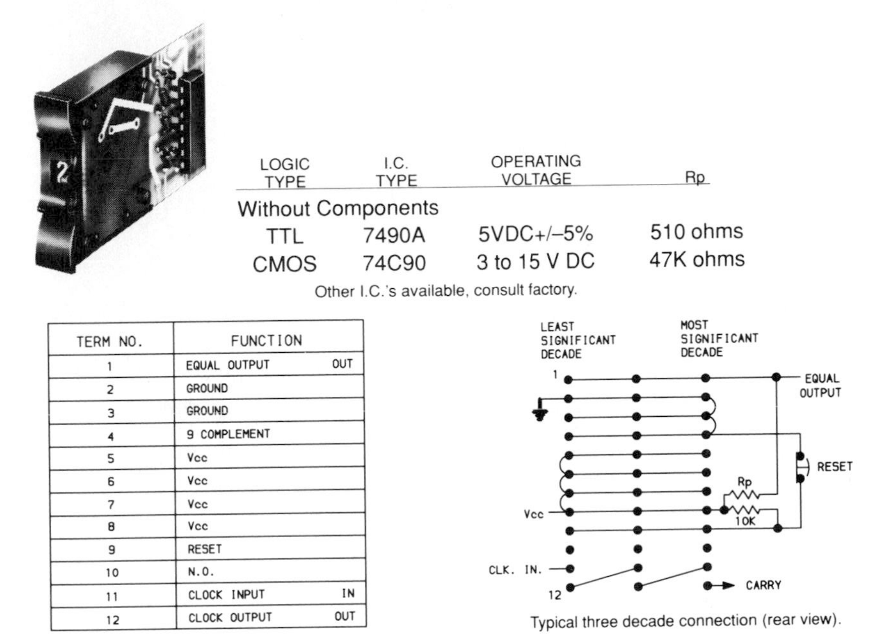

LOGIC TYPE	I.C. TYPE	OPERATING VOLTAGE	Rp
Without Components			
TTL	7490A	5VDC+/–5%	510 ohms
CMOS	74C90	3 to 15 V DC	47K ohms

Other I.C.'s available, consult factory.

TERM NO.	FUNCTION	
1	EQUAL OUTPUT	OUT
2	GROUND	
3	GROUND	
4	9 COMPLEMENT	
5	Vcc	
6	Vcc	
7	Vcc	
8	Vcc	
9	RESET	
10	N.O.	
11	CLOCK INPUT	IN
12	CLOCK OUTPUT	OUT

Typical three decade connection (rear view).

FIGURE 4.24 A typical counter-timer decade. (*Courtesy of C&K Components, Inc., Newton, Mass.*)

CAPACITANCE RANGE	MAXIMUM VOLTAGE
Without Components	
10 to 90.0 µf	10 V DC
1 to 9.0 µf	15 V DC
0.1 to 0.9 µf	25 V DC
0.01 to 0.09 µf	50 V DC
0.001 to 0.009 µf	100 V DC

Other capacitors available, consult factory.

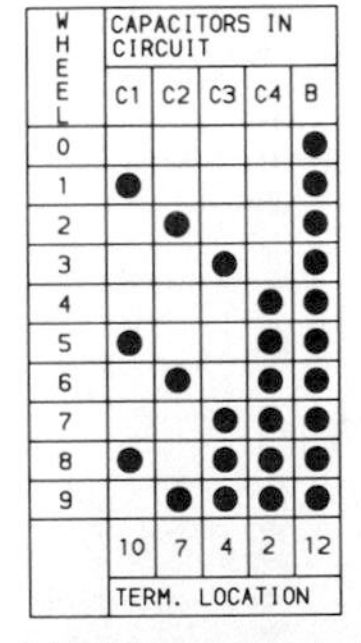

WHEEL	C1	C2	C3	C4	B
0					●
1	●				●
2		●			●
3			●		●
4				●	●
5	●			●	●
6		●		●	●
7			●	●	●
8	●		●	●	●
9		●	●	●	●
TERM. LOCATION	10	7	4	2	12

(Column group header: CAPACITORS IN CIRCUIT)

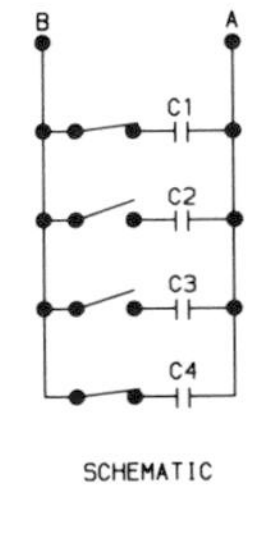

SCHEMATIC

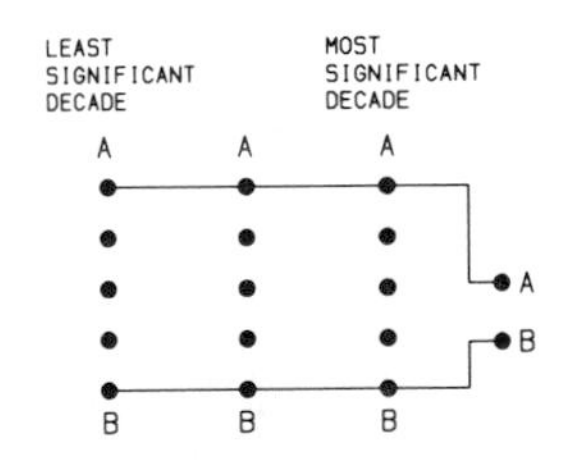

Typical three decade connection (rear view).

FIGURE 4.25 A typical capacitance decade on a thumbwheel switch. (*Courtesy of C&K Components, Inc., Newton, Mass.*)

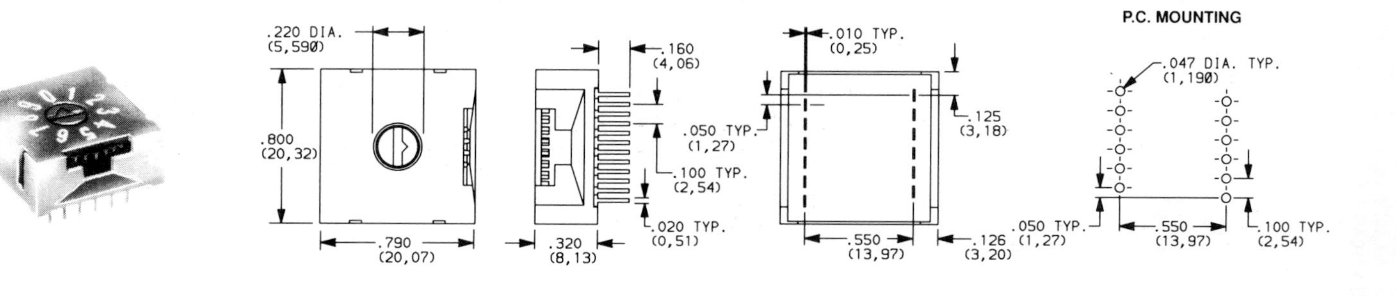

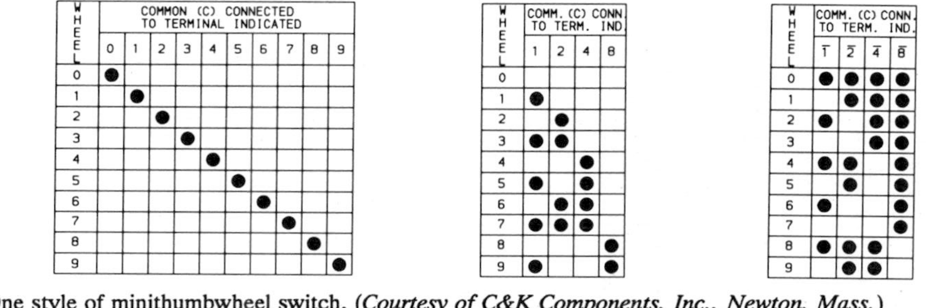

WHEEL	COMMON (C) CONNECTED TO TERMINAL INDICATED									
	0	1	2	3	4	5	6	7	8	9
0	●									
1		●								
2			●							
3				●						
4					●					
5						●				
6							●			
7								●		
8									●	
9										●

WHEEL	COMM. (C) CONN. TO TERM. IND.			
	1	2	4	8
0				
1	●			
2		●		
3	●	●		
4			●	
5	●		●	
6		●	●	
7	●	●	●	
8				●
9	●			●

WHEEL	COMM. (C) CONN. TO TERM. IND.			
	$\overline{1}$	$\overline{2}$	$\overline{4}$	$\overline{8}$
0	●	●	●	●
1		●	●	●
2	●		●	●
3			●	●
4	●	●		●
5		●		●
6	●			●
7				●
8	●	●	●	
9		●	●	

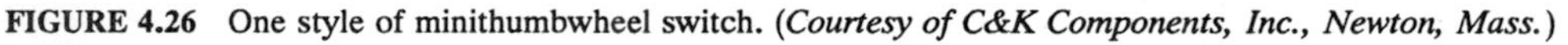

FIGURE 4.26 One style of minithumbwheel switch. (*Courtesy of C&K Components, Inc., Newton, Mass.*)

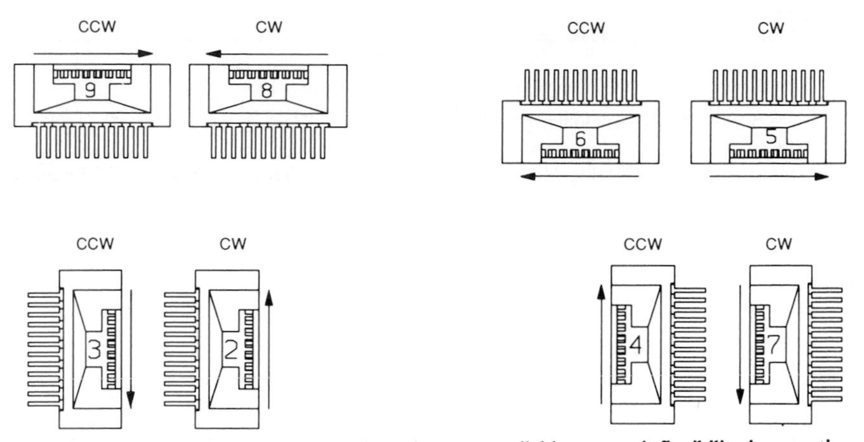

FIGURE 4.27 A variety of marking orientations are available to permit flexibility in mounting. *(Courtesy of C&K Components, Inc., Newton, Mass.)*

ings available (Fig. 4.27) to permit the user to orient the switch as needed on the PCB, allowing the dial to be read from any selected angle. (2)

GLOSSARY

Assembly. One or more switch modules with relevant spacers and end brackets held together by assembly studs or straps.

Bezel. The stationary face of the switch case that is visible from the front of the panel.

Dial positions. Normally refers to the total number of possible dial settings of an individual switch module, i.e., 2-, 8-, 10-, 12-, or 16-position switch.

Gloss finish. A reflective finish on the visible surfaces of the switch case and end brackets which allows the use of the switch parts in their as-molded state.

Hard mounted. An assembly which is mounted to the control panel through the use of screws or studs along with nuts and washers.

Matte finish. A special nonreflective finish applied to the visible surfaces of the switch module and end brackets after molding.

Module. One switch in a series of switches capable of being combined into an assembly of one or more switches with end brackets and assembly straps or studs.

Semimatte or semigloss finish. A semireflective finish on the visible surfaces of the switch case and end brackets.

Snap-in mounting. An assembly which is pressed through a precut mounting hole in the control panel and held in place by tangs integral to the assembly. Usually, no tools or additional hardware are required.

Station. Same as module, normally numbered from left to right.

Stops. Hardware that mechanically limits rotation of the dial to a certain number of switching positions.

REFERENCES

1. "Digital Switches," XCEL-Digitran Inc., Ontario, Calif., 1983.
2. "Catalog No. 9105, Switches—Newton Division," C&K Components, Inc., Newton, Mass., 1991.
3. "Selector Switches Catalog No. CE—815R1," Cherry Electrical Products, Waukegan, Ill., 1991.

CHAPTER 5
TOGGLE SWITCHES*

Toggle switches are the workhorses of switch technology. Favored for their rugged simplicity (Fig. 5.1), toggle switches are used extensively in a wide variety of applications ranging from consumer electronics to complex, multipole, space vehicle requirements. Consisting of (*a*) a mechanical operator which initiates switch operation, (*b*) a set of low-resistance metal (usually silver) contacts that make or break the electric circuit, and (*c*) the switch mechanism itself, linked to the mechanical operator, which opens and closes the contacts, toggle switches, with their visible mechanical actuator, provide an unambiguous indication of switch status to the operator.

FIGURE 5.1 Toggle switches are rugged, highly reliable devices. (*Courtesy of Micro Switch, Freeport, Ill.*)

*Numbers in parentheses indicate items in the References at the end of this chapter.

HOW TOGGLE SWITCHES FUNCTION

As depicted in Fig. 5.2, toggle switches utilize an actuator, the operator interface, to trigger the internal mechanism which changes the switch state. Different manufacturers design this internal mechanism in a variety of ways, but all devices operate in basically the same way. Some manufacturers choose to design the entire internal mechanism including contacts, while others attach one or more precision snap-action switch modules (Chap. 2) to an actuating mechanism all within the same package (Fig. 5.3).

For the reasons explained in Chap. 1, toggle switch manufacturers produce switches with two basic break mechanisms:

- Slow make–slow break
- Quick make–quick break

Slow Make–Slow Break

This mechanism is usually identified with ac applications because its relative slowness provides enough time delay for the ac wave to go through its zero energy level, at which point the contact arc is extinguished. The teeter-totter mechanism depicted in Fig. 5.2 is typically associated with the slow make–slow break type of mechanism. This type of mechanism usually has face-to-face contacts, referred to as *butt contacts.*

Quick Make–Quick Break

The quick make–quick break mechanism is significantly different from the slow make–slow break mechanism. The quick make–quick break style uses additional

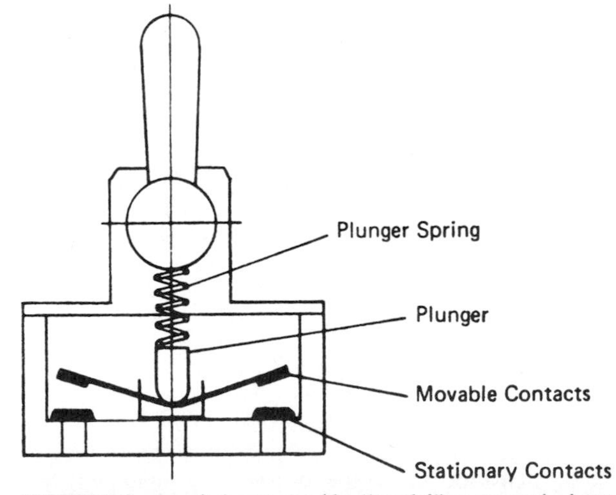

FIGURE 5.2 A switch actuator (the "toggle") operates the internal switch mechanism. (*Courtesy of Eaton Corporation, Milwaukee, Wis.*)

motive power, such as that provided by a compression-type spring (Fig. 5.4), to produce a snap-action response which quickly separates the contacts. This mechanism is particularly useful in dc applications where swift extinguishing of the arc is dependent on very fast contact separation. This type of mechanism is associated with *wiping-action* contacts. [Adapted from (1).]

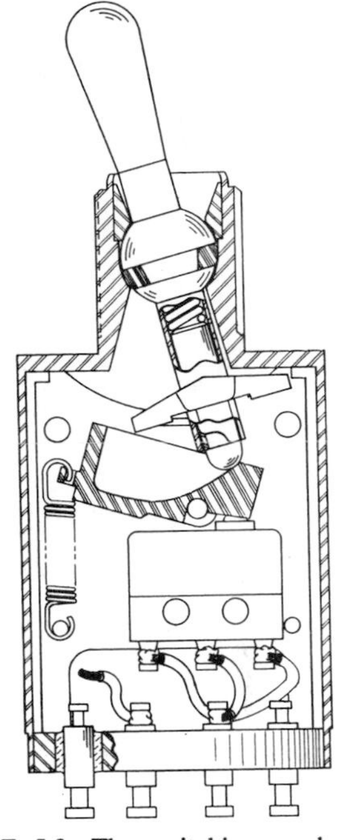

FIGURE 5.3 The switching mechanism of this special-design toggle switch is a precision snap-action module. (*Courtesy of Otto Engineering, Inc., Carpentersville, Ill.*)

Positive-Action Switches

A third type of break mechanism, although not as commonly used as the slow make–slow break and quick make–quick break types, is worthy of mention. Used primarily in military equipment, this type is referred to as the *positive-action switch* and is essentially a hybrid of the slow make–slow break and quick make–quick break mechanisms. Operation is by a toggle lever assembly (Fig. 5.5) which is mechanically linked to the movable contact member to *ensure* positive make and break.

What is the purpose of this positive-action mechanism? If the contacts

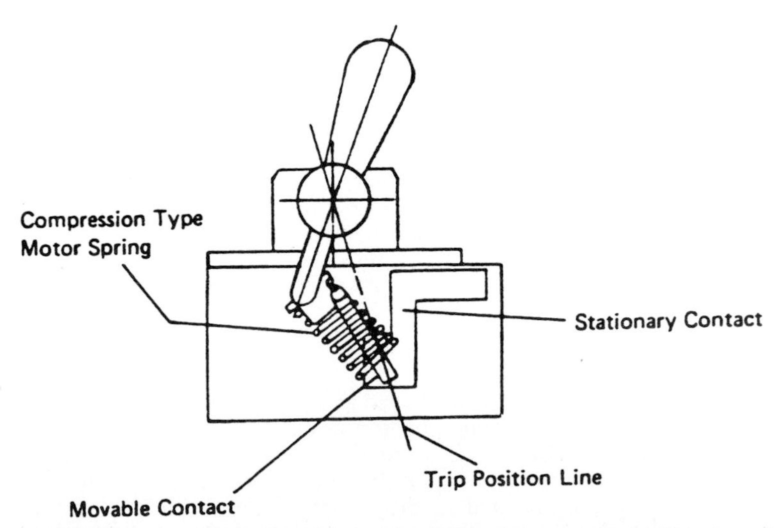

FIGURE 5.4 A quick make–quick break toggle switch mechanism. (*Courtesy of Eaton Corporation, Milwaukee, Wis.*)

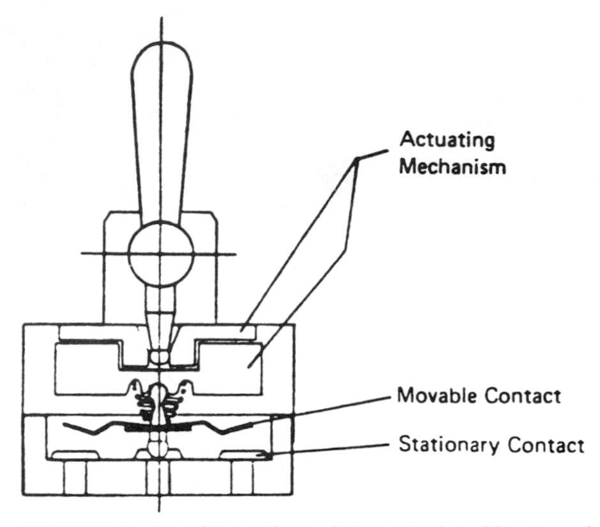

FIGURE 5.5 A positive-action switch mechanism. (*Courtesy of Eaton Corporation, Milwaukee, Wis.*)

weld during usage and it is of utmost importance that they be opened, the positive-action mechanism causes a mechanical break and lightly welded contacts are separated.

Like the quick make–quick break type of switch, the speed of the moving contacts is independent of the actuating mechanism. Once contact movement is started, the motion cannot be stopped before the switching action is completed. The switch mechanism combines both butt and wiping action contact, providing advantages for both ac and dc load applications. (1)

Poles

Toggle switches are available with from one to eight poles. Typical toggle switch terminal configurations and associated schematics are shown in Fig. 5.6. The figure depicts a style of toggle switch which uses precision snap-action modules as the basic switch technology and has one actuator design which is adaptable for use with either single or multiple snap-action modules.

Momentary Action

Although toggle switches are primarily used as an *alternate action* type of contact in which the switch is in either one position or another until changed by the operator, some toggle switches also feature *momentary action* positions in which switch contact is maintained as long as the operator holds the actuator lever in the maintained position. As soon as the actuator lever is released, an internal spring-loaded mechanism moves the actuator back to a neutral position and contact at the momentary position is broken.

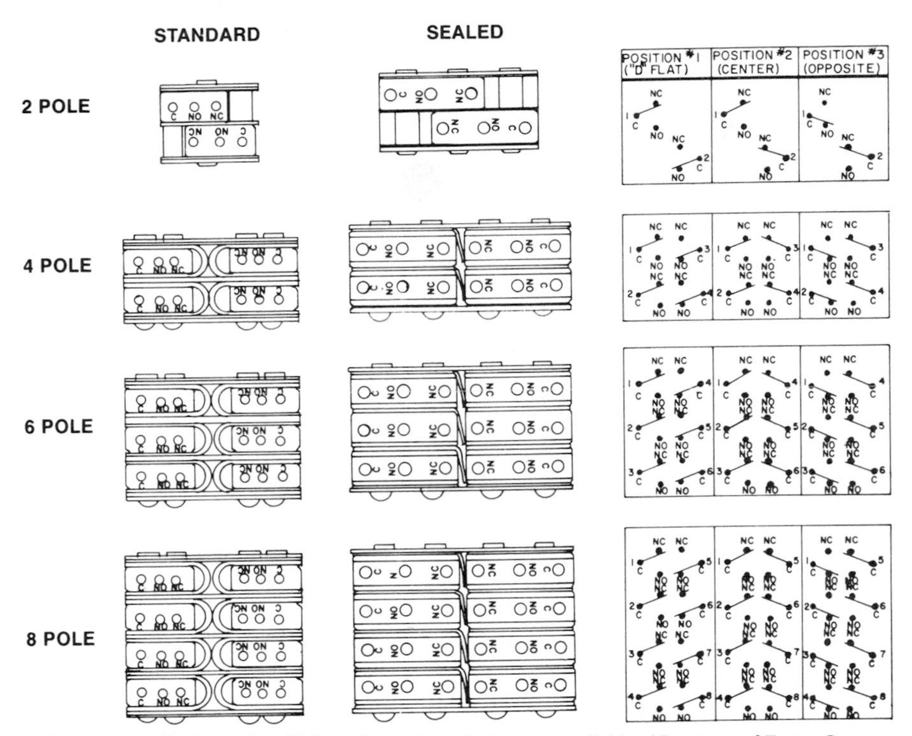

FIGURE 5.6 Single- and multiple-pole toggle switches are available. (*Courtesy of Eaton Corporation, Milwaukee, Wis.*)

Lever Lock Bushing Styles

One useful option available for toggle switches is the lever lock bushing. This mechanism provides a means of preventing accidental actuation of the toggle lever by incorporating a spring device internal to the switch package which requires a pull outward (toward the operator) on the specially configured toggle lever tab to bypass locking notches in the bushing. Figure 5.7*a* shows details of the notch-and-tab type of mechanical lockout. Figure 5.7*b* depicts the wide variety of lever lock bushing styles available. Table 5.1 illustrates a portion of a lever operation chart from a manufacturer's publication. The chart indicates direction of movement against which the lever is locked and any momentary action option available with a particular locking bushing. A lever locking configuration is available for almost any conceivable lock-and-contact combination. (2)

Surface Mount (SMT) Toggle Switches

In addition to standard-size toggle switches, some manufacturers are producing SMT toggle switches, small in size and designed to withstand the *IR* and vapor phase reflow soldering and cleaning processes. One such SMT device is depicted in Fig. 5.8. Switches such as this are often available from the manufacturers on tape and reel for automated placement. (3)

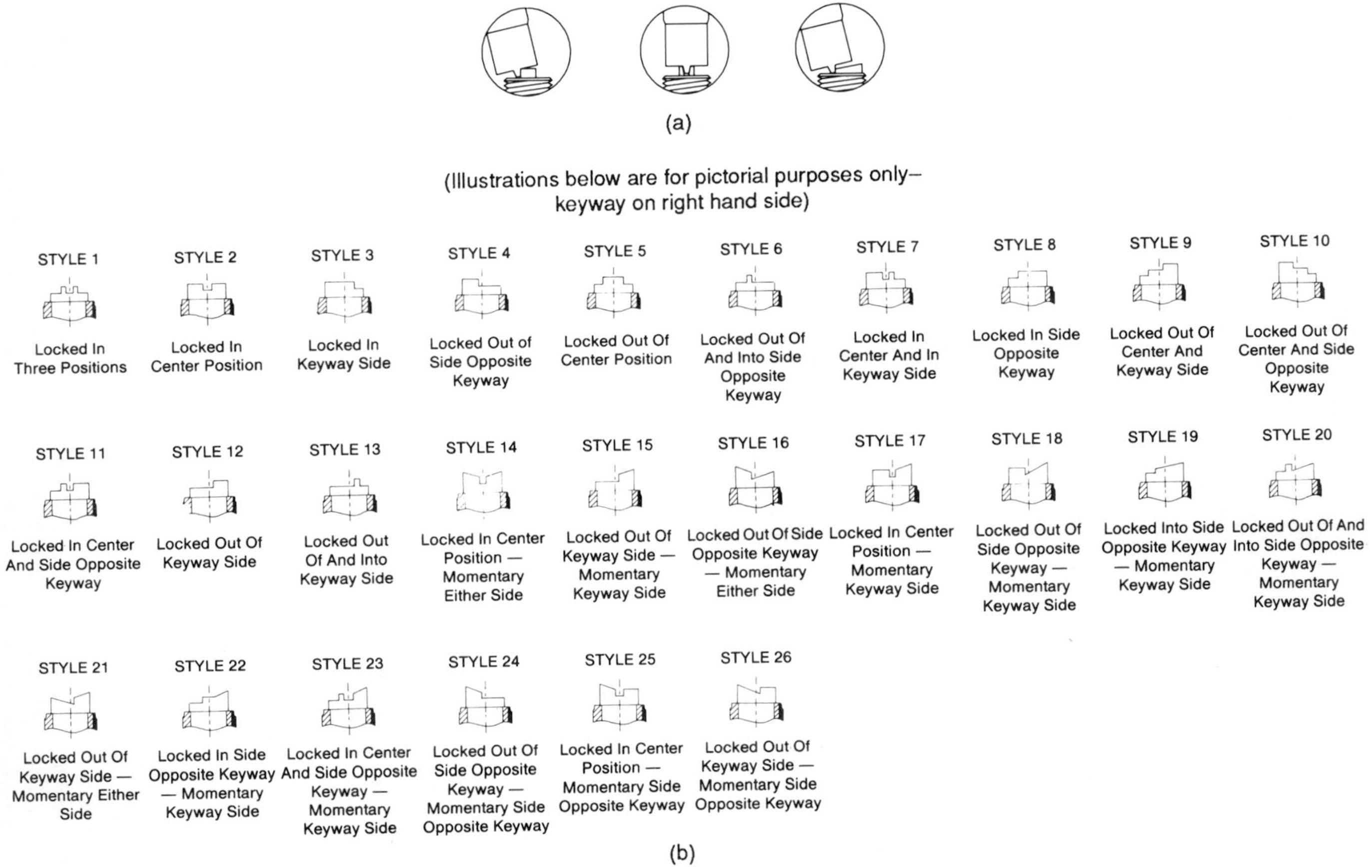

(a)

(Illustrations below are for pictorial purposes only–
keyway on right hand side)

STYLE 1
Locked In Three Positions

STYLE 2
Locked In Center Position

STYLE 3
Locked In Keyway Side

STYLE 4
Locked Out of Side Opposite Keyway

STYLE 5
Locked Out Of Center Position

STYLE 6
Locked Out Of And Into Side Opposite Keyway

STYLE 7
Locked In Center And In Keyway Side

STYLE 8
Locked In Side Opposite Keyway

STYLE 9
Locked Out Of Center And Keyway Side

STYLE 10
Locked Out Of Center And Side Opposite Keyway

STYLE 11
Locked In Center And Side Opposite Keyway

STYLE 12
Locked Out Of Keyway Side

STYLE 13
Locked Out Of And Into Keyway Side

STYLE 14
Locked In Center Position — Momentary Either Side

STYLE 15
Locked Out Of Keyway Side — Momentary Keyway Side

STYLE 16
Locked Out Of Side Opposite Keyway — Momentary Either Side

STYLE 17
Locked In Center Position — Momentary Keyway Side

STYLE 18
Locked Out Of Side Opposite Keyway — Momentary Keyway Side

STYLE 19
Locked Into Side Opposite Keyway — Momentary Keyway Side

STYLE 20
Locked Out Of And Into Side Opposite Keyway — Momentary Keyway Side

STYLE 21
Locked Out Of Keyway Side — Momentary Either Side

STYLE 22
Locked In Side Opposite Keyway — Momentary Keyway Side

STYLE 23
Locked In Center And Side Opposite Keyway — Momentary Keyway Side

STYLE 24
Locked Out Of Side Opposite Keyway — Momentary Side Opposite Keyway

STYLE 25
Locked In Center Position — Momentary Side Opposite Keyway

STYLE 26
Locked Out Of Keyway Side — Momentary Side Opposite Keyway

(b)

FIGURE 5.7 Lever lock mechanisms: (*a*) details of the notch-and-tab lock mechanism, and (*b*) a multitude of lever lock configurations are available. [*Courtesy of* (a) *C&K Components, Inc., Newton, Mass., and* (b) *Eaton Corporation, Milwaukee, Wis.*]

TABLE 5.1 A Lever Lock Chart Shows How Lock Mechanism Is Configured for Each Contact Arrangement

Circuit with lever in . . .		
Up position	Center position	Down position (keyway)
ON→	←OFF→	←ON
ON	←OFF→	ON
ON	←OFF	NONE
ON	OFF	←ON
ON	NONE	←OFF
ON	NONE	←ON
ON	←OFF	ON
ON→	OFF	←ON
ON→	NONE	←OFF
ON→	NONE	←ON
*ON	←OFF→	ON*
ON	OFF→	ON*
*ON	←OFF	ON*
ON→	←OFF	NONE
NONE	OFF→	ON*
ON	←OFF→	ON*
ON→	NONE	ON*
ON→	NONE	OFF*
OFF→	NONE	ON*
ON	←OFF	ON*
ON→	←OFF	ON*

*Momentary contact.
Arrows indicate direction of movement against which lever is locked.
Source: Eaton Corporation, Milwaukee, Wis.

USEFUL VARIATIONS FOR BASIC TOGGLE SWITCHES

Wiring for Three-Way Switches

Occasionally, an engineer may need a three-way function switch. Figure 5.9 depicts a method of providing this function using a toggle switch with an ON-ON-ON function and user-added jumpers.

Reversing Circuit

The reversing circuit is a popular variation which utilizes a two-pole switch and jumpers. Many switch applications are for series or universal motor control. In this type of motor, it is only necessary to reverse the direction of current in either the field windings or the armature to reverse the mechanical output direction of a motor shaft. As depicted in Fig. 5.10, adding diagonal jumpers to a two-pole switch provides the reversing effect. Assuming that the switch illustrated is making contact at terminals 1-5 and 2-6, current direction is from left to right or coun-

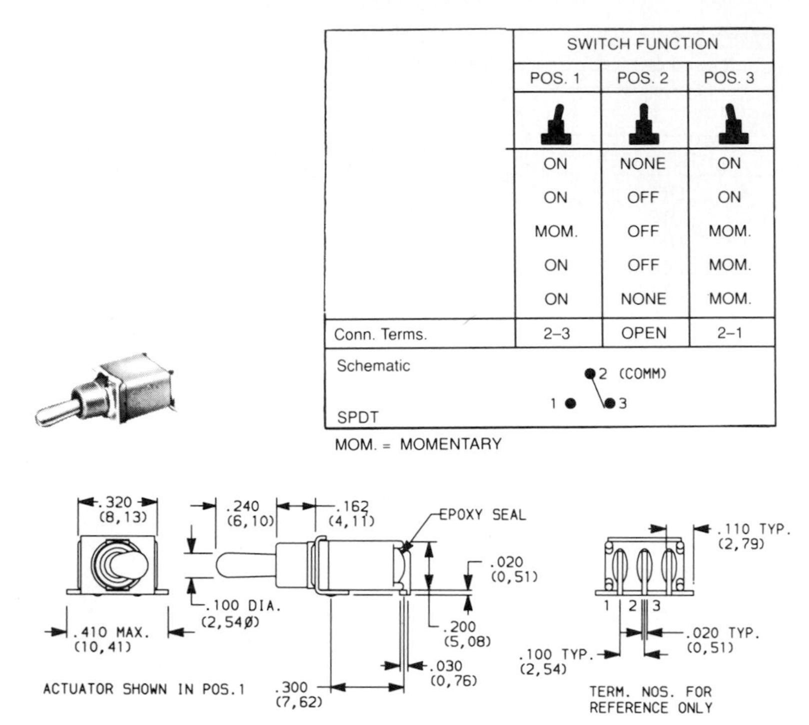

FIGURE 5.8 A typical miniature SMT toggle switch. (*Courtesy of C&K Components, Inc., Newton, Mass.*)

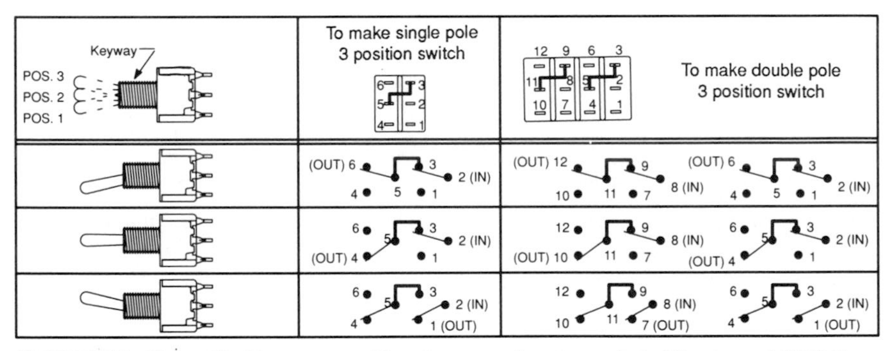

FIGURE 5.9 User-added jumpers provide a three-way function switch. (*Courtesy of C&K Components, Inc., Newton, Mass.*)

terclockwise through the armature. If we wish to reverse motor direction, we simply transfer the circuit so that L1 and L2 are connected to 1-3 and 2-4, respectively. Tracing the circuit with the aid of dotted lines, we can see how the jumpers serve to reverse the current. Typical applications for this type of reversing circuit are the "forward-reverse" on electric drills and screwdrivers and the "up-down" on automobile windows. (1)

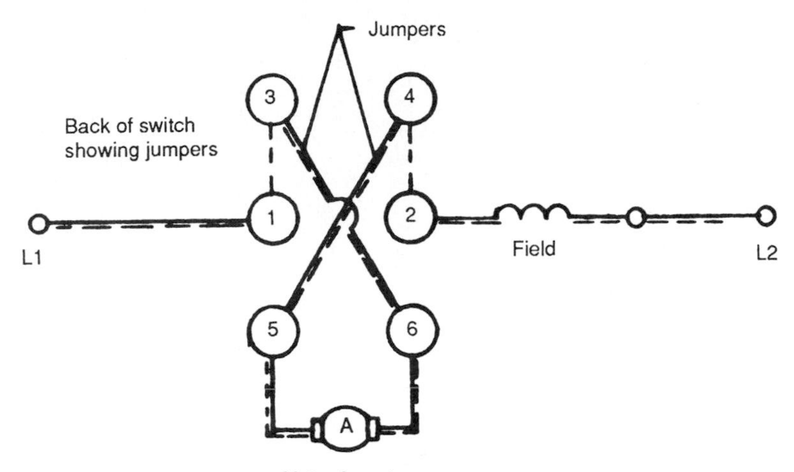

FIGURE 5.10 Jumpers on a two-pole switch provide a reversing circuit. (*Courtesy of Eaton Corporation, Milwaukee, Wis.*)

GLOSSARY

Alternate action. A switch mechanism which maintains a particular switch state until the operator physically changes that state. Once changed, the switch remains in that state until once again acted on by the operator.

Butt contact. A type of contact style in which the opposing contacts meet head-on. Opposite of *wiping contact.*

Contact. The points of electrical contact in a switch.

Momentary action. A switch mechanism that does not self-maintain the position to which it is actuated. Typically, a spring-return mechanism in the switch returns the contacts to their original position when the actuator is released by the operator.

Pole. The number of completely separate circuits that can pass through a switch at one time.

Quick make–quick break. A nonteasable, snap-action type of switch mechanism designed to provide fast contact separation. Primarily designed for dc operation.

Slow make–slow break. A switch mechanism with relatively slow operation to provide a slight time delay, permitting the ac wave to go through its zero energy level.

Wiping contact. Lateral travel of movable contact over fixed contact while pressure between the two exists. This action helps cleanse the contacts of contamination. See *Butt contact.*

REFERENCES

1. "Switch Fundamentals," Eaton Corporation, Cutler-Hammer Products, Aerospace and Commercial Controls Division, Milwaukee, Wis., 1991.
2. "Aerospace Switches," Eaton Corporation, Cutler-Hammer Products, Aerospace and Commercial Controls Division, Milwaukee, Wis., 1990.
3. "Catalog No. 9105, Switches—Newton Division," C&K Components, Inc., Newton, Mass., 1991.

CHAPTER 6
PUSHBUTTON SWITCHES*

Pushbutton switches are among the simplest of all types of switches. In a basic pushbutton switch, a spring holds a shorting bar in a position which causes a circuit condition to exist. This circuit condition can be either a closed path or an open path. The operator, by pressing the button, overcomes the spring tension and causes the opposite circuit condition. A simple type of pushbutton switch can have one of two conditions, either normally open or normally closed (Fig. 6.1). The term "normally" *always* refers to the switch before it is actuated. This is true even though in some applications the switch is actuated a greater percentage of the time.

Pushbutton switches can also be among the most complex of switch types (Fig. 6.2*a* and *b*). Illuminated pushbutton switches designed for use in commercial and military aircraft are intricate devices consisting of a precision switching package, an actuator device to trigger the switch action, and specially selected

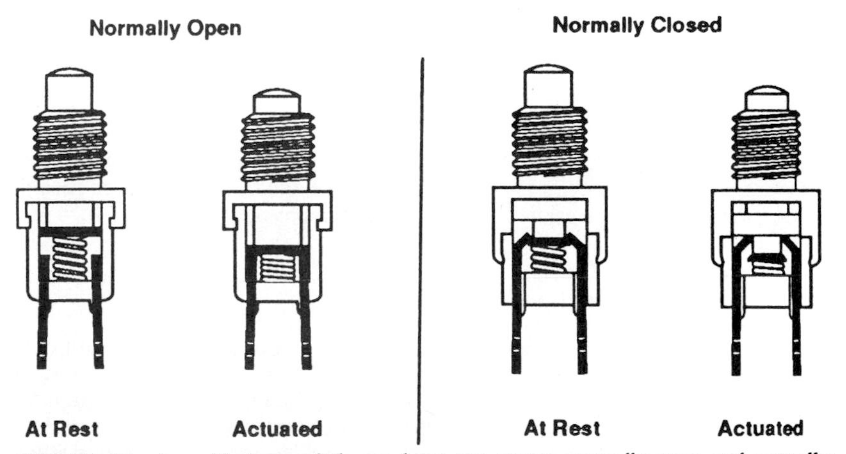

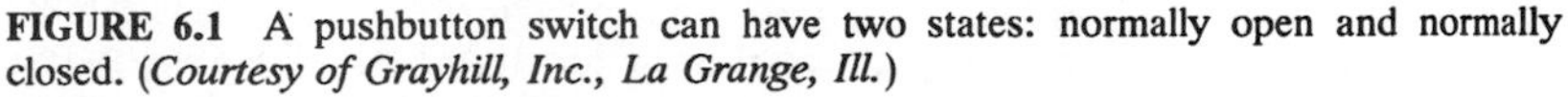

FIGURE 6.1 A pushbutton switch can have two states: normally open and normally closed. (*Courtesy of Grayhill, Inc., La Grange, Ill.*)

*Numbers in parentheses indicate items in the References at the end of this chapter.

FIGURE 6.2 Pushbutton switches range from the simple to the complex. [(*Left*) *Courtesy of Grayhill, Inc., La Grange, Ill., and* (*Right*) *©Korry Electronics, 1992.*]

illumination sources (lamps) which, when combined with optical filters, legend plates, and color filtering, produce consistent and precisely controlled illumination levels at the switch pushbutton face.

HOW PUSHBUTTON SWITCHES FUNCTION

Pushbutton switches can be classified by two characteristics: the *mechanical* action that occurs when the switch is opened or closed, and the circuit condition resulting from the *electrical* configuration or contact arrangement prior to actuation.

Mechanical Action of Pushbutton Switches

There are three basic types of mechanical action, each of which is a result of the physical construction of the switch. These types of mechanical action are *wiping action, butt action,* and *snap action* (Fig. 6.3). In a wiping-action switch, the moving member (shorting bar) wipes across the contacts of the switch terminals. In a butt-action switch, no wiping action occurs. The moving member merely touches or butts against the contacts of the switch terminals. In a snap-action switch, a spring mechanism causes the moving member to butt against the contacts of the terminals. The pushbutton of the switch merely determines when the spring is actuated. The switching in a snap-action type of switch is generally in-

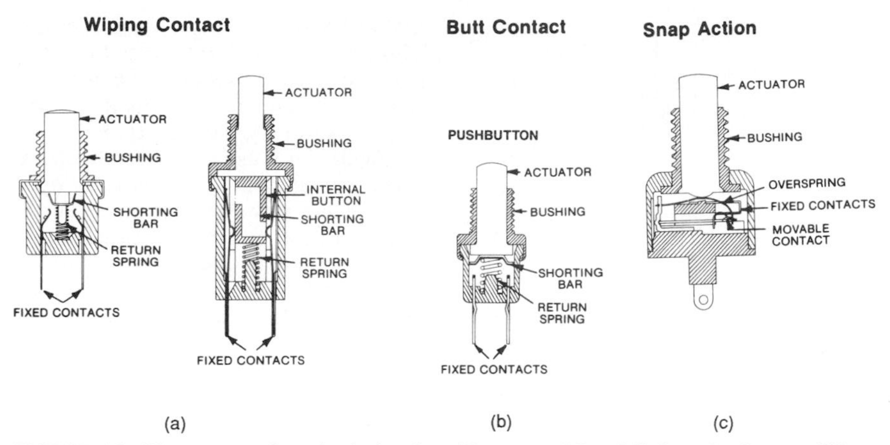

FIGURE 6.3 Three types of mechanical action. (*Courtesy of Grayhill, Inc., La Grange, Ill.*)

dependent of the rate of travel of the pushbutton. In fact, a number of pushbutton types use individual or ganged precision snap-acting switches (Chap. 2) as the switching mechanism, often contained in a housing with the switch actuator and pushbutton. The precision mechanism of a snap-acting switch is ideally suited for the accurate, reliable actuation required for this type of pushbutton switch when used in commercial or military aircraft.

The wiping-action switch cleans the contacts of the terminals as it is actuated. This feature is desirable since it maintains low contact resistance during the life of the switch. On the other hand, wiping action causes the contact surfaces to wear, shortening the life of the switch.

The butt-action switch, of course, does not have the wiping action; therefore, there is less wear in this type of switch. The self-cleaning feature is lost in a butt-action switch, and contact resistance will generally increase during the life of the switch. Because of the small amount of wear, the butt-action switch can be used for many cycles of operation if the increasing contact resistance can be tolerated in a circuit.

When very low voltages (approximately 0.1 V or less) are to be switched, the environment may cause a film to build up upon the contacts. Under these conditions, the butt-action pushbutton switch might be erratic in operation and a wiping-contact switch should be considered. Likewise, where the environment deposits contamination on the switch contacts, a wiping-action switch is usually a better choice than a butt-action switch.

The advantage of a snap-acting switch is that the spring mechanism controls the point of make or break. This mechanism, isolated from any influence by the human operator, causes rapid make and break of the contact. This allows more of the arcing to dissipate in air and permits larger currents to be switched. A snap-action switch is essentially a butt contact.

Electrical Configurations of Pushbutton Switches

The electrical configuration or contact arrangement of a pushbutton switch refers to the circuit condition caused by the switch before it is actuated, that is, before

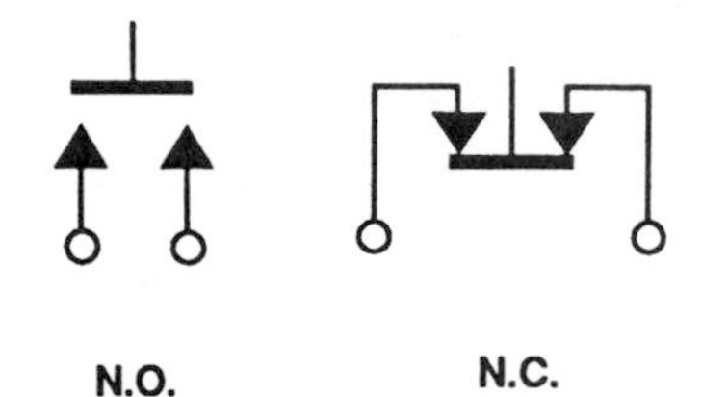

FIGURE 6.4 Two basic contact arrangements. (*Courtesy of Grayhill, Inc., La Grange, Ill.*)

the pushbutton (plunger) is depressed. This position is referred to as the "normal" position. The simplest contact arrangements are the "normally open" or the "normally closed" types.

Figure 6.4 illustrates these two contact arrangements. The normally open (NO) diagram shows that the circuit is open until the button is depressed and a shorting bar connects the terminals. The normally closed (NC) diagram shows that the circuit is closed until the button is depressed and the shorting bar is moved away from the terminals.

Most pushbutton switches are of the momentary-contact type, which means contact action occurs only while the button is depressed. Releasing the force on the button causes the contacts to return to their normal condition.

Sometimes a pushbutton switch is needed which will close one circuit and open another. For example, an airline pilot can interrupt the music in the cabin to make announcements. The switch opens (disconnects) the music circuit and closes (connects) the pilot's microphone circuit. When the button is released, the music circuit is reconnected. The circuit diagram for this arrangement is shown in Fig. 6.5. There are three terminals in this switch. One of them is common to both of the other two and is called a *common* (marked C). With the pushbutton at rest, the common is in contact with the normally closed (NC) terminal. When the button is pushed, that circuit is disconnected and contact is then made between the common terminal and the third, or normally open (NO), terminal.

FIGURE 6.5 A three-terminal pushbutton switch circuit. (*Courtesy of Grayhill, Inc., La Grange, Ill.*)

The terminal that is common to two or more other terminals is called a *common pole,* or most often, just a *pole.* The terminals that a pole may contact are called *throws.* The contact arrangement shown in Fig. 6.5 is single-pole double-throw (SPDT) since the common terminal may contact one of two electrically isolated circuits. Switches that can close or open only one current path are referred to as single-pole single-throw (SPST). A doorbell is a single-pole single-throw switch, as are the NO and NC switches shown in Fig. 6.4.

More than one pole can be incorporated in a pushbutton switch. The circuit diagram for such a switch is shown in Fig. 6.6. The figure indicates that there are two common terminals (C1 and C2) or poles. A normally open and a normally closed section is associated with each common terminal. The dashed line indicates that the button moves both poles at the same time.

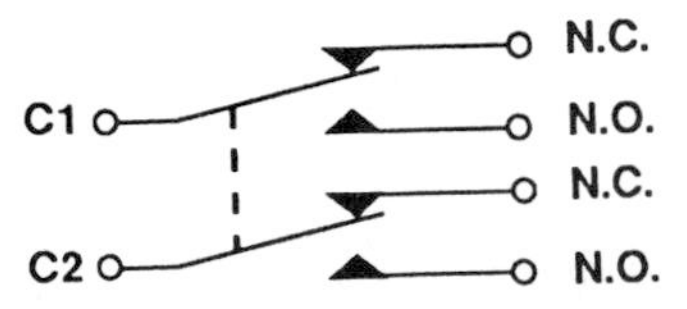

FIGURE 6.6 A switch with two common terminals. (*Courtesy of Grayhill, Inc., La Grange, Ill.*)

When the pushbutton is depressed, the contact is broken between C1 and its associated NC terminal and between C2 and its NC terminal. When the movement of the button causes C1 to contact its NO terminal, C2 will likewise contact its NO terminal. Since there are two poles in this switch and two throws associated with each, this type of switch is called a double-pole double-throw switch. This is usually abbreviated DPDT.

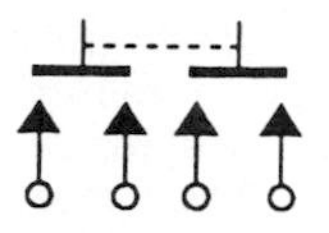

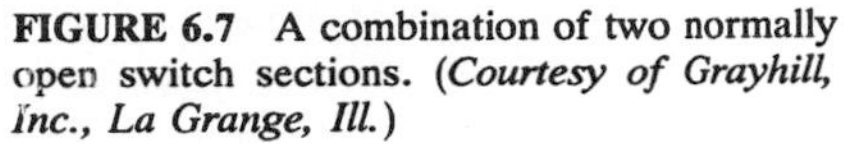

FIGURE 6.7 A combination of two normally open switch sections. (*Courtesy of Grayhill, Inc., La Grange, Ill.*)

Yet another contact arrangement is a combination of two normally open sections. The circuit diagram for this type of arrangement is shown in Fig. 6.7. A separate shorting bar is associated with each pair of terminals. The dashed line indicates that the button operates both shorting bars at the same time. This type of arrangement is called a double-pole switch. This contact arrangement may also be referred to as a double-pole single-throw switch, which is abbreviated DPST.

In switches that have more than one throw position, two types of contact action are possible. These are a *break before make* and a *make before break.*

To explain these types of contact action, consider the SPDT contact arrangement shown in Fig. 6.5. In the normal position of this arrangement, the common arm is touching the upper contact (NC) and the path is broken between the common and the lower terminal (NO). If the switch were a "break before make" type, the contact at the NC section would be broken before contact was made at the NO section. If the switch were a "make before break," contact would be made at the NO section before it was broken at the NC section.

Movement Characteristics of Pushbutton Switches

The movement characteristics of a pushbutton switch can be seen in Fig. 6.8, which shows a pushbutton and a bushing of a switch (not all pushbutton switches, of course, have bushings—the important item in this figure is the pushbutton itself). There are three positions of importance on the downstroke of the button. The *free position* is the pushbutton location with no force (except the force of gravity) exerted on it. The *operating position* is the pushbutton location when the contact action (opening or closing) takes place. The *total overtravel position,* or *bottom position,* is the *maximum* downward travel of the pushbutton.

Two distances are also defined. The distance between the free position and the operating position is called the *pretravel* distance. The distance between the operating position and the total overtravel position is called the *overtravel* distance. The

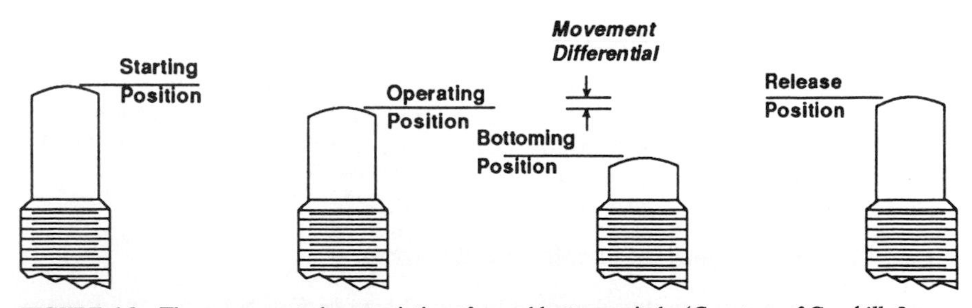

FIGURE 6.8 The movement characteristics of a pushbutton switch. (*Courtesy of Grayhill, Inc., La Grange, Ill.*)

release position is the position on the upstroke of the pushbutton where the contact action no longer takes place (i.e., "broken" for an NO switch or "closed" for an NC switch). Because of the construction of most switches, the operating and release positions are at different locations. The distance difference between these two positions is called the *differential travel.* [Adapted from (1)–(5).]

VARIATIONS WITHIN THE PUSHBUTTON FAMILY

There are literally thousands of variations of pushbutton switches. In this section, we examine a few of the many types available in order to get a perspective on the diverse categories of pushbutton switches.

Snap-Acting Momentary Pushbutton Switches

As mentioned earlier in this chapter, a snap-acting switch mechanism operates independently of the rate at which the pushbutton is actuated. The switches depicted in Fig. 6.9*a* and *b* are small snap-acting devices available with up to four poles (Fig. 6.9*b*). The package shown is a popular size, and many pushbutton switch manufacturers produce an equivalent model(s). (6)

Sealed Subminiature Snap-Acting Pushbutton Switches. Figure 6.10 depicts a series of subminiature pushbutton switches with integral seals on the pushbutton to seal the device against dust and moisture. The switches also feature a double-break design which offers wider switching possibilities in addition to providing longer life and greater capacity. The advantages of a double-break contact system were discussed in Chap. 1. (7)

Precision Snap-Acting Module Pushbutton Switches

As in some styles of toggle switches examined in Chap. 5, many different types of pushbutton switches are manufactured using precision snap-acting switch modules (Chap. 2) as the basis for the pushbutton package. Figure 6.11 illustrates two of these devices. The switch module packages are attached to retainer plates which align the modules under the actuator. Single and multiple modules can be packaged in this way to provide a variety of switch contact configurations. (7)

Wiping-Contact Environmentally Sealed Pushbutton Switches

The switch depicted in Fig. 6.12 is a wiping-contact pushbutton device with encapsulated leads and an O-ring seal on the pushbutton shaft for environmental resistance. This momentary action switch is a double-pole double-throw (DPDT) switch capable of operation from −40°F to 185°F (−40 to 85°C). (8)

Butt-Contact Momentary Pushbutton Switches

The tiny pushbutton switch depicted in actual size in Fig. 6.13 is a fully operational, normally open, single-pole single-throw pushbutton switch. The device

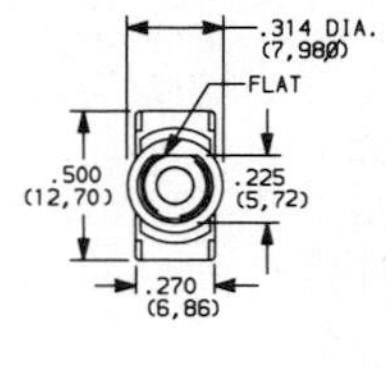

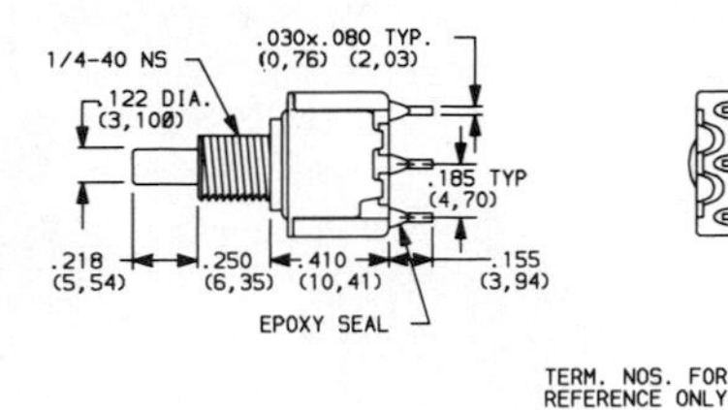

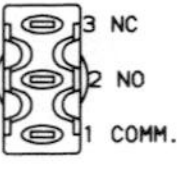

TERM. NOS. FOR REFERENCE ONLY

(a)

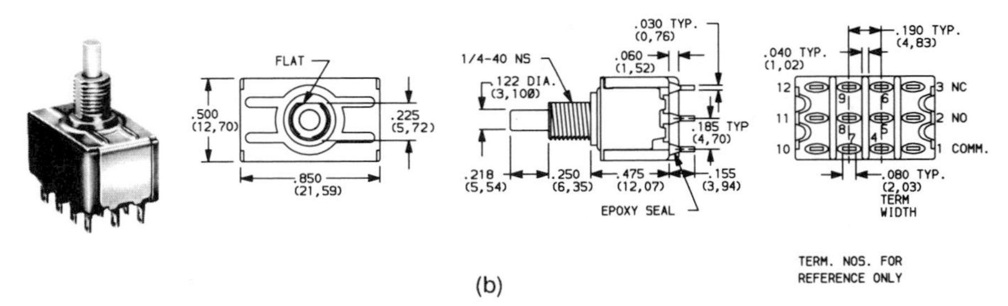

SWITCH FUNCTION		CONNECTED TERMINALS		SCHEMATIC
POS. 1	POS. 2	POS. 1	POS. 2	
ON	MOM.	1-3	1-2	SPDT 1 (COMM) 3 2
ON	MOM.			
ON	MOM.	1-3,4-6,7-9,10-12	1-2,4-5,7-8,10-11	4PDT 1 3 2 4 (COMM) 6 5 7 9 8 10 12 11
ON	MOM.			

FLAT .500 (12,70) .225 (5,72) .850 (21,59)

1/4-40 NS .030 TYP. (0,76) .060 (1,52) .122 DIA. (3,100) .185 TYP (4,70) .218 (5,54) .250 (6,35) .475 (12,07) .155 (3,94) EPOXY SEAL

.040 TYP. (1,02) .190 TYP. (4,83) 3 NC 2 NO 1 COMM. .080 TYP. (2,03) TERM WIDTH

TERM. NOS. FOR REFERENCE ONLY

(b)

FIGURE 6.9 Snap-acting momentary pushbutton switches: (*a*) an SPDT switch, and (*b*) a DPDT switch. (*Courtesy of C&K Components Inc., Newton, Mass.*)

shown incorporates a sealed bushing and pushbutton as well! The switch contacts are capable of carrying ½ A resistive at 115 V ac, and it is rated for 250,000 operations at rated load. (8)

Square-Base 0.50-in (12,7 mm) Pushbutton Switches for PCB Mounting

The miniature-base pushbutton switch depicted in Fig. 6.14*a* and *b* is designed for placement beneath flexible membrane overlays or custom keycaps as shown in Fig. 6.14*b*. This small package contains a metal dome (Chap. 9) as the switching element. Variable-height buttons are available to provide the user some variety in mounting the device. (9)

FIGURE 6.10 Sealed subminiature snap-acting pushbutton switches. (*Courtesy of Otto Controls, Inc., Carpentersville, Ill.*)

Surface Mount (SMT) Pushbutton Switches

Two styles of SMT pushbutton switches are shown in Fig. 6.15*a* and *b*. The ultraminiature key switch depicted in Fig. 6.15*a* is only 0.245 in (6,22 mm) square and is sealed to withstand *IR* and vapor-phase reflow soldering and cleaning processes. The device is sealed by the use of an internal silicone rubber actuator seal and insert molded terminals. Figure 6.15*b* illustrates one of five different SMT terminal styles available in this ultraminiature switch. It, too, is sealed to withstand the SMT process.

Like most SMT devices, the two switches described above are available on tape and reel for automated placement. (6)

Integrated Switch Panels

Integrated switch panels (Fig. 6.16) are designed to provide illuminated pushbutton switch functions in a low-profile integrated panel. Typically mounted

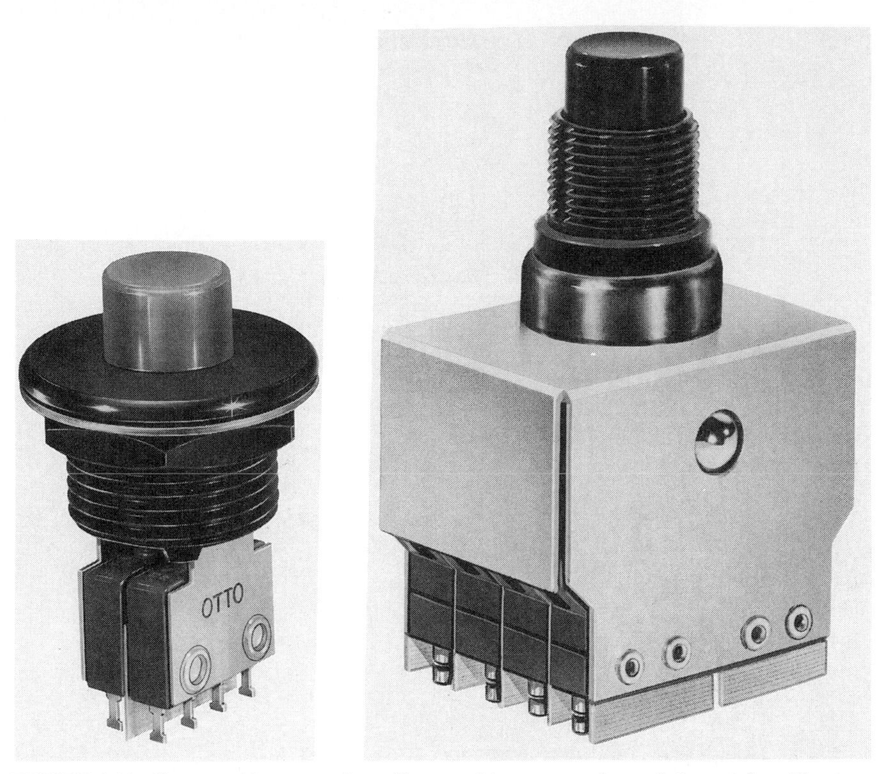

FIGURE 6.11 Some pushbutton styles utilize precision snap-action switch modules. (*Courtesy of Otto Controls, Inc., Carpentersville, Ill.*)

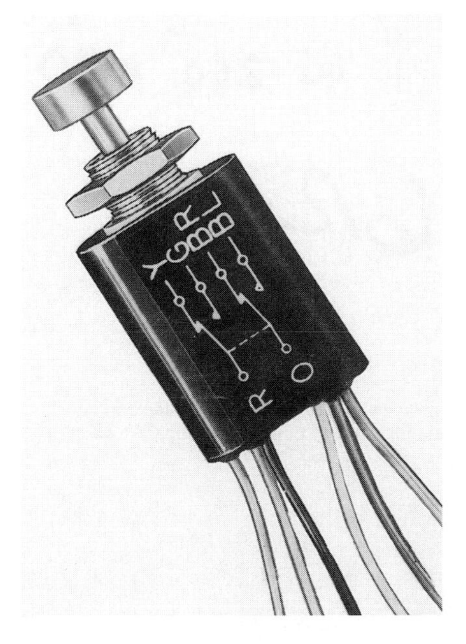

FIGURE 6.12 Wiping-contact environmentally sealed pushbutton switch. (*Courtesy of Grayhill, Inc., La Grange, Ill.*)

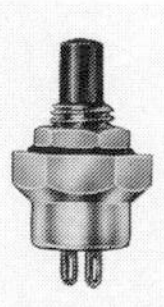

FIGURE 6.13 Butt-contact momentary pushbutton switch (shown actual size). (*Courtesy of Grayhill, Inc., La Grange, Ill.*)

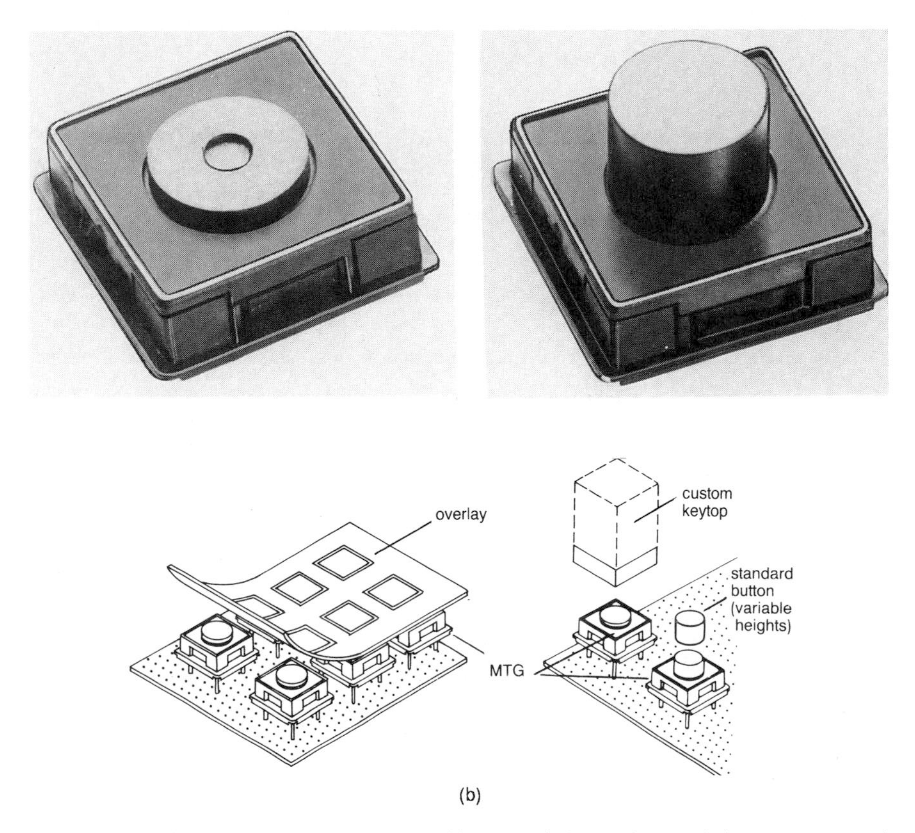

FIGURE 6.14 (*a*) PCB-mount square-base pushbutton switch contains metal-dome contacts, and (*b*) square-base switch mounts under flexible membrane or custom keycaps. (*Courtesy of Schurter, Petaluma, Calif.*)

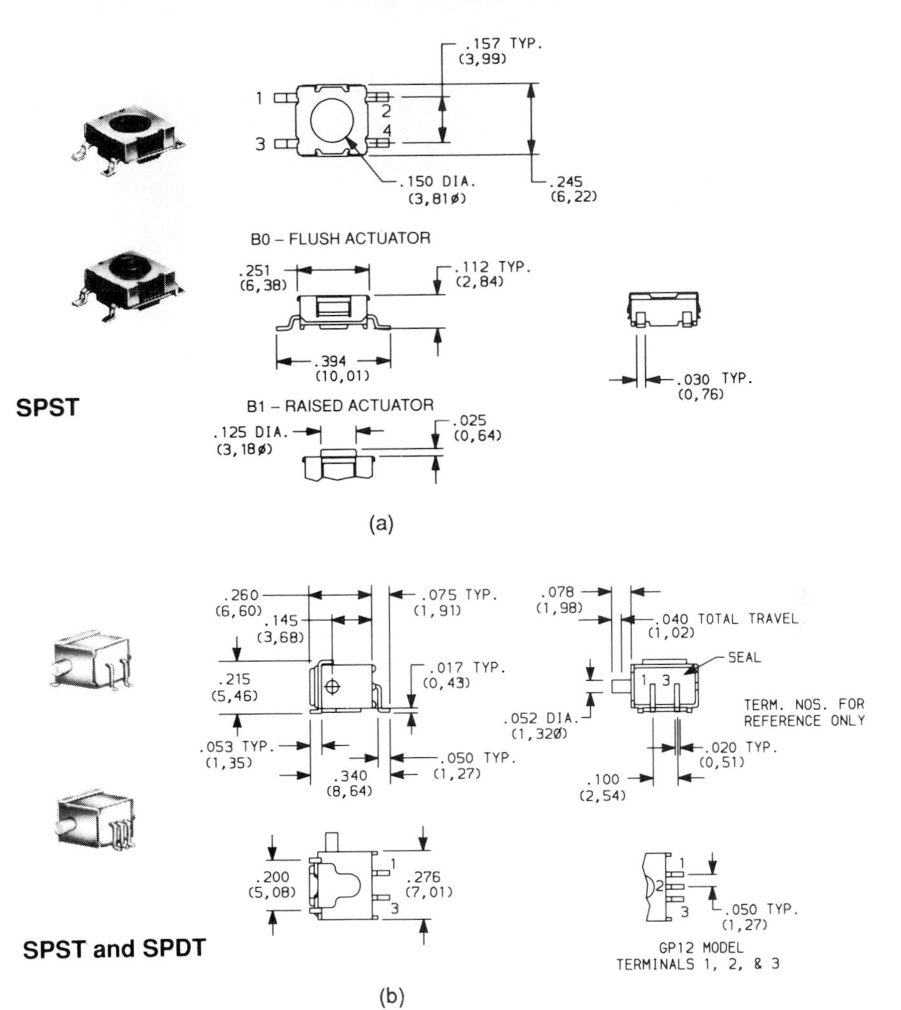

FIGURE 6.15 (*a*) A sealed ultra-miniature SMT key switch, and (*b*) an SMT switch with a side-actuated pushbutton. (*Courtesy of C&K Components, Inc., Newton, Mass.*)

over LCD or CRT displays to provide switch input interaction with the displayed information, the panels are manufactured using one of three basic switch techniques (Fig. 6.17*a, b,* and *c*). Figure 6.17*a* utilizes a metal dome switch contact (Chap. 9), while the panel illustrated in Fig. 6.17*b* uses an encapsulated metal dome switch similar to the square-base pushbutton switch for PCB mounting described earlier in this chapter. The panel in Fig. 6.17*c* utilizes a garter-spring switch proprietary to Korry Electronics Company, Seattle, Wash.

The pushbuttons in these panels can be illuminated to meet sunlight readability and NVIS requirements (more on NVIS later in the chapter).

Table 6.1 depicts the features available for each of the three switching tech-

FIGURE 6.16 A low-profile illuminated integrated key panel. (*Courtesy of Interface Products Inc., Oceanside, Calif.*)

niques as offered by Korry Electronics Company in their Chromarray™ line of integrated panels. (10)

Illuminated Pushbutton Switches

Illuminated pushbutton switches are used in a wide variety of applications, from machine tool controls to sophisticated "stealth" aircraft. While commercial and industrial-grade illuminated devices can prove to be just as interesting, we choose to examine the type of illuminated pushbutton switch designed for high-reliability applications such as commercial and military aircraft. This examination can provide insight to the user or potential user who may be considering use of this imaginative device.

The lighted pushbutton switch is similar in operation to an unlighted pushbutton switch from an electrical standpoint; the operator pushes on the switch face, and this motion is transferred through the pushbutton–lamp holder module to the snap-action switch mechanism, initiating the switching action. The switch may be alternate-action or momentary-action, single-pole or multiple-pole and is often designed using precision snap-acting switch modules (Chap. 2) as the switching mechanism. Some manufacturers design their own switching mechanism for use in their pushbutton switches, but the fundamental switch action is

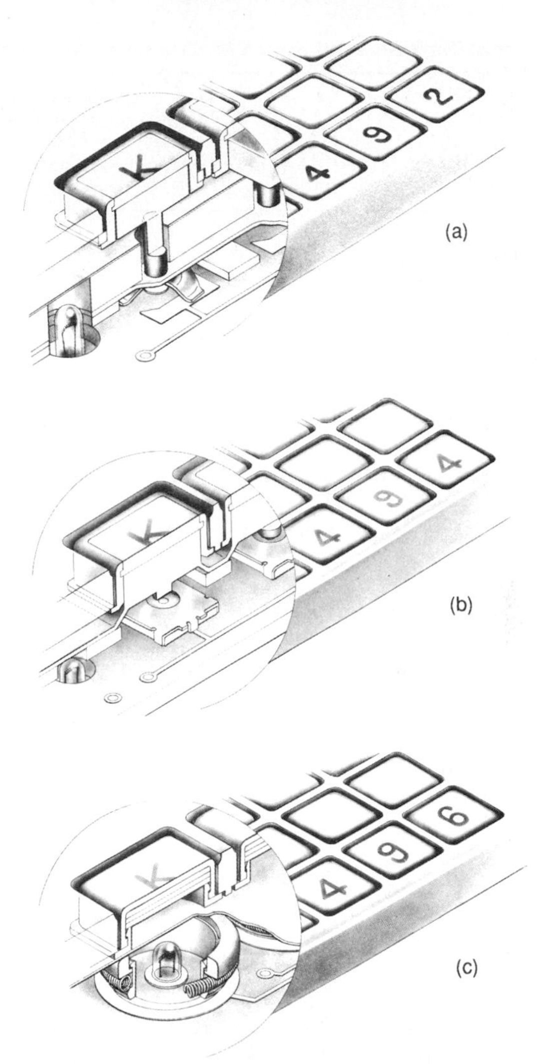

FIGURE 6.17 Three basic switch techniques for integrated panels: (*a*) the metal dome, (*b*) the encapsulated-key switch, and (*c*) the garter-spring switch. (©*Korry Electronics, 1992.*)

similar to the snap-acting module design. Figure 6.18 depicts schematically the component configuration in a typical illuminated pushbutton switch.

If the switch is an alternate-action device, the design of the actuating mechanism will include a latch or other holding mechanism to hold the external pushbutton module down until a second actuation returns the module to its original position. The term alternate-action describes a pushbutton that is latched

TABLE 6.1 Features of Three Styles of Switch Mechanism for Integrated Switch Panels

	Chromarray™		
Feature	492 Snap dome	494 Encapsu-lated dome	496 Garter spring
Lighting:			
Incandescent lamps	*S*	*S*	*S*
Area lighting	*S*	*S*	*S*
Unlit panel indicia	*S*	*S*	*S*
Individually lit keycaps	*N*	*N*	*S*
Sunlight-readable hidden legend keycap	*N*	*N*	*S*
Lit panel indicia	*O*	*O*	*O*
LED lighting	*O*	*O*	*O*
NVIS compatibility	*O*	*N*	*O*
EL lighting	*N*	*O*	*O*
Hot-spot indicator on keycap	*N*	*N*	*O*
Split-legend cap with annunciator	*N*	*N*	*O*
Front relampability	*N*	*N*	*O*
Mechanical:			
Front panel of molded or machined plastic or aluminum	*S*	*S*	*S*
Panel thickness (min 0.30–0.60 in; depends on features)	*S*	*S*	*S*
Keycaps ⅜ to ¾ in	*S*	*S*	*S*
Dripproof sealing	*S*	*S*	*S*
Standard travel (0.001 in)	20 ± 5	20 ± 5	30 ± 5
Standard actuation force (oz)	22 ± 6	24 ± 10	20 ± 5
Choice of actuation force	*O**	*N*	*O*
Submersible sealing	*N*	*O*	*O*
Electrical:			
Momentary switching	*S*	*S*	*S*
Contact resistance <100 mΩ	*S*	*S*	*S*
EMI/RFI shielding to 40 dB	*S*	*S*	*S*
Current switching up to 100 mA	*N*	*N*	*S*
Alternate switching	*O*	*O*	*O*
Two-pole single-ground switching	*N*	*N*	*O*
EMI/RFI shielding to 80 dB	*N*	*N*	*O*
TEMPEST shielding	*N*	*N*	*O*

Source: ©Korry Electronics, 1992.
S = standard feature, *O* = optional feature, *N* = not currently available.
* = Multiples of 11 oz.

down every other actuation (the device has two normal button positions). [Adapted from (5).]

If the switch is a momentary-action device, the switch contacts "make" when the operator pushes the pushbutton down and "break" when the operator releases the pushbutton.

Figure 6.19 provides a more detailed look at a high-reliability illuminated pushbutton switch. The device depicted is manufactured by Korry Electronics Company, Seattle, Wash. The figure illustrates the precision components required to produce a switch approved for use in critical applications. Several features of the switch are readily apparent; the removable pushbutton–lamp holder module which permits replacement of the lamps, the snap-acting switch modules packaged at the base of the switch, and the actuator mechanism designed to provide momentary or alternate action while permitting positive tactile response after switch closure. What may not be apparent is the intricate optical filtering and lighting enhancement contained in the pushbutton module. Most aircraft applications require that the lighted legend information be visible in a sunlit cockpit; ambient illumination can approach 10,000 fc at flying altitudes. Special optical design techniques, closely guarded by individual switch manufacturers, provide the necessary light output to meet the "sunlight readable" requirement (Fig. 6.20). Table 6.2 details minimum luminance and sunlight readability measurements from one manufacturer of illuminated pushbutton switches. The special requirements of Night Vision Imaging Systems (NVIS) for military pilots have complicated further the light filtering needs within the pushbutton module. An introduction to NVIS technology is found later in this chapter.

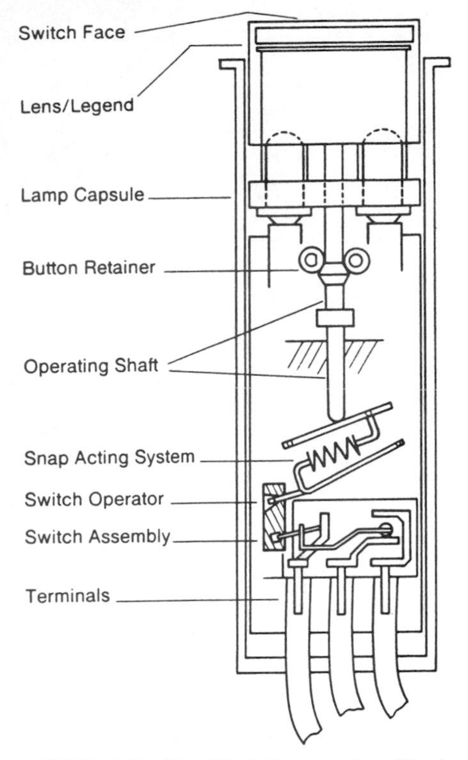

FIGURE 6.18 Simplified diagram of an illuminated pushbutton switch. (*Courtesy of Eaton Corporation, Milwaukee, Wis.*)

Figure 6.21 depicts the configuration of a two-lamp illuminated pushbutton switch. Note again the use of precision snap-acting modules for electrical switching and the actuating mechanism design. Figure 6.22 illustrates typical mechanical dimensions for the device shown in Fig. 6.21. The switch offers two options for termination as shown in Fig. 6.22, solder and PCB terminals. (11)

Another method of providing an interconnect to the switch module is depicted in Fig. 6.23*a* and *b*. A Common Termination System (CTS) is utilized in which a separate connector module is prewired (Fig. 6.23*a*) using standard snap-in contact sockets. The connector module then fastens to the rear of the switch module (Fig. 6.23*b*) to complete the assembly. (12)

Using the foregoing as a very fundamental introduction to illuminated

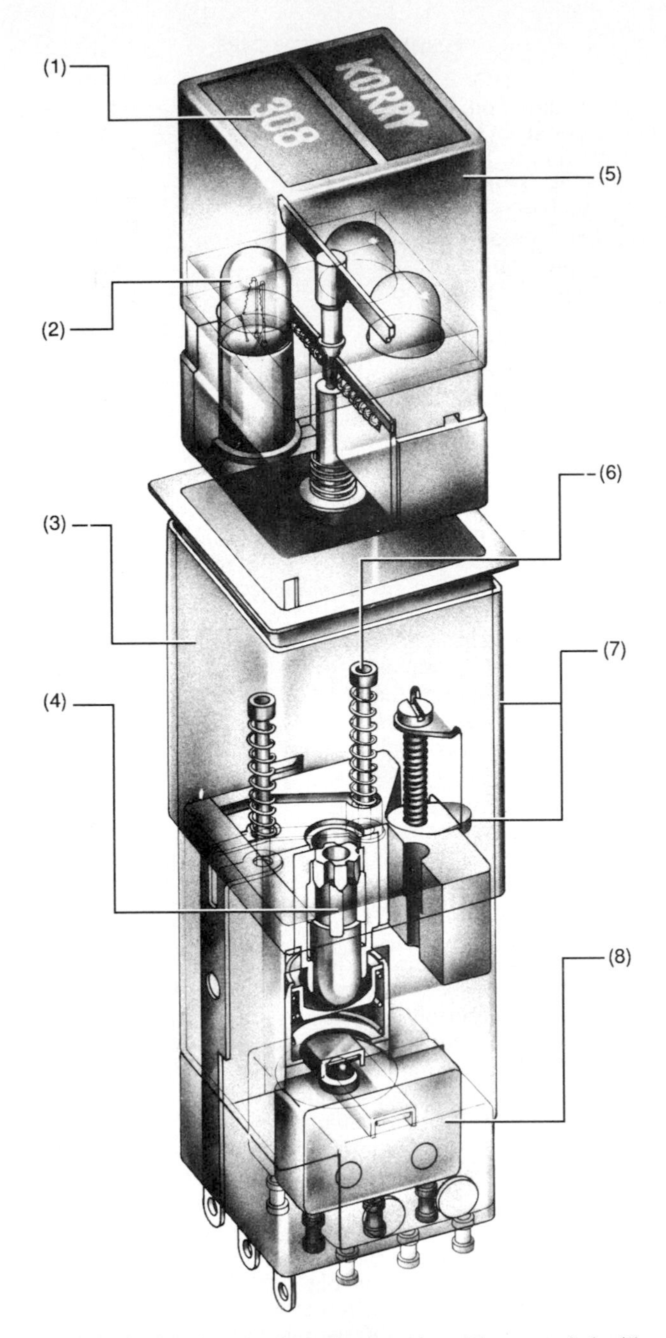

FIGURE 6.19 Details of an illuminated pushbutton switch: (1) legend design provides sunlight readability, (2) lamp holder accepts incandescent or LED lamps, (3) rugged housing encloses switch actuation mechanism and precision switches, (4) actuator provides momentary or alternate action, (5) lamp holder extracts for ease of lamp replacement, (6) spring-loaded lamp contacts provide secure electrical contact, (7) cam-shaped lugs mount switch against front panel, and (8) one to four precision snap-acting switch modules are mounted to base of switch. (*©Korry Electronics, 1992.*)

FIGURE 6.20 Sunlight readable switches are a necessity in a cockpit. (*Courtesy of Aerospace Optics, Inc., Fort Worth, Tex.*)

TABLE 6.2 Minimum Luminance and Sunlight Readability

Legend type	Color	Brightness, fL* 0.30 MSCP lamp	Brightness, fL* 0.15 MSCP lamp	Sunlight readability contrast†
S	*Yellow*	350	350	0.6
	Red	185	185	0.6
	Green	185	185	0.6
	White	185	185	0.6
	Blue	150	150	0.4
B	*Yellow*	300	300	
	Red	150	150	
	Green	150	150	N/A
	White	150	150	
	Blue	100	100	
W	*Yellow*	300	300	
	Red	120	120	
	Green	110	110	N/A
	White	175	175	
	Blue	90	90	
N	*Yellow*	75	75	
	Red	35	35	
	Green	25	25	N/A
	White	35	35	
	Blue	25	25	
C	*Yellow*	300	300	
	Red	120	120	
	Green	110	110	N/A
	White	175	175	
	Blue	90	90	

Source: *© Korry Electronics, 1992.*

*Minimum average brightness of each legend character at rated voltage.

†Contrast ratio = legend-background ÷ background in 10,000 fc per MIL-S-22885D measured at 30° included angle.

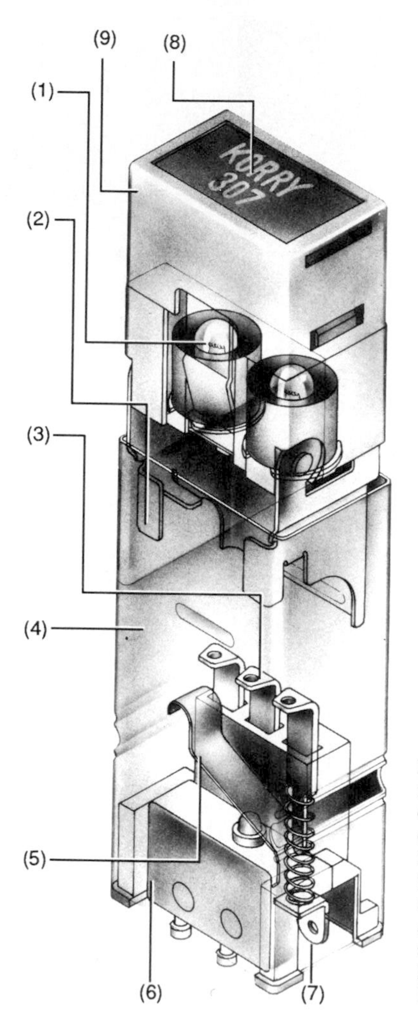

FIGURE 6.21 Design details of a two-lamp illuminated pushbutton switch: (1) adaptable to either T-1 or T-1¾ incandescent lamps, (2) cap retainer holds cap during relamping, (3) spring-loaded lamp contacts maintain contact pressure throughout cap travel, (4) rugged welded housing, (5) leaf spring actuator provides overtravel, (6) precision snap-acting switch module, (7) sealed lamp terminals, (8) sunlight readable legend with NVIS-compatible lighting available, and (9) molded thermoplastic cap houses legend. (©*Korry Electronics, 1992.*)

pushbutton switches, we now also provide an equally fundamental introduction to NVIS-compatible pushbutton switches. NVIS-compatibility is becoming increasingly important in military aircraft, and illuminated pushbutton switch manufacturers have been leaders in defining and refining the complex optical filtering requirements to meet NVIS specifications.

INTRODUCTION TO NIGHT VISION IMAGING SYSTEMS (NVIS)

Military interservice use of NVIS has increased significantly over the last few years. These goggles (Fig. 6.24) enhance and multiply the effects of light at night, allowing the crew to see ground targets with amazing clarity. (13)

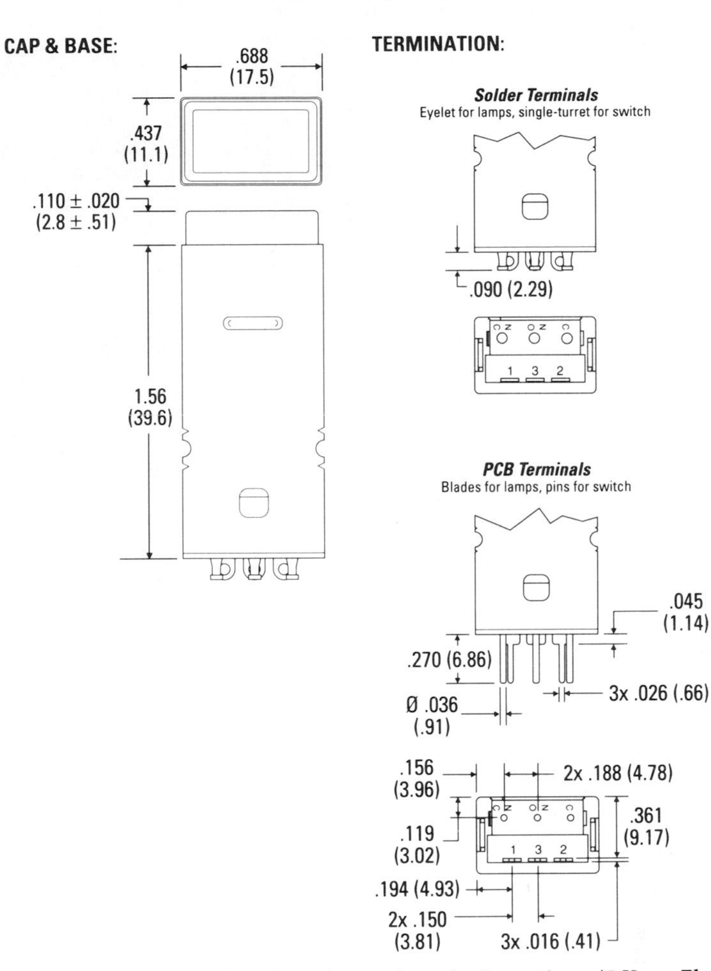

FIGURE 6.22 Mechanical dimensions and termination options. (©*Korry Electronics, 1992.*)

Principles of Night Vision Imaging. Figure 6.25 illustrates the electromagnetic spectrum, one portion of which is the visible spectrum, encompassing wavelengths between 400 and 700 nanometers (nm). Our eyes respond only to wavelengths in the 400- to 700-nm region, which we normally experience as colors. The less light, however, the less we distinguish color and detail. On a dark night, we lose color perception entirely, and objects become shadowy, somewhat brighter or dimmer shapes.

There are two ways of improving night vision. The first is to increase the amount of light reaching the eye, as with a telescope or flashlight. The second is through imaging technology, by creating a visible, phosphor-screen image from normally imperceptible radiation, as in a night vision imaging system (NVIS) or forward-looking infrared (FLIR) camera.

An NVIS usually looks like a pair of goggles or binoculars, is worn by the viewer, and responds to visible and near-infrared wavelengths up to 930 nm.

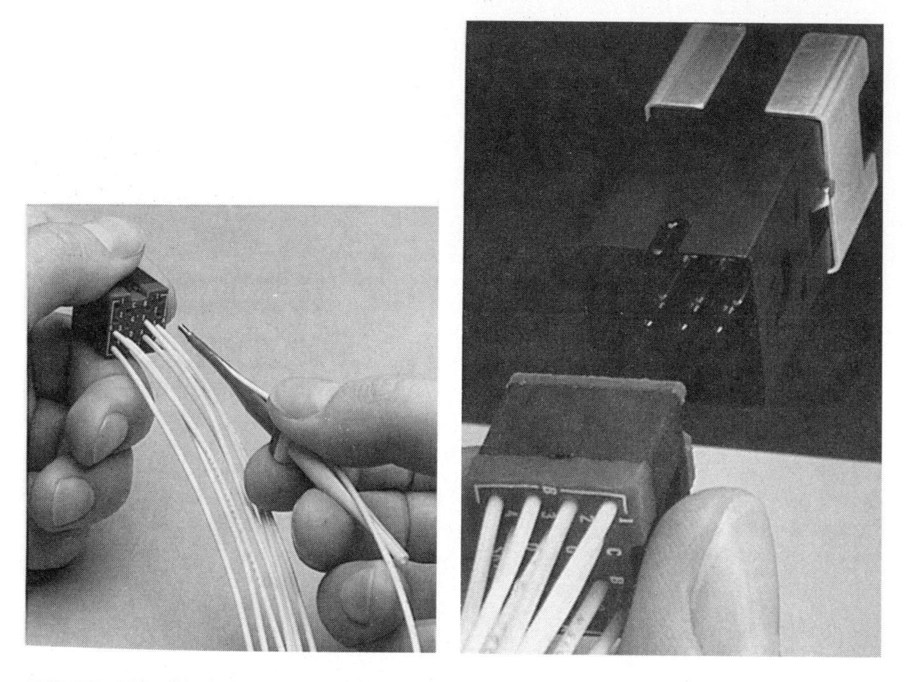

FIGURE 6.23 (*a*) Connector module is prewired, and (*b*) connector module is attached to pushbutton switch. (*Courtesy of Aerospace Optics, Inc., Fort Worth, Tex.*)

FLIR is built into aircraft, with special sensors below the nose cone and a display screen in the crew station. It responds to wavelengths in the 8000- to 16,000-nm range, forming a visual representation of heat. Compared with NVIS, FLIR has two disadvantages: it doesn't track with the pilot's head movements, and it is not real-time imaging.

Of the two types of imaging technologies, NVIS and FLIR, NVIS is directly applicable to illuminated pushbutton switches. Therefore, all discussion regarding imaging technology in this section is limited to NVIS and NVIS-compatible lighting.

Besides perceiving only one part of the spectrum, our eyes respond to some wavelengths more intensely than to others. For example, we see green light as much brighter than the same amount of red light. In daylight, our visual response describes a bell curve—known as the *photopic* or standard observer curve—that peaks in the green area, at 555 nm (see Fig. 6.26). When our eyes adapt to the dark, their response curve shifts toward blue and is known as the *scotopic* curve (also illustrated in Fig. 6.26).

Just like the eye, every type of NVIS has a response curve describing its sensitivity to different wavelengths. Early forms of NVIS (Gen 2 and earlier) were more sensitive to visible light than to infrared (IR). More recently, Gen 3 was developed with greater response to IR, in both relative and absolute terms.

The response, or sensitivity, of NVIS to IR makes it far more useful for night vision than a device that only amplifies visible light. At night there is much more available IR than visible light (Fig. 6.27). Also, many materials—foliage, concrete, or

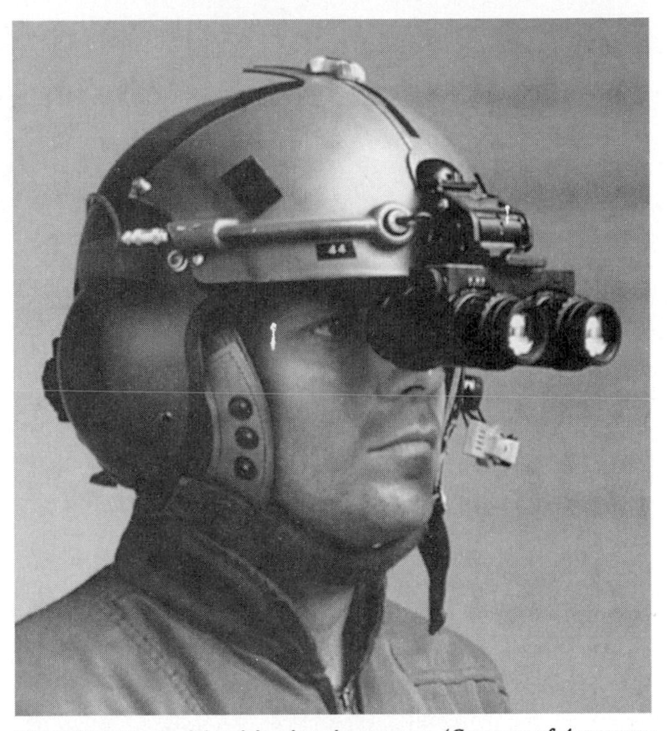

FIGURE 6.24 A night vision imaging system. (*Courtesy of Aerospace Optics, Inc., Fort Worth, Tex.*)

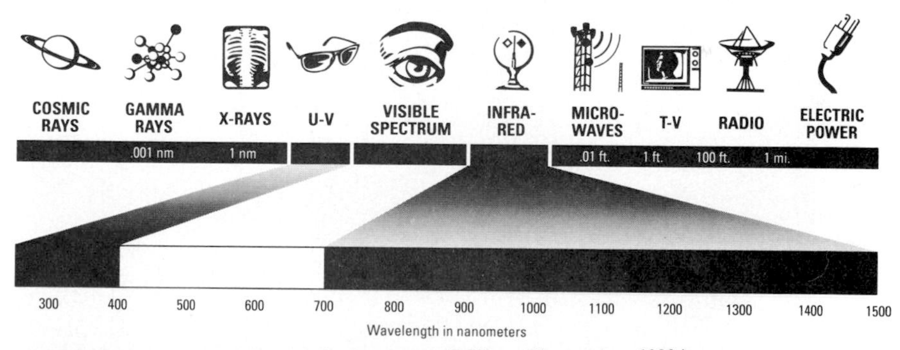

FIGURE 6.25 The electromagnetic spectrum. (©*Korry Electronics, 1992.*)

stone, for example—reflect IR more efficiently than they do visible light. Foliage, in particular, is such a high reflector of red and IR that if human visual sensitivity to red were higher, trees would appear red, not green (Fig. 6.28).

The combination of NVIS IR response, night sky irradiance, and foliage reflectivity is very important for low-flying aircraft, especially helicopters, because it gives the pilot a very sharp image of nighttime terrain.

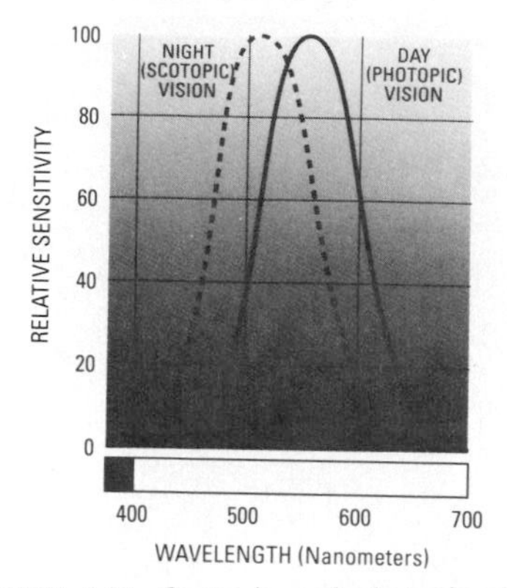

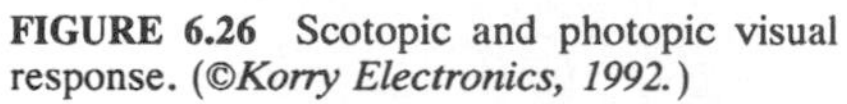

FIGURE 6.26 Scotopic and photopic visual response. (©*Korry Electronics, 1992.*)

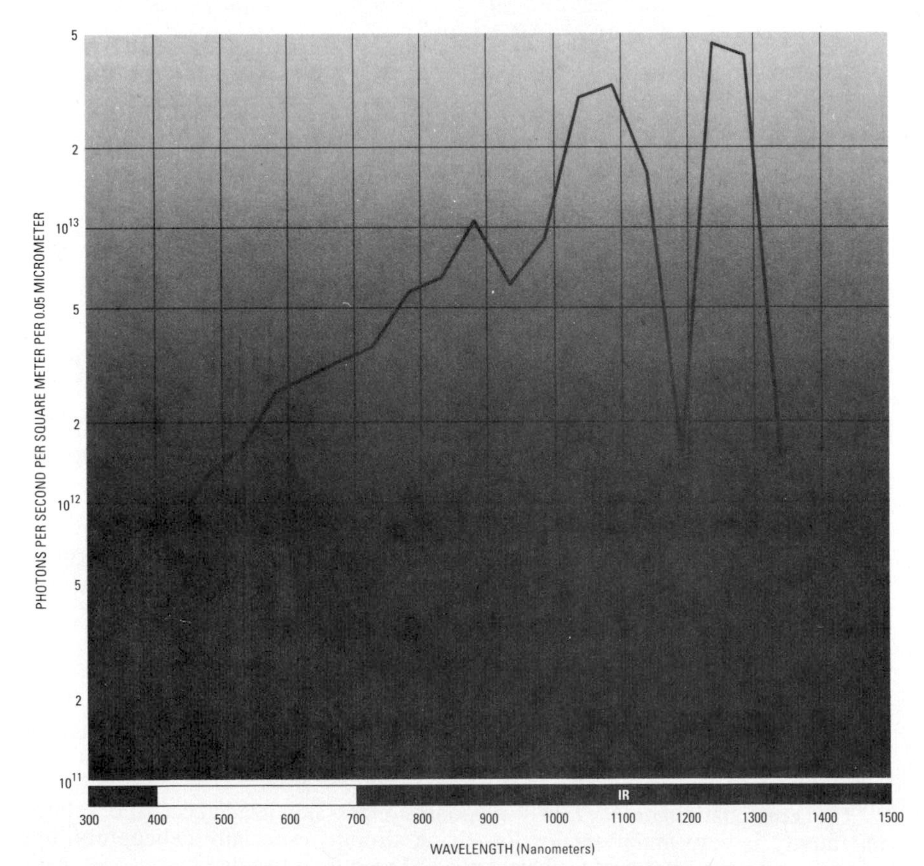

FIGURE 6.27 Natural night-sky spectral irradiance. (©*Korry Electronics, 1992.*)

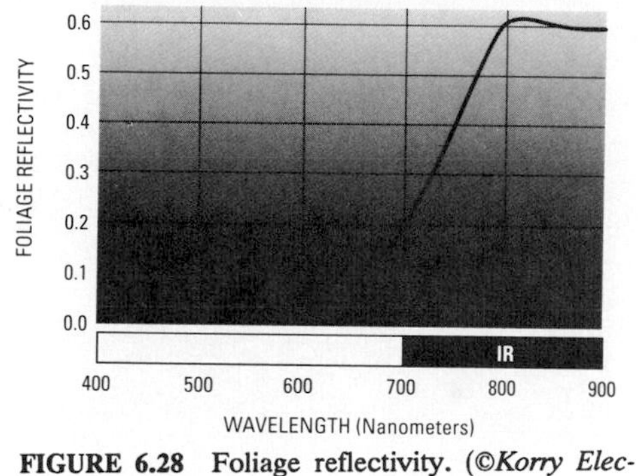

FIGURE 6.28 Foliage reflectivity. (©*Korry Electronics, 1992.*)

NVIS Technology. NVIS goggles consist of sometimes one, usually two, image-intensifier tubes and look much like binoculars mounted to a head strap or helmet (Fig. 6.24). An NVIS intensifier tube (Fig. 6.29) is a vacuum-tube device similar to a miniature video camera and screen packaged together. The stages in the intensifier tube are:

- A photocathode receptor, which converts visible and IR light energy into electrons
- A microchannel plate, which multiplies the number of electrons emitted by the photocathode
- A green phosphor screen, which converts the electrons into a visible range

All three elements are packaged close to each other to minimize blurring of the converted visible image. An electron accelerating voltage is applied between each of the three stages to produce amplification and maintain image quality.

The two main types of intensifier tube in use are classified as Gen 2, which uses a *multialkali* photocathode, and Gen 3, which uses the far more sensitive *gallium arsenide* photocathodes developed in the 1970s. The improvement is obvious by comparing typical Gen 2 and Gen 3 response curves (Fig. 6.30). Gen 3

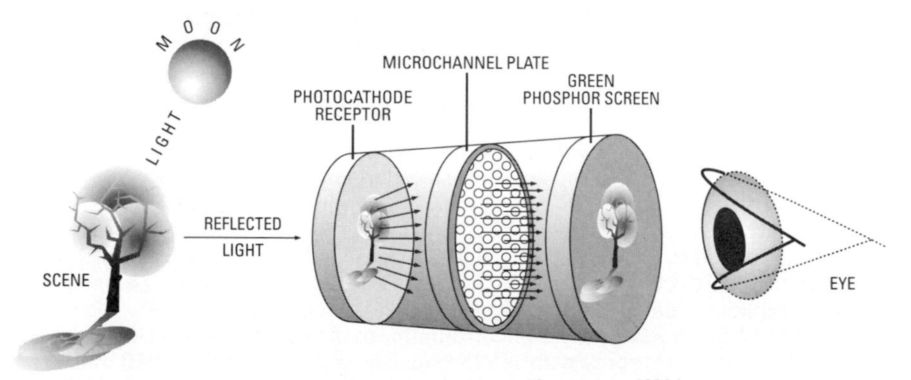

FIGURE 6.29 Diagram of an intensifier tube. (©*Korry Electronics, 1992.*)

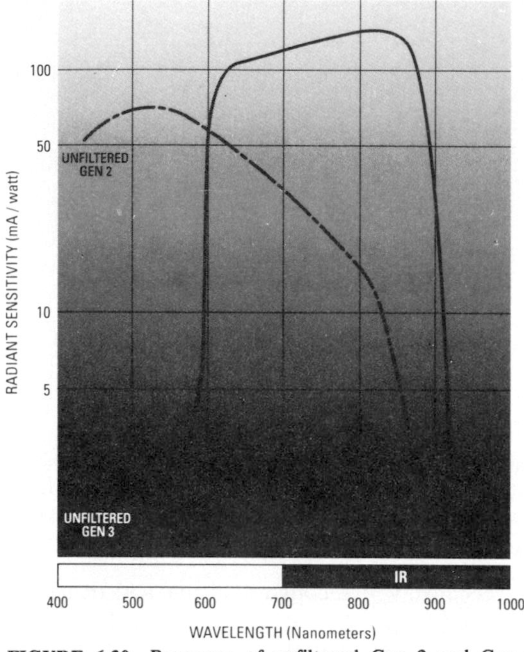

FIGURE 6.30 Response of unfiltered Gen 2 and Gen 3 NVIS. (*©Korry Electronics, 1992.*)

tubes cost approximately three times more than Gen 2, but their useful life is also three times longer.

Even at low light levels, a Gen 3 tube produces a very crisp image on the phosphor screen with little loss of resolution.

NVIS-Compatible Lighting. A pilot flying with NVIS must also be able to read the illuminated displays in the crew station with unaided normal vision. Likewise, an infantryman must still be able to read and operate communication or control equipment. Displays must be visible to the unaided eye while emitting very low levels of IR. If a display emits small amounts of radiant energy within the response range of the IR-sensitive photocathode, it may degrade the performance of the NVIS.

NVIS-compatible illuminated displays do not interfere with NVIS and remain readable to the unaided eye. NVIS compatibility also means that enemy NVIS find it difficult or impossible to detect the light source at a distance.

The success of an NVIS application, particularly in aviation, depends as much on the performance of the NVIS-compatible displays as on the NVIS itself. Switch manufacturers involved in NVIS technology must extend extra care and attention to detail regarding filter design, legends, light sources, and mechanical packaging in order to attain NVIS compatibility. The following discussion highlights just one of the many complex and interlocking intricacies of NVIS technology.

A normal incandescent, EL (electroluminescent), or LED (light-emitting diode) light source interferes with NVIS because it emits significant IR radiation within the response range of NVIS. This excess IR causes interference in several ways:

1. IR reflections from the crew station canopy produce ghost images in the NVIS.
2. IR can bloom the goggles, creating a halolike glow around the displays and reducing visibility.
3. High levels of IR activate the automatic gain control (AGC) and further degrade sensitivity and resolution.

The AGC feedback circuitry optimizes the gain or level of NVIS intensification. A sudden increase in radiant energy within the NVIS response increases the electron current flow in the intensifier tube. The AGC detects this increase and compensates by reducing the voltage acceleration, lowering intensification so that the phosphor image dims and loses contrast. This protects the NVIS user from being blinded and disoriented by the flash of an explosion or flare. As soon as the radiant energy decreases, intensification goes back up. AGC also increases intensification when ambient irradiation is low, as on a moonless night.

An unfiltered incandescent bulb being turned on nearby has the same visual effect through the NVIS as a camera flashbulb to the unaided eye, followed by loss of image definition and poor visibility. The goggles actually shut down after the initial burst.

An improperly designed or manufactured NVIS-compatible display may still activate the AGC and reduce sensitivity and target resolution, or cause persistent ghosting or blooming.

NVIS for Aviation. Pilots first experimented with NVIS by wearing Gen 2 AN/PVS-5 goggles in helicopters (Fig. 6.31). Intended for low-speed ground applications, the sensitivity and resolution of these goggles proved inadequate for the high speed of aircraft, which leaves no room for hesitation or error, especially at low altitude. Another disadvantage was the AN/PVS-5's face mask, which restricts peripheral vision. Interference from crew station lighting also hampered visibility, and pilots at first turned off or taped over any nonessential displays. The problems of Gen 2 NVIS were resolved in three ways:

1. Switching to Gen 3 NVIS for sharper, more detailed imaging
2. Adding *minus-blue filters* to reduce NVIS response to visible light from crew station displays
3. Redesigning crew station lighting to make it compatible with the response of the filtered Gen 3 NVIS

Redesigning crew station lighting required definition of NVIS compatibility and verification of methods. MIL-L-85762, a triservice specification, filled this need by setting standards for aircraft interior lighting compatible with the Gen 3 Aviator's Night Vision Imaging System (ANVIS), also known as AN/AVS-6 (Fig. 6.32).

Designed for helicopter pilots, ANVIS filtering transmits red, to take full advantage of radiation reflected from foliage and optimize visibility at treetop level. At the same time, the ANVIS minus-blue filter cuts out most of the low night-sky radiance in the 500- to 600-nm band (Fig. 6.33), which does not impair visibility through ANVIS, because foliage has relatively low reflectivity in this band.

FIGURE 6.31 Gen 2 AN/PVS-5 binocular. (©*Korry Electronics, 1992.*)

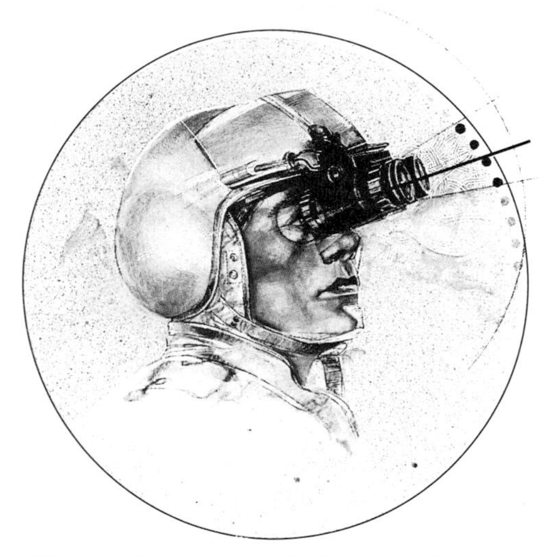

FIGURE 6.32 The AN/AVS Aviator's Night Vision Imaging System (ANVIS) Gen 3 type 1, class A NVIS. (©*Korry Electronics, 1992.*)

These goggles provide excellent peripheral vision and allow the pilot to glance below the eyepieces to view instrumentation but exclude the use of red displays in the crew station.

Revision A of MIL-L-85762 expanded the specification to address new Gen 3 NVIS refinements for fixed-wing aircraft (aircraft other than helicopters). The

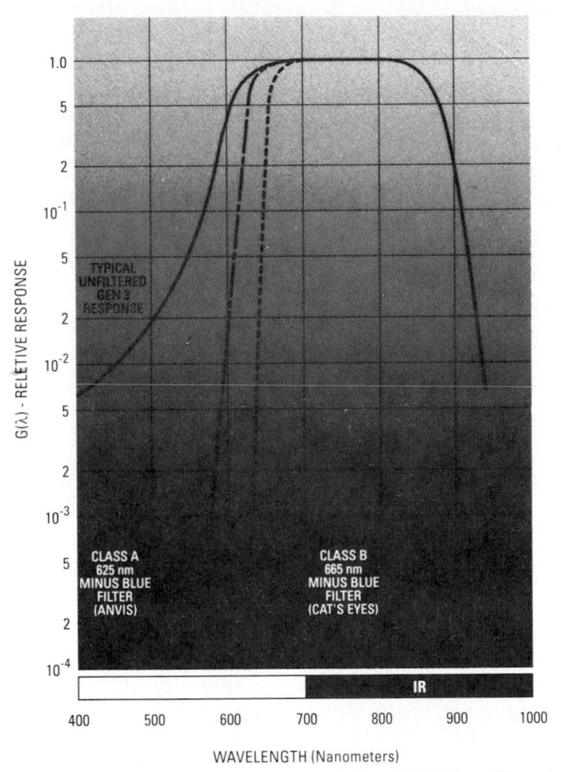

FIGURE 6.33 Spectral response of Gen 3 NVIS unfiltered and with class A and class B minus-blue filters. (©*Korry Electronics, 1992.*)

specification also formally introduced the general term "NVIS" and replaced the term "ANVIS radiance" (AR) with "NVIS radiance" (NR).

MIL-L-85762A classifies Gen 3 NVIS-compatible lighting according to NVIS type and class:

Type I Direct View Image. The view through the goggle eyepiece includes only the NVIS phosphor-screen image. The pilot must glance down to see the instruments. This is primarily used in helicopters. (14)

Type II Projected Image. Addition of a combiner lens to the eyepiece allows the pilot to simultaneously view the NVIS phosphor-screen image and crew station instrumentation. This is used primarily in fixed-wing aircraft and is a real advantage in low-flying jets. Combiner lenses also provide improved compatibility with head-up displays (HUD). (14) When a HUD is employed, both the outside scene and the HUD are viewed through the goggles. The HUD display information is collimated to near-infinity so that it is in focus when viewed through the goggles and is projected in a green color. The goggles are able to detect the HUD and image it to the user because the cutoff filter is constructed with a "window" in the visible green region at 545 nm. (15)

Class A. Addition of a 625-nm minus-blue cut-on filter to the objective lens produces the response shown in Fig. 6.33. The 1 percent relative response is at

595 nm, which allows crew station illumination in blue, green (advisory), and yellow (caution, master caution, warning). To avoid interference from orange and red displays, MIL-L-85762A excludes them from class A crew stations. Class A is used primarily in helicopters.

Class B. Addition of a 665-nm minus-blue cut-on filter to the objective lens produces the response shown in Fig. 6.33. Here, the 1 percent relative response is at 625 nm, which excludes most of the red from the NVIS response to allow crew station illumination in blue, green (advisory), yellow (caution), and red (master caution, warning). Class B is used in fixed-wing aircraft.

This scheme defines four categories of NVIS (Fig. 6.34), of which only two are widely used (type I, class A and type II, class B). Type I, class A is the same as the older designation ANVIS.

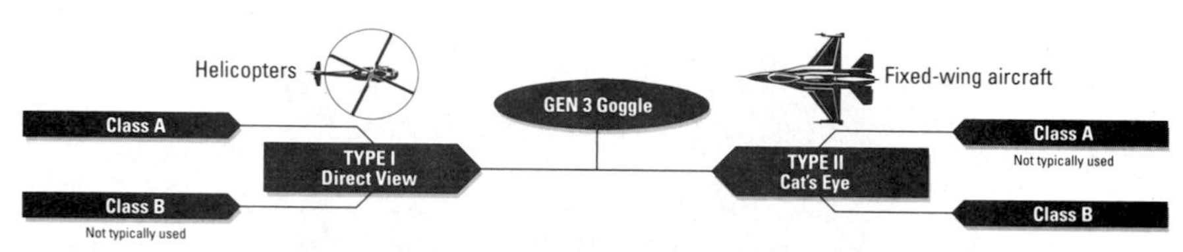

FIGURE 6.34 MIL-A-85762A NVIS classification. (©*Korry Electronics, 1992.*)

NVIS Radiance. To be NVIS-compatible, an illuminated display should emit a minimum of radiation within the response range of the NVIS and should not cause any degradation of the NVIS image. NVIS radiance (NR) is a figure of merit defined by MIL-L-85762A to measure the degree of NVIS interference of a display at a specified luminance. As a figure of merit, NR can be used to compare the relative NVIS compatibility of displays. NR is calculated from the measured spectral radiance of a display. The spectral radiance curve of a typical filtered incandescent green display is shown in Fig. 6.35. The NR calculation can be broken down into two steps:

Step 1. Compute the overlap radiance $\mathcal{R}$. $\mathcal{R}$ is the area under the curve obtained by multiplying the display radiance (Fig. 6.35*a*) and the NVIS response function (Fig. 6.35*b*). The resulting curve is shown in Fig. 6.35*c*.

Step 2. Luminance L is the area under the curve obtained by multiplying the display radiance (Fig. 6.35*a*) and the photopic response (Fig. 6.35*d*). The resultant curve is shown in Fig. 6.35*e*.

Although not separately defined by MIL-L-85762A, the overlap radiance $\mathcal{R}$ is a useful quantity for simplifying the calculation of NR:

$$\text{NR} = \frac{\mathcal{R}(0.1)}{L} \qquad \text{NR} = \frac{\mathcal{R}(15)}{L}$$

where the normalizing luminance level is the constant 0.1 for green and 15 for yellow and red, as specified in MIL-L-85762A. MIL-L-85762A sets minimum and maximum limits for NR. Because yellow (caution) and red (warning) displays must attract pilot attention in both daylight and NVIS missions, MIL-L-85762A sets minimum NR levels for yellow and red high enough to attract the pilot's immediate attention. Because there are two NVIS response curves, class A and B, there are two corresponding classes of NR: NR_A and NR_B.

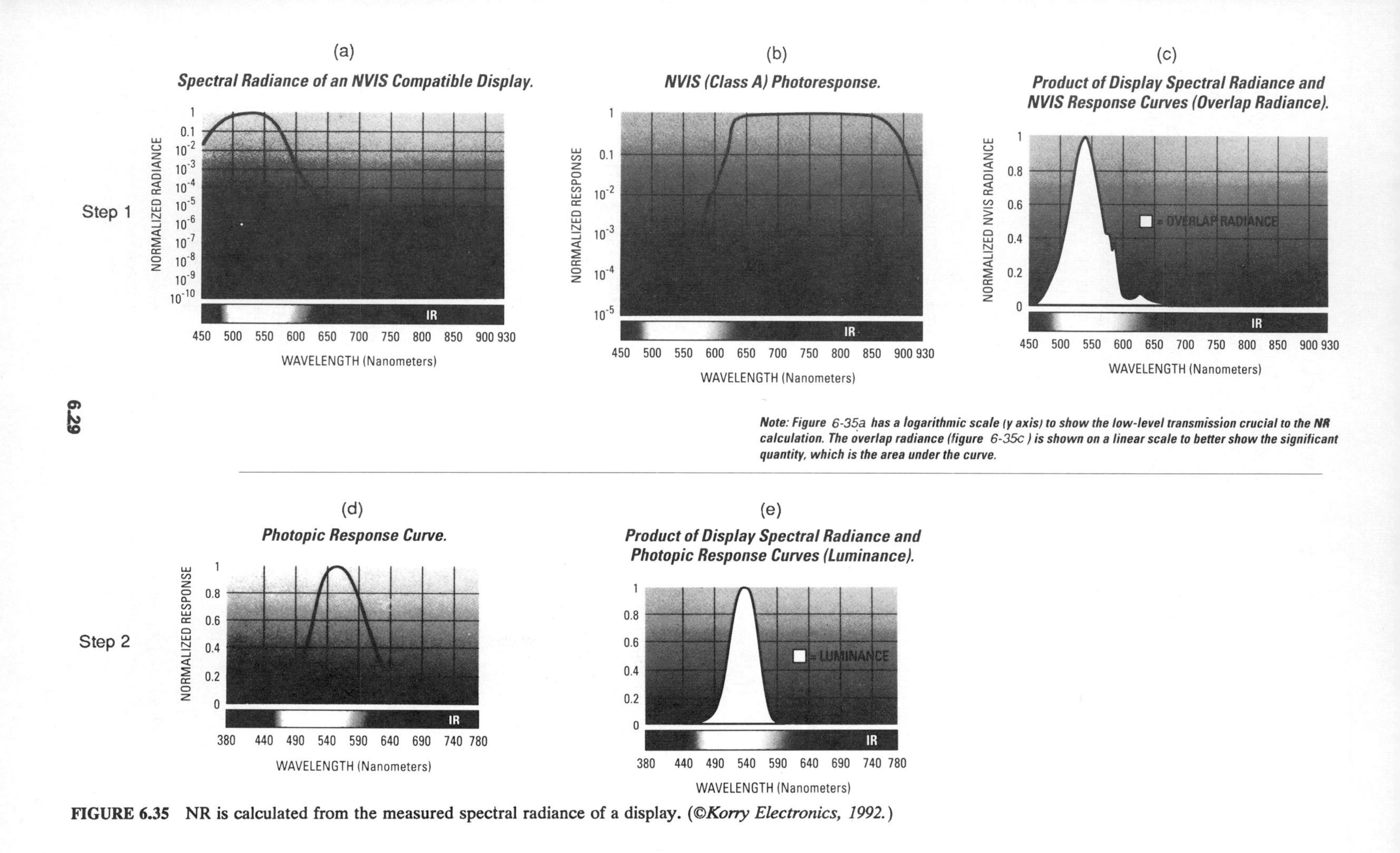

FIGURE 6.35 NR is calculated from the measured spectral radiance of a display. (©*Korry Electronics, 1992.*)

When a pushbutton switch manufacturer involved in NVIS technology produces an NVIS-compatible switch, switch panel, or keyboard, the filter assembly is a critical part of the product, because more than any other single component it ensures NVIS-compatible illumination. Korry Electronics Co., Seattle, Wash., publishes technical documentation attesting to proprietary NVIS filters designed to meet the NR limits of MIL-L-85762A by safety factors of 1.5 to more than 5, depending on the color and light source. (14)

In addition, a conscientious NVIS pushbutton switch manufacturer will always pay particular attention to the mechanical packaging of every NVIS-compatible switch specifically to prevent unfiltered light leaks. The best filter in the world will not meet NR specifications if the mechanical seal permits any light leaks.

Design Issues. There are three critical design objectives NVIS switch manufacturers must use to satisfy MIL-L-85762A:

1. Minimize NR within specification and prevent leaks of unfiltered light.
2. Meet color specifications.
3. Maximize visible light transmission for the best possible luminance, contrast, and daylight readability.

These design objectives are interrelated. Filter and legend design (Fig. 6.36) requires accurate testing of materials, complex analysis of trade-offs, as well as strict control of variability in production. The application (user) engineer should make careful evaluation of a potential NVIS switch supplier by fully understanding how the supplier intends to meet the design objectives.

Color. MIL-L-85762A defines UCS coordinates for four colors as shown in Fig. 6.37. NVIS colors are not standard aviation colors but are particular to NVIS because of the limits on NR. Class A red is not allowed because red falls within the class A NVIS response and would cause unacceptable NR. Table 6.3 shows typical UCS color coordinates from one manufacturer.

Green A and B for both class A and B NVIS are the simplest to design because the green wavelengths are far enough from the Class A and B NVIS response curves that they do not incur an NR penalty (Fig. 6.38). At the same time, they are in the band of maximum photopic response, so it takes relatively less energy to produce high luminance. There are few conflicts between meeting color, luminance, and NR specifications. In general, the same situation holds for class B yellow.

Class A yellow, however, presents a dilemma. True yellow color is a broadband emission extending from the green to the red. As a result, the desired color band is right against the steeply rising class A NVIS response curve (Fig. 6.38). Any energy in the redder wavelengths carries an exponentially increasing NR penalty. The only solution compliant with MIL-L-85762A is to limit emissions to the green side of the yellow band.

The same design and production dilemma holds for class B red, with the added difficulty that the photopic response to red is lower than to green, as shown in Fig. 6.39. This means approximately 50 percent more light output is required in the red band to meet luminance and contrast requirements. Again, the only acceptable solution per MIL-L-85762A is to hold the red to the orange side of the band.

Luminance and Contrast. MIL-L-85762A specifies luminance and contrast values so that all illuminated components will be sunlight readable (at full voltage in 10,000 fc of ambient light) and usable in daytime as well as at night. For discrete

FIGURE 6.36 Filter and legend design for NVIS compatibility requires complex interaction of a number of engineering disciplines. (©*Korry Electronics, 1992.*)

switches, MIL-L-85762A adopts the contrast sections of MIL-S-22885, but relaxes the on/background and on/off contrast values from 0.6 to 0.4 minimum. MIL-L-85762A recognizes it must sacrifice some of the available luminance, just as it sacrifices color, to maintain its NR requirements.

Illumination Sources and NVIS Compatibility. Table 6.4 summarizes the NVIS compatibility of various illumination sources, in terms of which colors can meet MIL-L-85762A requirements for NR.

NVIS for Ground Operations. The U.S. Army uses both Gen 2 and Gen 3 NVIS as goggles and weapon sights for ground operations. Because the peak sensitivity of Gen 2 goggles is in the visible green band, with relatively low sensitivity in the

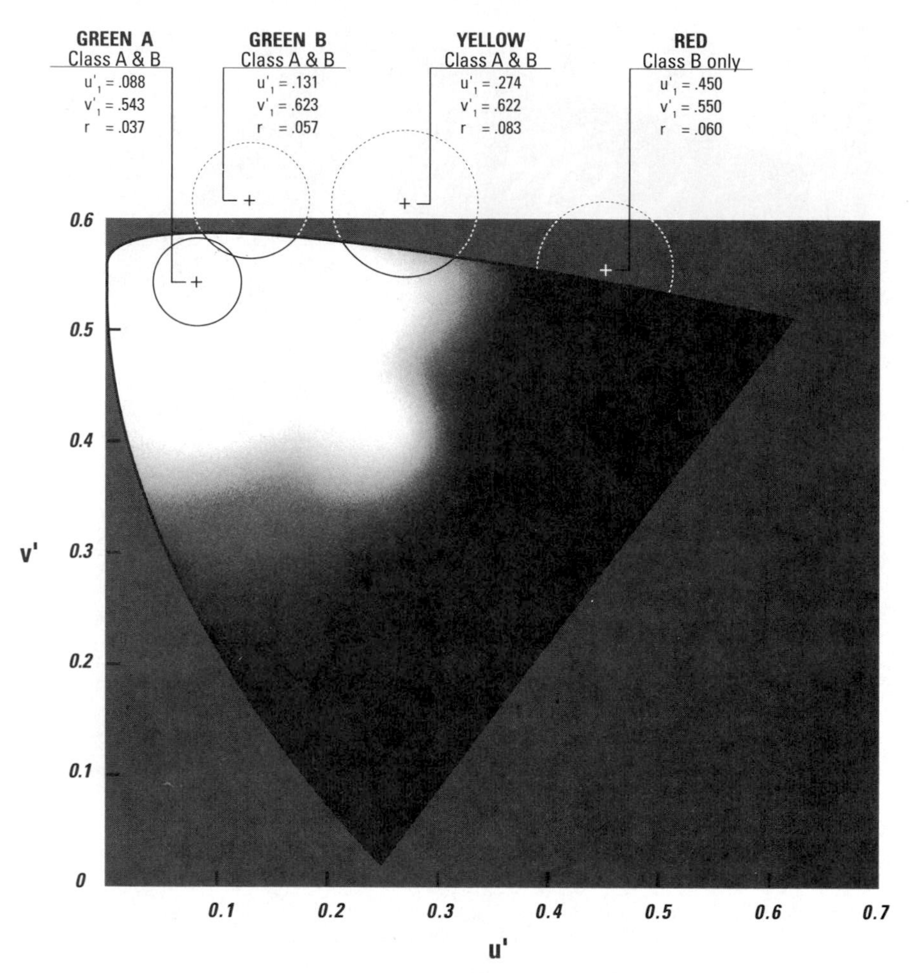

FIGURE 6.37 CIE 1976 UCS diagram. (©*Korry Electronics, 1992.*)

near-infrared (Fig. 6.30), they depend on an abundance of moonlight for best visibility. The best example of this NVIS type is the U.S. Army's AN/PVS-5 (Fig. 6.31), the most common goggles for ground operations. The U.S. Army also uses unfiltered Gen 3 NVIS, better suited to moonless conditions, when higher infrared (IR) sensitivity is preferred. The unfiltered Gen 3 response range still includes visible light down to 450 nm (Fig. 6.30). NVIS for ground operations remains sensitive to low levels of visible light in order to maximize sensitivity to any available nighttime radiation. NVIS for aviation takes a different approach and filters out most of the visible light to help avoid interference from crew station lighting.

Secure Lighting. Secure lighting is a form of NVIS-compatible illumination for ground-based applications. It is defined by the U.S. Army CECOM (Communications—Electronics Command) at Fort Monmouth, N.J., in a Secure Lighting Statement of Work. Secure lighting involves the modification of light sources to meet two goals:

TABLE 6.3 Comparison of Class A and Class B Crew Station Display Lighting

Korry type I, class A NVIS crew station displays—rotary-wing aircraft				
	1976 UCS color coordinates			
Color	u'	v'	Radius	NR_A*
Green A	0.088	0.543	0.037	$<1.7 \times 10^{-10}$ at 0.1 fL
Green B	0.131	0.623	0.057	$<1.7 \times 10^{-10}$ at 0.1 fL
Yellow	0.274	0.622	0.083	$<1.5 \times 10^{-7}$ and $>5.0 \times 10^{-8}$ at 15 fL

Korry type II, class B NVIS crew station displays—fixed-wing aircraft				
	1976 UCS color coordinates			
Color	u'	v'	Radius	NR_B*
Green A	0.088	0.543	0.037	$<1.7 \times 10^{-10}$ at 0.1 fL
Green B	0.131	0.623	0.057	$<1.7 \times 10^{-10}$ at 0.1 fL
Yellow	0.274	0.622	0.083	$<1.5 \times 10^{-7}$ at 15 fL
Red	0.450	0.550	0.060	$<1.4 \times 10^{-7}$ at 15 fL

Source: ©Korry Electronics, 1992.
*Excludes jump lights.

1. To minimize detectability by enemy NVIS while preserving enough luminance to be usable
2. To maximize compatibility with Gen 2 NVIS

The Statement of Work sets three design priorities to meet these goals:

Priority 1, Wavelength Restrictions. Total energy above 700 nm is to be no more than 0.5 percent of the total energy emitted between 350 and 930 nm. The 0.5 percent cutoff is to be between 600 and 700 nm, and as close to 600 nm as possible. This minimizes red and near-infrared light emission but retains wavelengths required to maintain visibility to the unaided eye.

Priority 2, Luminance Restrictions. Light sources are to be dimmed down to 0.05 fL or less at night, just enough to be visible at close range. They must still provide full brightness in daylight.

Priority 3, Viewing Angle Restrictions. The viewing angle is to be as small as possible, consistent with the application to curtail enemy observation. A viewing angle of ±10° is desirable whenever possible. These design priorities are not specifications, and allow flexibility in designing and retrofitting the large quantity and variety of Army equipment. The Army Statement of Work was never coordinated with MIL-L-85762A and does not use NVIS radiance (NR) to quantify performance. Similarly, it does not specify color coordinates for blue, green, or yellow.

Spectroradiometry. To calculate NR for a display, MIL-L-85762A requires radiance measurement from 450 to 930 nm with a spectroradiometer. A spectroradiometer splits the light emitted by the tested source into its constituent wavelengths, much as a prism spreads daylight into a rainbow of color. In turn, it focuses each narrow band of wavelengths onto a high-sensitivity photodetector, which converts radiant energy into an electrical signal. The spectroradiometer's computer stores these measurement signals and calibration data in memory, calculates absolute radiance, then derives NR, luminance, and color coordinate val-

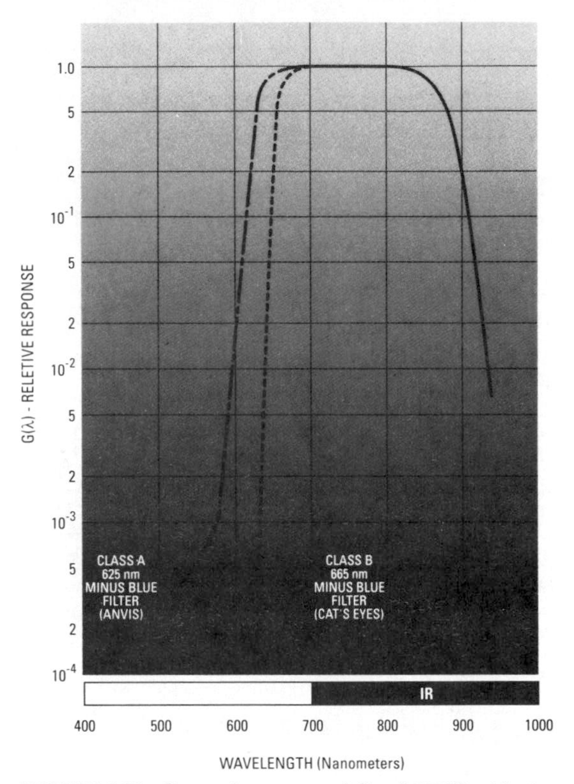

FIGURE 6.38 Spectral response of Gen 3 NVIS with class A and class B minus-blue filters per MIL-A-85762A. (©*Korry Electronics, 1992.*)

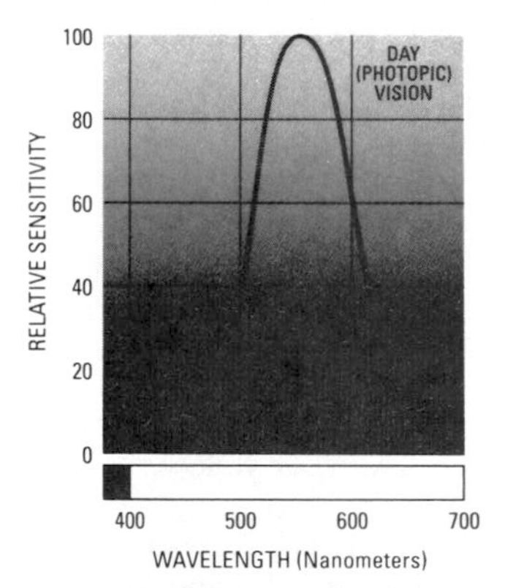

FIGURE 6.39 Human visual response. (©*Korry Electronics, 1992.*)

ues. (14) Compliance of illuminated pushbutton switches to MIL-L-85762A requirements is certified by spectroradiometry.

MIL-L-85762A's requirement for a scanning spectroradiometer to verify radiance for first-article and production inspection reflects the state of light-measurement technology when the specification was developed. A scanning spectroradiometer is a precision laboratory instrument well suited for qualification and first-article testing but not adapted to the part quantities and time constraints of production inspection.

NVIS pushbutton switch suppliers such as Korry Electronics Co. (Seattle, Wash.), Aerospace Optics, Inc. (Fort Worth, Tex.), and Eaton Corp. (Costa Mesa, Calif.) have in-house facilities and test equipment to certify compliance to

TABLE 6.4 Illumination Sources and NVIS Compatibility

Illumination source	NVIS color
Incandescent	Blue,* white,* green A and B, yellow, red
LED backlight	Close to green B
LED programmable display	Close to green B
EL	Blue,* green A and B

Source: ©Korry Electronics, 1992.
*Not defined by MIL-L-85762A. All other colors per MIL-L-85765A.

MIL-L-85762A. The first-time NVIS user should *not* try to venture into the NVIS arena without the assistance and advice of switch manufacturers who have significant, relevant experience in this highly complex and critical field. (14)

NVIS Worldwide. Foreign governments and aircraft and avionics manufacturers are familiar with MIL-L-85762A. Many are developing NVIS with different response curves than MIL-L-85762A's class A or B, and often borrow its system of NVIS classification, expanding it with class C, D, etc.

International manufacturers often assign different priorities to color, luminance, contrast, and NVIS radiance (NR). For example, they may accept higher NR values in exchange for increased luminance and improved sunlight readability.

The British use a convention of quantifying NVIS compatibility for radiance with "NVG gain," which is the ratio of red and infrared to green energy in the display. This has been defined in a provisional specification (Working Paper 6, 1986) by the Royal Aerospace Establishment. This paper describes a 13(λ) response curve for helicopters and a 13W(λ) curve for fast jets.

The 13W(λ) curve specifies a response window in the green region for viewing head-up display projections. A new British Defense Standard 00970 is in progress which will formally document the 13(λ), 13W(λ), and MIL-L-85762A specifications to allow crew station designers a choice of conventions. (14)

GLOSSARY

General

Actuator. The part of the switch to which an external force is applied to operate the switch.

Alternate action. A switch in which the operable position is maintained after the first actuation, and then disengaged with the second operation.

Momentary action. A switch in which the actuator returns from its operating position to its normal or free position when the actuating force is removed.

Normally closed (NC). Switch in which the circuit is closed without actuation (with actuator in the "normal" position).

Normally open (NO). Switch in which the circuit is open without actuation (with actuator in the "normal" position).

NVIS-Specific Terms

AGC (automatic gain control). AGC feedback circuitry optimizes the gain, or level of NVIS intensification. A sudden increase in radiance reduces or shuts down intensification so that the phosphor image dims and loses contrast. This protects the NVIS user from being blinded and disoriented by the flash of an explosion or flare. As soon as the radiance goes down, normal intensification resumes. AGC also increases intensification when ambient irradiation is low.

AN/AVS-6 goggles. See *ANVIS*.

AN/PVS-5 goggles. A Gen 2 NVIS for ground operations.

AN/PVS-7 goggles ("Cyclops"). A Gen 2 or unfiltered Gen 3 NVIS (with a single intensifier tube) for ground applications.

ANVIS (Aviator Night Vision Imaging System). Known more precisely as AN/AVS-6 goggles, ANVIS is a Gen 3, type I, class A NVIS for use in helicopters and slow fixed-wing aircraft. The original MIL-L-85762 addressed only ANVIS and lighting compatible with it.

ANVIS compatible. Describes illumination compatible with ANVIS or the AN/AVS-9 goggles, as defined by MIL-L-85762. ANVIS compatibility is equivalent to type I, class A NVIS compatibility, as defined by MIL-L-85762A.

Blooming. Refers to the halolike flaring around a display seen through the NVIS when the display emits a critical amount of radiant energy in the NVIS response range. MIL-L-85762A specifies a minimal amount of blooming in caution, master caution, and warning displays to get the pilot's attention.

Cat's Eyes. The Mark III Cat's Eyes are a Gen 3, type II, class B NVIS developed by GEC in the United Kingdom. They are the standard for high-speed fixed-wing aircraft.

Class A. Refers to a 625-nm minus-blue filter added to a Gen 3 NVIS to reduce its sensitivity to visible light below the red band. Class A NVIS photoresponse is 1 percent at 595 nm.

Class B. Refers to a 665-nm minus-blue filter added to a Gen 3 NVIS to reduce its sensitivity to visible light, including most of the red band. Class B NVIS photoresponse is 1 percent at 625 nm.

Gallium arsenide. The photosensitive material used in the photocathodes of Gen 3 intensifier tubes. It is far more sensitive than the multialkali cathodes in Gen 2 tubes.

Gen 2. NVIS intensifier tube technology that uses a multialkali photocathode and multichannel plate (MCP) intensifier. Gen 2 NVIS is now used only for ground operations.

Gen 3. NVIS intensifier tube technology that uses a gallium arsenide photocathode and multichannel plate (MCP) intensifier. Gen 3 tubes are more sensitive than Gen 2 and are used in NVIS for aviation.

Image intensifier tube. See *Intensifier tube.*

Infrared (IR). Electromagnetic energy outside the visible band with wavelengths between 700 and 1500 nm.

Intensifier tube. A vacuum tube device that converts visible and infrared light to electrons, multiplies the electrons, and then converts them to a visible image on a phosphor screen.

Luminance. A unit of lighting measurement identical to radiance except that it is based on the visual (photopic) brightness of a light source rather than the total brightness. It is expressed in lumens (rather than watts) per steradian per surface unit. A commonly used alternative unit is the footlambert (fL). See *Radiance.*

Microchannel plate (MCP). An MCP is a thin disk of tiny glass tubes or microchannels (about 10 μm in diameter) fused together like a wafer. Inside, the channels are coated with an electron emission film connected to input and output electrodes. An electron striking the input side of a channel causes a cascade of electron emissions, multiplying the electron output. Tubes with an MCP intensifier (Gen 2 and 3) are often called wafer tubes.

MIL-L-85762A. A triservice specification that defines NVIS-compatible illumination for aircraft interiors.

Minus-blue filter. A filter that cuts off any visible light below a certain wavelength. See *Class A* and *Class B.*

Multialkali. The photosensitive coating used on photocathodes for Gen 2 intensifier tubes.

NR. See *NVIS radiance.*

NVG (night vision goggles). An older term for head- or helmet-mounted NVIS. Now superseded by NVIS.

NVIS (Night Vision Imaging System). A general term to describe night imaging systems containing intensifier tubes sensitive to the visible and near-infrared bands.

NVIS compatible. A display is NVIS compatible if it does not activate the NVIS AGC circuit and interfere with the visibility of the NVIS image.

NVIS radiance (NR). A nondimensioned measure of the radiance detected by an NVIS.

Photocathode receptor. A device which converts visible and IR light energy into electrons.

Photopic. Refers to the normal response of the human eye in daylight. The photopic or standard observer curve is a bell-shaped curve that describes this response.

Radiance. The radiant flux emitted by a light source per unit solid angle and per unit projected area of radiating surface (expressed in watts per steradian per surface unit).

Response. The sensitivity of a light detector to specific wavelengths. The photopic response curve describes the relative sensitivity of the human eye over the visible spectrum. The NVIS response curve describes the sensitivity of that NVIS (expressed in milliamperes per watt) over the visible and IR spectrum.

Scotopic. Related to night vision. The scotopic curve describes the human eye's response to light once it has adapted to the dark.

Secure lighting. A form of NVIS-compatible lighting for ground applications defined by the U.S. Army CECOM Secure Lighting Statement of Work.

Spectral radiance. Radiance emitted by a source over a band of the spectrum. It is measured with a spectroradiometer and described by a curve. See *Radiance.*

Spectroradiometer. A device for measuring the spectral radiance of a display. To calculate NVIS radiance, MIL-L-85762A requires a scan from 450 to 930 nm, the response range of unfiltered Gen 3 NVIS.

Type I. Refers to NVIS with a direct-view image: the view through the goggle eyepiece includes only the NVIS phosphor-screen image. The pilot must glance down to see the instruments.

Type II. Refers to NVIS with a projected image: the addition of a combiner lens to the goggle eyepiece allows the pilot to simultaneously view the NVIS phosphor-screen image and crew station instrumentation.

REFERENCES

1. "Pushbutton Switch Terminology," Grayhill, Inc., La Grange, Ill., 1989.
2. "Butt Action and Snap Action Pushbutton Switches," Grayhill, Inc., La Grange, Ill., 1981.
3. "Wiping Action Pushbutton Switches," Grayhill, Inc., La Grange, Ill., 1987.
4. "The Switch in an Electrical Circuit," Grayhill, Inc., La Grange, Ill., 1991.
5. "Switch Fundamentals," Eaton Corporation, Milwaukee, Wis., 1991.
6. "Catalog No. 9105," C&K Components Inc., Newton, Mass., 1991.

7. "Otto Precision Switches, Catalog No. 104," Otto Controls, Inc., Carpentersville, Ill., 1990.
8. "Engineering Catalog," Grayhill, Inc., La Grange, Ill., 1990.
9. "The Complete Switch Line—Catalog No. 1/91S," Schurter, Petaluma, Calif., 1991.
10. "Chromarray™ Integrated Switch Panels," Korry Electronics Company, Seattle, Wash., 1991.
11. "Chromalux 307 Switchlight," Korry Electronics Company, Seattle, Wash., 1991.
12. "VIVISUN Series 95™ 4 Pole Solderless QUIK-CONNECT™ Options," Aerospace Optics, Inc., Fort Worth, Tex., 1991.
13. "VIVISUN Series 95™ Sunlight Readable Switches, Data Sheet No. 95-1-86-3 Rev. 1," Aerospace Optics, Inc., Fort Worth, Tex., 1991.
14. "An Introduction to NVIS," Korry Electronics Company, Seattle, Wash., 1991.
15. "Crew Station Lighting for Night Operation," Eaton Corporation, Costa Mesa, Calif., 1991.

CHAPTER 7
PROGRAMMABLE SWITCHES*

As computer controlled systems increase in complexity, their designs are limited by fixed legend pushbutton switches (Chap. 6) which perform only one function. This situation has created a need for new capabilities not possible with dedicated-function switches. For certain applications, a programmable switch system may answer this need. A *programmable switch* provides multifunction switching and multilegend capabilities in a single pushbutton system (Fig. 7.1). The unique aspect of this concept is that different legends can be displayed at different times on a single pushbutton display surface.

Programmable switches provide interactive communications between the operator and the system computer. The system computer presents information to the operator by sending messages which appear as legends on the pushbutton displays. The operator can send a reply to the system computer by actuating the pushbutton displaying the desired action. The actuation signals the system computer to perform the appropriate action. New information is then presented to the operator, who once again can send a reply by actuating the pushbutton displaying the operator's selection.

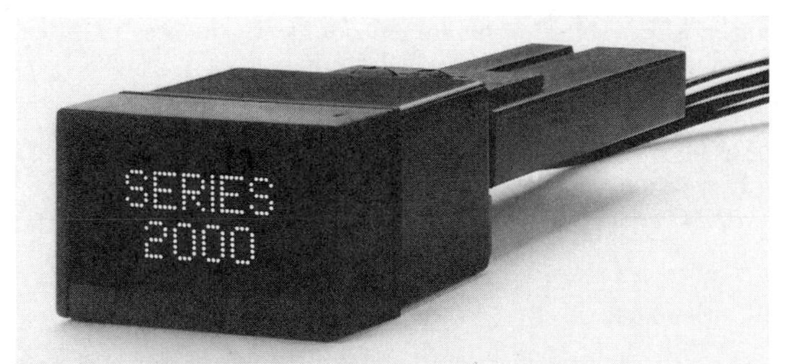

FIGURE 7.1 A programmable switch provides multifunction switching and multilegend capabilities in a single pushbutton module. (*Courtesy of Aerospace Optics, Inc., Fort Worth, Tex.*)

*Numbers in parentheses indicate items in the References at the end of this chapter.

Programmable switches can be programmed through the system computer to lead an operator through any sequence of events. The operator can be guided through a complete sequence of instructions or any combination of operations in a predetermined order. Complex manuals and checkout procedures can be eliminated. For example, in an aircraft application the programmable switch could be configured to take the pilot through each mode of flight in sequence beginning with the preflight instrument checklist and ending with engine shutdown procedures.

A programmable switch array can also significantly reduce the number of dedicated controls needed in a system (Fig. 7.2). Since each programmable switch can perform multiple functions, a separate switch is no longer required for each system function. Seldom used controls can be eliminated. Tens or even hundreds of individual controls can be replaced by a small array of programmable switches. As an example, a 4 × 4 matrix containing 16 programmable switch modules could be used in a control panel where each programmable switch module might have five different legends in four modes (16 × 5 × 4) taking the place of 320 dedicated function switches. (1)

HOW PROGRAMMABLE SWITCHES FUNCTION

In order to simplify illustration of how programmable switches function, two such devices have been selected as examples. One device is manufactured by Aerospace Optics, Fort Worth, Tex., and the other by C.Itoh Technology, Inc., Irvine, Calif. The first device (Fig. 7.3), designated the VIVISUN Series 2000, is a complete, ready-to-use system consisting of four programmable multifunction pushbutton (PMP) switch modules with LED displays, one refresh processor unit (RPU), and four cables which connect each PMP to the RPU (Fig. 7.4). (1)

The second device (Fig. 7.5), called a D880 GLM (graphic LCD module) multifunction programmable switch, consists of a single, PCB-mounted programmable switch with a liquid crystal display (LCD) featuring two-color LED backlighting. (2)

The VIVISUN Series 2000 is designed primarily for high-reliability applications requiring survivability in severe environments, such as those encountered in military or commercial aircraft, naval equipment, and ground tactical equipment (Fig. 7.6).

The D880 GLM is designed for commercial or industrial environments encountering typical operating temperatures of 32°F (0°C) to 104°F (40°C) and an operating humidity of 10 to 80 percent.

VIVISUN Series 2000

Our first example, the VIVISUN Series 2000 switch module, contains a 560 pixel dot matrix LED display with drive electronics, a solid-state Hall-effect switch, and a mechanism to provide tactile feedback when the pushbutton is depressed. The Hall-effect switch signals the RPU that the PMP has been actuated. The operator receives information from the system computer by legends presented on the PMP display. The operator initiates action by depressing the pushbutton surface displaying legends representing the operator's selection.

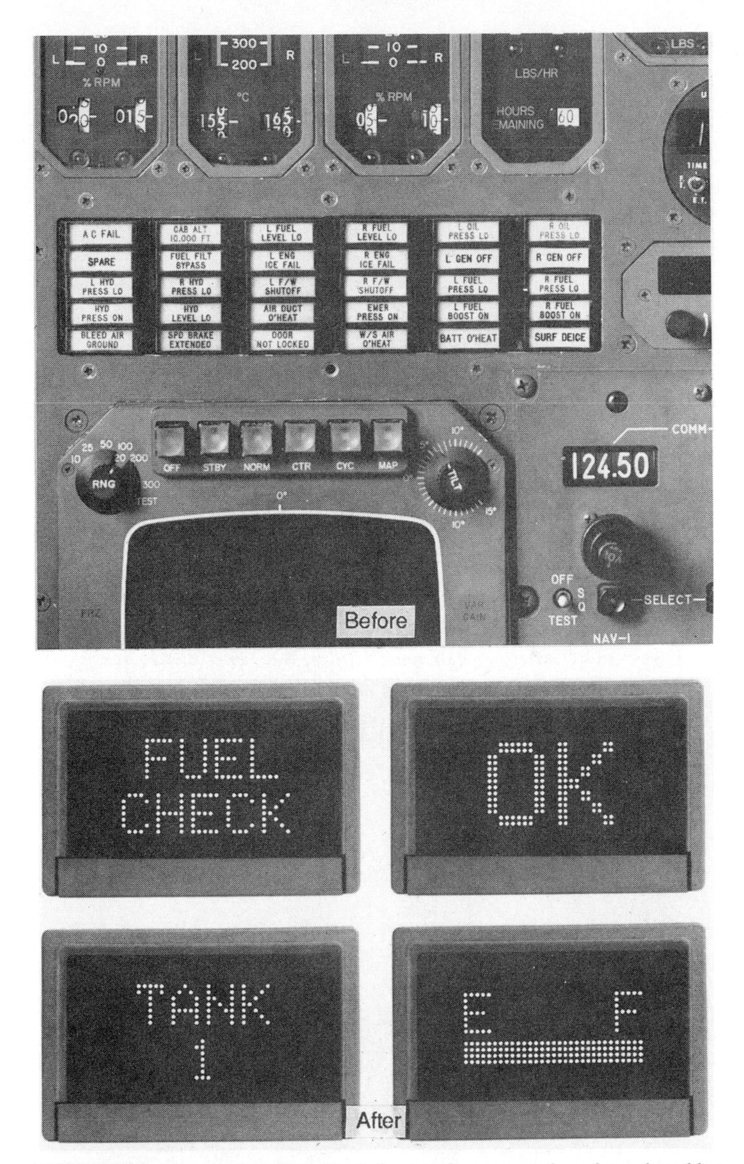

FIGURE 7.2 The number of dedicated controls on a panel can be reduced by substituting an array of programmable switches. (*Courtesy of Aerospace Optics, Inc., Fort Worth, Tex.*)

The RPU contains the refresh and processing electronics necessary to function as an intelligent interface between the PMP and the system computer. It receives coded messages from the system computer and converts these messages to dot patterns which are displayed as legends on the PMP. When a PMP is actuated, the RPU sends a coded message to the system computer identifying which PMP was actuated. The RPU independently handles routine functions

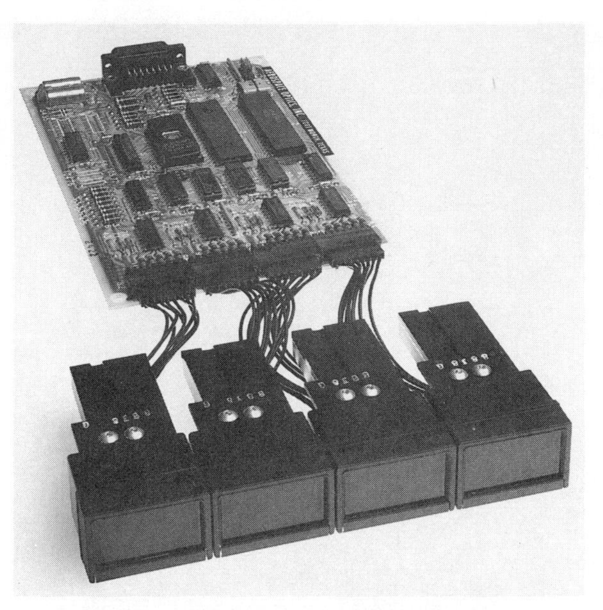

FIGURE 7.3 The VIVISUN Series 2000 programmable switch system. (*Courtesy of Aerospace Optics, Inc., Fort Worth, Tex.*)

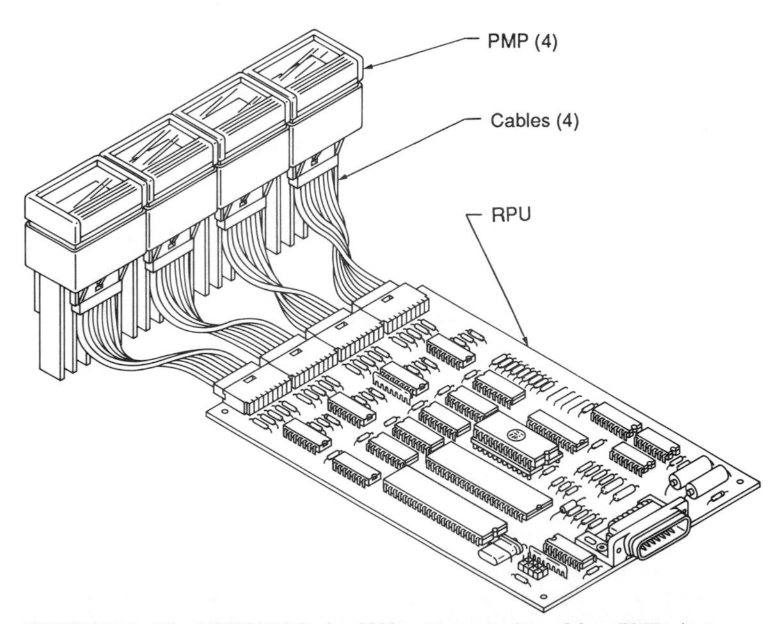

FIGURE 7.4 The VIVISUN Series 2000 system consists of four PMPs (programmable multifunction pushbutton), one RPU (refresh processor unit), and four interconnect cables. (*Courtesy of Aerospace Optics, Inc., Fort Worth, Tex.*)

FIGURE 7.5 The D880 GLM multifunction programmable switch features a 24 × 36 matrix LCD display. (*Courtesy of C.Itoh Technology, Inc., Irvine, Calif.*)

Temperature:
Operating −40 to +160°F (−40 to +71°C)
Nonoperating −65 to +185°F (−55 to +85°C)
High Temperature: In accordance with MIL-STD-810C, method 501.1, procedure II, +160°F (+71°C) maximum, operating
Low Temperature: In accordance with MIL-STD-810C, method 502.1, procedure I, −40°F (−40°C) operating
Thermal Shock: In accordance with MIL-STD-810C, method 503.1, procedure I, −65 to +185°F (−55 to +85°C) nonoperating
Shock: In accordance with MIL-STD-810C, method 516.2, procedure I, 30 g
Vibration: In accordance with MIL-STD-810C, method 514.2, procedure I

Frequency, Hz	Acceleration
5–14	0.100 in DA
14–23	1.0 g
23–74	0.036 in DA
74–2000	10.0 g

when the RPU interface connector is rigidly mounted to the fixture and the circuit board is supported using edge retainer guides. (Consult Aerospace Optics for details.)
Relative Humidity: In accordance with MIL-STD-810C, method 507.1, procedure I, 85 to 95 percent RH for 10 days
Salt Spray: In accordance with MIL-STD-202, method 101, test condition A, when only the front of the panel mounted display is exposed to the test environment
Fungus: In accordance with MIL-STD-454H, requirement 4
Explosion: In accordance with MIL-STD-202, method 109

FIGURE 7.6 Environmental characteristics of the VIVISUN Series 2000 programmable switch system. (*Courtesy of Aerospace Optics, Inc., Fort Worth, Tex.*)

such as self-testing, display refresh, luminance, and blinking control so the system computer can concentrate on system priority functions.

Figure 7.7*a* illustrates how a legend is presented to the operator. A coded message containing the legend information is transmitted from the system computer to the RPU via an RS-422 or RS-232C serial data link. This coded message is highly structured and contains parity information. Parity gives the RPU the ability to check the message for validity. Once the message has been validated, the RPU transmits an acknowledgment message back to the system computer. If the message is declared invalid, the RPU transmits a retry message back to the sys-

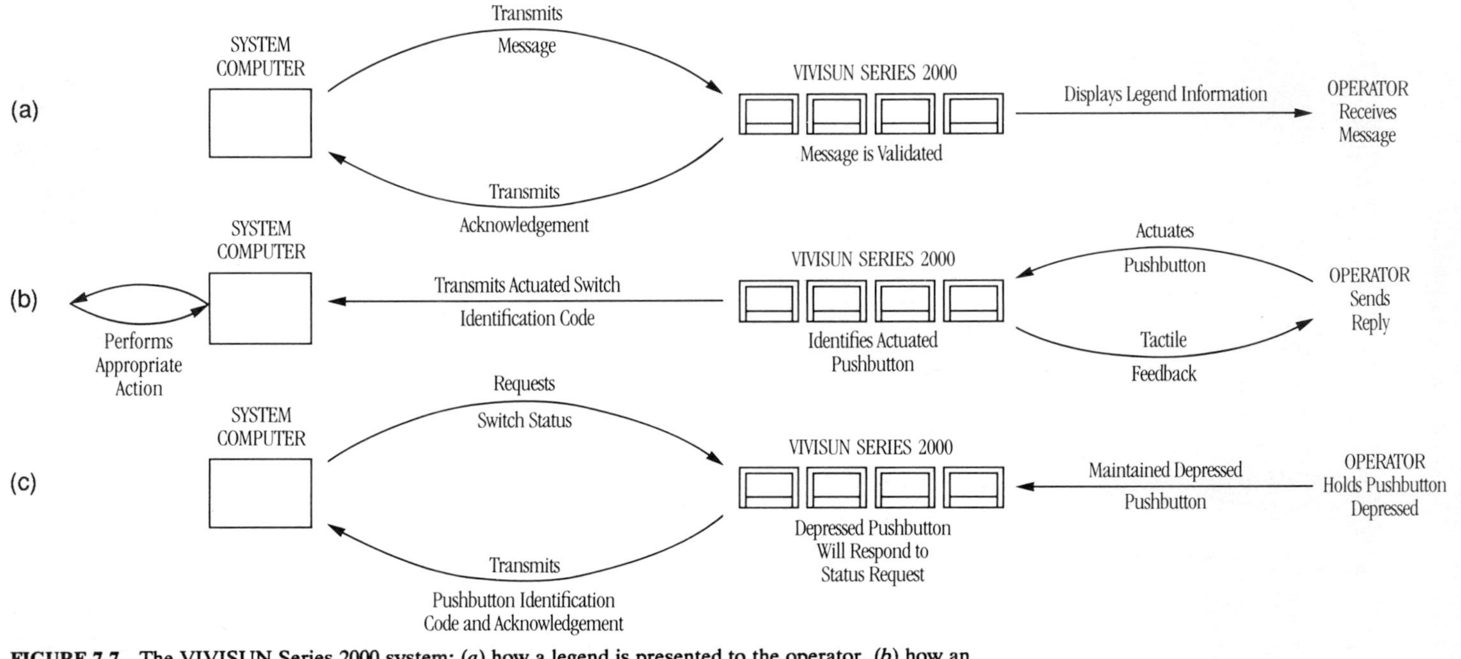

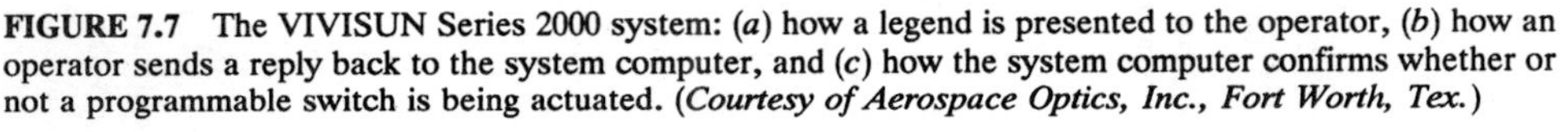

FIGURE 7.7 **The VIVISUN Series 2000 system: (*a*) how a legend is presented to the operator, (*b*) how an operator sends a reply back to the system computer, and (*c*) how the system computer confirms whether or not a programmable switch is being actuated. (*Courtesy of Aerospace Optics, Inc., Fort Worth, Tex.*)**

tem computer. No further action is taken that would destroy any existing legends until a valid message is received. This "handshaking" assures the system computer that the legend information is always true and correct. Once a message is validated, the RPU converts the message into a dot pattern and turns on the appropriate pixels in the selected PMP, which then displays the legend information to the operator. The RPU then awaits either an operator response or a new message from the system computer.

Figure 7.7*b* shows how an operator can send a reply back to the system computer. The operator selects the PMP displaying the chosen legend and actuates that PMP. The PMP notifies the operator of actuation by providing an acknowledgment in the form of a tactile feedback force. At the same time the PMP sends a signal to the RPU that it has been actuated and then the RPU transmits a message to the system computer containing the identification code for that PMP. This code is unique to the actuated PMP switch so the system computer knows exactly which PMP switch was actuated by the operator.

Figure 7.7*c* depicts how the system computer can confirm whether or not a PMP switch is being held depressed. When the system computer transmits a "switch status request" command to an RPU, the identification codes of all PMP switches being held depressed are transmitted back to the system computer. This is immediately followed by an acknowledgment message notifying the system computer that the "switch status request" command has been executed. This command can be used at any time by the system computer to confirm whether or not any PMP switch is being maintained in a depressed position.

LED Display. The VIVISUN Series 2000 PMP utilizes LED technology. The display is comprised of 560 LED dice and support electronics assembled into a ceramic hybrid microcircuit. The LEDs are organized into a matrix of 16 rows by 35 columns. Under RPU control it can display 5 × 7 or 10 × 14 format text, graphics, or any combination of text and graphics, anywhere on the display surface. The LEDs provide typical solid-state long life and reliability and are provided in three colors: green, red, and amber.

Loop Networking Capabilities. In order to reduce the number of serial ports required by the system computer, each RPU is equipped with networking capabilities. Instead of requiring one serial port for each RPU, the loop networking feature allows up to four RPUs to be connected to one serial port. This means that up to 16 PMPs and 4 RPUs can be handled by one serial port instead of only four PMPs and one RPU. Each of the four RPUs within a network has its own unique address code. Figure 7.8 shows the serial data line interconnections and the RPU address codes used to communicate with up to 4 RPUs and 16 PMP switches from one system computer serial port.

Each RPU is jumper configured with a unique address code from address 0 to address 3 prior to network installation. Legend commands sent from the system computer are coded with address information showing the destination RPU and PMP switch. Messages not addressed for a particular RPU are ignored and simply retransmitted down the loop to the next RPU. As an example of this process, the system computer would transmit a message to the network coded for RPU address 3. This message first arrives at RPU address 0. RPU address 0 would compare the message address with its jumper setting and find that the message is coded for RPU address 3. RPU address 0 then ignores the message and retransmits the entire message to RPU address 1. This compare and retransmission process occurs again from RPU address 1 to RPU address 2 and from RPU address

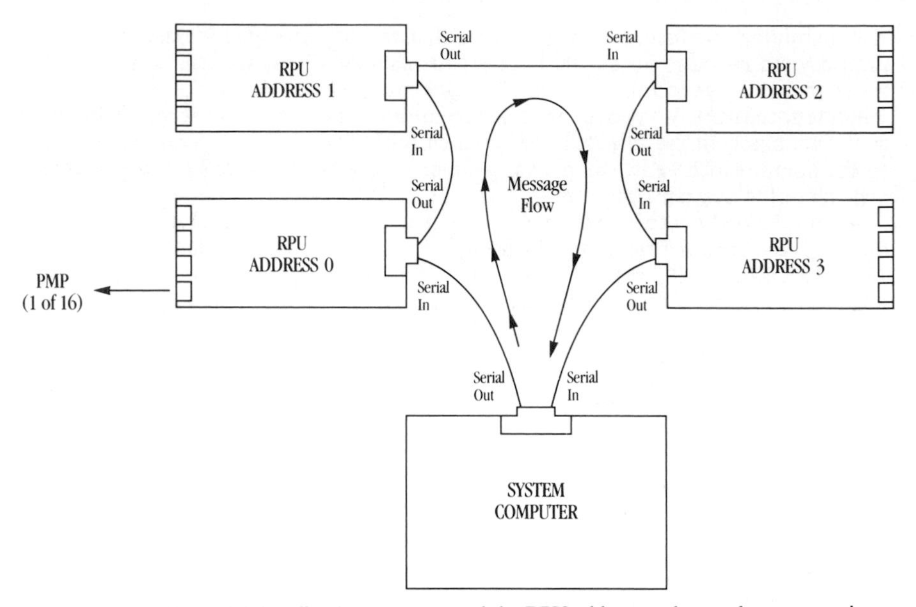

FIGURE 7.8 The serial data line interconnect and the RPU address codes used to communicate with up to four RPUs and 16 PMP switches from one system computer serial port. (*Courtesy of Aerospace Optics, Inc., Fort Worth, Tex.*)

2 to RPU address 3. When RPU address 3 checks the message, it finds that the message address corresponds to its jumper address setting. RPU address 3 now not only echoes the command back to the system computer but checks the message for validity, sends an acknowledge back to the system computer, and executes the command. All other RPUs and PMPs within the network remain unchanged.

When one of the 16 PMP switches is actuated, the RPU controlling it is notified. That RPU then transmits an actuation message containing the PMP identification code to the system computer. The PMP identification code is determined by the RPU address code and the RPU connector to which it is connected. The PMPs have no individual coding, and any PMP can be connected to any RPU.

Software Features. The software command structure of the RPU is designed to make legend generation flexible and efficient. The command set is divided into two groups, global and specific. Global commands affect all RPUs and PMPs within the network while specific commands apply only to one RPU or PMP.

Global commands are designed to be a "shortcut" for commands of a general nature. They allow attributes such as blink timer synchronization, luminance control, switch status checks, and clear all displays. The blink synchronization command assures that all PMP switches within a network will blink in unison. The switch status check is used to determine whether or not any PMP switch is being held depressed. The clear all displays command allows a quick initialization of all network RPUs and PMPs during system power up. The luminance control command allows the intensity level of all network PMP switches to be adjusted simultaneously.

Specific commands are directed at a single RPU or PMP. These commands control such attributes as self-test, all-pixels-on, blinking, text formats and graphics modes. The self-test feature performs a comprehensive RPU self-test on all RAM and ROM memory, the refresh controller, and the PMP Hall-effect switches. The results of the self-test are transmitted to the system computer. The all-pixels-on command enables the operator to visually verify proper operation of the display on the selected PMP. The blinking and clear display commands are also directed to a specific PMP. In the text mode, the RPU provides both a 5 × 7 and a 10 × 14 character format. In graphics mode, PMP pixels can be controlled either individually or in horizontal or vertical line groups.

The RPU allows legends to be presented on the PMP display in more than one mode simultaneously. For example, text legends can be displayed and later updated with an underscore or surrounded by a box.

All commands and legend control statements are within the ASCII (American Standard Code for Information Interchange) character set. ASCII coding allows commands to be entered on a standard computer keyboard. Legend control statements will also be visible within source code listings. The ASCII command structure helps programmers become familiar with the command set sooner because messages can be developed and sent to the programmable switch using a personal computer. (1)

D880 GLM

In our second example, the D880 GLM integrates a custom IC driver and a low-power graphic supertwist LCD with multicolor backlighting into a keycap of a single-pole single-throw (SPST) momentary conductive rubber contact switch, all within a dimension of less than 1 in^2 (25,4 mm^2) as viewed from the operator's position (Fig. 7.9).

LCD Display. The graphic LCD consists of 864 pixels in a 24 × 36 matrix which provides full screen graphics; 18 characters (3 lines × 6 characters) can be displayed using a 5 × 7 font, in a total display area of 0.592 in (15,94 mm) × 0.492 in (12,50 mm).

Backlighting. Red and green backlighting is available on the D880 GLM. Color can be changed by reversing the 5 V applied to the LED terminals. An amber backlighting option is also available and is obtained by using an ac voltage across the LED terminals.

Driver Circuitry. The custom IC driver internally generates the driving signals for display on the LCD after latching the display data, which are serially transmitted from the microcomputer or LCD controller. The IC driver also provides all the interfacing functions to allow a "daisy chain" of D880s. In this way, a number of switch modules (up to 48 when using the Hitachi HD 61830B) can be controlled by the microcomputer or LCD controller. Pin assignments for the module are depicted in Fig. 7.10.

Intelligent Controller. A universal controller is also available from C.Itoh Technology, Inc., for controlling up to 48 switches. The controller can communicate with any computer or can be used in a stand-alone mode. The controller has the following features:

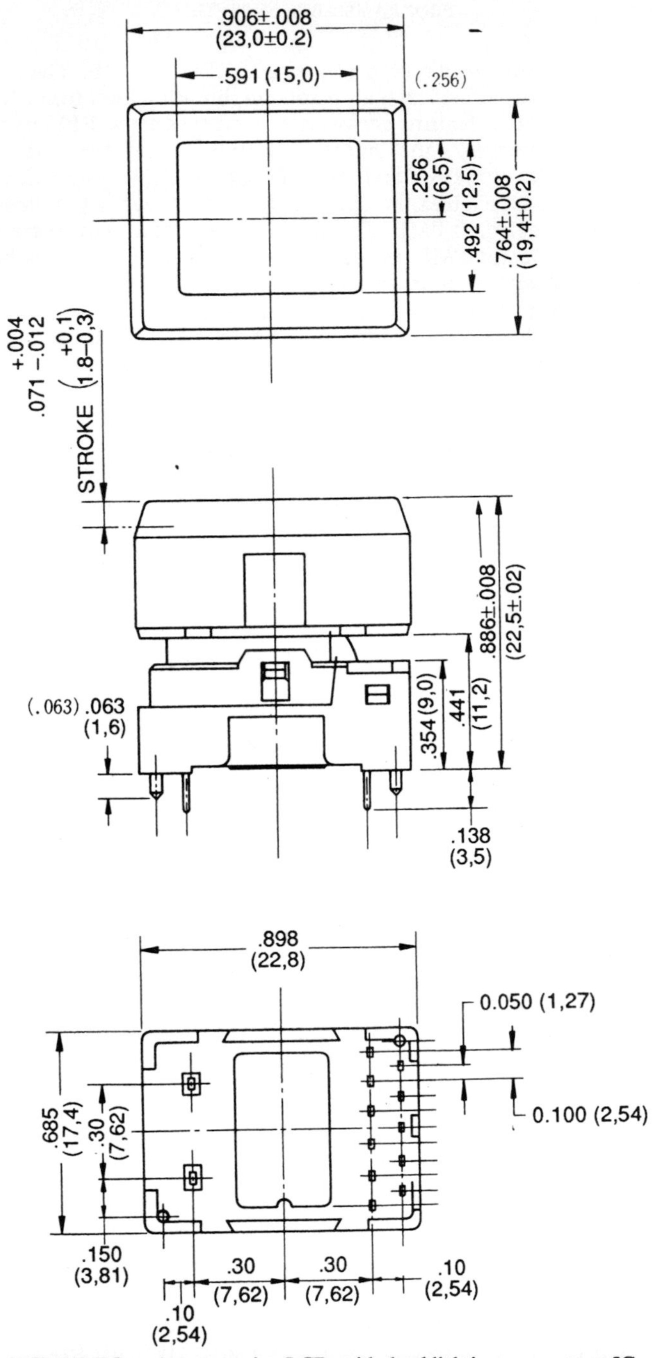

FIGURE 7.9 A supertwist LCD with backlighting, a custom IC driver, and an SPST switch are integrated into the D880 GLM package. The keycap takes up less than 1 in² (25,4 mm²). (*Courtesy of C.Itoh Technology, Inc., Irvine, Calif.*)

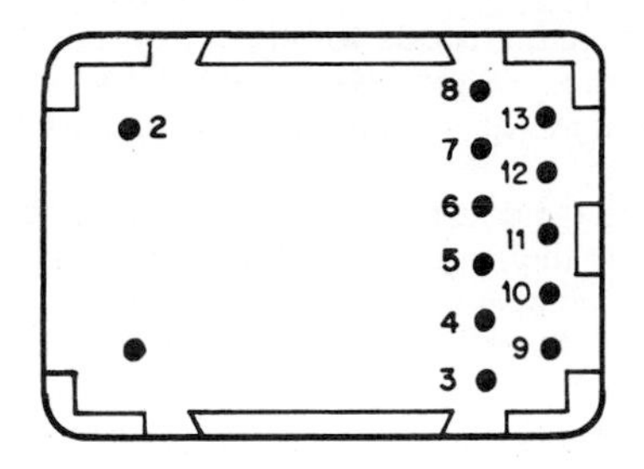

No.	Pin	Function	Connection
1	SW1	switch	user defined
2	SW2	switch	user defined
3	VLC	supply voltage for LCD	power supply
4	GND	ground (±OV)	power supply
5	DIN	data input	DOUT or controller
6	SCP	serial clock pulse	controller
7	LED K	LED cathode	power supply
8	FLM	first line marker	controller
9	VDD	supply voltage for logic +5V	power supply
10	LED A	LED anode	power supply
11	DOUT	data output	DIN
12	LP	latch pulse	controller
13	RST	reset signal	controller

FIGURE 7.10 Pin assignments for the D880 GLM. (*Courtesy of C.Itoh Technology, Inc., Irvine, Calif.*)

- Memory with battery backup for 255 user-defined legends and attributes.
- Attributes attached to each legend so when a switch is pressed, the attribute which corresponds to the legend being displayed on that switch will be executed, causing some or all of the legends on the switches to be changed, depending on the content of attributes. This allows stand-alone operation.
- Trimmer adjustments for LCD contrast and LED backlight brightness.
- A TTL switch output.
- Software for computer.
- Communication via RS-232/422 to any computer or terminal. (2)

TECHNIQUES FOR USING PROGRAMMABLE SWITCHES

The full potential of a programmable switch system will not be realized if it is used to simply replace a single dedicated function switch. To use the system's full capabilities, careful planning needs to be performed early in the application design process, allowing the programmable switch to be part of a totally integrated system. Always contact the switch manufacturer as early as possible in the process.

Planning and organizing the control of a system is made more efficient by the use of a flowchart, enabling the application engineer to visualize the control flow easily without the need to program the system computer.

Control planning can be divided into three simple steps:

1. *Identify* all system functions that are to be controlled with the programmable switch system.
2. *Group* these functions into functionally related control groups or menus.
3. *Organize* the control groups for a control scheme.

When a control system is planned properly, the control flow will be straightforward and easy to understand.

APPLICATIONS EXAMPLE

Again using the VIVISUN Series 2000 as a model, the following application example illustrates how only one Series 2000 system could be used to control all major systems on an advanced aircraft. Each row of Fig. 7.11 shows the legend information displayed on each of the four PMP switches as it is presented to the pilot. Solid lines originating from a PMP switch indicate actuation of that PMP switch. Dotted lines originating from a PMP switch indicate that the legend information changes as the result of actuation of another PMP switch. Arrows are used to indicate the direction of the control flow.

The example shown incorporates both multilegend and multifunction modes of operation. PMP switches are used to display system status, perform interactive checklists, and directly control system functions. When the pilot first activates the system computer, the legend information in row 1 is displayed. The legends indicate that the battery, fuel, and breathing oxygen levels are within system limits and that the engine can now be started. The pilot then actuates the "ENGINE START" switch, and once the engine is started, the system computer performs an automatic test of all engine, hydraulics, and power systems. Since all systems are within safe limits, the pilot actuates the "CHECKLIST" switch, causing the checklist shown in row 3 to be presented. The pilot acknowledges each requested check by actuating the appropriate PMP switch. When actuated, each switch face goes blank until the entire checklist is complete. In this example, only wing ice, flaps position, and brakes are included in the checklist. System status is also shown during the checklist operation, the result of continuous system diagnostics. If any fault were to occur, legends would appear indicating the type of fault and suggested remedies. In this example, all systems are "OK." Once complete, the "FLIGHT MENU" legend is displayed, allowing the pilot to enter the aircraft systems control menu. When actuated, the flight menu is displayed, shown on row 5.

The "FLIGHT MENU" is the origin for all electronic warfare, weapons, system control, and diagnostic functions. By actuating the "EW MENU" switch in row 5, the pilot can select active jamming, passive radar warning receiver, and

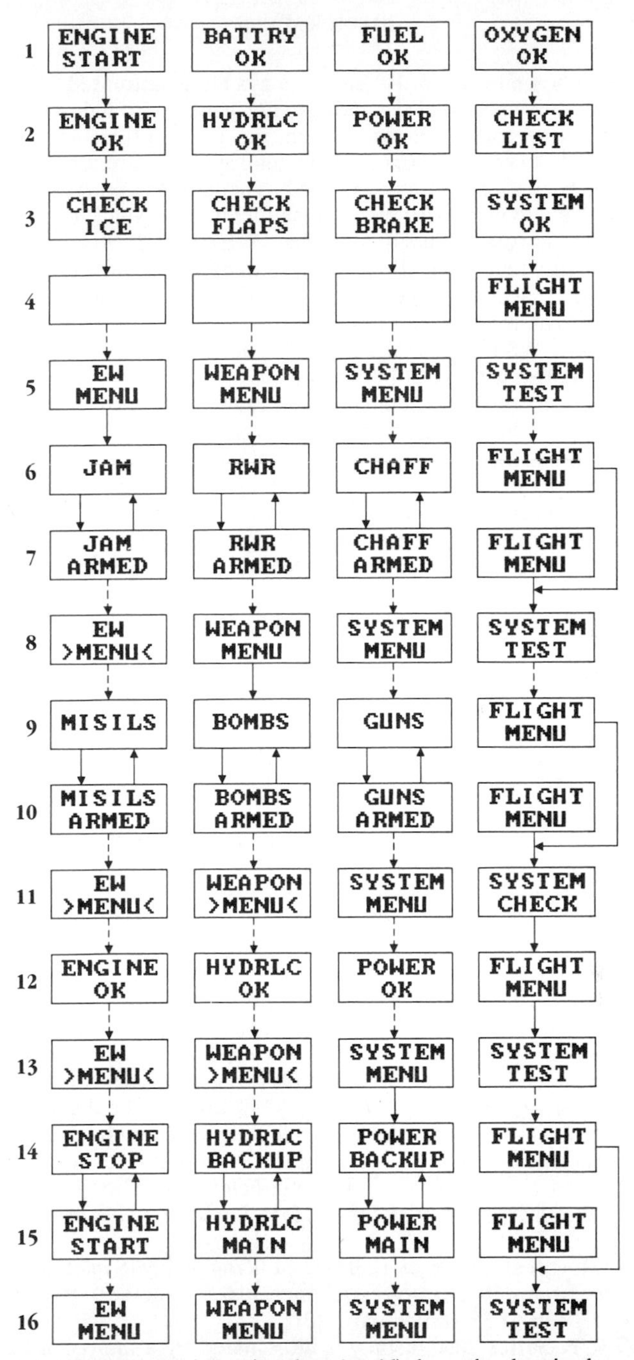

FIGURE 7.11 Each row (numbers 1 to 16) shows the changing legend information on each of four PMPs as it is presented to a pilot in an example of programmable switch application. (*Courtesy of Aerospace Optics, Inc., Fort Worth, Tex.*)

defensive chaff shown in row 6. Whenever a system is activated, the "ARMED" legend appears. When actuated, legends are designed to "toggle" between armed and unarmed states for simplicity. When the pilot returns by actuating the "FLIGHT MENU" switch in row 7, arrowhead symbols are used to show that at least one of the menu's selections is armed. This lets the pilot know at a glance that a selection has been made within that menu. The same control scheme is used for weapons in rows 9 and 10 after the "WEAPONS MENU" switch is actuated in row 8. In a real application, the system computer would be programmed to recognize the attached weapons and provide all necessary launch sequence and targeting instructions to the pilot.

Actuating the "SYSTEM CHECK" switch in row 11 causes a manual systems diagnostics of major systems to be shown in row 12. This can be performed at any time to confirm status of the engine, hydraulics, and power systems. In row 13, the pilot actuates the "SYSTEM MENU" switch. In this menu, the pilot can stop and start the aircraft engine as well as manually switch between main and backup hydraulics and power. In row 15, the pilot has actuated the "FLIGHT MENU" switch to return to the flight menu shown in row 16. Note that no electronic warfare or weapons system is any longer armed. To prevent accidental weapons release, the system computer disarmed these systems automatically when the engine was stopped in row 14. This feature, although only an example, can be used to increase system safety by using the system computer to prevent unsafe operating modes. (1)

HUMAN FACTORS CONSIDERATIONS IN SELECTING PROGRAMMABLE SWITCHES

In the process of selecting and specifying programmable switches, the human factors considerations most often influencing evaluation include the switch size, display emission characteristics, display resolution, character height, refresh rate, scratch resistance, and tactile feedback. Software features and attributes are also important to provide flexibility in legend presentation.

It is important that the overall programmable switch size be large enough to be seen and actuated easily. Text characters and graphics patterns must be large enough to be visible at a typical viewing distance of 28 to 30 in (711,2 to 762 mm). Extremely small characters may be difficult to read at this distance and should be avoided. Again, referring to our two examples, the VIVISUN Series 2000 has an overall size (Fig. 7.12) of 1.0 × 1.5 in (25,4 × 38,1 mm). The display resolution is optimized at 40 pixels per inch (25,4 mm), yielding a minimum character height in the 5 × 7 format of 0.162 in (4,15 mm). Figure 7.13 illustrates the character *sizes* for both 5 × 7 and 10 × 14 formats. The character sizes shown in Fig. 7.13 generate, in turn, the character *styles* for the 5 × 7 and 10 × 14 formats depicted in Fig. 7.14. (1)

The D880 GLM features an overall size of 0.764 × 0.906 in (19,4 × 23,0 mm). In a 5 × 7 font style, 18 characters (3 lines × 6 characters) can be displayed. Pixel size is 0.014 in (0,35 mm) square. (2)

The VIVISUN Series 2000 display is illuminated by a matrix of light-emitting diodes (LEDs), yielding the following display emission characteristics:[1]

[1]Display emission characteristics for the Series 2000 VIVISUN are courtesy of Aerospace Optics, Fort Worth, Tex.

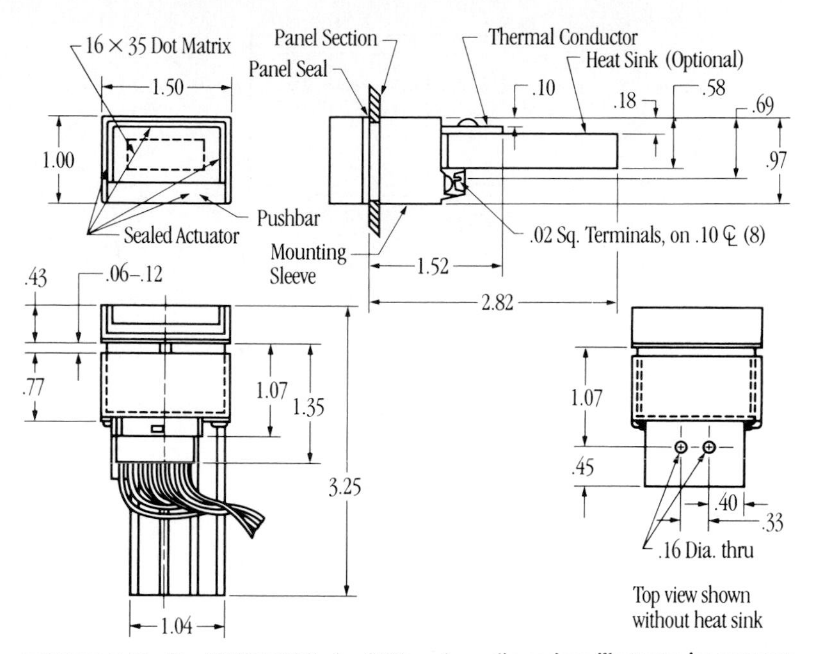

FIGURE 7.12 **The VIVISUN Series 2000 package dimensions illustrates its compact size. (*Courtesy of Aerospace Optics, Inc., Fort Worth, Tex.*)**

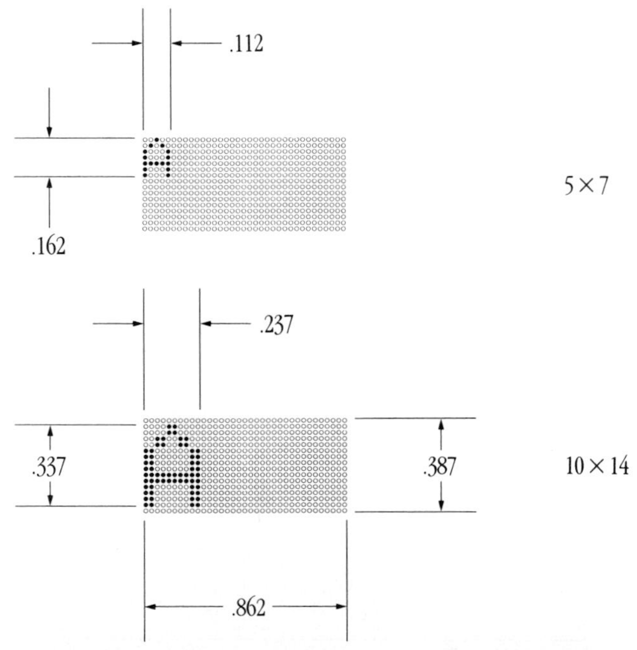

FIGURE 7.13 **Character sizes for the 5 × 7 and 10 × 14 formats for the Series 2000. (*Courtesy of Aerospace Optics, Inc., Fort Worth, Tex.*)**

5 × 7 FORMAT

10 × 14 FORMAT

FIGURE 7.14 Character styles generated from the character size formats shown in Fig. 7.13. (*Courtesy of Aerospace Optics, Inc., Fort Worth, Tex.*)

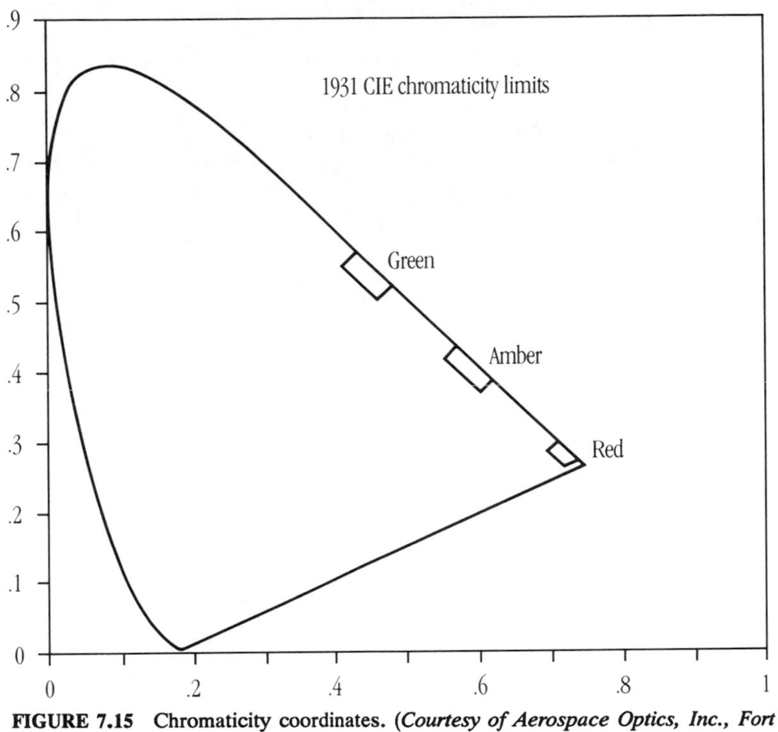

FIGURE 7.15 Chromaticity coordinates. (*Courtesy of Aerospace Optics, Inc., Fort Worth, Tex.*)

Chromaticity Coordinates. See Fig. 7.15 for limits.

Color	*x*	*y*
Green	0.459	0.541
Red	0.724	0.276
Amber	0.592	0.408

Peak Wavelength. Figure 7.16 shows the spectral radiance for each color.

Color	Wavelength, nm
Green	572
Red	650
Amber	593

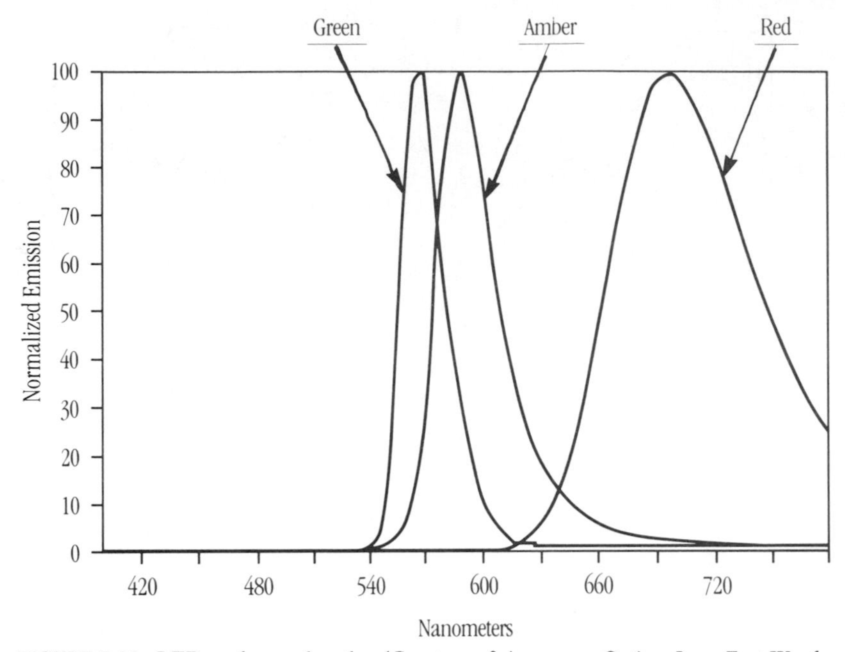

FIGURE 7.16 LED peak wavelengths. (*Courtesy of Aerospace Optics, Inc., Fort Worth, Tex.*)

Average Luminance. Minimum values at 77°F (25°C) ambient temperature.

Color	Luminance, fL
Green	150
Red	40
Amber	60

Additionally, the display luminance can be adjusted in 35 brightness levels by software control.

Display Contrast. Minimum contrast when measured with the test setup shown in Fig. 7.17.

Color	Ambient illumination, fc	Contrast*
Green	8000	3.0
Red	4000	3.0

*On/background.

Additional Human Factors Considerations

If the programmable switch module will be used in an aircraft or vehicular application, the specifying engineer should consider that displays requiring periodic

$$\text{ON/BACKGROUND contrast } C_L = \frac{B_2 - B_1}{B_1}$$

where:
B_1 = Average background luminance
B_2 = Average pixel luminance, lighted

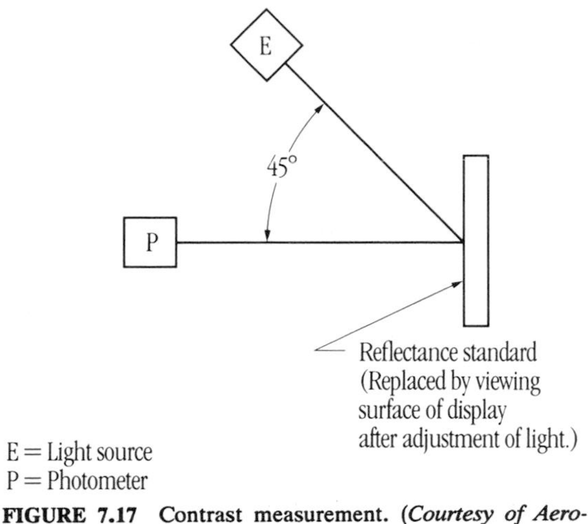

FIGURE 7.17 Contrast measurement. (*Courtesy of Aerospace Optics, Inc., Fort Worth, Tex.*)

refresh must be refreshed at a rate in excess of 400 frames per second. This will prevent stroboscopic "flickering" of the display during the vibration often encountered in these applications.

Since the face of a programmable switch is designed to be touched during actuation, it must be resistant to scratches and fingerprints. The specifying engineer should ascertain if the programmable switch under consideration has been treated with a process to resist scratches and fingerprints.

The actuating surface of a programmable switch must provide a positive tactile feedback for operator acknowledgment when actuated. It must also provide a degree of protection against accidental actuation. The tactile feedback should remove any need for the operator to visually read the display to determine whether or not it was actuated. The programmable switch should provide both a positive tactile feel and integral actuation barriers. The sides and top of the display should form a raised area to form an actuation barrier. This barrier helps prevent multiple actuation should the operator's finger slip off the programmable switch during actuation.

The specifying engineer should consider the software attributes that provide flexibility in legend presentation. Consideration should be given to the device which permits text and graphic patterns to be displayed together on the same display face. This gives the engineer the ability to present text legends and later update these legends with graphics attributes such as an underscore or surrounding box. This enables the operator to know at a glance when a legend representing a system feature is selected or available. The switch should also offer a blinking attribute for purposes of prompting and warning. (1)

GLOSSARY

LED. Light-emitting diode. A solid-state semiconductor device that emits light when current passes through it.

Pixel. Picture element. The smallest area or dot on a display that can be accessed individually.

Programmable switch. A device which provides multifunction switching and multilegend capabilities in a single pushbutton system. A unique feature of this device is that different legends can be displayed at different times on a single pushbutton display surface. Additionally, programmable switches provide interactive communications between the operator and the system computer.

REFERENCES

1. "VIVISUN Series 2000 Programmable Display Switches," Aerospace Optics, Inc., Fort Worth, Tex., 1988.
2. "D880 GLM Graphic LCD Module Multifunction Programmable Switch," C.Itoh Technology, Inc., Irvine, Calif., 1992.

CHAPTER 8
MEMBRANE SWITCHES*

Membrane switches consist of circuitry printed with silver conductive ink on sheets of clear polyester substrates. The substrates are then assembled, using a sheet of pressure-sensitive adhesive (PSA) between the two layers, bonding the printed circuitry substrates together with the conductive layers facing each other but separated by the thickness of the PSA layer. The shape and location of the conductive ink circuitry define the position and style of each switch contact. As the operator pushes the top, or "flex," layer downward toward the bottom, or "stable," layer, the conductive circuitry pattern printed on the inside surface of the top polyester substrate comes in contact with the conductive pattern on the inside surface of the bottom polyester substrate, electrically joining one conductive layer to its opposite on the other layer for as long as the contact is held by the operator. With the proper design of the conductive patterns on each substrate, this temporary electrical joining can be read as a switch closure.

Typically, the conductor pattern is brought out to flexible tails formed as part of the extension of the polyester substrates. This assembly constitutes the basic membrane switch construction (Fig. 8.1).

Often, a colorful graphic panel layer is added to the top, or operator's side, of the basic membrane switch in order to direct the operator's touch to the proper switch location and to enhance the visual appeal of the membrane switch itself and the control panel or enclosure to which it is attached. Additionally, the graphics panel may contain embossed ridges around each key position to tactilely direct an operator's finger to the correct switch area. Embossment of the graphic panel is also used as a decorative embellishment to add visual impact, for example, by incorporating a raised corporate logo or symbol. Attachment of the graphics panel to the membrane switch is normally by a layer of pressure-sensitive adhesive.

Outwardly, most flexible membrane switches appear to be the same. However, their construction must change depending on the environment in which the switch must operate. The construction must be chosen with attention to all factors of its operating environment including maximum temperature, quality of atmosphere, and method of interconnection. Most membrane switch manufacturers use one of the following methods to construct the switch assembly. The three distinct types of flexible construction offered by the W. H. Brady Co., Milwaukee, Wis., are used as examples.

*Numbers in parentheses indicate items in the References at the end of this chapter.

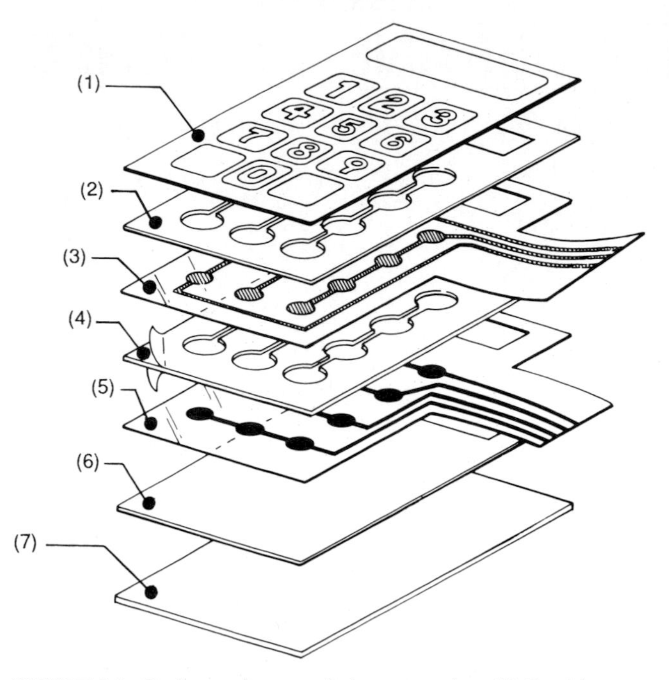

FIGURE 8.1 Basic membrane switch construction: (1) Graphic overlay, (2) pressure-sensitive adhesive layer, (3) top switch circuit, (4) pressure-sensitive adhesive switch spacer, (5) bottom switch circuit, (6) pressure-sensitive adhesive layer, (7) backer material. Note venting in (2) and (4). (*Courtesy of 3M Identification and Converter Systems Division, St. Paul, Minn.*)

Type F65

This is the most popular type of construction of the three examples. Designated by the Brady Co. as their type F65, it consists of three layers laminated together to form the switch assembly. These three layers consist of an upper circuitry layer called the "flex," a lower circuitry layer called the "stable," and a third layer, the spacer (this is the construction shown in Fig. 8.1), that separates the two layers. All three layers are composed of polyester substrates which are held together by pressure-sensitive adhesives. This style of switch is internally vented; that is, a continuous channel in the spacer layer connects all switch cavities together to create a passage for air to travel between switch positions when a switch is actuated. In this fashion, no external venting is necessary. The switch package is totally sealed from outside air and any contamination contained in the atmosphere. The one important disadvantage of this construction is an upper temperature limit of 149°F (65°C). Operation above this temperature, especially thermal cycling, will cause the polyester to "breathe." When returned to ambient temperature, a pressure differential will exist causing sensitive, or in the worst cases, collapsed switches. Unfortunately, this failure mode is permanent and can be avoided only by externally venting the switch.

Type F85

If operation above 149°F (65°C) is a necessity, the type of construction the Brady Co. designates as its type F85 is preferred. The switch is constructed in a manner similar to the type F65, but the switch is externally vented in order to eliminate the collapsed switch failure mode. Any time a switch is vented, certain precautions must be taken to protect the electrical integrity of the switch. One of the greatest concerns in a vented switch is the risk of contaminants entering the switch cavity. Among possible contaminants is condensed moisture which, upon infiltrating the switch cavity, can cause silver migration (this phenomenon is discussed further in this chapter). In order to minimize this occurrence, the type F85 (and similar constructions by other switch manufacturers) employ several design techniques. First, the circuitry is separated by polarity so that circuits of different potential are each on different circuitry substrates. Second, exposed silver is encapsulated in a carbon or graphite overcoat which inhibits the migration process. In order to accomplish these design parameters, the switch must employ two flex tails, one exiting from each of the two circuit layers. These can be overlapped to form a single tail to minimize connection costs, but the type F85 style of construction is more expensive than the type F65.

Type FNC

A final construction technique offered by the Brady Co. is designated the type FNC construction. It is less costly than either the F65 or F85 techniques since it employs a single-layer substrate which is folded back on itself and adhered together by the spacer layer to form the switch. Switches of this type of construction are generally limited to a noncondensing environment. They are externally vented, but typically do not have the protection parameters built into the construction as do the type F85 versions. Type FNC switches are recommended for use where the switch edge is captured by a bezel. They are specifically designed to perform in applications where a low-cost switch will be used in a controlled, relatively benign environment. [Adapted from (1).]

Substrates for Membrane Switches

In general, polyester is the material of choice as a substrate on which is printed the conductive silver circuitry. This is due to polyester's unique characteristics which make it an ideal material for switch use. Polycarbonate is also available but has limitations.

Polyester. Polyester is the most widely used substrate for membrane switches because of its availability, cost, and properties. Polyester is a strong, flexible plastic that has enough rigidity to retain its original shape even after bending, flexing, folding, and repeated actuation of switches.

All polyester material used in switch layers is stabilized, providing a final product that will not shrink during high-temperature use or storage. Polyester can be obtained in varying degrees of opacity and color. Most material selected for use in membrane switches is clear (Fig. 8.2). Cloudy or colored material would not affect the performance of the switch, although the material would be noticeable in a display window area.

FIGURE 8.2 A technician inspects a polyester substrate membrane layer. (*Courtesy of GM Nameplate, INTAQ Division, Seattle, Wash.*)

Since polyester has excellent chemical and solvent resistance along with good tensile strength, it is the material of choice in applications where harsh chemicals or variation in temperature occur.

Polycarbonate. While polycarbonate can be used in membrane switch manufacturing, it is not the usual choice as a substrate. The basic advantage to using polycarbonate is that the material is more thermally stable than polyester at elevated temperatures. At lower temperatures, polycarbonate tends to expand. Additionally, if the switch is to be folded, polycarbonate will crease easily at the fold line. The primary disadvantage of polycarbonate is that it is not as solvent-resistant as polyester. In many cases, just the presence of solvent vapors will be detrimental to the life of a polycarbonate-based switch. This is due to the permeability of polycarbonate to solvent vapors. Polycarbonate is also prone to work hardening, causing it to crack completely after approximately 1000 flex cycles, so it should not be specified for applications where there is a possibility of the tail's being flexed.

A comparison of the physical properties of polyester and polycarbonate materials as the basic conductor-carrying substrates for membrane switches is shown in Table 8.1. [Adapted from (2).]

Material Incompatibility Problems

The majority of membrane switch designs commonly incorporate silver inks and pressure-sensitive adhesives. Conductive silver ink is a cost-effective way to deliver printed traces capable of handling the low power requirements of microprocessors. Pressure-sensitive adhesive systems are readily available and require no special equipment for use in assembly operations.

TABLE 8.1 Physical Properties of Substrates for Membrane Switches

Property	Units	Polyester* (1 mil)	Polycarbonate† (5 mil)
Density or specific gravity	g/cm^3	1.39	1.20
Distortion temperature	°F	275	275
Maximum usage temperature	°F	150	260
Minimum usage temperature	°F	−40	−40
Tensile strength	lb/in^2	25,000–30,000	9000–11,000
Stress to produce 5% elongation	lb/in^2	15,000	7250
Ultimate elongation	%	120–150	100–150
Fold endurance	Number of folds	100,000	250–400
Tear strength	lb/in	1800	1150–1570
Bursting strength	lb/in^2	66	30–45
Shrinkage at 300°F	%	0.05†	0.28
Dielectric strength	V/mil	7500	2200
Surface resistance	Ω/square	10^{16}	10^{15}
Water absorption	%	0.6	0.35
Switch life cycles	Number of cycles	10,000,000	10,000,000
Thickness available‡	mil	4,5,7	5,7

Source: Ref. 2.
*These values are typical for the thicknesses specified. For specific applications, the material being used should be tested.
†Value for prestabilized film.
‡Values given are the usual values for Brady switches.

However, for experienced users of membrane switches, it is common knowledge that both ink and adhesive can contribute to membrane switch failure in a variety of ways, including the chemical interaction of the materials used and the resultant effect on reliability. This section discusses the modes of failure which can be caused by these characteristics and suggest ways that the user-engineer can avoid potential problems by frank discussion with the potential switch supplier(s) concerning the supplier's methods of dealing with materials incompatibility.

Chemical Compatibility. An analysis of chemical compatibility is an important starting point in the selection of silver ink and pressure-sensitive materials. Whether pure silver ink or another conductive filler is used, the principle is the same.

Silver ink consists of finely divided conductive particles which have been combined with a polymeric binder. The binder serves to bond the silver ink to the substrate surface in a flexible matrix in such a way that the electrical conductivity remains a useful property.

The polymer is selected for its ability to adhere to the membrane switch substrate (polyester is the most commonly used substrate) and for its ability to maintain flexibility during the life of the switch. The polymer must be soluble in solvent systems which are compatible with screen process printing operations. Frequently, nonconductive filler is added to enhance the screen process printing properties of the ink. In most silver ink compounds, these binder resins are proprietary and little technical information about the chemical nature of their performance is provided by the manufacturer.

As might be expected, a wide range of binder resins are available, each offering different performance and compatibility characteristics. There is no right or wrong formulation of ink and binder, but the user has a right to expect a membrane switch supplier to carefully select an ink formulation compatible with the adhesive system the supplier is using. Many pressure-sensitive systems contain a variety of chemical components—residual solvents, unreacted short-chain polymers, tacifiers, detacifiers, etc.—which can attack the silver ink, and use of such adhesive systems must be avoided by the potential supplier.

In a series of tests conducted by the Berquist Switch Co., Minneapolis, Minn., evidence of this mechanism was demonstrated by accelerated aging tests at 212°F (100°C). In these tests, samples of pressure-sensitive adhesive were placed in intimate contact with samples of conductive ink processed according to the manufacturer's instructions. The assembly was then aged and inspected at varying time intervals.

The tests showed that the degree of silver ink transfer to the pressure-sensitive adhesive surface is a function of the exposure time under these environmental conditions. In other words, longer exposures caused more silver transfer to the pressure-sensitive adhesive surface. This is the result of "plasticization" of the silver ink binder resin. Materials migrating from the pressure-sensitive adhesive attacked the silver ink binder resin and reduced its cohesive strength sufficiently to permit transfer to the adhesive surface. The attack of the binder resin will, in time, reduce the overall bond strength and eventually reach the silver ink substrate interface. At this point, switch failure will occur.

Mechanical designs which place potential electrical shorting components in vertical proximity to one another on a pressure-sensitive surface are particularly susceptible to failure caused by adhesive "creep." Under conditions where compressive loading forces are placed toward the center of the membrane switch, the resulting shear action will cause the pressure-sensitive adhesive to flow into the switch area. "Plasticized" silver ink will flow with the pressure-sensitive adhesive and be displaced toward the negative conductor. The resulting failure occurs when sufficient creep has taken place to transfer conductive ink down the sidewall of the space and onto the negative surface.

In many cases, this motion may be accelerated when switch designs incorporate improper pressure accommodation techniques which can result in relatively high pressures when the switch is activated. Such air pressure contributes to the overall tendency of the adhesive to creep in an effort to relieve the high-pressure area.

A major concern to the user, of course, is contamination of the actual contact surface by pressure-sensitive adhesives. Failures occur when a small volume of pressure-sensitive adhesive, for whatever reason, has been deposited at the interface of the switch contact. When the switch is activated, the pressure-sensitive material may serve to bond the two switch surfaces together in a shorted condition or may serve to act as an insulator, preventing contact and proper switch functions.

A good membrane switch design requires considerable care regarding the silver ink–pressure-sensitive interface. Compatibility of materials is paramount since no mechanical design can compensate for chemical attack and degradation of the functioning component. In practice, it is probably easiest to select a highly professional supplier that clearly understands which silver ink is most desirable and then tests for compatibility with pressure-sensitive systems. Mechanical designs can minimize electrical failures resulting from silver ink transfer by locating conductive traces properly. This technique, however, is not a solution to the elimination of the pressure-sensitive adhesive migration and plasticization of the silver polymer. Proper mechanical design can also reduce substantially the mobility of the pressure-sensitive adhesive system. Proper use of design techniques such as air reservoirs and venting paths (see following section) to control pressure buildup are extremely helpful. Assembly techniques which eliminate small "balls" of pressure-sensitive adhesive in vent channels and along spacer walls will also help prevent this problem. The user should insist (and indeed, specify) that the supplier select a pressure-sensitive adhesive system which is not mobile. [Adapted from (3).]

Pressure Buildup in Nonvented Switches

Air pressure is an important and often overlooked element in the design and mounting of a membrane switch. What follows is a general discussion about the effects of pressure changes on membrane switches. Designers are usually more concerned with chemical resistance and other more familiar hazards. However, when naturally occurring pressure changes inside a membrane switch are ignored, they can lead to field failures. This section reviews the changing pressures inside a membrane switch and the designs used to accommodate those changes.

Pressure. The air pressure changes which affect membrane switches can be divided into two categories: (1) natural causes such as changes in altitude and temperature, and (2) the mechanical effects of actuating the keys in the switch.

Environmental Pressure Changes. During storage, transportation, and use, a membrane switch may encounter a wide range of temperatures and, sometimes, altitudes. An informal standard has evolved for storage and transportation which requires a switch to withstand temperatures ranging from −40°F (−40°C) to +185°F (+85°C) and altitudes up to 30,000 ft (9144 m). While geographic altitude changes are not severe, the altitudes reached in air transport can be significant. Most airplane cargo holds are now pressurized to some degree, but the possibility of shipment in an unpressurized hold cannot be ignored.

The hazard is not simply high or low air pressure but the differential pressure which can develop between the air inside and outside the switch (Fig. 8.3). A membrane switch constructed of plastic films is not a rigid vessel. The layers which surround the key cavities are flexible and capable of accommodating some change in air volume. This capacity is limited, however, and expansion puts stress on the bond between the layers. The switches are designed so that compression, as in pressing a key, closes the contacts. This closure is desirable only when a key is actuated. It is not desirable when it is caused by differential air pressure.

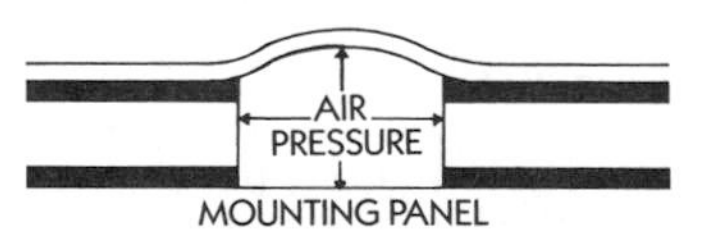

FIGURE 8.3 Differential pressure can put stress on the bond between layers of a membrane switch. (*Courtesy of Berquist Switch Co., Minneapolis, Minn.*)

Altitude. Referring to Table 8.2, an altitude of 30,000 ft (9144 m) can cause a pressure differential of 12.3 lb/in^2. Another common specification, 15,000 ft (4572 m), results in a differential of 6.4 lb/in^2. Denver, Colo., the "mile-high city," has a differential of 2.7 lb/in^2 from sea level. The differential resulting from altitudes of 30,000 ft (9144 m) or 15,000 ft (4572 m) could be expected to have a time span in hours. The differential experienced at Denver could be a *lifetime* condition for a membrane switch.

TABLE 8.2 Air Pressure at Various Altitudes

Altitude, ft	Barometric pressure, in Hg	Pressure, lb/in^2	Pressure differential, lb/in^2
Sea level	29.9	14.7	0
1,000	28.9	14.2	0.5
5,000	24.9	12.2	2.5
10,000	20.6	10.1	3.6
15,000	16.9	8.3	6.4
20,000	13.8	6.8	7.9
25,000	11.1	5.4	9.3
30,000	8.9	4.4	12.3

Source: Ref. 4.

Temperature. Changes in temperature are a more pervasive condition during the operating life of a membrane switch. Storage and transportation conditions are more severe than operating conditions, but the effects of pressure differentials are not dependent on operation. An unrestrained volume of air would increase in volume at the rate of 1/273 of its volume at 0°C for each degree rise in temperature. In a substantially restrained system such as the key cavities of a membrane switch, a large part of the potential expansion is translated into internal pressure. The potential expansion of a membrane switch cavity in accommodation of increased air volumes is design-dependent. It is determined by the physical dimensions of the cavity, the materials which make up the layers of the switch, and the number of those layers. The compensatory expansion of the key cavity through flexure of the layers is assumed to be at 10 percent of the original volume. Stated another way, the first 10 percent increase in volume is assumed to be accommodated by flexure in the layers of the membrane switch. At 1/273 volume increase per °C rise in temperature, this covers a temperature rise of 49°F (27.7°C), i.e., from +77°F (+25°C) (room temperature) to +126°F (+52.2°C). The remainder, +59°F (32.8°C) to the industry maximum of +185°F (+85°C), is assumed to be converted to pressure. This temperature rise is equivalent to a 12 percent pressure increase since pressure and volume vary inversely.

Mechanical Pressure Changes. When a key is pressed in a membrane switch, the top conductive layer is deflected to contact the bottom conductive layer, compressing the air in the cavity (Fig. 8.4). The volume of air in the cavity can easily be reduced to 1/10 of the original volume, increasing the internal air pressure by a factor of 10. The formula, based on Boyle's law and used to estimate increases in pressure due to reductions in volume, is

$$P_1 \times V_1 = P_2 \times V_2$$

where P = pressure and V = volume. If the temperature remains constant, the pressure is inversely proportional to the volume. The reduction in volume and concomitant increase in pressure are dependent on the ratio of the area open in the spacer layer and the thickness of the spacer layer.

Using the guidelines discussed above, it becomes evident that substantial pressures can build within a membrane switch. The most extreme pressures are the rapid fluctuations resulting from actuation, but altitude and temperature can create milder long-term conditions. All these situations must be addressed in the physical design of the switch.

Designing around Pressure. The first design element used to accommodate pressure changes is venting. Venting channels between keys in the spacer layer allow air exchanges from key to key during actuation. Without these channels the volume reduction during actuation creates a back pressure, raising actuation force and putting severe stresses on the bonds between layers. The stress on the bonds can lead to delamination over time in switches constructed with pressure-sensitive adhesives.

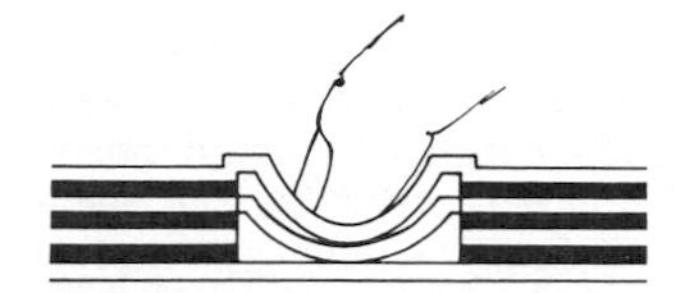

FIGURE 8.4 Actuating a membrane switch compresses the air in the switch cavity. (*Courtesy of Berquist Switch Co., Minneapolis, Minn.*)

The venting channels must be sized to handle the air exchange rapidly. This becomes crucial as the opening diameter increases in proportion to the spacer thickness. In a switch with large key openings and thin spacer material, it is possible to encounter "sympathetic actuations." As one key is released, the movement of air into the cavity of the released key can draw enough air out of an adjacent key to close it momentarily.

Venting between keys relieves the pressure from actuation, but pressure changes resulting from altitude and temperature affect the entire switch at once. Opening the venting pattern to the atmosphere allows the internal and external pressures to equalize, eliminating the pressure differentials. A closed venting pattern (internal venting only) contains the pressures within the switch and depends on the strength of the bond between the layers to prevent eventual delamination.

Venting to Atmosphere. The drawback to designs which are vented to atmosphere is that the atmosphere and all its chemical agents and contaminants will be pumped into the switch through venting channels by the air exchange involved in actuating the keys (Fig. 8.5). However, when a switch is designed to be vented to atmosphere the location of the venting can be chosen to minimize the danger of contamination. The switch can be designed to vent into the case or mounting to avoid spills, sprays, and condensation. Venting patterns can also be designed to scavenger pads of metal or convoluted paths to put distance between the atmosphere and the vulnerable metals inside the switch.

Internal Venting. The drawback to internal venting designs in membrane switches bonded with pressure-sensitive adhesives is that internal pressures can build to the point where the air makes channels through the adhesive to the at-

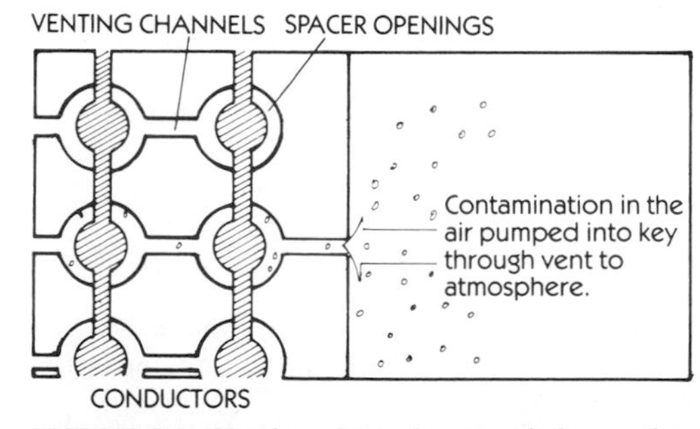

FIGURE 8.5 Venting of membrane switches, unless properly designed, can be a pathway for contaminants. (*Courtesy of Berquist Switch Co., Minneapolis, Minn.*)

mosphere. When the situation which caused the pressure buildup is reversed, the air will not seep back into the switch. The classic example is membrane switches shipped by air which arrive closed owing to lack of pressure in the airplane cargo hold. Internal pressures can also cause eventual delamination.

Heat-Seal Switches. Another approach is to build an internally vented switch which can withstand the internal pressures by eliminating the pressure-sensitive adhesives. Adhesives which are *thermally* activated can be used. This process is referred to as heat sealing. Testing done with a vacuum chamber showed that heat-sealed membrane switches with a 0.125-in (3,17-mm) perimeter seal withstood a simulated altitude of 40,000 ft (12,192 m) for 7 days with no discernible loss of bond. Under the same conditions samples constructed with pressure-sensitive adhesives and a 0.250-in (6,35-mm) perimeter bond failed within hours.

In summary, a membrane switch will be subject to changes in internal pressure due to altitude, temperature, and actuation. Various designs can be used to accommodate these pressures, but each has limitations. The application engineer must discuss specific applications with the manufacturer to decide which design is suitable. [Adapted from (4).]

Membrane Switch Life

The purpose of life testing is to estimate the useful life of a membrane switch employed in the field. Basically, life testing consists of exposing the switch to field conditions in accelerated form. When the exposure is carried on to failure, it can serve to identify the weakest factor in the switch design. The factor may be a physical design element such as the venting pattern, or it may be a faulty material combination or an incorrect manufacturing operating condition.

This section discusses life testing of completed membrane switches. In life testing, the field conditions are laboratory simulations and the switch is actuated mechanically. Since the field conditions and the actuation are artificial, the resulting data are not a guarantee but rather an estimate of performance. By combining test data and aging formulas with the history of similar designs, it is possible to project the field performance of a membrane switch with some statistical confidence. Additionally, some of the elements involved in setting up life testing will be surveyed along with a discussion of the uses and limitations of the resulting information.

Test Setup: What (and How) to Test and Measure. There are two general types of life testing to consider: (1) nonoperating (static) and, (2) operating (dynamic). Nonoperating testing is aimed at environmental conditions which may be found in storage, transportation, and the workplace. This testing identifies the capabilities and sensitivities of the individual materials which make up the switch and their performance in combination with each other. Operating (dynamic) testing is aimed at the final application of the switch, and variables in this testing include type of actuator, current, actuation force, and cycle time, as well as the environment in which the switch will operate.

For both types of testing, the samples should be constructions approximating the actual switch in order to develop data on the behavior of switch materials in specific design configurations.

Whenever possible, the sample switches should be manufactured on production machinery using production tooling. The use of production equipment is important in creating samples which will reflect the influence of the manufacturing processes and perform like production parts. The best test samples are production prototypes of the membrane switch.

Environmental Test Equipment. The temperature and relative humidity ranges experienced by switches during storage and transportation can be extreme. Uninsulated warehouses, truck trailers, and airplane cargo holds are some of the places where these extreme environments are encountered. The geographic location of the workplace, seasonal weather extremes, and type of industry will also affect the environment to be considered in testing. These situations have led to an informal industry standard of −40°F (−40°C) to +185°F (+85°C) and 0 to 95 percent relative humidity.

It is not difficult to create high temperature and humidity in the laboratory with an oven and humidity cells. However, cycling temperature, humidity and chemical environments, simulating exposure to sunlight, and creating changes in air pressure all require more sophisticated equipment. Environmental chambers which combine chemical sprays or vapors and gases, ultraviolet light exposure, temperature, humidity, and barometric pressure cycling are much more complex and expensive pieces of equipment.

To gain adequate background data on the effects of environments on membrane switches, the manufacturer must make a commitment to acquire the required equipment and to allocate time and personnel to conduct the testing. However, without significant input from the user community, especially on new designs, the testing may not be conducted under the correct environmental conditions to provide useful data.

Nonoperating (Static) Testing. By accelerating the effects of time with elevated temperature in the presence of oxygen, thereby aging the switch, it is possible to identify long-range potential failures due to engineering errors which may include material incompatibility, improper manufacturing processing conditions, and design flaws. Once the correspondence between actual time and hours at elevated temperatures is established, samples can be "aged" before testing for chemical resistance, actuated life, and other elements of the expected environment.

One of the primary concerns in life testing is resistance to chemicals found in the workplace. Environmental chemical resistance is critical in the top surface of the switch, whether the graphic film itself or coatings on its surface, as well as in the conductive materials inside. While the manufacturer of the materials and the switch supplier can provide chemical resistance information, the user must determine which chemicals and combinations will be present in the application and unambiguously detail these chemicals in the membrane switch specification. Contact testing, by immersion, spraying, or wiping, can be carried out at ambient temperatures rather simply as long as the chemicals are neither extremely volatile nor toxic in concentration. Chemical resistance testing at ambient conditions, however, has the drawback of not reflecting the accelerated activity of chemical agents at elevated temperatures. Testing at elevated temperatures requires sophisticated equipment as previously mentioned.

When a switch which is vented to the atmosphere is to be used in an environment containing chemical agents, the chemical resistance testing must include the materials inside the switch since the keys act as pumps drawing the environment inside. When the key is closed, it pushes air out; as the key opens, it draws air back in. Whether as vapor or condensed or applied solution, chemical agents present at the opening of the venting to the atmosphere will be pumped in with the action of the keys. Since the keys will be operated under load, the presence of current must be included as a possible agent in internal chemical reactions.

Operating (Dynamic) Testing. Testing becomes more complicated as it progresses to operating testing. The factors of the actuator, current, mounting angle, cycle time, and actuation force enter the picture. If operating life testing is to yield useful numbers, it must be accelerated. Acceleration can be accomplished by using harsher versions of the actuator, current, cycle time, and temperature than are expected in the field.

In life testing, the actuator is a simulation of the operator. In most cases this means the human finger, although some switches are actuated mechanically. The durometer of the human finger has been estimated at 45 shore A scale. A person pushing a key does not strike it at a true right angle. While pushing down, the finger is at an angle to the surface, resulting in a slight wiping action. The force required to actuate a switch is design-dependent and the testing equipment should be adjusted accordingly.

The angle at which a switch will be mounted is frequently overlooked. Few switches are designed to be operated in a horizontal position, and as the angle increases, so do the effects of gravity. This is most evident in a pressure-sensitive adhesive bonded switch at elevated temperatures.

Cycle time is the most obvious factor in dynamic life testing. In order to have data in time to be useful, life testing is run at rates over time which are never seen in the field. Even a key used to jog a machine will not be operated at that rate for hours and days as in life testing.

The operating current for a switch is determined by the electronic device(s) it will trigger. In life testing, the current can be controlled to suit the application or may be dependent on devices used to count actuations. Counters triggered by the switch itself are preferable, since they will indicate cycles at failure as well as multiple or intermittent actuations—actuations which are as much failures as opens and shorts.

Measurement. Measurable changes in characteristics, such as flexibility, peel strength, light transmittance, actuation force, electrical resistance, bounce time, and dimensional stability, can be as important in defining success or failure as opens or shorts. The definition of failure must be provided to a large extent by the switch user, who must decide which characteristics are the most important for the application. A 10 percent rise in contact resistance over 10,000 actuations may be significant in one application but not in another. The degree of change in a given characteristic which is acceptable over time is dependent on the application. Visible change in the graphic layers should also be included in the test results. Visible evidence of change or degradation, although subjective, can be valuable information. Table 8.3 depicts the measurable and subjective characteristics involved in switch testing.

Successful Failures. When a switch fails, it is important to determine the cause of failure. Failures due to limitations of the testing equipment or improper sample preparation are not indicative of life in the field. However, if the design or materials are the cause of failure, the information can be used to improve the switch and extend its life. (5) It is important, as is stressed a number of times in this book, to contact the switch manufacturer with details of the switch failure and to

TABLE 8.3 Measurable and Subjective Characteristics Involved in Switch Testing

Measurable characteristic	Measurement tool
Contact resistance	Ohmmeter
Insulation resistance	Megohm bridge
Actuation force	Force gauge
Bounce time	Oscilloscope
Bond (layer to layer)	Peel strength tester
Dimensional stability	*X-Y* coordinate measuring device
Light transmittance (lens area)	Transmitter densitometer, IR scan
Brittleness	Flex tester
Subjective characteristic	**Evaluative tool**
Surface abrasion	Glossimeter
Tactile feel	Strain gauge, oscilloscope
Delamination	Visual
Curl and warping	Visual, dual indicator, *X-Y* coordinate
Fading	Colorimeter, visual
Ink adhesion (graphic layer)	Crosshatch-tapepull, visual
Color shift	Colorimeter, visual

send the failed switch back to the supplier, unopened, so that determination can be made of the root cause of the failure and corrective action can be taken by the manufacturer.

Actuation Force

Actuation force is design-dependent and usually predictable. The application engineer, in specifying actuation force, must be aware of the constraints that specification places on the actual design process in producing a membrane switch.

The basic function of a membrane switch is to complete a circuit by pressing on the top (graphic) layer, thereby making contact between separate conductive elements. The pressure required to make contact in a membrane switch is called actuation force. Actuation force is a measurable characteristic frequently used as an acceptance criterion.

Design factors which affect actuation force in membrane switches are:

1. The type of materials used and the thickness of the layers
2. The size of the openings in the internal layers
3. The venting pattern
4. The embossing pattern of the graphics
5. Additional components, such as domes, shielding for EMI and RFI, and backlighting devices

Using a basic, standard-construction membrane switch (Fig. 8.6) as an example, the following discussion illustrates the effect of several of the above noted factors. In this example, the graphic layer and top switch layer must be depressed to come in contact with the bottom switch layer in order to actuate the switch. As

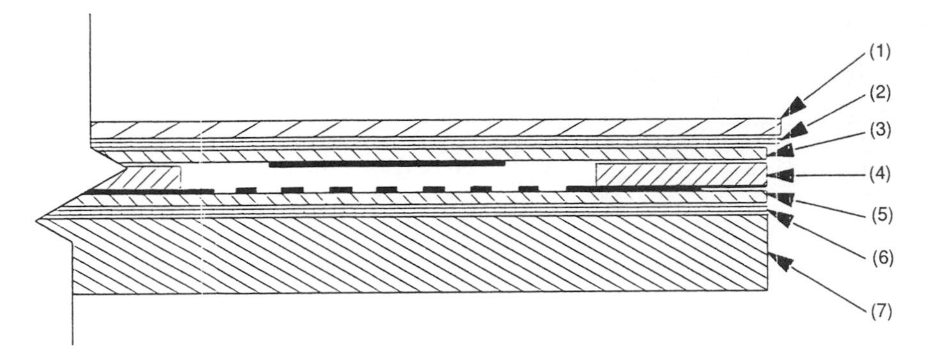

FIGURE 8.6 Each layer of a membrane switch affects the characteristics of the switch: (1) graphic overlay, (2) adhesive layer, (3) top switch layer, (4) spacer layer, (5) bottom switch layer, (6) adhesive layer, (7) backer panel. (*Courtesy of GM Nameplate, INTAQ Division, Seattle, Wash.*)

the intervening layers become thicker the distance the top layers must travel becomes greater and the required force increases. The effects of changing spacer *thickness* can readily be seen in Fig. 8.7*a*. If the size of the opening in the spacer layer is decreased, the angle of deflection of the top layers becomes sharper and the force required to make contact becomes greater. Figure 8.7*b* illustrates the changes in actuation force as the spacer hole *diameter* changes. If the spacer openings are not vented to the atmosphere or to a reservoir of air, the air trapped in the spacer openings will resist the pressure being applied to the top of the switch, requiring greater force to make contact, as well as causing a stiff feel.

Some embossing patterns stiffen the graphic layer, like corrugations in cardboard, and increase the required force. Component layers installed above the conductive layers, such as EMI/RFI shielding, will add to the actuation force because they must be deflected along with the graphic and top switch layers. Just as all these elements can be manipulated to increase the actuation force, they can also be manipulated to decrease the actuation force or to compensate for other design requirements which run contrary to the desired result.

A tactile design (Fig. 8.8*a* and *b*) uses a plastic or stainless-steel dome. The effects of material layer thickness, opening size, and venting remain the same, but a new element is added. The pressure is applied to the face of the switch until the inertia of the dome is overcome and it collapses, making contact. The sensation of collapse can be used as a form of tactile feedback, but the design must be carefully controlled. If the force required to collapse the dome is not sufficient to carry the center of the dome all the way through to electrical contact, the operator will receive a false message ("the tactile feel through my finger indicates I made switch contact") when, in fact, no contact was made. A poor design using domes can also result in multiple actuations caused by physical vibration of the domes or by edge actuation followed by center actuation.

Embossed Graphics. An embossing pattern in the graphics layer is usually chosen for switch (finger) location and cosmetic reasons, but it can create unpredictable effects on actuation force and feel. Embossing is a heat-deformation process which can leave latent stresses in a plastic film. Dense embossing patterns, keys

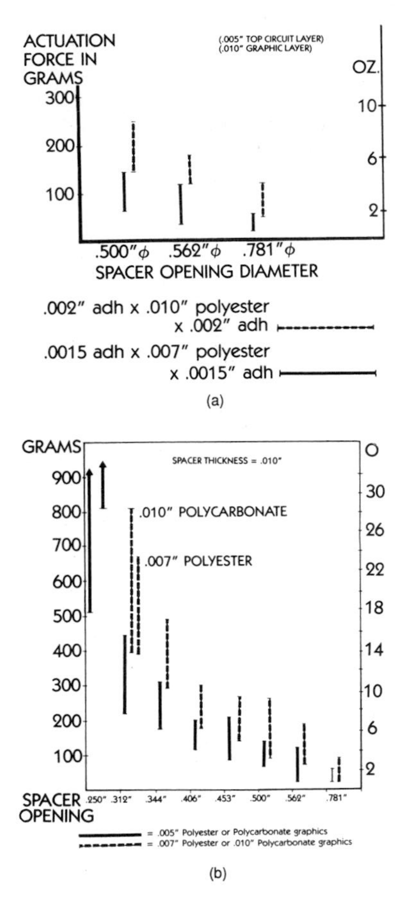

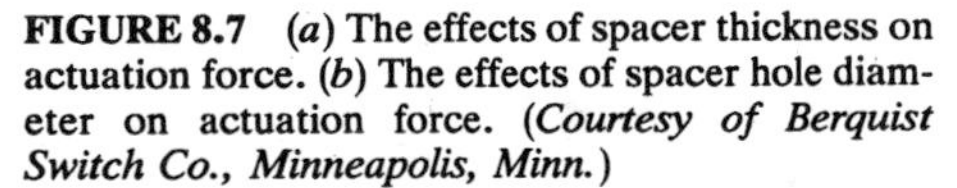
FIGURE 8.7 (*a*) The effects of spacer thickness on actuation force. (*b*) The effects of spacer hole diameter on actuation force. (*Courtesy of Berquist Switch Co., Minneapolis, Minn.*)

close together, or images embossed inside a key perimeter can lead to "oil canning." This is the tendency for the center of the key to bow either up or down. When the key bows down over a snap dome, the feel can be deadened because it is not free to travel through its entire range. When the key bows up, there can be a tendency for the graphics to lift. The tendency of a given embossing pattern to "oil can," or affect the actuation force and feel in some other way, can be predicted to some degree. For this reason "rim embossed" and "whole pad embossed" keys are the most common. The effect of other patterns should be decided on the basis of prototypes with the option of compensating for undesirable effects by modifying other layers in the switch when unusual embossing is necessary. See Chapter 10 for details of embossing styles.

Measuring the Force. Actuation force is usually measured using a "push-pull" or strain gauge and some indicator of electrical contact. These gauges read force only and do not indicate the elusive and important quality of "feel." In order to

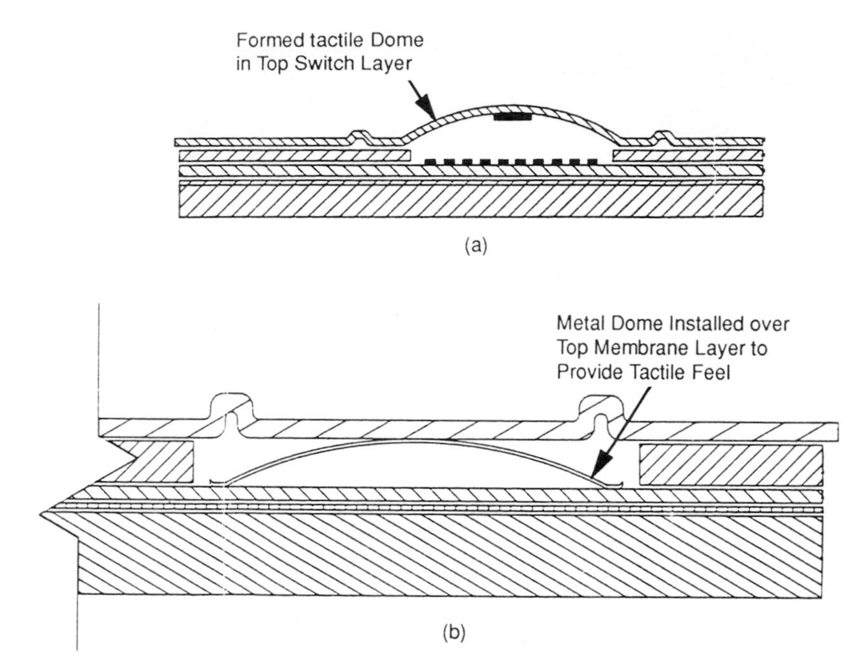

FIGURE 8.8 A membrane switch can be designed to incorporate tactile feel (*a*) by formed domes in the polyester overlay, or (*b*) by utilizing a metal dome (Chap. 9) as a tactile device. (*Courtesy of GM Nameplate, INTAQ Division, Seattle, Wash.*)

use actuation force readings as acceptance criteria, the following factors must be taken into consideration:

1. Different gauges will give different readings, especially if the probes are not the same shape and durometer. A conical metal probe, by concentrating the applied force in its point, will give a significantly lower reading than a spherical rubber probe.
2. Readings are operator-dependent when a hand-held gauge is used.
3. Speed and angle of application, flexure of the support layer, and positioning of the gauge within a key can also result in variations.

It is important to note that when actuation force is used as an acceptance criterion, the measuring tools and method of use *must* be correlated between user and supplier! [Adapted from (6).]

Substrates for Membrane Graphic Panels

Two widely used substrates for membrane graphic panels are polyester and polycarbonate. Each has very different characteristics and surface treatments. The following section details some of the unique properties of these materials, their advantages and disadvantages, and other rarely utilized plastic materials. Table 8.4 depicts a comparison of physical properties of the most popular membrane graphic panel substrates.

TABLE 8.4 Physical Properties of Plastics for Faceplates

Property	Units	Shiny polyester*	Nonglare polyester*	Shiny polycarbonate*	Nonglare polycarbonate*	Textured polycarbonate†
Gloss	Gloss units	+191‡	+50	+186‡	+53	−1.8–4.6§
Haze	%	60	20–25	<0.7	20	100
Distortion temperature	°F	155	155	275	275	275
Maximum usage temperature	°F	150	150	200	200	200
Minimum usage temperature	°F	−40	−40	−40	−40	−40
Tensile strength	lb/in^2	25,000	25,000	9000	9000	9000
Stress to produce 5% elongation	lb/in^2	15,000	15,000	7000	7000	7000
Ultimate elongation	%	120	120	100	100	100
Tear strength	lb/in	1800	1800	1150	1150	1150
Burst strength	lb/in^2	66	66	30	30	30
UV exposure,¶ sunlight	h	180	200	100	200	100
Weatherability (QUV¶)	h	>1100	25	1400	>336	1400
Humidity, 100°F, 95%	h	720	200	720	200	720
Abrasion resistance at 1000 g	cycles	6000	—	45,000	—	45,000
Thickness available	mil	4–7	4–7	5–30	5–30	5–20

Source: Ref. 8
*Values typical for 1-mil film.
†Values typical for 5-mil film.
‡Measured at 20°.
§Measured at 85°.
¶1 h equivalent to approximately 100 h actual exposure.

Polyester. Polyester is available in gauges up to 0.008 in (0,20 mm), with a shiny, slightly wavy surface which allows for very sharp colors and good definition of printed graphics. In gauges above 0.008 in, the material is naturally cloudy in appearance. Since the shiny material can produce an undesirable glare in certain applications, a nonglare finish can be applied overall or selectively (as part of a hardcoating process discussed later in this chapter), or a brushed polyester can be used, which has a built-in nonglare surface as the material comes from the factory. Brushed polyester, however, has several drawbacks, including the fact that it is available only in thin gauges. Use of the brushed polyester usually results in a part that can become brittle over time and is easily dented. Additionally, the clarity is not as good as it is using clear polyester as a graphic substrate, followed by a top hardcoat application which imparts nonglare qualities.

Polyester holds up extremely well to weathering and normal usage. The material will not yellow or become brittle under sunlight or ultraviolet light. Polyester, even without a protective hardcoat finish, is resistant to many chemicals found in industrial settings, household chemicals, and petroleum products (Table 8.5).

As a general rule, polyester does not emboss as well as polycarbonate because sharp breaks cannot be easily achieved in the material. Integral tactile domes can be formed in polyester if the form tooling provides a gradual break from the flat to the formed surface. Polyester has excellent memory and will return to its original flat state when exposed to temperatures above 149°F (65°C). These factors need to be considered when embossing or forming is done.

Between −40°F (−40°C) and 149°F (65°C) polyester remains dimensionally stable. Polyester graphic panels exposed to temperatures above 149°F (65°C) will begin to shrink and distort. Above 149°F there will be measurable shrinkage. The

TABLE 8.5 Chemical Resistance of Materials for Faceplates

Solvent	Shiny polyester	Nonglare polyester	Shiny polycarbonate	Nonglare polycarbonate	Textured polycarbonate
Distilled water	N	N	N	N	N
5% detergent solution	N	N	N	N	N
5% soap solution	N	N	N	N	N
Windex (with ammonia)	VS	N	N	N	N
Isopropyl alcohol	VS	N	M	N	M
Hydrogen peroxide	N	N	N	N	N
Acetone	M	VS	E	E	E
10% sodium hydroxide	S	VS	VS	VS	VS
Acetic acid 10%	N	N	N	N	N
Sulfuric acid 10%	N	N	S	VS	S
Hydrochloric acid	N	N	—	—	—
Methyl ethyl ketone	N	N	E	M	E
Toluene-xylene	N	N	E	M	E
Hexane	N	N	VS	N	VS
Naphtha (petroleum)	N	N	VS	N	VS
20 weight SAE oil	N	N	VS	N	VS
Gasoline	VS	N	VS	N	VS
1,1,1-Trichloroethylene	S	VS	E	M	E

Key: Description refers to amount of attack. N = none, VS = very slight, S = slight, M = moderate, E = extreme.

Source: Ref. 8

glass transition temperature of polyester is 158°F (70°C), and the material starts to flow at this temperature.

In summary, the attributes of polyester that affect membrane switch graphic panel design are:

- Polyester is mechanically superior to polycarbonate.
- Polyester can be expected to perform well for at least 4 to 5 million actuations.
- Pretextured surfaces are available.
- Chemical resistance is superior to polycarbonate.
- Polyester is more difficult to die-cut and emboss, compared with polycarbonate.
- Silk-screen printing is more difficult with polyester, and the resulting colors are not quite as vivid as colors on polycarbonate.
- Polyester holds up well under close-proximity fluorescent lighting for extended periods of time. (7)

Polycarbonate. Polycarbonate is a popular choice for membrane graphic panels for a variety of reasons. Polycarbonate is not as naturally solvent-resistant as polyester, and this fact is both an asset and a liability when using polycarbonate as a membrane graphic panel. Since polycarbonate is less solvent-resistant it is relatively easy to print on and, in fact, can be processed in standard printing presses. The liability, of course, is that the polycarbonate, unless protected by a hardcoat, is attacked by a number of relatively common chemicals as shown in Table 8.5. Another factor making polycarbonate a popular choice is that it is available not only with a clear, shiny surface but also with integral matte or textured surfaces. Each surface treatment results in a different "look" to the graphic panel. With the advent of hardcoated surfaces capable of being textured to produce a variety of surfaces, the integral textures available on polycarbonate have become less of a factor but nonetheless are a consideration, especially to the industrial designer who may wish to match or contrast the surface of a membrane graphic panel with the surface of a housing or enclosure.

Polycarbonate is easier to form thermally than polyester. Crisp edges around embossed areas can be readily obtained. However, it is generally not good practice to emboss in areas which will be flexed frequently since polycarbonate will stress, causing fractures in these areas.

Over extended periods of time, in direct sunlight, clear areas of polycarbonate will tend to yellow slightly. Because the glass transition temperature of polycarbonate is much higher than that of polyester, the material is stable up to 275°F (135°C).

If polycarbonate is used as a graphic panel substrate, it is always prudent for the user to specify hardcoating on the top surface of the panel. The slight increase in unit cost is more than offset by the enhanced survivability of the coated polycarbonate.

To summarize, the attributes of polycarbonate which can affect membrane graphic panel design are:

- Polycarbonate embosses easily.
- Polycarbonate provides an excellent surface for application of graphics and reproduction of color.
- Polycarbonate has a good resistance to many chemicals.

- Pretextured surfaces and hardcoating enhance the durability of the graphics panel face.
- Polycarbonate tolerates fluorescent lighting quite well but does not hold up well in extended periods of sunlight (UV).
- Polycarbonate can withstand up to 1 to 2 million actuations. (7)

Other Substrate Materials. Although very few, if any, switch manufacturers use the following materials any longer, they are worth mentioning in case special circumstances warrant their consideration:

Vinyl (polyvinyl chloride) tends to be too flexible for the vast majority of applications and has a slower recovery rate than either polyester or polycarbonate, resulting in multiple switch closures. Vinyl is also plasticized, which eventually causes any pressure-sensitive adhesive attaching the graphic panel to the basic membrane switch to release, and the graphic panel detaches from the switch.

Styrene has good optical characteristics but is prone to scratching and tends to tear easily. Styrene starts to distort above 140°F (60°C), which limits its use in any environmental situation exceeding that temperature.

Polysulfones have excellent high-temperature performance characteristics but have not progressed to the stage where the material can be produced optically clear and suitable for graphics printing. [Adapted from (8).]

Colors for Graphic Panels

One of the outstanding features of graphic overlay panels is the ability to choose virtually any color or combination of colors for the printing of the panel graphics.

Color Standards. Color chips provide the most accurate duplication of desired colors and are the preferred method of color specification. The Pantone color matching system (PMS) may also be used for initial color selection. Federal Standard 595 may also be used as a color standard as most membrane switch manufacturers are now familiar with this document.

The specifying engineer should always insist that the selected switch manufacturer submit actual color samples, on the correct material and with the specified texture, for approval prior to the start of any manufacturing. This is most important because texturing can cause the color to change slightly.

The following should be considered when choosing colors for the graphics panel:

1. Avoid large areas or backgrounds of white or beige, if possible. All colors are created by silk screening ink onto the transparent poly material. Opacity is more difficult to achieve with white or beige colors, compared with darker colors.
2. Avoid similar shades of colors that will be printed together. For example, dark red lettering on a black background will not be as readable as a lighter red on black. The same is true of light blue lettering on a light gray background instead of, perhaps, a dark blue.
3. Avoid combining colors that clash or colors that will visually distract the membrane switch user.
4. Colors must overlap at the edges of LED windows, and depending on the choice of colors, this often will produce a "ghosting" of the overlapped color.

If, because of color requirements, this ghosting cannot be avoided, it is recommended that a thin black border be printed over the ghosting to improve appearance.

Nonfunctional Graphics. Nonfunctional graphics such as company logos and product or model names may also be included as part of the graphic panel design and printing. For an even more dramatic effect, these special graphics may be embossed as well.

Windows. Windows for displays and annunciators are easily incorporated into graphic panels. Since the base material is transparent anyway, *not* printing in a certain area will leave a window of the desired shape. Reflections off displays beneath the window can be minimized by selective hardcoating and texturing on the window itself, or treating the window with a glare-reducing material, while retaining nearly all the transparency of the base material.

Textures. Both polyester and polycarbonate are available with a variety of texture finishes. A gloss graphics panel finish should be avoided because, more than others, it will show scratches, smears, etc. A velvet-textured surface works well in most applications. Durability of the polycarbonate surface can be enhanced through the use of a clear hardcoat. This hardcoat can be textured during the application process to produce excellent results. (7)

The hardcoat can be applied in multiple passes using different silk-screen artwork configurations and each pass can be textured selectively, to produce a graphics panel with several different texture appearances.

How Pressure-Sensitive Adhesives Attack Graphic Inks

Graphic inks used in screen printed graphic panels are usually printed on the back side of plastic films to protect them from chemical degradation and abrasion. This so-called second-surface printing protects the graphic ink layers from harsh environmental conditions but brings the inks into contact with the pressure-sensitive adhesives used to bond the graphic panel to the basic membrane switch. In some cases the pressure-sensitive adhesives will attack the graphic ink layer and lead to color changes or delamination. Pressure-sensitive adhesives contain solvents, tacifiers, and plasticizers which enable them to "wet out" the surface to which they are applied. "Wetting out" refers to intimate contact achieved by a liquid on a solid surface and is the mechanism by which the adhesives bond. The solvent and plasticizer component of an adhesive also allows the adhesive to remain pliable enough to maintain adhesion to flexible materials such as plastic films. When the graphic inks are printed on the back side of the plastic film, the *inks* become the surface to which the pressure-sensitive adhesive bonds, not the substrate itself.

Composition of Graphic Inks. Screen process graphic inks consist of pigment particles or dyes, a binder resin, fillers, modifiers, and solvents. Graphic inks can be cured in several ways, but "conventional curing" (solvent drying) and ultraviolet curing are the most common methods. Conventional inks are cured by removing the solvents with heated currents of air. The resin system may crosslink to some degree depending on the type of resin. When dried, these inks consist of particles of pigment and fillers bound in resin. The structure is porous, and active solvents in pressure-sensitive adhesives can penetrate the graphic ink layer. Ultraviolet-cured inks contain little or no solvent and cure by crosslinking on exposure to the proper intensity and wavelength of light. Because these inks crosslink extensively, they are less susceptible to solvent attack. The disadvantage to ultraviolet curing materials is that the pigment loading must be kept low enough to allow light penetration in order to cure. The resulting cured inks are not opaque enough for most graphic applications. Ultraviolet-cured inks are frequently used for display windows in graphic panels where light transmittance is desirable.

Opacity is one reason that conventionally cured inks are used in process printing for graphic panels. Another reason is that the solvent component in the wet inks will clean and slightly etch the surface of many plastic films, resulting in excellent adhesion. The solvents also enhance layer-to-layer adhesion in multicolor printing by redissolving the surface of the dried ink on which the wet ink is printed.

Solvent Attack. Since the surface of a dried conventional ink can be redissolved by solvents in a printing process, it may also be redissolved by the solvents in a pressure-sensitive adhesive. In the printing process, the solvents in the wet layer of ink are removed in the curing oven before they penetrate the dry layer far enough to cause damage. The solvent content of a pressure-sensitive adhesive is slight compared with a wet graphic ink, but the pressure-sensitive adhesive will be in contact with the graphic ink for the life of the membrane switch, and such solvents are highly active.

When a graphic ink has been softened, it may lose adhesion to the plastic film, change color, bleed, or be deformed with mechanical stress. The severity of the attack depends on (1) the solubility of the ink resin and the nature of the solvents in the adhesive, (2) the degree of cure of the graphic ink, and (3) the environmental conditions where the graphic panel is stored and used. (9)

It is recommended that the user consult with a potential membrane switch supplier to determine the manufacturer's understanding of pressure-sensitive adhesives and their role in solvent attack and the methods used by the manufacturer to prevent this mode of failure.

Hardcoated vs. Uncoated Surfaces on Membrane Graphic Panels

Hardcoating is a material, usually in liquid form, coated or printed on the top surface of a plastic film and then cured. One purpose of the coating is to improve the chemical and abrasion resistance of the graphic layer. The applied hardcoating can be glossy, matte, or textured, depending on the material and the processing. With some processes, such as screen printing, the coating can be applied selectively. Selective texturing is the second purpose of hardcoating because it allows areas such as display windows to be glossy or nonglare while keys and other areas can be textured. When a window is created directly on the graphic panel by applying hardcoating, it eliminates cutting a hole through the graphics and laminating a separate window sheet, thus preserving the environmental integrity of the graphic panel.

The success of a hardcoat, as a means of extending the life of a graphic panel through improved chemical and abrasion resistance, depends on several things:

the chemistry of the hardcoating material, compatibility with the film to which it is applied, and correct processing.

Another area which deserves consideration is the edge of the graphic layer or membrane switch. The protection provided by a hardcoat does not extend over the sides. Protection for the sides of the assembly must be provided in the mounting design.

Chemistry. The chemistry of a hardcoating material will dictate its curing methods, possible chemical and abrasion resistance, as well as possible texture variations. Any hardcoating material is resistant to some chemicals and susceptible to others, so the specifying engineer has an obligation to inform the switch manufacturer of the chemicals expected in the application. In this reference, the term "chemicals" includes cleaners, solvents, gases, pollutants, slurries, and any other materials in the workplace which are expected to come in contact with the graphic panel. These hostile materials are not always obvious but may be quite active at elevated temperatures. For example, some of the chemicals which have been found to be reactive include cocoa butter, antistatic treatments, some household chemical cleaners, and sewer sludge.

Data on the chemical resistance of hardcoating materials may be available from the membrane switch manufacturer, but the test data must be used judiciously. Testing is conducted in a given set of conditions. Any change in peripheral conditions, such as humidity or temperature, may cause a different reaction. Where field conditions differ substantially from test conditions, it may be necessary to conduct new testing at expected field conditions to establish a correlation.

Compatibility. Compatibility between a hardcoating material and graphic films required similar values for characteristics such as thermal coefficient of expansion and flexibility. Clarity and the ability of the hardcoating to adhere to the graphic plastic film are also important considerations. Incompatibility between the hardcoating material and the film shows up as a loss of adhesion in most cases. The loss of adhesion may show up at once or become evident in further processing or in the field. Lack of adhesion can be the result of a difference in the flexibility of the two materials, a chemical reaction between the materials, or a poor bond initially.

Process Control. Process control is vitally important in achieving the full potential of the coating-film combination. The first concern in the control of a hardcoating process is cleanliness. Particulates, oils, or plasticizer on the surface of a graphic film will interfere with the bond at the hardcoat-film interface. The hardcoat will stick to the dirt, but the dirt will not stick to the film. Fingerprints are a particular hazard because they introduce oils, salts, and acids to the film surface. Particulates in the hardcoating material itself reduce the integrity of the cured hardcoating and become stress points reducing the film's chemical resistance and flexibility.

The second element in process control is to assure the correct cure level. Hardcoating materials may be cured several ways, although conventional and ultraviolet curing are the most common. An improperly cured material will not reach its potential in chemical resistance or other characteristics. An inert atmosphere is required for proper cure or to create textures with some hardcoating materials. Undercure results in a soft coating containing excess solvents or unreacted components; overcure creates a brittle coating.

Further processing, such as embossing and die cutting, stresses the hardcoat-film combination. Embossing can cause cracks in the hardcoating layer because the film must expand under heat and pressure to conform with the embossing die. The amount of stress increases as the embossing pattern becomes more dense and as the draw becomes deeper. Die cutting can chip the hardcoating on the cut line. Brittle or overcured hardcoatings are prone to this type of failure. Care must be taken to control processing stresses to maintain the integrity of the hardcoating so that it can withstand conditions in the field.

Chemical Resistance. Table 8.6 depicts results from tests conducted by Berquist Switch Co., Minneapolis, Minn., in which hardcoated polycarbonate or polyester materials were immersed in or wiped with a variety of chemicals, both nonvolatile and volatile.

Test 1 was performed using nonvolatile chemicals. The hardcoated samples were placed in a beaker containing the chemical solution and were checked daily.

Test 2 involved volatile chemicals. Samples of the hardcoated materials were placed on a glass coupon with a pad saturated in the chemical solution placed on top of the materials and held down with a second glass coupon. The samples were checked daily, and the solution was replenished as necessary.

Test 3 was performed by wiping the hardcoated samples ten times (back and forth) with a saturated pad under a 14-oz (400-g) weight. This testing was directed at the resistance of the hardcoating layer and the graphic film beneath it. These tests were conducted at ambient conditions and may not reflect the behavior of the materials at elevated temperatures. (10)

The INTAQ Division of GM Nameplate, Seattle, Wash., has conducted similar tests on their commercial-grade hardcoat (Series GM 1-030, 035, 045, and 055), the results of which are shown in Table 8.7. The "double rub" test column reports the hardcoat's durability when subjected to a 7.7-lb (3500-g) load, 200 double rub test performed with industrial cloth saturated with each liquid noted. The "soak" test column reports the durability of the hardcoat when subjected to the more severe "puddle" or "soak" test in which a cotton swab, saturated with the liquid noted, is left on the hardcoat surface for 24 h. (11)

INTAQ also has developed a hardcoat specifically for medical and laboratory electronics applications. GM 1-060 (clear finish) and GM 1-065 (textured finish) withstand normal industrial solvents plus strong disinfectants commonly used in hospitals and laboratories, including KODAN Tinktur Forte, a chemical disinfectant which severely attacks normal hardcoating. KODAN is used in the European medical environment. (12)

Abrasion Resistance. Abrasion can take several forms: particulated, rubbing, scratching, and ordinary abuse. Particulates are dirt, sand, and other hard particles rubbed against or blown onto the graphic layer of a switch. Rubbing abrasion is usually encountered either in cleaning the switch or when the operator wears gloves. Scratching occurs when objects, such as pens or other tools, are used to actuate a switch or when objects glance off the surface of the switch. Abuse is obvious when it occurs, but it cannot be predicted.

Abrasion resistance is tested using several measurable characteristics: weight, light transmittance, and reflectance. Abrasion resistance tests are conducted with abrasive wheels, falling sand, weighted pads, pencils with varying degrees of lead hardness, and other types of abusive items that may come in contact with the switch. The object of the testing, and therefore the test method, varies with the

TABLE 8.6 Chemical Resistance Tests

Chemical	Hardcoat 1		Hardcoat 2		Commercial coated
	Polycarbonate	Polyester	Polycarbonate	Polyester	
Test 1. Immersion test in nonvolatile chemicals					
Motor oil	No reaction	No reaction	No reaction	No reaction	
3 in 1 oil	No reaction	No reaction	No reaction	No reaction	
Transmission fluid	No reaction	No reaction	No reaction	No reaction	
Fertilizer (plant food)	White blush	White blush	Slight orange stain	Slight orange stain	
Coffee	Stained	Stained	Stained	Stained	
Pepsi	Stain	Stain	Stain	Stain	
"409" cleaner	Wrinkled	Dissolved	Wrinkled	Dissolved	
Test 2. Immersion test in volatile chemicals					
Diesel fuel	No reaction	Slight loss of texture	No reaction	No reaction	
Battery acid	Turned to white powder	Blistered	Turned to white powder	Blistered	
Turpentine	Increased transparency	Transparent border	White blush	No reaction	
Methylene chloride	Film melted	Hardcoat blistered	Film melted	White blush	
Isopropyl alcohol	Contact edge clearer	Contact edge clearer	Slight blush	Slight blush	
Hydrogen peroxide	No reaction	No reaction	No reaction	No reaction	
Phosphoric acid (20%)	White blush	White blush	White blush	No reaction	
Sodium hydroxide (10%)	Dissolved hardcoat	Chalky	Wrinkled	Chalky	Removed finish
Gasoline	No reaction	No reaction	No reaction	No reaction	
Test 3. Rubbing test with various chemicals					
Motor oil	No reaction	No reaction	No reaction	No reaction	No reaction
Gasoline	No reaction	No reaction	Milky streaks	No reaction	No reaction
Sodium hydroxide (10%)	Streaky	No reaction	No reaction	No reaction	No reaction
Phosphoric acid (10%)	Slightly transparent	Slightly transparent	No reaction	No reaction	No reaction
Methylene chloride	No reaction	No reaction	Slight white blush	Slight white blush	Film melted
Alcohol	No reaction	No reaction	No reaction	No reaction	No reaction
Hydrogen peroxide	No reaction	No reaction	No reaction	No reaction	No reaction
Diesel fuel	No reaction	No reaction	No reaction	No reaction	No reaction

Source: Ref. 10.

TABLE 8.7 Rub and Soak Testing of Commercial-Grade Hardcoat

Liquid	Double rub test	Soak test
Isopropyl alcohol	No detectable effect	Moderate attack after 1 h
Tetrahydrofuran	No detectable effect	No detectable effect
Methyl chloride	No detectable effect	Moderate attack after 1 h
Methyl sulfide	Slight deterioration	Blisters quickly
Dimethylformamide	Not tested	Blisters quickly
Acetone	No detectable effect	Attacks immediately
Deoxane	No detectable effect	Moderate attack after 24 h
Ethyl acetate	Slight deterioration	Blisters quickly
Point 1 normal HCl	No detectable effect	Moderate attack after 1 h
Carbon disulfide	No detectable effect	No detectable effect
Ether	No detectable effect	No detectable effect
Caustic etch solution 6%	No detectable effect	No detectable effect
Score detergent 6%	No detectable effect	No detectable effect
Skydrol	No detectable effect	No detectable effect
Chloroform	No detectable effect	Moderate attack after 24 h
Benzene	No detectable effect	No detectable effect
X-4 recorder ink	No detectable effect	Light stain after 1 h
Gasoline (80 octane)	No detectable effect	No change or blemish
Ammonia (household)	No detectable effect	No change or blemish

Source: Ref. 11.

application. The primary concern is usually the point at which the abrasion becomes noticeable and objectionable. Matte or textured surfaces hide abrasion by scattering light. A smooth surface, such as a mirror, reflects details clearly; a matte or textured surface does not. Bare glossy plastic films scratch visibly with slight abrasion. The abrasion resistance of hardcoating is most evident in display window areas. The abrasion resistance does not, however, require a sacrifice in light transmittance. The light transmittance chart in Table 8.8 shows light transmittance readings for four degrees of texture on clear polyester film, clear polycarbonate film, and commercially textured polycarbonate film.

Matte, semigloss, or textured surfaces have the added advantage of cutting glare, allowing an operator to see the keys or the display behind the window clearly in strong light. As with chemical resistance, it is important that the user-engineer specify the type of abrasion expected in the field so that appropriate tests can be performed prior to production. [Adapted from (10).]

TABLE 8.8 Percent of Light Transmitted through Four Degrees of Texture on Three Plastic Films

	Clear polyester	Clear polycarbonate	Textured polycarbonate
Bare film	89	90	88
Gloss	89	90	87
Nonglare	88	89	87
Matte	88	88	86
Velvet	73	73	75

Source: Ref. 10.

Silver Migration

Silver migration is a phenomenon that can occur in membrane switches since the conductive medium of this switch technology is primarily silver. When silver migration occurs, the results are generally undesirable because adjacent conductors will eventually be connected by the migrating electrical path, causing a short. It should be noted that silver migration under actual conditions is somewhat rare since the conditions required for migration to occur are not usually available in the correct intensity or combination. Improper design, however, is the typical cause of silver migration.

How Silver Migration Occurs. Silver migration is an electrolytic process, which, under certain circumstances, may produce insulative failure in membrane switches. The elements necessary for silver migration to occur are:

- A silver-containing anode
- An electrolyte such as water
- Electrical potential

In a simplified occurrence in which water is the electrolyte, the metallic silver at the anode reacts with the water and goes into solution. Under the influence of an applied potential, the ions in the solution migrate toward the cathode and are electrolytically deposited in the form of metallic silver. The formation that begins at the cathode and "grows" toward the anode is called dendrites. Eventually, a complete path "grows" to join the cathode to the anode, creating a low-resistance conductive path that will short the two potentials together. If high current is passed, the short may be temporarily destroyed by burning away. However, the shorting path will quickly re-form.

Variables in Silver Migration. Numerous variables in the mechanism of silver migration can slow the process or even completely stop it. The following is a partial list of these variables:

1. Distance of separation between anode and cathode
2. Anode metal composition
3. Substrate on which the silver is used
4. Presence of an electrolyte
5. Magnitude of dc potential
6. Elapsed time
7. Ionic contaminants in the conductor
8. Temperature
9. Conductive ink binder

By controlling one or more of the variables, the probability of silver migration can be reduced or eliminated. To simplify the variables, we can look at silver migration as a triangle (Fig. 8.9). Eliminate any one of the three major variables and silver migration does not occur.

Normally, any one of the three major variables cannot be removed completely and therefore membrane switch manufacturers who understand the silver migration process will attempt a compromise with the three, as well as considering the nine variables shown in the list above. For example, a switch manufacturer might protect against condensing humidity by encapsulating all electrical conductors with a coating that resists moisture. The switch manufacturer may also consider running conductors of opposite potential on separate layers or encapsulating the conductor in the contact areas with a protective nonmigrating conductor. (13)

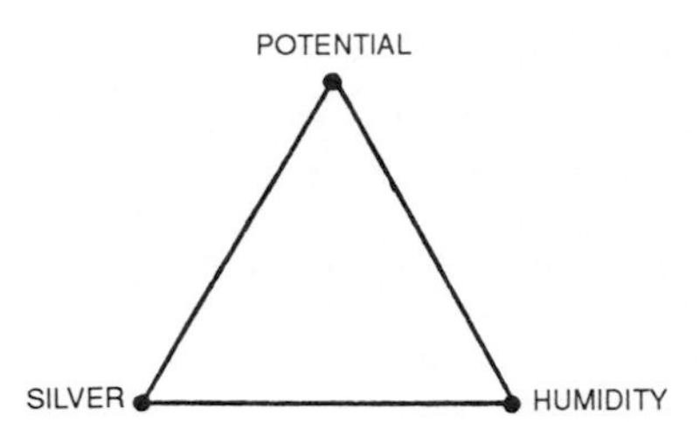

FIGURE 8.9 The three major variables which cause silver migration; remove any one and the probability of silver migration is reduced or eliminated. (*Courtesy of W. H. Brady Co., XYMOX Division, Milwaukee, Wis.*)

Planar vs. Perpendicular Flex Tails

The various circuit configurations on any membrane switch panel are typically routed to a single location for termination to circuit boards containing interface electronics. This exit point is the beginning of what is commonly called the flex tail. The flex tail is simply a section of one of the circuit substrates, containing parallel conductive traces, which join the membrane switch circuitry to a connector. The length of this tail will vary from design to design but will typically be from 2 to 6 in (51 to 152 mm) in length. This section discusses how this flex tail is designed to exit the membrane switch.

Planar Tails. A common form of tail construction in the membrane switch industry is simply a continuation of the flex tail from a given edge of the membrane switch itself. Because it is a continuation of the circuit substrate and in the finished construction exists in the same plane as the circuit, it is called a "planar" tail (Fig. 8.10*a*). This construction is easily manufactured and is acceptable for a wide variety of applications. However, when a part is mounted to a housing from

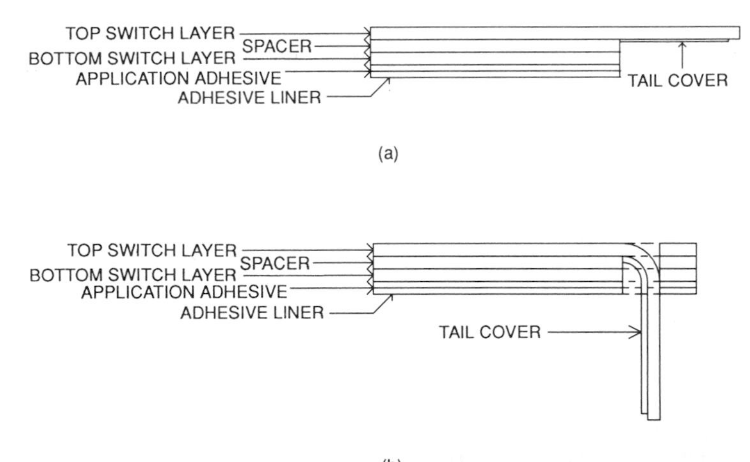

FIGURE 8.10 There are two basic designs for exiting flex tails from membrane switches: (*a*) the planar tail design, and (*b*) the perpendicular tail design. (*Courtesy of W. H. Brady Co., XYMOX Division, Milwaukee, Wis.*)

the front, there are several potential problems associated with the planar tail construction. One potential problem is that of the finished appearance of the product. It is difficult to conceal a hole in the housing for the tail when that hole is at the edge of the part. In many cases, a trim bezel is necessary. Second, a sealing problem could exist in that liquids spilled on the surface can leak into the interior of the product through the flex tail hole in the housing. An alternative to the planar flex tail construction which overcomes the problems of appearance and leakage is called a perpendicular tail.

Perpendicular Tails. The perpendicular tail is appropriately named since the tail section of the circuit substrate is inserted through a slot in lower layers of the switch lamination during construction of the switch (Fig. 8.10*b*). This constrains the tail to a plane that is perpendicular to the plane of the circuit substrate. The tail now exits from the back of the part instead of the side and can be located at virtually any point on the part not obstructed by particular key locations. This type of tail breakout can be accomplished from either the upper or lower circuit substrate. This offers a great deal of flexibility for the location of the exit point in that it no longer is at the edge of the part and therefore is not susceptible to the disadvantages of the planar tail construction. The variability in tail location gives greater freedom to the overall package design and can decrease the tail length requirement, making the switch package more efficient to produce. More importantly, the presence of a perpendicular tail provides an adhesive seal around the entrance to the interior of the product, protecting the product from environmental spills that could otherwise damage the electronics. (14)

Connectors for Membrane Switches

Membrane switches, in general, cannot be directly soldered because of temperature limitations of the substrate materials. As a result, connectors have been developed that will provide the mechanical and electrical interconnect between the flex tail of the membrane switch and the printed wiring board of the host electronics (or other device). These connectors fall into three major categories: crimp-on; low insertion force, and zero insertion force. Each type has its own advantages and disadvantages.

Crimp-on Connector. Crimp-on connectors, as their name implies, are mechanically crimped onto the tail of a membrane switch. This is a permanent connection that cannot be removed. These connectors must be applied by equipment usually designed by the connector manufacturer. Some styles of this connector incorporate an insulated housing which is installed after all conductive tracks have been terminated. Other connectors have the contacts preloaded into the housing and all contacts are applied at the same time (mass terminated). The advantages of this type of connector are:

- Connection to the membrane switch is usually gastight.[1]
- The connector can be applied through most flex tail cover materials, leaving no conductors exposed to the environment.

[1]A gastight connection is one in which the connector contact "seals" to the conductor track, blocking the entry of gaseous contamination on the contact surfaces.

- Once installed, the connector may be inserted and removed from the mating connector without damage to the tail. This is important if the switch connection must be frequently removed for service to the final product.
- This type of connector is generally a male-female type of connector, mating with widely available 0.025-in (0,63-mm) square posts or sockets.
- Some housings can be provided with strain relief.

The major disadvantage of this type of connector is that it tends to be the most expensive termination, since its application requires an additional mating connector to be installed on the control board.

Zero Insertion Force Connectors. Zero insertion force connectors form a connection between the conductive ink of the switch tail and the control circuitry by pinching the flex tail conductors between the connector contacts. No force is required to insert the tail into the connector and no reliable connection is made until the connector is closed. This pressure connection is subject to failure due to vibration and stress; therefore, this type of connector generally has some strain relief. This is usually accomplished by punching holes through the flex tail, through which locking pins on the connector are inserted, preventing the flex tail from being removed unless the connector is opened. These types of connectors are not an integral part of the membrane switch and are therefore procured separately by the end user and usually installed on the control circuit PCB.

The advantages of this type of connector are:

- This type of connector provides a reasonable connection without substantial damage to the flex tail conductors.
- Usually a method of strain relief is provided.
- This type of connector is generally lower-cost than the crimp-on type.

The disadvantages to this type of connector are:

- Not all connectors are gastight.
- The pressure at the contact point will encourage some silver-polymer conductive inks to weld to the connector, particularly in high-temperature applications. This could be of concern if the tail were to be removed from the connector.
- The additional strain relief holes in the tail generally increase the cost of the membrane switch since their placement tolerance requires special tooling.
- Contacts can be contaminated with flux when prepared for wave soldering to the PCB.
- After a number of closures, the compression force of the contacts is reduced, applying less force to the flex tail connection.

Low Insertion Force Connectors. Low insertion force connectors rely on the spring force of the contact to form the connection to the conductors on the flex tail and to mechanically hold the flex tail in place. As the tail is inserted into the connector, the contacts apply force on the conductive ink and scrape the conductors. This abrasive action limits the number of insertions and removals before the conductive ink fails. The number of insertions is highly dependent on the membrane switch conductors and connector manufacturer.

The advantages of this type of connector are:

- It is generally the lowest cost of the three types.
- Usually a gastight connection is formed.
- The area required for mounting to a PCB is generally less than the area required for mounting zero insertion force connectors.

The disadvantages of this type of connector are:

- The tail can be removed and inserted a limited number of times.
- No strain relief is provided to hold the flex tail in place.
- Contacts can become contaminated with flux when prepared for wave soldering to a PCB. (15)

The choice and qualification of any connector, of course, is up to the end user, and these or similar manufacturers should be contacted for additional information about specific characteristics of membrane connectors. Any contact with potential suppliers should be done as early in the design process as feasible.

Backlighting of Membrane Switches

There are several methods of backlighting membrane switches. Each method has its advantages and disadvantages; therefore, the user should thoroughly investigate each method carefully before deciding on a particular type for the application in question. Contact either the membrane switch manufacturer offering backlighting options or the backlight supplier. Ask to see illuminated samples backlighting the style of membrane switch under consideration. Since backlighting is very subjective, it is always best to get a consensus from the ultimate user community as to what constitute acceptable backlighting levels. Once this is accomplished, agreement should be reached between the specifier and the supplier as to how the backlighting should be specified and how it should be measured.

Woven Fiber Optics. One of the latest methods of backlighting membrane switches (including membrane metal dome switches, Chap. 10) is woven fiber optics (Fig. 8.11) in which plastic optical fibers are woven together into a flat, flexible light-emitting panel. The optical fibers extend from the woven portion of the panel and connect to a remote light source. As in other fiber-optic applications, light enters the woven panel through each highly polished fiber end. Unlike other applications of fiber optics, however, "microbends" in the optical fibers cause the transmitted light to be emitted from the fiber through the side of the fiber. The light hits each bend and leaves *through the cladding* without physically interrupting the cladding surface. The fibers are not hot-stamped, cracked, or etched. The panel is woven with variable cross-fiber spacing that causes all available light to be emitted uniformly.

Woven fiber-optic panels, such as those manufactured by Lumitex, Inc., Cleveland, Ohio, are constructed of one to six woven layers. Multiple layers allow more efficient use of the light source, enhancing brightness or extending source life. In some applications a reflective layer is attached to the side of the woven panel away from the part to be backlit in order to provide unidirectional

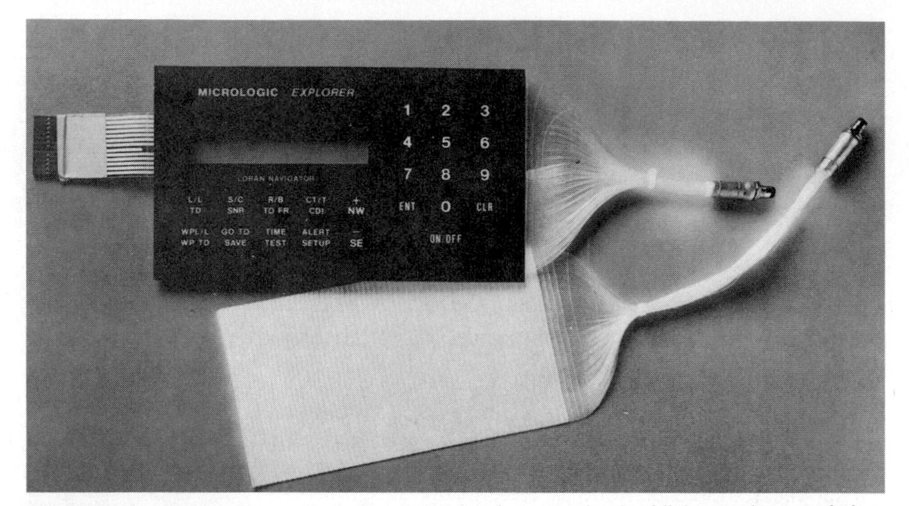

FIGURE 8.11 Woven fiber-optic light-emitting panels are used to backlight membrane switches. (*Courtesy of Lumitex Inc., Cleveland, Ohio.*)

light emission and to increase light intensity. Depending on the placement of the fiber-optic panel in relation to the switch mechanism on the membrane assembly, the reflective layer, manufactured of metallized polyester, may also function as an EMI/RFI shield. A typical construction is shown in Fig. 8.12. Lumitex uses a special construction for most membrane switch applications. Consisting of a single woven layer, this construction results in a 0.040-in (1,0-mm) thick, very pliable, thin backlight.

Woven panel sizes are unlimited within a range of 0.5 in^2 (12,7 mm^2) to 1152 in^2 (29,260 mm^2), and up to 24 ft (7,3 m) in length. The fiber cable connecting the panel and source lamp can be up to 30 ft (9,1 m) long. In applications utilizing a basic, flat membrane switch, the fiber-optic panel can be installed beneath the switch assembly itself. Depending on the lamp selected, the intensity of light

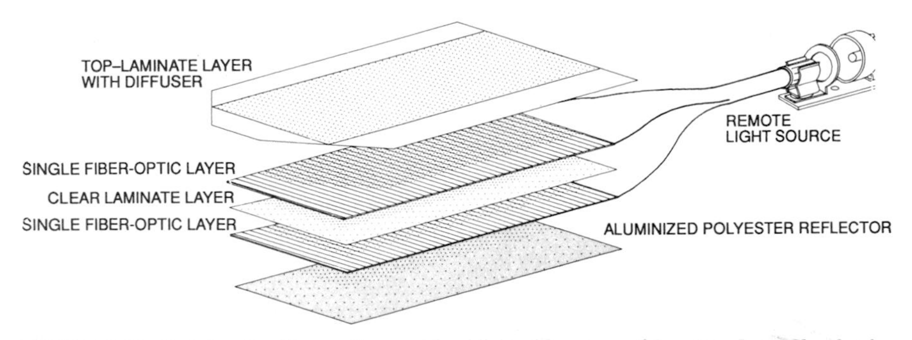

FIGURE 8.12 Details of a fiber-optic panel backlight. (*Courtesy of Lumitex Inc., Cleveland, Ohio.*)

emitted from the fiber-optic panel is generally high enough to provide adequate illumination through all switch layers.

A variety of light sources are available, from single LEDs to 150-W halogen lamps. Most sources are low-voltage dc, and lamp life is typically available to 100,000 h with either LEDs or halogen sources. An LED is the most common light source. A single LED can illuminate a panel up to 4 in (101,6 mm) wide. Wider panels require multiple LEDs. For white light, low-wattage incandescent or halogen lamps may be used.

Membrane switch backlighting panels may be die-cut to allow placement of displays and other components. In some applications, multiple fiber-optic panels can be illuminated by a single source. This can be done to illuminate both a switch and an accompanying LCD, even if the panels have a different number of layers.

Some advantages of woven fiber-optic panels are:

- Light intensity is not reduced over time.
- Low-voltage, low-current dc sources eliminate inverter noise, prolong battery life.
- The panels eliminate hot spots and uneven lighting. Placement *over* metal domes eliminates shadowing. (16)

Electroluminescent (EL) Backlighting. Electroluminescence is a solid-state phenomenon that uses phosphors, not heat, to generate light. An electroluminescent lamp is constructed in a manner similar to a capacitor—two conducting surfaces with a dielectric in between. An EL panel is a cold, uniform light source specifically designed to be thin and flexible.

The basic building block for the EL lamp is aluminum foil. Serving as the back electrode, the foil has a dielectric coating applied to give the lamp high-voltage integrity. A layer of light-emitting phosphor is the transparent front electrode, which defines the active (or lighted) area. On top of the electrode an opaque bus bar is screened to provide a more uniform electric field across the entire lamp surface, ensuring more uniform luminance. Front and back leads, specially treated to ensure the moisture barrier is not compromised, are applied to provide power access, and then a moisture-absorbing desiccant is laminated to the assembly. Finally, the entire sandwich is enclosed in an environmental barrier to protect the lamp from exposure to the atmosphere, especially moisture, and to provide electrical insulation. Figure 8.13*a* and *b* illustrates the construction of a typical EL lamp and the range of colors available.

A number of factors, including cost and reliability, affect the options with respect to lamp dimensions. Lamps with a continuously illuminated area can be produced in a variety of shapes, in sizes as small as 0.5 × 1.0 in (12,7 × 25,4 mm) up to 12 × 20 in (304,8 × 508,0 mm) or up to 6 × 34 in (152,4 × 863,6 mm). Most flexible lamps are produced in the form of rectangles or "strips." Lamps with curvilinear forms or with holes, cutouts, and notches can be made, although there are limits on minimum feature size. (17)

When selecting and specifying fiber-optic or EL lamps for membrane switch backlighting, the lamp manufacturer should be contacted to determine what restrictions are imposed for limitations on lighted areas, borders, bus bars, and lead-connector options.

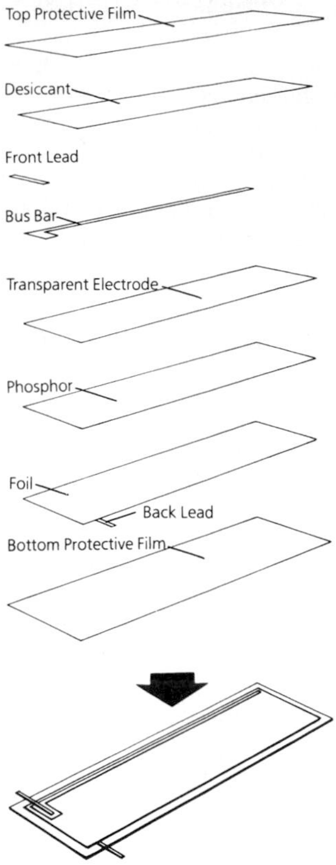

Top Protective Film
Desiccant
Front Lead
Bus Bar
Transparent Electrode
Phosphor
Foil
Back Lead
Bottom Protective Film

(a)

TYPICAL PERFORMANCE DATA (AT 115 VOLTS)

COLOR	FREQUENCY (Hz)	LUMINANCE		RADIANCE (WATTS/ft²)	SPECTRAL PEAK EMISSION	CHROMATICITY	
		(ft.L)	(Cd/M²)			X	Y
AVIATION GREEN	60	4.5	15.4	.010	514	.225	.540
	400	24.0	82.2	.066	510	.205	.485
BLUE-GREEN	60	4.5	15.4	.010	510	.215	.500
	400	22.5	77.1	.066	506	.195	.430
GREEN-WHITE	400	23.0	78.7	.069	566	.330	.420
	1000	40.0	136.8	.135	562	.305	.365
YELLOW-GREEN	400	9.0	30.8	.018	546	.375	.575
STANDARD WHITE	400	16.0	54.8	.053	n/a	.330	.350
YELLOW-WHITE	400	23.0	78.7	.050	n/a	.410	.430
YELLOW-ORANGE	400	5.0	17.1	.011	590	.520	.450
RED	400	4.0	13.7	.019	618	.655	.337

SPECIAL COLORS, INCLUDING NVG COMPATIBLE, AND IR-EMITTING ARE AVAILABLE

(b)

FIGURE 8.13 EL (electroluminescent) lamps are commonly used to backlight membrane switches: (*a*) construction details of a typical EL lamp, and (*b*) a variety of colors are available for EL lamps. (*Courtesy of Loctite Luminescent Systems, Lebanon, N.H.*)

GLOSSARY

Abrasion resistance. The amount of scratching that a material can withstand without showing deleterious effects.

Adhesion. The ability of conductive ink to adhere to a substrate.

Curing. The process of thoroughly drying conductive ink for maximum reliability.

Die cut. To cut thin material, such as polyester, by means of sharp knife blades secured in a holding tool.

Flexible tail (flex tail). The circuit termination exit extending from the flexible circuit to which is typically attached a clincher-style connector.

Migration. The leaching out of suspended particles in a conductor when exposed to a high-humidity environment.

PSA. Pressure-sensitive adhesive.

Resistivity. The volume resistivity of the conductive ink recorded as ohms per square per mil thickness.

Selective texture. A surface treatment used on membrane graphic overlays to selectively texture areas on the overlay.

Specular gloss. The amount of light reflected off the surface of a material at a specific angle. Usually recorded in "gloss units."

Switch resistance. The total resistance, in ohms, that is recorded in a closed switch. The measurement is made from the tail of the switch for membranes.

Venting. An air channel, or breather, cut in the spacer layer of a membrane switch, which connects groups of switches for the purpose of equalizing air pressure during switch closure.

REFERENCES

1. "Xymox Application Note No. 1, Switch Comparison," W. H. Brady Co., Milwaukee, Wis., 1986.
2. "Xymox Application Note No. 4, Substrates for Membrane Switches," W. H. Brady Co., Milwaukee, Wis., 1986.
3. "Membrane Switch Failures Due to Materials Incompatibility," Berquist Switch Co., Minneapolis, Minn., 1991.
4. "Analysis of Pressure Build-up in Nonvented Switches," Berquist Switch Co., Minneapolis, Minn., 1991.
5. "Membrane Switch Life, A Critical Testing Comparison," Berquist Switch Co., Minneapolis, Minn., 1991.
6. "Actuation Force," Berquist Switch Co., Minneapolis, Minn., 1991.
7. "Design Guide," Square D Company, Data Entry Products, Loveland, Colo., 1990.
8. "Xymox Application Note No. 5, Substrates for Faceplates," W. H. Brady Co., Milwaukee, Wis., 1984.
9. "How Pressure Sensitive Adhesives Attack Graphic Inks," Berquist Switch Co., Minneapolis, Minn., 1991.
10. "Graphic Panel Life, Hardcoated vs. Uncoated Surfaces," Berquist Switch Co., Minneapolis, Minn., 1985.

11. "Test Report, Commercial Grade Hard Coat," GM Nameplate, INTAQ Division, Seattle, Wash., 1984.
12. "Test Report, Laboratory Grade Hard Coat," GM Nameplate, INTAQ Division, Seattle, Wash., 1989.
13. "Xymox Application Note No. 2, Silver Migration," W. H. Brady Co., Milwaukee, Wis., 1986.
14. "Xymox Application Note No. 3, Planar vs. Perpendicular Tails," W. H. Brady Co., Milwaukee, Wis., 1990.
15. "Xymox Application Note No. 6, Connectors," W. H. Brady Co., Milwaukee, Wis., 1990.
16. "Lumitex®, Creators of Woven Light," Lumitex, Inc., Cleveland, Ohio, 1991.
17. "The Design Guide," Loctite Luminescent Systems, Lebanon, N.H., 1990.

CHAPTER 9
METAL DOMES*

Metal domes, also commonly known as snap domes,[1] are used as the basic switch element in a number of different types of switches. This chapter explains the fundamentals of metal domes, their electrical and mechanical characteristics, various shapes and sizes available (Fig. 9.1), and how to design the matching circuitry. How metal domes are incorporated into the switches themselves is the subject of Chap. 10.

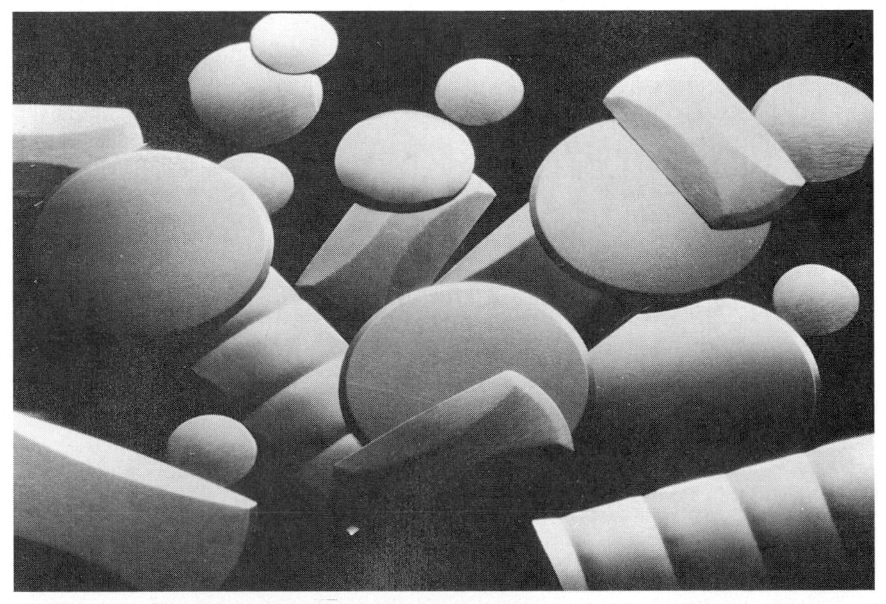

FIGURE 9.1 There is a metal dome for almost every conceivable application. (*Courtesy of Square D Company, Data Entry Products, Loveland, Colo.*)

*Numbers in parentheses indicate items in the References at the end of this chapter.
[1]The term Snap Dome is trademarked by the Square D Company, Data Entry Products Division, Loveland, Colo.

Interestingly, metal domes are sometimes used only as a tactile providing medium (particularly in some styles of membrane switch) and in this type of application are not an electrically active part of the switch circuit.

HOW METAL DOMES FUNCTION

Metal domes are exactly as the name implies, dome-shaped disks of metal designed to "snap" over when pressure is applied to the top of the dome. The design is such that when pressure is released, the dome snaps back to its original shape, awaiting the next pressure application and release cycle. The tactile feel present in the snap-down, snap-up sequence makes the metal dome an ideal mechanism for use as a switch contact. A combination of dome shape and material composition and thickness gives each type of dome unique characteristics: actuation force, actuation travel, tactile feel, bounce, contact resistance, and mating circuit design.

OPERATING CHARACTERISTICS OF METAL DOMES

Switch Life

The mechanical characteristics of a typical metal dome are shown in Table 9.1. It may be of interest to the user-engineer that even though the life of a metal dome is considered to end when the dome no longer exhibits tactile feel, usually the metal dome will continue to function electrically for thousands or even tens of thousands of actuation cycles. (1)

Tactile Feel

Although Chap. 1 discusses tactile feel in depth, one useful definition is the finger sensation experienced by the user when the switch is actuated. As pointed out in Chap. 1, the evaluation of this sensation is quite subjective and therefore difficult to quantify, but few characteristics are so important to the perception of performance and acceptance of the final switch configuration as tactile feel. It is pos-

TABLE 9.1 Mechanical Specifications for a Typical Metal Dome

Dome material	Stainless steel
Total travel	See Table 9.2
Storage temperature	−40 to +185°F (−40 to +85°C), specials up to 105°C
Operating temperature	−4 to +185°F (−20 to +85°C)
Humidity	0 to 95% (no condensation)
Gold plating (optional)	Type I, 99.7% gold, minimum Grade C, Knoop hardness 130–200 Class 1, 0.00005 in thickness, min. MIL-G-45204C
Actuation life	100,000–3,000,000 cycles

Source: Ref 1.

sible to compare the tactile feel of a single switch with that of other switches and rate one against the others. Chapter 1 discusses the relative values assigned to the sensation of tactile feel. It can be attenuated by the material used in the final switch configuration. In a membrane–metal dome switch (Chap. 10), the material, material thickness, any embossment of the material, and material layers can significantly change the perception of tactile feel. The same set of material characteristics can increase the measurement of actuation force from the bare metal dome to the final keyboard or keypad assembly. What this indicates to the prudent design engineer is to use actuation force specifications as a starting place for the application and to consider the effects of the materials used in conjunction with the dome in defining the actual force specified. When deciding on a final actuation force, actuation travel, tactile feel combination, it is *always* best to build a mock-up of, at the very least, a single key station using materials anticipated to be used in the final configuration. Metal dome manufacturers are generous with a variety of samples to assist engineers in producing these mock-ups (or breadboards). Always ask for assistance when necessary. If a metal dome manufacturer seems hesitant about supplying samples, consider taking them off your supplier list. If the supplier finds it difficult to provide a few samples to an engineer, how will this same supplier be able to deliver production lot quantities once a purchase order has been placed?

Trip Force

Trip force is the amount of force required to press the metal dome switch into a flattened shape to bridge two contacts. Square D Data Entry Products measures their snap domes with an Ametek gauge, using a 0.062-in (1,57-mm) flat-bottom plunger. Standard trip forces for the company's metal domes are 6 oz (170 g), 7.75 oz (220 g), and 10.5 oz (300 g). Table 9.2 depicts the variety of trip forces available from this manufacturer. (1)

The user-engineer should also be aware of the possibility of *stacking* metal domes in a keypad or keyboard application to dramatically increase the trip force. As many as three metal domes have been successfully stacked in special assemblies. The final trip force will again depend on the materials used for dome seals and over the stacked domes, but as a general rule of thumb (and discounting the effects of materials), add the trip force of each dome used to get a basic, rough-order-of-magnitude trip force, and use this as a starting point for building the all-important mock-up.

Typically, metal dome manufacturers maintain a maximum tolerance of ±1 oz (30 g) on their standard products, but it is always wise to check with the manufacturer of choice to determine what standard tolerances the manufacturer supports. If the user-engineer requires a special trip force and/or tolerance, it is advisable to check with a number of metal dome suppliers since they are often called upon to produce specials, and the special dome needed may already be tooled.

LAYOUT REQUIREMENTS FOR PRINTED CIRCUIT BOARDS

PCB Patterns and Dimensions

Figure 9.2*a* depicts typical PCB patterns and dimensions suggested for use with the various types of metal domes portrayed in Table 9.2. While Fig. 9.2*b* illus-

TABLE 9.2 Metal Domes in a Variety of Sizes, Trip Forces, Tactile Feel, and Life Characteristics Are Available

Snap dome size, in/mm	Approx free height, in	Trip force, g	Tactile feel	Actuation life, cycles
Round snap dome switches				
0.156/3.96	0.008	150 ± 50	Medium	100,000
0.220/5.6	0.012	220 ± 30	Medium	500,000
0.270/6.86	0.012	170 ± 30	Medium	3,000,000
0.270/6.86	0.012	220 ± 30	High	500,000
0.270/6.86	0.012	220 ± 30	Medium	500,000
0.270/6.86	0.014	300 ± 30	High	100,000
0.350/8.9	0.015	170 ± 30	High	3,000,000
0.350/8.9	0.016	220 ± 30	High	1,000,000
0.350/8.9	0.017	300 ± 30	High	100,000
0.350/8.9	0.015	220 ± 30	High	3,000,000
0.425/10.8	0.017	220 ± 30	Medium	1,000,000
0.425/10.8	0.019	300 ± 30	High	100,000
0.750/19	0.029	220 ± 30	Medium	100,000
Rectangular snap dome switches				
0.35 × 0.70/8.9 × 17.8	0.025	220 ± 30	High	100,000
0.50 × 1.0/12.7 × 25.4	0.050	220 ± 30	High	100,000
Special application domes				
0.350/8.9	0.015	220 ± 30	Medium	1,000,000
0.350/8.9	0.015	300 ± 30	High	500,000
0.500/12.7	0.022	300 ± 30	High	1,500,000
Linear switch bars				
0.40 × 1.0/10.2 × 25.4	0.032	220 ± 30	High	100,000
0.40 × 1.0/10.2 × 25.4	0.032	300 ± 30	High	100,000
0.50 × 1.0/12.7 × 25.4	0.032	220 ± 30	Medium	100,000
0.50 × 1.0/12.7 × 25.4	0.032	300 ± 30	High	100,000

Source: Ref. 1.

trates the minimum spacing requirements for the Square D Data Entry Products Snap Dome switch, other manufacturers will have their own, perhaps different, requirements. It is important to note that the dimensions shown in Fig. 9.2*b* are minimum dimensions and more space should be allowed, if possible.

Venting. Venting of metal dome switch PCBs is done to assure the unrestricted transfer of air which may be trapped inside the cavity formed by the dome–dome seal assembly process. The venting essentially allows the metal dome to "breathe" during actuation. While not all metal dome switch applications need to be vented, unless there is already a level of experience in designing nonvented switches, it is recommended that vents be made part of the design. Most metal dome manufacturers recommend either double traces, feed-through holes under the switch, or special back seals with built-in air displacement tracks (Fig. 9.3). Check with the metal dome manufacturer for recommendations on hole diameters and air displacement track dimensions. [Adapted from (1).]

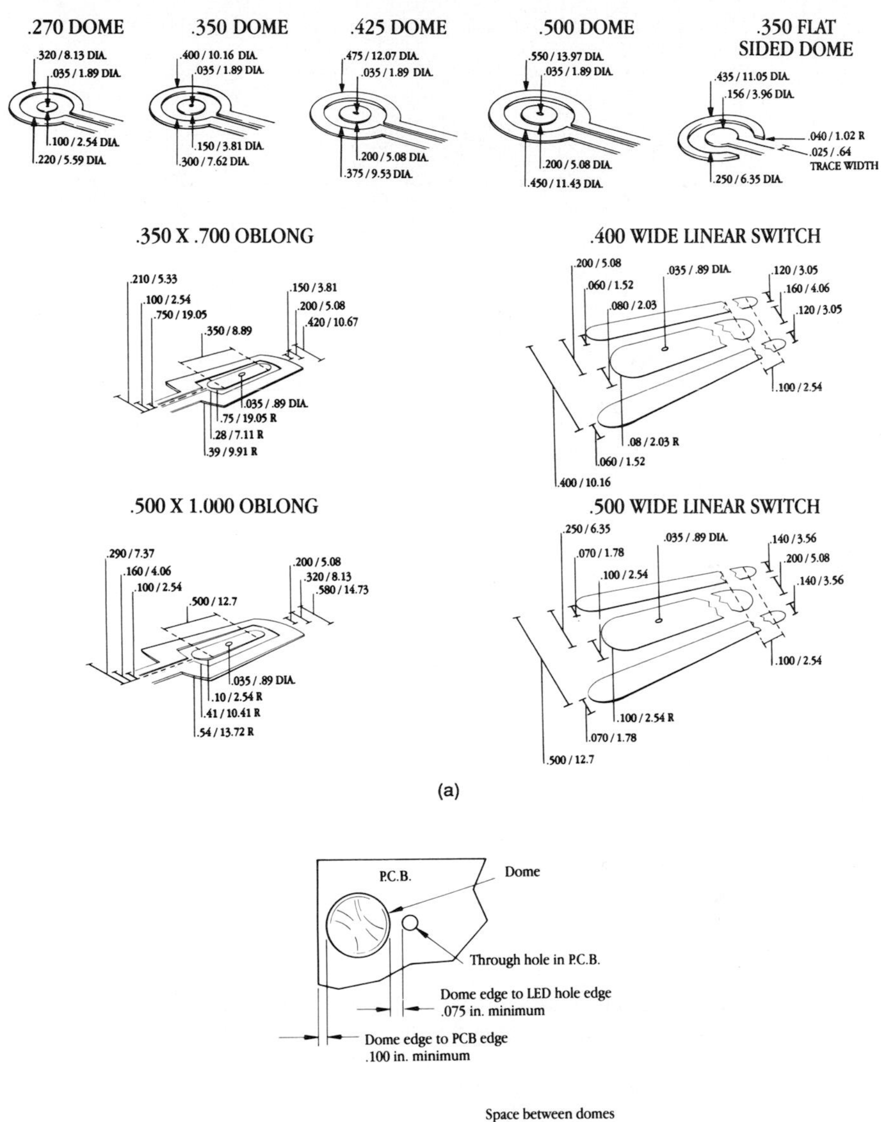

FIGURE 9.2 (*a*) Typical PCB patterns and dimensions for use with various metal domes, and (*b*) minimum recommended spacing requirements. (*Courtesy of Square D Company, Data Entry Products, Loveland, Colo.*)

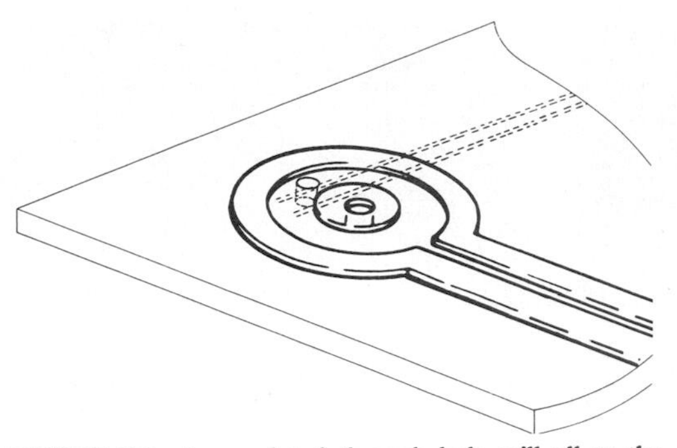

FIGURE 9.3 A nonplated through hole will allow the dome to vent to the backside venting tracks. (*Courtesy of Square D Company, Data Entry Products, Loveland, Colo.*)

ASSEMBLY CONSIDERATIONS

When seeking a metal dome supplier, always inquire how the parts are handled through the various manufacturing processes. Preference should be given to the manufacturer whose processes do not require any direct human touch on the domes, as the oils and salts present on the human skin can contaminate the surface of the dome and either drive up contact resistance or render the dome electrically unusable.

The domes should be placed in tightly sealed containers after manufacture. If the user will be installing the metal domes on PCBs or flexible circuits at his or her own factory, it is essential that the domes be handled as little as possible and then only with vacuum-pickup-type equipment (Fig. 9.4). If the domes are to be handled directly, the operator *must* wear clean cotton gloves and must minimize contamination of the domes by moisture (breath), food particles, tobacco smoke, or other contaminants in the assembly area. The amount of care taken to avoid contamination will be directly reflected in the quality and consistency of the metal dome's performance in its final assembly.

Assembly considerations are so important to the proper performance of metal dome assemblies that one manufacturer, Square D Company Data Entry Products, has defined the following recommendations for handling of the domes during assembly.

Figure 9.5 shows the steps which are typical for the assembly of metal dome switches. In Fig. 9.5*a* the domes are aligned and registered in a magnetic fixture that matches the pattern of the PCB to which it is to be assembled. The dome seal, a sheet of die-cut polyester with pressure-sensitive adhesive, is placed on the mating fixture (Fig. 9.5*b*).

The magnetic fixture is then aligned with the tooling pins of the other half of the fixture, and when the two halves are pressed together and then separated, the domes adhere to the adhesive side of the polyester seal sheet (Fig. 9.5*c*).

At this point it is important that all soldering on the PCB is completed, and that the PCB has been thoroughly cleaned with flux remover, then a mild detergent, and then rinsed and dried. Care must be taken to ensure that no res-

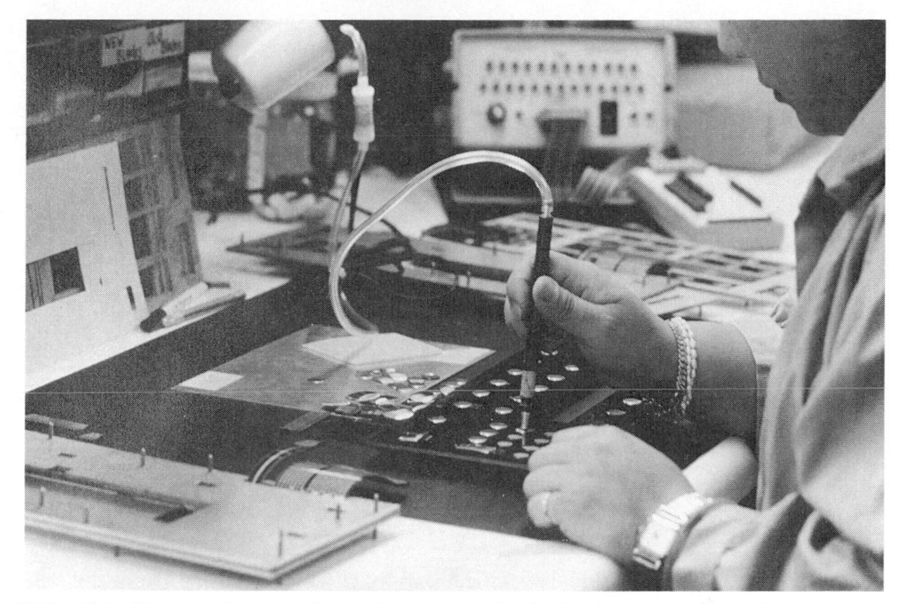

FIGURE 9.4 Vacuum-pickup equipment is recommended for handling metal domes in an assembly environment. (*Courtesy of GM Nameplate, INTAQ Division, Seattle, Wash.*)

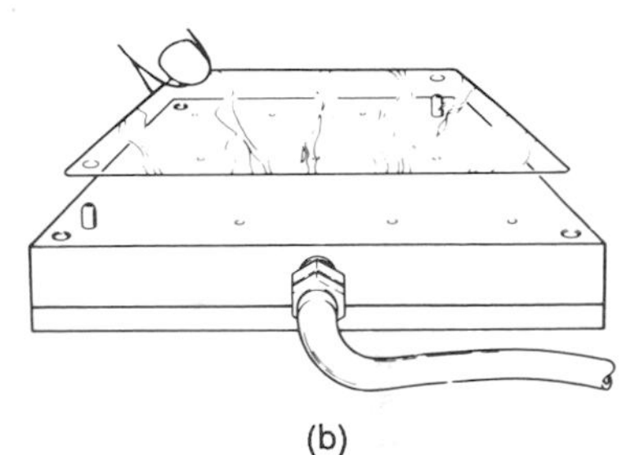

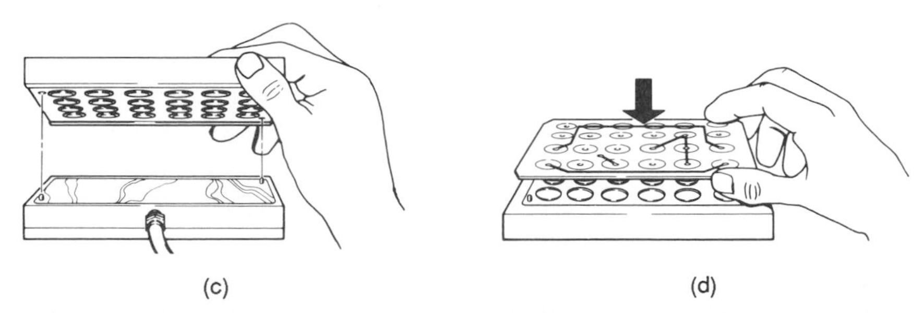

FIGURE 9.5 Typical metal dome assembly steps: (*a*) align and register domes in precision fixture, (*b*) place dome seal on the mating fixture, (*c*) press two halves together—domes adhere to the adhesive side of the dome seal, and (*d*) align and adhere PCB to metal dome–dome seal subassembly. (*Courtesy of Square D Company, Data Entry Products, Loveland, Colo.*)

idue is left on the PCB, as this could cause possible contamination of the switch contacts.

The PCB is then aligned, PCB switch contact patterns facing the concave side of the domes, with the same tooling pins and pressed onto the polyester seal sheet (Fig. 9.5*d*). The PCB–metal dome assembly is removed from the tool and the seal is burnished down to the PCB surface to remove wrinkles and ensure adhesion. (1)

MECHANICAL, ELECTRICAL, AND ENVIRONMENTAL SPECIFICATIONS

Mechanical Specifications

As depicted in Table 9.1, metal dome switches are usually produced using stainless steel as the base material. Quality-conscious metal dome manufacturers will passivate the domes after manufacturing.

Electrical Specifications

Table 9.3 shows the electrical specifications for Square D's Snap Dome switches. The specifications show results when the domes are used with hard-nickel-plated PCB assemblies.

TABLE 9.3 Electrical Specifications for a Typical Snap Dome

Contact configuration	Form A, SPST, normally open
Current	10 mA typical, 40 mA maximum
Voltage	24 V dc into resistive load only
Breakdown voltage	Over 250 V dc
Contact resistance	1.0 Ω typical, 20 Ω max*
Contact bounce	Make 1 ms typical, 10 ms max. Break up to 20 ms

Source: Ref. 1.
*For lower contact resistance, gold plating is an option on the domes and on the PCB.

Contact Bounce. *Contact bounce,* discussed in Chap. 1, is generally described as the successive "making" or "breaking" of switch contacts during actuation or release. With some circuits, this kind of chatter can be picked up as multiple entries. If a circuit is particularly sensitive to contact bounce, a debounce, or delay, can usually be added to the circuitry without difficulty. Contact bounce in metal domes is affected by outside influences such as:

- Speed of actuation
- Angle of actuation
- Amount of force applied
- Contact plating

"Worst case" situations should therefore be considered taking into account these various influences.

Square D's Snap Dome switches are rated at 1 ms *typical,* or 10 ms *maximum,* on the "making" of a circuit and up to 20 ms *maximum* on the "breaking" of a circuit.

Environmental Specifications

Metal dome switches will operate in temperature ranges that are broader than those of the other materials which typically are used to construct the completed keypad or keyboard. These temperature limits will dictate the operating temperature range of the metal domes.

When metal domes are mounted on a PCB, a *back seal* is recommended on the PCB in all applications because the seal keeps out dust, dirt, moisture, etc. Additionally, the back seal provides insulating properties. Typical back seal material is 0.002 in (0,051 mm) thick transparent polyester with a 0.0008 in (0,020 mm) thick adhesive on one side.

Metal dome switches should not be used in an environment where the switch may be pressed directly by a sharp, pointed object, as this may cause hyperextension of the dome, permanently affecting its electrical and mechanical characteristics. In applications where hyperextension of the dome is of concern, use of a dimple dome is recommended.

PREASSEMBLED METAL DOME SWITCHES

A convenient way for the user-engineer to specify and procure metal dome switches consists of an array of domes preattached to the adhesive side of die-cut dome seal material. This array can be tailored to exactly fit the requirements specified by the end user.

The domes are precision registered, using tooling similar to that described earlier in the chapter, on the polyester dome seal by the metal dome manufacturer, complete with pressure-sensitive adhesive backing, tooling holes, and any other die-cut designs the user requires. The protective release liner of the dome seal, removed to install the domes, is reapplied to the dome seal, to protect the domes and the adhesive during shipping and handling.

Once received at the user's facility, the entire part can be mounted quickly and easily to the user's PCB or flexible circuit assembly in one operation (by simply removing the release liner), saving assembly time and enhancing quality control (Fig. 9.6). [Adapted from (1).]

GLOSSARY

Back seal. A thin sheet of material, usually polyester, with adhesive on one side. The material is adhered to the rear of a PCB on which are assembled metal domes to protect the subassembly from the environment.

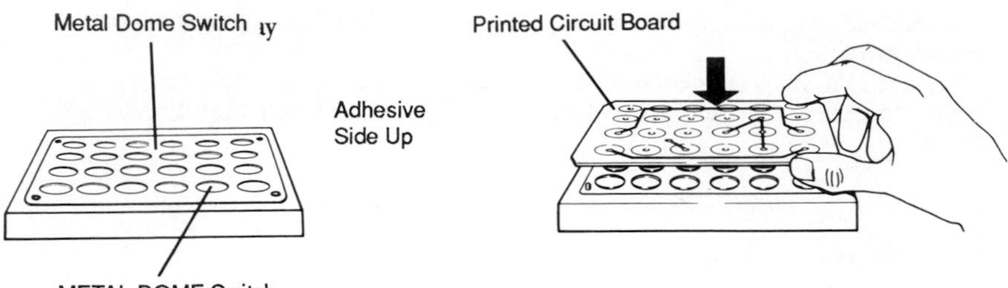

FIGURE 9.6 Metal domes preattached to the dome seal by the dome manufacturer can enhance assembly-line efficiency. (*Courtesy of Square D Company, Data Entry Products, Loveland, Colo.*)

Contact bounce. The successive "making" or "breaking" of switch contacts during actuation or release.

Metal dome. A dome-shaped disk of metal (usually stainless steel) designed to "snap" over when pressure is applied to the top of the dome.

Trip force. The amount of force required to press a metal dome into a flattened shape to successfully bridge two contacts.

Venting. Breathing holes put into a PCB or channels die-cut into spacer material to allow the unrestricted transfer of air which may be trapped inside the cavity formed by the metal dome–dome seal.

REFERENCE

1. "Design Guide," Square D Company, Data Entry Products Division, Loveland, Colo., 1990.

CHAPTER 10

MEMBRANE–METAL DOME AND RUBBER–METAL DOME SWITCHES*

Two distinct, but related, types of switches will be discussed in this chapter. Using the basic metal dome discussed in Chap. 9, we examine keypads using membrane–metal dome and rubber–metal dome construction techniques. In the definition of the two techniques we use in this chapter, the *membrane–metal dome* style switch (also called *graphic overlay* style) utilizes metal domes attached to either a PCB or flexible circuit which is then covered with a membrane graphics panel similar to the type discussed in Chap. 8. The domes are actuated by the operator's finger pressing directly on the membrane graphics layer.

The *rubber–metal dome* style switch utilizes the same metal domes, attached to a PCB in this case, covered by a molded elastomer boot. The elastomer boot not only protects the metal dome–PCB subassembly from the effects of the environment but also acts as the mechanism through which the domes are actuated.

Each style has its advantages and both styles provide unique solutions to the usual, as well as unusual, keypad design applications.

HOW MEMBRANE–METAL DOME SWITCHES ARE CONSTRUCTED

Metal Dome Subassembly

Two types of metal dome–substrate subassemblies are commonly used in membrane–metal dome switches:

- The metal dome–PCB subassembly
- The metal dome–flexible circuit subassembly

These two types of subassemblies are the fundamental electric circuit portion of the membrane–metal dome switch; the membrane graphic overlay acts as the legend panel and, with embossing, provides an exact target area over the metal dome switch for the operator's finger.

*Numbers in parentheses indicate items in the References at the end of this chapter.

Metal Dome–PCB Subassembly. The membrane–metal dome assembly shown in Fig. 10.1 shows the PCB–metal dome–*dome seal* subassembly. The graphic overlay is added to this subassembly to produce the final keypad configuration. The advantages of this type approach are:

- Metal-to-metal switch reliability
- Consistent, positive tactile feel
- The ability to add components to the PCB
- A broad operating temperature range (1)

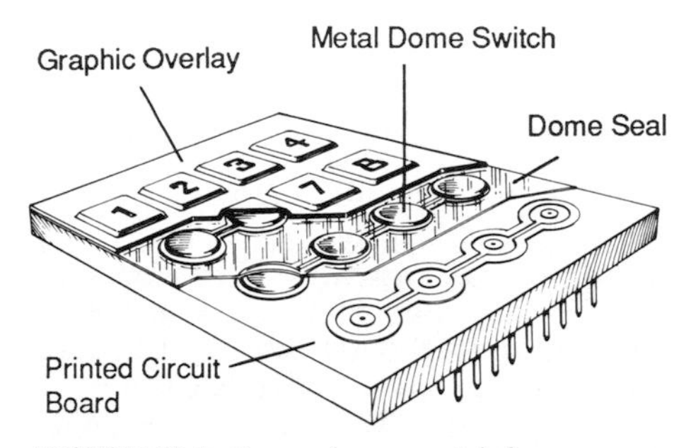

FIGURE 10.1 A membrane–metal dome assembly. (*Courtesy of Square D Company, Data Entry Products, Loveland, Colo.*)

Metal-to-Metal Switch Reliability. The metal dome contact on the metal plating (nickel, gold, etc.) of the PCB contact provides the lowest contact resistance when compared with the metal dome–flexible circuit subassembly style.

Tactile Feel. The metal-to-metal hard surface contact provides consistent tactile performance over the life of the switch.

Adding Components to the PCB. The ability to add components to the PCB (special connectors, decode electronics, debounce circuitry, etc.) can be of real value to the user-engineer. Additionally, as will be discussed later in the chapter, lamps for backlighting can be easily added to the PCB.

Broad Operating Temperature Range. On a strict comparison basis, the operating temperature range of a PCB-based switch will be higher than that of a flexible circuit-based switch. However, the weakest link in this chain is the lowest rated temperature part of the total construction, which is usually the temperature limit of the graphic overlay, depending on the material selected.

See Chap. 8 for a detailed comparison of graphic overlay materials and their environmental characteristics.

PCB Materials. In a Design Guide (1) published by Square D Company, Data Entry Products, Loveland, Colo., the company recommends the following parameters for PCBs used in metal dome–PCB subassemblies:

- The base material for the PCB should be FR4 glass base, 1 oz copper, double-sided. Standard board thicknesses are 0.015 in (0,38 mm), 0.031 in (0,78 mm), 0.062 in (1,57 mm), 0.093 in (2,36 mm), and 0.125 in (3,17 mm).

- Plating with hard tin-nickel or hard nickel will be 0.000150 to 0.000250 in (0,0038 to 0,0063 mm) thick, with a hardness of RC40. (1) Gold plating is an option available from most manufacturers, but consultation is encouraged to determine thickness and hardness parameters.

PCB Layout

A schematic of the PCB should be incorporated in the switch specification. If the user-engineer is going to procure the entire switch from the switch manufacturer (and unless the user has experience in manufacturing switches of this type, it is best to let the experts do it!), the specification (including mechanical dimensions of the final keyboard configuration and schematic) is the basic information needed by the manufacturer to proceed with circuit artwork.

It is at this point that the user-engineer, in consultation with the switch manufacturer, needs to determine if additional components will be required on the PCB so that the necessary additions to the artwork can be implemented. *The ideal time to install any additional components is before the keypad is assembled,* since there is a good chance the heat encountered during soldering would adversely affect a completed switch assembly.

Terminations

There are three basic choices for PCB terminations for the metal dome–PCB style:

- Pins
- Headers
- Connectors

Pins. Square pins, 0.025 in (0,63 mm) per side, are press-fit and soldered to the PCB (Fig. 10.2). Pin length is typically measured from the back of the board to the end of the pin and ranges from 0.125 in (3,17 mm) to 0.750 in (19,0 mm). Pins are generally placed on 0.100-in (2,54-mm) centers. Pins are tip plated, with gold plating optional. (1) Contact the switch manufacturer for any special dimensional or plating requirements.

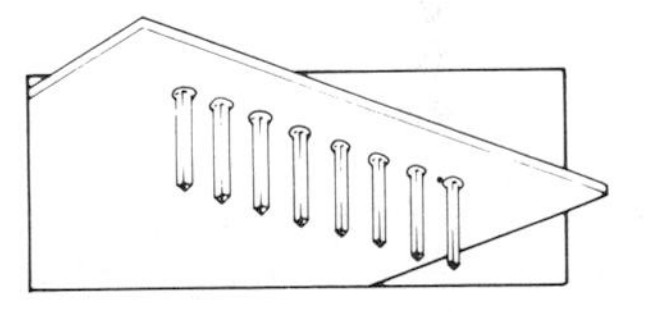

FIGURE 10.2 Square pins are press-fit and soldered to the PCB. (*Courtesy of Square D Company, Data Entry Products, Loveland, Colo.*)

Headers. It is not possible to list all the many shapes and sizes of header terminations available. A typical standard header is depicted in Fig. 10.3.

All standard headers are 0.025 in (0,63 mm) square and are placed on 0.100-in (2,54-mm) centers. If a nonstandard header is dictated by design requirements, always consult with the switch manufacturer prior to incorporating the device. (1)

Connectors. Switch manufacturers can incorporate a wide variety of connectors onto the PCB itself. Restrictions vary as to location, physical dimensions, and electrical requirements, but a typical connector style is shown in Fig. 10.4 that most manufacturers can accommodate.

.025 (0,64)
L
.100 (2,54)
• determined by PCB thickness
.100 (2,54)

PCB THICKNESS inch (mm)	PIN LENGTH (L) inch (mm)
.031 (0,78)	.250 (6,35)
.031 (0,78)	.500 (12,7)
.031 (0,78)	.750 (19,05)
.031 (0,78)	1.000 (25,4)
.062 (1,57)	.250 (6,35)
.062 (1,57)	.500 (12,7)
.062 (1,57)	.750 (19,05)
.062 (1,57)	1.000 (25,4)
.094 (2,38)	.250 (6,35)
.094 (2,38)	.500 (12,7)
.094 (2,38)	.750 (19,05)
.094 (2,38)	1.000 (25,4)
.125 (3,17)	.250 (6,35)
.125 (3,17)	.500 (12,7)
.125 (3,17)	.750 (19,05)
.125 (3,17)	1.000 (25,4)

FIGURE 10.3 A typical PCB header. (*Courtesy of Square D Company, Data Entry Products, Loveland, Colo.*)

Metal Dome–Flexible Circuit Subassembly. Figure 10.5 depicts a typical metal dome–flexible circuit subassembly. The flexible circuit consists of, essentially, the bottom half of a membrane switch (Chap. 8). The switch half is manufactured in the same manner as basic membrane switches, that is, conductive ink printed on a transparent, flexible polyester or polycarbonate substrate. This half-switch is attached to a backer plate for rigidity. At this point in the construction, the half-switch–backer subassembly emulates the PCB in the metal dome–PCB subassembly discussed earlier. The metal domes are installed on the switch-backer subassembly with a dome seal and this subassembly is then covered with the graphic overlay. The advantages to this construction are:

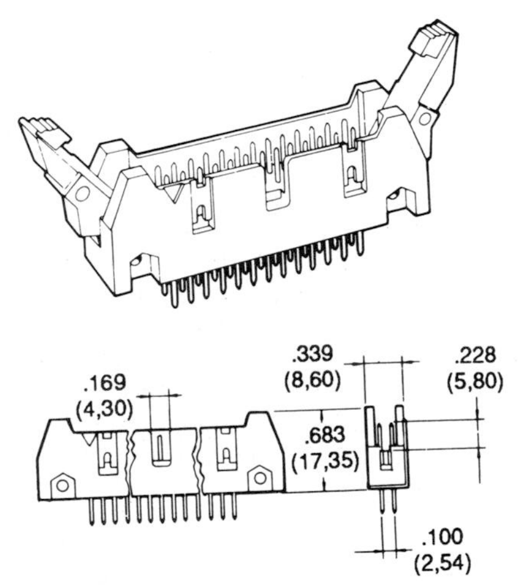

FIGURE 10.4 One of many different connector styles available for use with metal dome–PCB subassemblies. (*Courtesy of Square D Company, Data Entry Products, Loveland, Colo.*)

- Elimination of soldering for connectors
- Cost efficiencies through reduced manufacturing costs

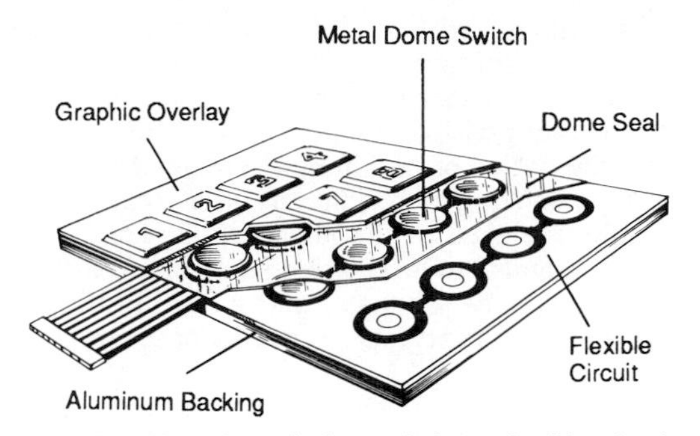

FIGURE 10.5 A typical metal dome–flexible circuit subassembly. (*Courtesy of Square D Company, Data Entry Products, Loveland, Colo.*)

- Very thin finished assembly
- Somewhat faster turnaround time from order to delivery (1)

Elimination of Soldering. Note the flex tail exiting from the flexible circuit in Fig. 10.5. By using a crimp-type connector on the end of the flex tail, soldering operations are eliminated.

Cost Effectiveness. Generally speaking, a half-membrane switch attached to a backer board is less costly than an equivalent PCB, particularly when the PCB requires terminals, header, or other type of interconnect.

Thin assembly. Discounting backer board thickness, which is a function of keyboard size and rigidity requirements, the complete metal dome–flexible circuit assembly could be as thin as 0.025 in (0,63 mm).

Faster Turnaround. Usually, flexible circuits are faster to produce than PCBs, but this is really a function of the manufacturer's vertical integration. Check with various manufacturers to determine actual delivery time of one technique vs. the other. (1)

Substrate Materials

The base materials are identical to those discussed in Chap. 8. A material typically recommended is 0.007 in (0,17 mm) thick polyester. The circuit artwork is silk-screened to the base material using conductive ink and carbon.

The connector flex tail for the flexible circuit is actually a ribbonlike extension, or "tab," of the base material with silk-screened connections. The connector is crimped directly to this flex tail and requires no solder.

As with the PCB style, a schematic should be made part of the basic keypad specification so that the switch manufacturer can generate the circuit artwork.

Generally, components (other than LEDs, discussed in Chap. 8) are not installed on flexible circuits used in switches. (1)

Mounting

The metal dome–flexible circuit subassembly must be mounted to a rigid mounting surface or backer material. The backer should be flat and not deflect when the switch is actuated. Aluminum, with a minimum thickness of 0.060 in (1,52 mm), is recommended by most manufacturers, but optional thicknesses are available. It is also recommended that this rigid backing be attached during manufacture of the subassembly to prevent possible damage during shipping and handling.

If the user prefers the option of mounting the metal dome–flexible circuit keyboard to his or her own backing plate (directly to a control panel, for example), keep in mind that adhesive used at the keyboard-backer interface will perform best when attached to metals, such as aluminum, or plastics with high surface energy, such as polyester, polycarbonate, and ABS. Plastics to avoid, owing to low surface energy, include polystyrene, polyethylene, and polypropylene. (1)

Terminations

Termination techniques used on basic membrane switches (Chap. 8) are applicable here also. A standard clincher-type connector offered by one manufacturer is depicted in Fig. 10.6. Consult with the switch manufacturer to determine which connectors are standard with that manufacturer. Choosing a standard the manufacturer is familiar with always results in maximum economies and fastest turnaround. (1)

Whichever style of metal dome subassembly is selected, PCB or flexible circuit, the next consideration is for the graphic overlay which will be attached to the subassembly. The reader is referred back to Chap. 8 for a detailed discussion on graphic overlays since the same principles and considerations apply here. However, it is well worth reinforcing some graphic overlay standards which are important to the efficient functioning of membrane–metal dome assemblies.

Graphic Overlays

In addition to a review of graphic overlay fundamentals in Chap. 8, the user-engineer should be aware of the following considerations when specifying graphic overlays for membrane–metal dome keyboards.

Embossing. Although *embossing* is not absolutely necessary on a graphic overlay designed for use on a membrane–metal dome keyboard, the nature of the con-

No. of positions	DIM A inch (mm)	DIM B inch (mm)	DIM C inch (mm)
8	.700 (17,7)	.990 (23,2)	.915 (23,2)
10	.900 (22,8)	1.190 (30,2)	1.115 (28,3)
12	1.100 (27,9)	1.390 (35,3)	1.315 (33,4)
14	1.300 (33,0)	1.590 (40,3)	1.515 (38,4)
16	1.500 (38,1)	1.790 (45,4)	1.715 (43,5)
18	1.700 (43,1)	1.990 (50,5)	1.915 (48,6)
20	1.900 (48,2)	2.190 (55,6)	2.115 (53,7)

FIGURE 10.6 A typical clincher-type connector for flexible circuit termination. (*Courtesy of Square D Company, Data Entry Products, Loveland, Colo.*)

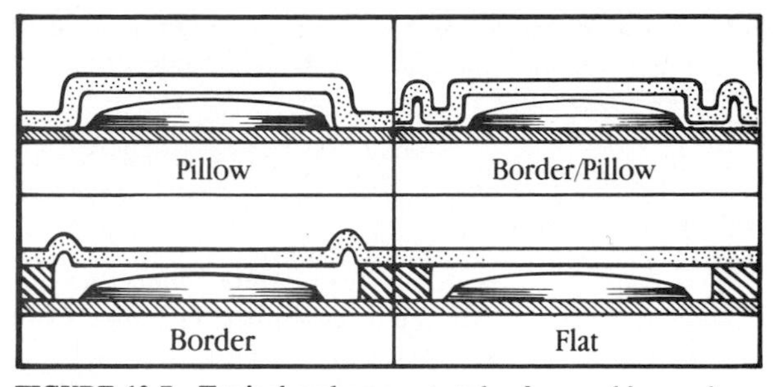

FIGURE 10.7 Typical embossment styles for graphic overlays. *(Courtesy of Square D Company, Data Entry Products, Loveland, Colo.)*

struction of the switch makes embossing a complement to the style, appearance, and feel of the keyboard. Figure 10.7 depicts typical embossment offered by a number of membrane–metal dome keyboard manufacturers and, for comparison, a nonembossed, flat design.

Pillow Embossing. Pillow embossing (sometimes called regular embossing) produces a raised key, giving the look and feel of a keycap keyboard. The metal dome switches are situated directly under the embossed keys. Note that a pillow embossed keyboard assembly requires the least amount of internal layers, which may provide an economic benefit to the user's design.

Border Embossing. Border embossing creates an outlined key. This style not only helps in centering the operator's fingers on the keys but, when combined with contrasting color graphics, it creates a visual, decorative definition to the keys.

Border and Pillow Embossing. Essentially a combination of the first two embossing styles, border and pillow embossing should be discussed with the switch manufacturer before this style is specified. It is somewhat more difficult to produce than border or pillow embossing alone.

Flat (No Embossing). A flat surface provides the most economical approach to graphic overlay keyboards. Definition and targeting of the operator key functions are determined strictly by the printed graphics on the flat, top membrane layer. The keyboard is smooth and easy to clean, and retains the excellent tactile feel. (1)

One manufacturer has outlined specific requirements for embossing produced at the company's facility:

- The *minimum* spacing from the edge of one key to the edge of another is 0.125 in (3,17 mm).
- The *maximum* height of an embossed key is 2.5 times the thickness of the material itself. For example, using a material that is 0.007 in (0,17 mm) thick, the maximum key height would be 0.017 in (0,43 mm) high.
- The *minimum* width for border embossing must be 5 times the thickness of the material. For example, using a material that is 0.007 in (0,17 mm) thick, the minimum width would be 0.035 in (0,89 mm) wide.
- The *minimum* radius of any inside corner must be 4 times the thickness for pillow embossing and 5 times the thickness for border embossing.
- The minimum distance between the edge of an embossed key and the edge of the keyboard is 0.250 in (6,35 mm). (1)

Different switch manufacturers may have their own individual embossing requirements, so consult with the manufacturer(s) of choice prior to beginning the design and specifying process.

Key Annunciation

The use of a PCB as the substrate to which the metal domes are sealed makes it quite easy to install LED lamps for *key annunciator lighting.* Commonly called *hotspot* lighting, key annunciation involves installing an LED directly on the PCB, beneath a window in the graphic overlay, as shown in Fig. 10.8*a*. Key annunciation is commonly used as an operator prompt to indicate that a key must be actuated, or as feedback to the operator that a key actuation has been received by the system processor. The annunciator window is normally positioned in one upper corner of a key switch (Fig. 10.8*b*) and can be made as transparent as the graphic layer base material or can be made in various levels of *deadfront* to keep the window hidden until illuminated by the LED. Incandescent lamps are not commonly used for this purpose because heat from the lamp might cause warping of the internal materials of the switch. The high-bright LEDs available from a number of sources are usually sufficient for most applications. Figure 10.9 shows a typical metal dome–PCB switch with LED key annunciation and embossing.

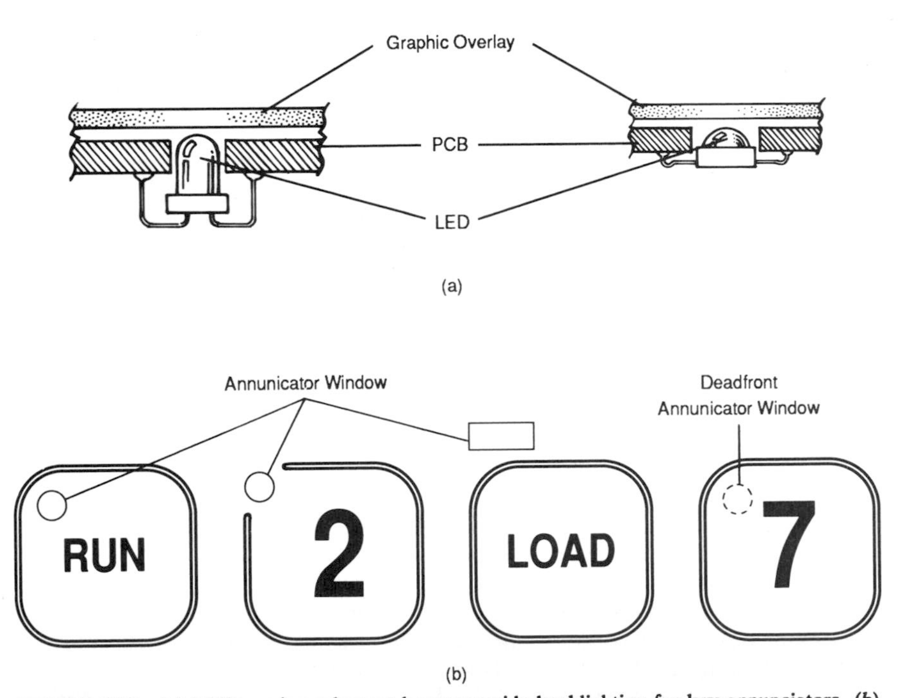

FIGURE 10.8 (*a*) LEDs or incandescent lamps provide backlighting for key annunciators, (*b*) annunciator windows are usually placed in one upper corner of a key switch. (*Courtesy of Square D Company, Data Entry Products, Loveland, Colo.*)

4

Switch manufacturers who normally integrate LEDs into their keyboard assemblies should be able to demonstrate to the user-engineer samples of various annunciator styles for evaluation purposes. If annunciator light levels are critical, measurements can be taken from selected samples and added to the switch specification.

Adhesives

When a graphic overlay is assembled to either a metal dome–PCB subassembly or a metal dome–flexible circuit subassembly, a high-performance acrylic adhesive is generally specified. Acrylic adhesives offer high shear strength and chemical resistance, as well as performance at temperatures from −40°F (−40°C) to 400°F (204°C). (1)

Mounting

Mounting configurations for graphic overlay keypads (using both PCB and flexible circuit–backer subassemblies) fall into three typical categories:

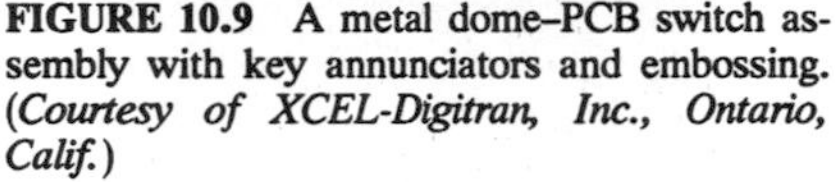

FIGURE 10.9 A metal dome–PCB switch assembly with key annunciators and embossing. *(Courtesy of XCEL-Digitran, Inc., Ontario, Calif.)*

- Recessed
- Back mounted
- Top mounted

Recessed. As shown in Fig. 10.10*a*, the housing for the keypad has a recessed edge. The keypad fits into the recess, resulting in a flush-mounted keyboard.

The keyboard may be attached to the housing with studs that are a part of the PCB. The PCB is counterbored so that, on installation, the head of the stud is flush with the top surface of the PCB. The PCB, with studs installed, can then go through its normal manufacturing procedure, including installation of the domes and dome seal.

If stud mounting of a flexible circuit graphic overlay keypad is desired, the studs are installed in the aluminum backer plate prior to the flexible circuit-backer plate assembly. The studs are mounted in the backer plate in a fashion similar to mounting in a PCB.

An alternative method of keypad attachment is to use a high-performance acrylic adhesive. The adhesive is placed between the bottom of the keypad and the recessed surface of the housing, as shown in Fig. 10.10*b*. (1)

Back Mounting. Figure 10.11*a* depicts a back-mounted keypad in which the keypad is attached to the back of an opening in the housing. The keypad is

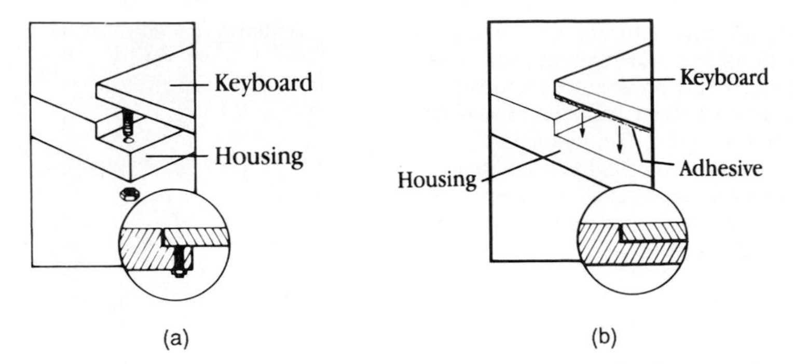

FIGURE 10.10 (*a*) Keyboard installation in a recessed panel using mounting studs, (*b*) keyboard installation in a recessed panel using adhesive. (*Courtesy of Square D Company, Data Entry Products, Loveland, Colo.*)

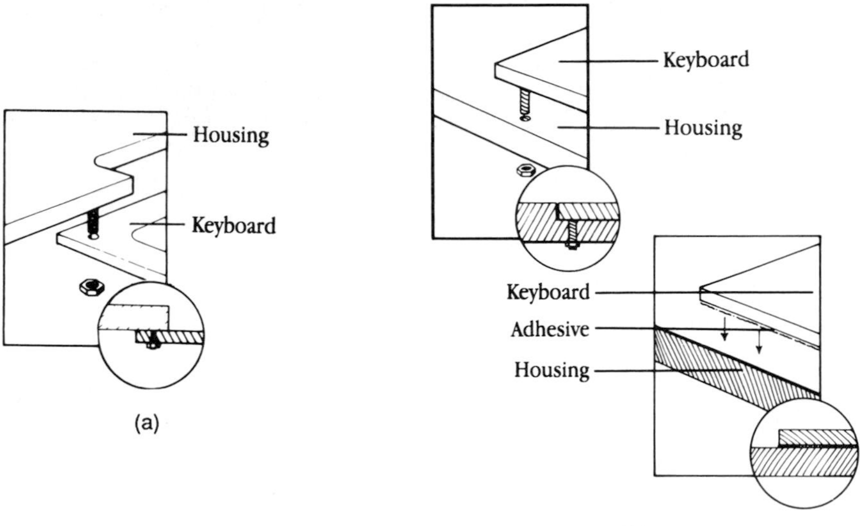

FIGURE 10.11 (*a*) Back-mounted keyboard mounts to back side of panel, (*b*) top mounting can be done but exposes the keyboard edges to the environment. (*Courtesy of Square D Company, Data Entry Products, Loveland, Colo.*)

secured with screws or studs in the housing that line up with the holes or slots in the keypad.

When this mounting method is selected, the typical distance from the edge of the hole to the edge of the keyboard is 0.125 in (3,17 mm). (1)

Top Mounting. The keypad may be mounted directly to the top of the housing as illustrated in Fig.10.11*b*. This is not a preferred method of mounting because it leaves the outside edge of the keypad exposed and vulnerable to damage caused by the penetration of liquids and chemicals or by the edge's getting caught and the keypad layers pulled apart. In some select applications, conformal coating can be applied to the edges of the keypad for added protection in this type of mounting. Top mounting using studs in the keypad, similar to the recess mounting techniques discussed above, can also be accomplished. (1)

Shielding

Many electronic components are sensitive to static discharge, and if the design in question includes such devices, protective shielding is recommended. Typically, ESD (electrostatic discharge) shielding is provided by aluminum foil laminated to an insulating polyester layer with acrylic pressure-sensitive adhesive backing.

HOW RUBBER–METAL DOME SWITCHES ARE CONSTRUCTED

When a keypad application requires operation in harsh conditions, while retaining an enhanced tactile feel, rubber–metal dome switches should be considered. Consisting of a metal dome–PCB subassembly covered by a molded elastomer boot, the final product is rugged and environmentally resistant, ideal for hot, cold, humid, dirty, and wet environments. Properly designed, rubber–metal dome keypads may even be submerged in water and expected to survive. A typical rubber–metal dome construction is shown in Fig. 10.12.

Metal Dome–PCB Subassembly

The metal dome–PCB subassembly is manufactured in an identical manner to the subassemblies used in membrane–metal dome keyboards explained earlier; that is, the metal dome is sealed to the PCB with a dome seal material.

Rubber Boot

Materials. The silicone rubber used in rubber-booted keypads is highly resistant to water, free steam, oxidation, high temperatures, and most chemicals, solvents,

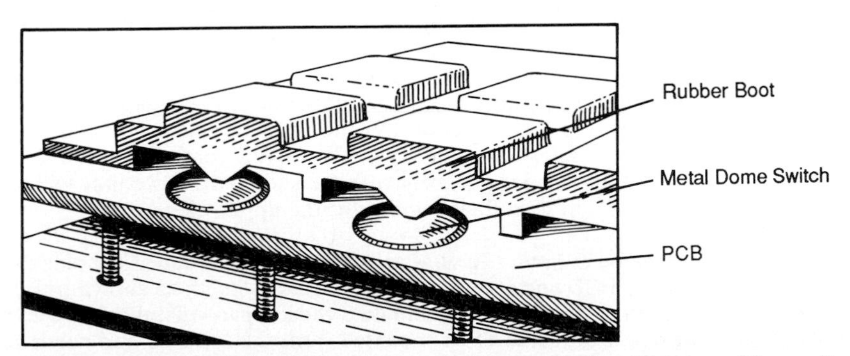

FIGURE 10.12 A typical rubber–metal dome switch construction. (*Courtesy of Square D Company, Data Entry Products, Loveland, Colo.*)

fuels, and oils. Chapter 11 gives more detailed descriptions of what environments silicone rubber can withstand.

Sizes. The keyboard boot is generally compression molded of silicone rubber which acts both as an actuator and as a gasket seal. The keys are molded into the rubber itself. Recommended parameters for the keys as suggested by one manufacturer are:

- A minimum size of 0.300 × 0.300 in (7,62 × 7,62 mm) and no larger than 0.750 × 0.750 in (19,05 × 19,05 mm)
- At least 0.125 in (3,17 mm) from the edge of one key to the edge of another
- At least 0.250 in (6,35 mm) from the edge of the key to the edge of the keypad assembly

Legends. Legends are applied to the boot in a manner similar to legends on conductive rubber switches (Chap. 11), and the reader is directed to that chapter for a detailed discussion on legend application techniques for silicone rubber.

Backlighting

When the boot is molded from *translucent silicone rubber,* a number of methods of backlighting the keypad are available. The keys may be lighted using LED, EL (electroluminescent), woven fiber optics, or incandescent components (Fig. 10.13*a*). Figure 10.14 depicts a translucent silicone rubber–metal dome switch assembly backlit by woven fiber-optic mats. Note the reverse legend markings used to produce a lighted legend–dark background character array. Chapter 8 contains complete details on EL and fiber-optic backlighting. The concepts contained in that chapter are also valid for backlighting translucent rubber. A black rubber or metal "egg crate" (Fig. 10.13*b*) works well to prevent hot spots in a backlighted keypad, particularly where LED or incandescent light sources are used. (1) Additional discussion of backlighting for silicone rubber switches is found in Chap. 11.

Mounting

Rubber-booted keypads are usually flange mounted or top mounted with a wraparound seal.

Flange Mount. In a flange mount design, the rubber-booted keypad extends to the edge of the PCB and is "sandwiched" in the product when mounted (Fig. 10.15*a*). The width of the flange is typically 0.500 in (12,7 mm), but there is no hard-and-fast rule for this. Working closely with the switch manufacturer will result in agreement on an optimum flange width for the application.

Top Mount with Wraparound Seal. In wraparound design, the rubber is molded to cover the edges of the PCB and to wrap around the PCB itself, as shown in Fig. 10.15*b*. This type of mounting is recommended if the keypad will be subjected to direct splashes of liquid or if it is to be mounted from the top side and needs to seal around the bottom edge of the keyboard. (1) This eliminates the need for an additional bottom gasket on the keypad. The user should perform a cost trade-

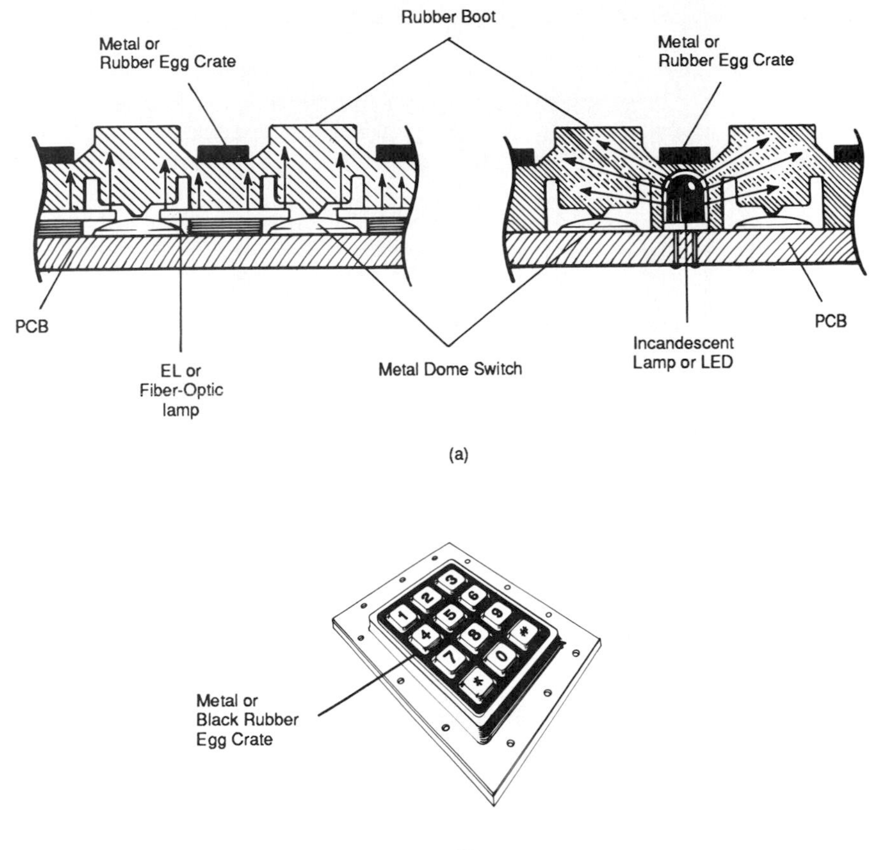

FIGURE 10.13 (*a*) Translucent silicone rubber keypads can be backlit using LED, EL, fiber-optic, or incandescent, (*b*) metal or rubber "egg crate" eliminates hot spots from keypad illuminators. (*Courtesy of Square D Company, Data Entry Products, Loveland, Colo.*)

off, however, to determine if the additional tooling cost (and perhaps additional individual unit cost also, if molding of the boot is more labor-intensive because of difficulties removing it from the tool) is offset by the cost of a separate gasket.

GLOSSARY

Deadfront. A method of finishing *key annunciator* windows which prevents the backlight means, an LED or incandescent lamp, from being seen until lighted. Without this treatment, ambient light reflecting off the lamp capsule might be perceived by the operator as the lamp being energized.

Dome seal. A thin polyester (or similar) sheet, with acrylic adhesive on one side, which is used to retain metal domes to a substrate such as a PCB or flexible circuit.

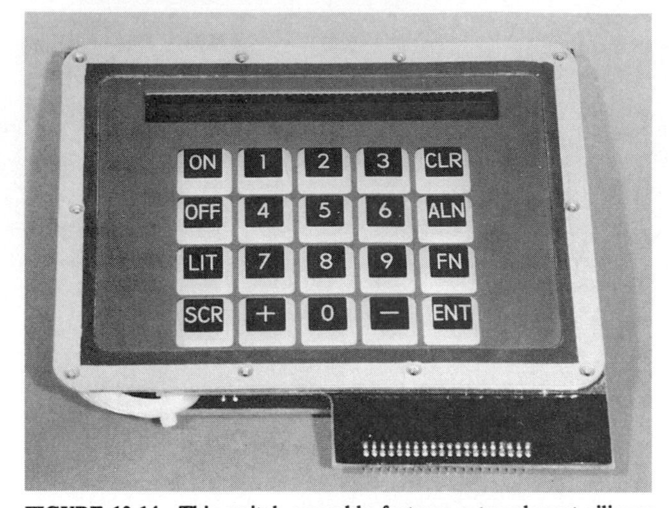

FIGURE 10.14 This switch assembly features a translucent silicone rubber–metal dome keypad with fiber-optic backlighting. (*Courtesy of XCEL-Digitran, Inc., Ontario, Calif.*)

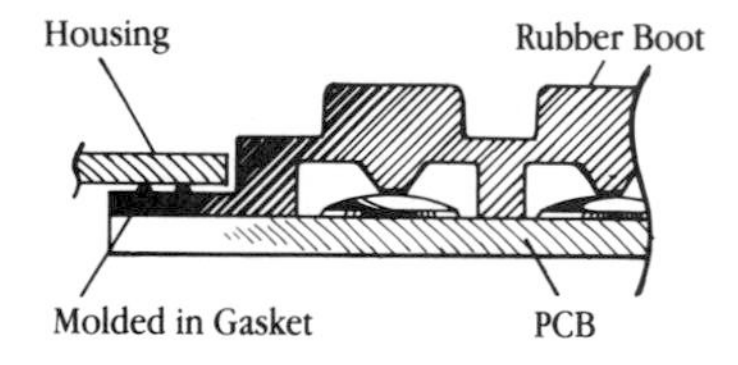

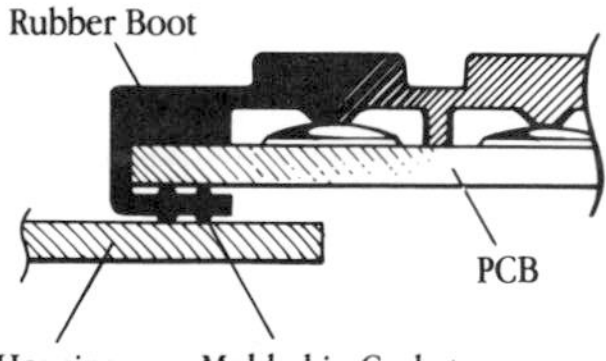

FIGURE 10.15 Two mounting styles are available for rubber–metal dome keypads, (*a*) flange mount and (*b*) top mount. (*Courtesy of Square D Company, Data Entry Products, Loveland, Colo.*)

Egg crate. A gridlike bezel, or frame, mounted over the top of keypads. In backlighted silicone rubber switches, the egg crate acts as an opaque light block to eliminate annoying bright spots of light created by LEDs or incandescent lamps.

Embossing. To produce a raised surface in a substrate material. Graphic overlays are often embossed to produce raised key areas, to simplify targeting by the operator. Embossing can be functional or decorative or both.

Graphic overlay. See *Membrane–metal dome.*

Hotspot. See *Key annunciator lighting.*

Key annunciator lighting. Small backlighted windows situated inside a key target area on a graphic overlay-style switch. The windows are usually lighted by high-bright LEDs or, less often, by incandescent lamps to produce a bright point of light indicating feedback from the system computer of a successful switch closure or an indication that the operator is required to actuate the associated key switch. Also called *hotspots.*

Membrane–metal dome. A style of switch which uses a metal dome, mounted on a PCB or flexible circuit, over which is mounted a graphic overlay. Also called a *graphic overlay switch.*

Metal dome–flexible circuit. A style of switch in which metal domes are installed directly on a flexible circuit and usually covered by a graphic overlay.

Metal dome–PCB. A style of switch in which metal domes are mounted onto a PCB containing etched and plated switch circuitry. The resulting subassembly can then be covered with a graphic overlay or a molded elastomer boot.

Rubber–metal dome. A style of switch in which a *metal dome*–PCB subassembly is covered with a molded elastomer boot.

Translucent silicone rubber. Silicone rubber compounded without added dyes, resulting in a translucent, light-transmitting material. Translucent silicone is ideal for producing backlit keypads.

REFERENCE

1. "Design Guide," Square D Company, Data Entry Products, Loveland, Colo., 1990.

CHAPTER 11
CONDUCTIVE RUBBER SWITCHES*

Although first introduced in the mid-1970s, conductive rubber switches are only now reaching their full potential as a reliable, low-cost, sealed switch technology. Conductive rubber switches are so called because the switch element is silicone rubber filled with conductive particles, usually carbon. Occasionally, silver or gold particles are used when minimal contact resistance is demanded by electrical requirements. A conductive rubber element is molded into a single sheet of silicone rubber, one conductive element for each switch position. The single rubber sheet holds the conductive elements in place beneath integrally molded bell-shaped domes (Fig. 11.1). The bell shapes impart the particular tactile feel re-

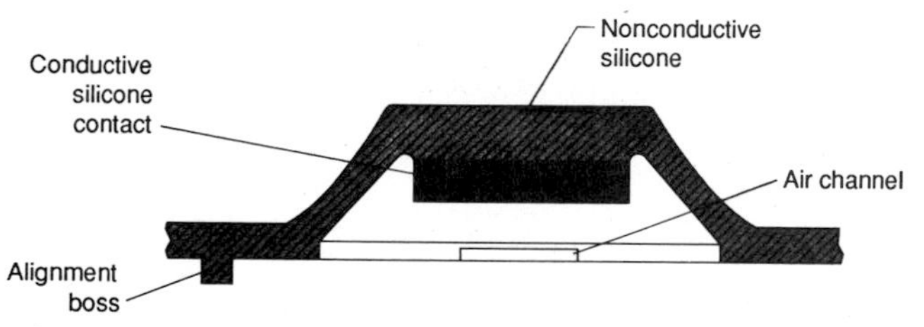

FIGURE 11.1 The conductive contacts are molded into the base rubber. (*Courtesy of Shin-Etsu Polymer America, Inc., Union City, Calif.*)

quired by the user. The sheet is placed over a printed wiring board which carries the necessary switch circuitry, backlighting means (if required), and connection device. The sheet also acts as a shield to prevent contaminants and spilled liquids from reaching the switch contacts and printed wiring board (Fig. 11.2).

How Conductive Rubber Switches Are Constructed

Conductive rubber switches are mechanical-type switches made of a single molded piece of silicone rubber with conductive contacts (often called "pucks"

*Numbers in parentheses indicate items in the References at the end of this chapter.

FIGURE 11.2 The one-piece molded rubber sheet provides sealing for the switch contacts. (*Courtesy of General Silicones Co., Arcadia, Calif.*)

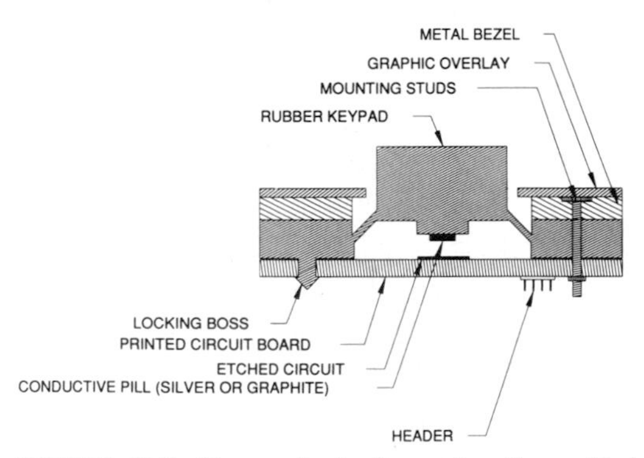

FIGURE 11.3 One method of securing the molded switch matrix to the PCB. (*Courtesy of Memtron Technologies, Inc., Frankenmuth, Mich.*)

for their resemblance to hockey pucks) integrally molded into the thin rubber keypad. This sheet, or switch matrix, is assembled between integral or separate keytops and a printed wiring board (Fig. 11.3). Each conductive rubber contact is directly over an etched interdigitated contact on the printed wiring board (Fig. 11.4). When pressed into contact with the printed wiring board etched contacts, the conductive rubber contact bridges the gap between the etched circuit and completes the switch contact.

FIGURE 11.4 The conductive rubber contact is located directly above the interdigitated contact on the PCB. (*Courtesy of Shin-Etsu Polymer America Inc., Union City, Calif.*)

DESIGN CONSIDERATIONS

The following design considerations are meant only to educate the potential conductive rubber switch user in the wide variety of design options available in producing a completed switch and are not an encouragement to design and specify a switch in a figurative vacuum. The reason for this cautionary note is the relatively easy access to a large number of switch manufacturers producing high-quality custom conductive rubber switches. These switch manufacturers have broad experience in producing the exact switch required to meet customer specifications. It would be unwise, as well as uneconomical, for the user-engineer to attempt to design and specify a conductive rubber switch without consultation with these manufacturers. Indeed, most of them have direct relationships with molding facilities in the Far East (Taiwan, in particular) where, except for a few exceptions, almost all conductive rubber molding is done. Much of the labor involved in producing conductive rubber switches is tedious hand labor, hence the nearly universal use of more cost-effective sources such as can be found in the Far East. With this continuous molding experience has come insight into the "black magic" of silicone processing, leading in turn to a thorough understanding of complex combinations of silicone materials, legend printing and overcoating, rubber durometer, tactile force and stroke designs, PCB materials and circuit plating, color matching, and conductive material selection. This vast experience is a resource the user-engineer should take advantage of as early in the design stage as possible in order to yield the most cost-effective conductive rubber switch feasible within the constraints of the state-of-the-art. The penalty for waiting too long to bring in expert consultation from these switch manufacturers could be anything from a complete redesign of a system front panel to the expensive retooling of injection molded components designed to interface with the conductive rubber switch. The user-engineer should also note the lead times involved in producing these molded switches. The wise engineer will keep in mind that the manufacturer must have time to review the engineer's specification, fabricate the mold for the silicone processing, develop artwork for the legend printing (if required), and mold first articles for the engineer's approval.

With this in mind, we can now proceed to a discussion about conductive rubber design considerations, highlighting those details necessary for the user-engineer to consider as he or she develops a specification which is relevant, useful, and attentive to cost constraints.

Design Details of the Conductive Rubber Contact

The heart of the conductive rubber switch is the conductive rubber contact itself. The contact is silicone rubber material filled with a conductive material to allow the contact to carry an electrical current. The vast majority of conductive rubber switches produced utilize molded silicone rubber filled with carbon particles to impart conductivity to the silicone rubber base material. This combination of silicone and carbon provides a conductive contact with a contact resistance of approximately 200 Ω. The carbon contact is very reliable and is available in a variety of sizes. Carbon contacts can be supplied as small as 0.080 in (2,0 mm) diameter and are usually available up to a maximum of 0.315 in (8,0 mm) diameter. (1) Conductive contacts are produced from a higher durometer material than the basic keypad and are matte-finished to promote a cleansing action of the PCB contact pad. The conductive contact should never be larger than the contact pad on the PCB, but it can be smaller without causing operational problems. (2)

One drawback to the carbon contacts is that they are usually available in circular shapes only, although oval contacts can be supplied at slightly higher prices. This is a minor inconvenience, however, since the cylindrical carbon contact will suffice for the vast majority of applications.

The second most common type of contact for rubber keypads is screened, or printed, contacts. In this technique, contact material consisting of conductive particles suspended in a silk-screenable liquid is applied to the bottom of each key position normally devoted to the conductive rubber contact. After application, the screened conductive material is cured. Screened contacts are available in all shapes and sizes (limited only by the screening or printing mechanism's ability to access the bottom of each key) and allow design flexibility and cost effectiveness because of the way the contact is applied to the rubber keypad. Unfortunately, this versatility is offset by the eternal engineering "trade-off," in this case, the disadvantage that screened contacts exhibit considerably shorter life when compared with carbon-filled conductive rubber contacts. Screened contacts also demonstrate significantly higher contact resistance. It is not uncommon for screened contacts to have contact resistance in excess of 1000 Ω, so careful attention must be given to the keypad's electrical and life requirements when the contact type is chosen. (1)

In instances where lower contact resistance is required, however, precious metal particles can be incorporated into the silicone rubber, providing contact resistances of approximately 5 Ω (gold) and 25 Ω (silver), but at a substantial increase in cost. For applications requiring high current, such as POWER ON switches, several conductive rubber switch manufacturers mold a metal wafer inside the rubber dome in place of the standard conductive rubber contact, again at an increase in cost. (3) The metal contact is molded to the base rubber during the manufacturing process and once attached is almost impossible to remove. Depending on the metal selected and the type of plating used, if any, contact resistance is typically less than 25 Ω. Metal contacts are not frequently used in rubber keypads because of the higher costs and longer lead times associated with the process, but they should be considered for special applications. (1)

TABLE 11.1 Specifications for Conductive and Insulating Silicone Rubber

Conductive rubber (carbon-filled):	
Specific gravity	1.05–1.10
Hardness, H°	50–65
Tensile strength, kg/cm^2	50–65
Elongation, %	150–300
Compression set, %	15–30
Volume resistivity, Ω-cm	2.5–4
Insulating rubber:	
Specific gravity	1.07–1.38
Hardness, H°	30–80
Tensile strength, kg/cm^2	50–80
Elongation, %	150–800
Compression set, %	13–21
Volume resistivity, Ω-cm	3×10^{14}–4×10^{15}

Source: Shin-Etsu Polymer America, Inc., Union City, Calif.

Table 11.1 shows physical properties for both the conductive silicone rubber and the insulating, or sheet, silicone rubber.

Design Details of the Molded Rubber Sheet

The sheet rubber that incorporates the bell-shaped tactile domes and the conductive elements is usually compression-molded as a single piece of silicone. The design of bell-shaped domes determines tactile feel and key travel. Figure 11.5*a* and *b* illustrates a few of the basic dome shapes which, when properly designed, can provide a nearly infinite variety of force and stroke combinations. The result is significant control over the finished keypad or keyboard's tactile feel, a very important *subjective* requirement in a switch.

Silicone rubber has extreme resistance to high-voltage ionization, remains flexible at −22°F (−30°C), and retains its shape and operating characteristics to 356°F (180°C). (4)

One of the most important considerations when designing and specifying molded silicone switches is the tolerances of the molded rubber sheet, in particular as it relates to potential mechanical interfaces on the control panel to which the switch ultimately mounts. Table 11.2 illustrates these standard dimensional tolerances.

Silicone Rubber. Silicone rubbers are complex polymeric structures which effectively combine the most desirable properties of both organic and inorganic polymers and are unique within the large family of elastomeric materials owing to properties intrinsic in their chemical composition. The silicone rubber used in conductive rubber switches is an industrial-grade material available from a number of raw material manufacturers, such as Shin-Etsu Polymer, General Electric, and Dow Chemical.

Silicone rubber is available in many different formulations, and the product should dictate the material chosen in any given situation. Silicone rubbers are

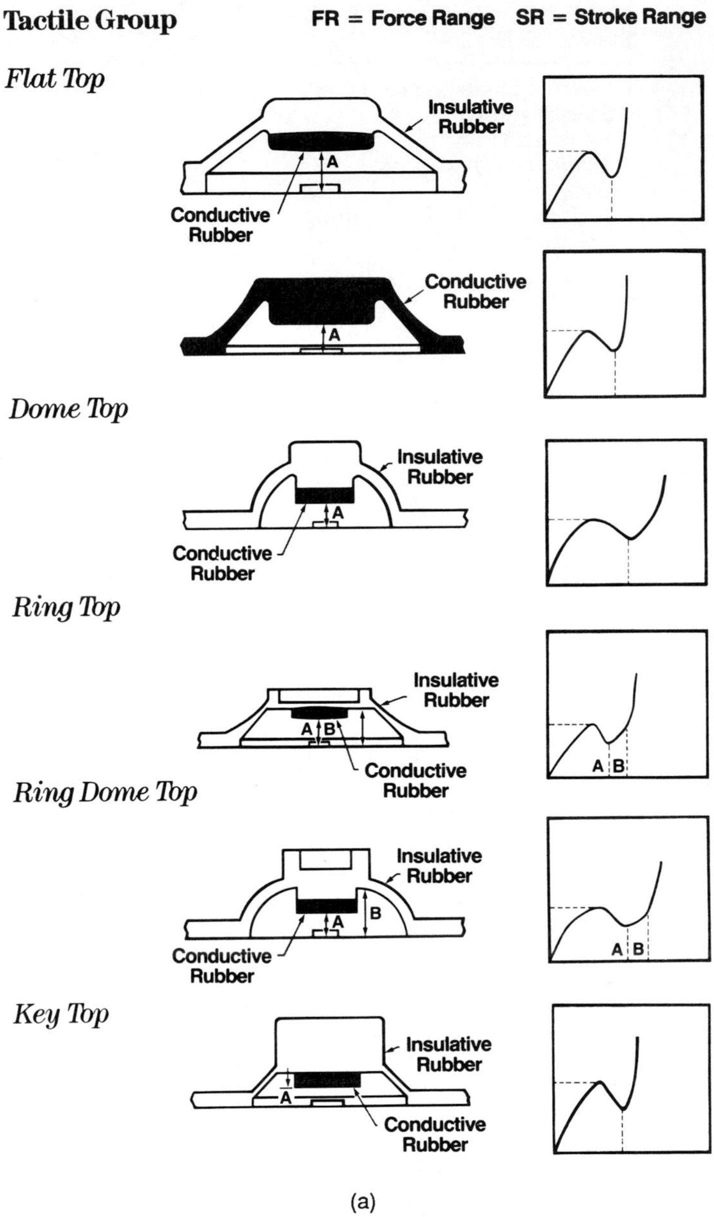

Tactile Group

FR = Force Range SR = Stroke Range

Flat Top

Insulative Rubber
A
Conductive Rubber

Conductive Rubber
A

Dome Top

Insulative Rubber
A
Conductive Rubber

Ring Top

Insulative Rubber
A B
Conductive Rubber
A B

Ring Dome Top

Insulative Rubber
B
A
Conductive Rubber
A B

Key Top

Insulative Rubber
A
Conductive Rubber

(a)

FIGURE 11.5 Basic dome shapes and their associated force-deflection curves. (*Courtesy of Shin-Etsu Polymer America, Inc., Union City, Calif.*)

Non-Tactile Group

FR = Force Range SR = Stroke Range

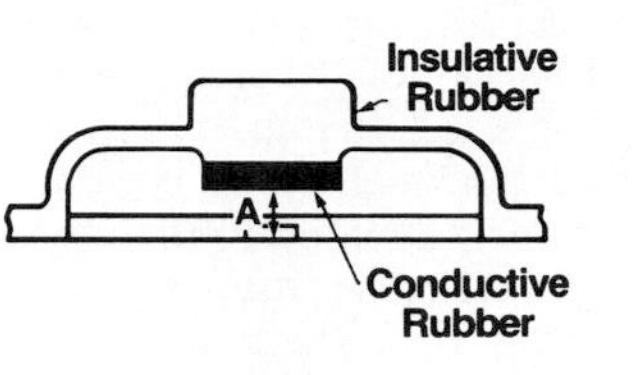

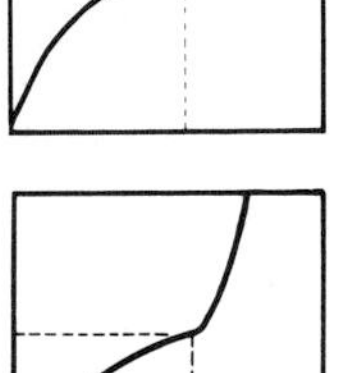

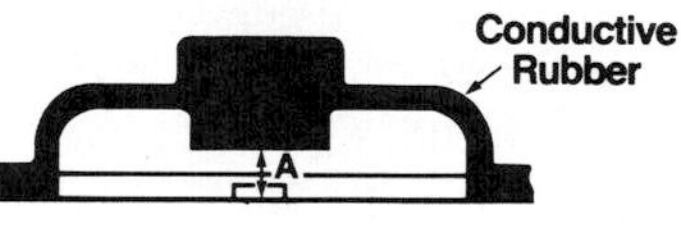

Ring Dome Top

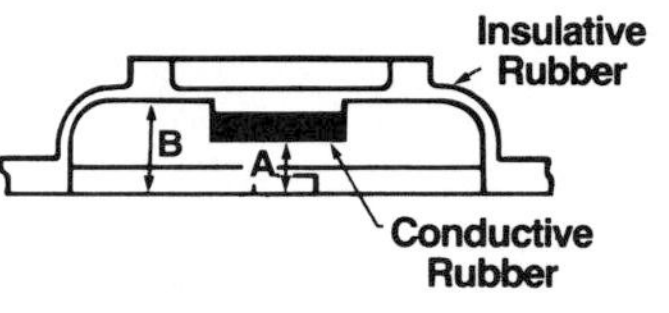

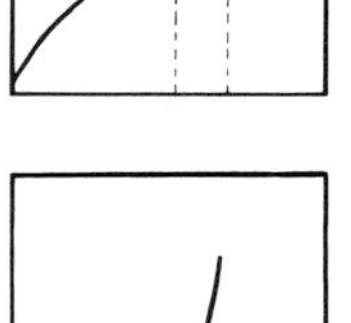

Flat Sheet

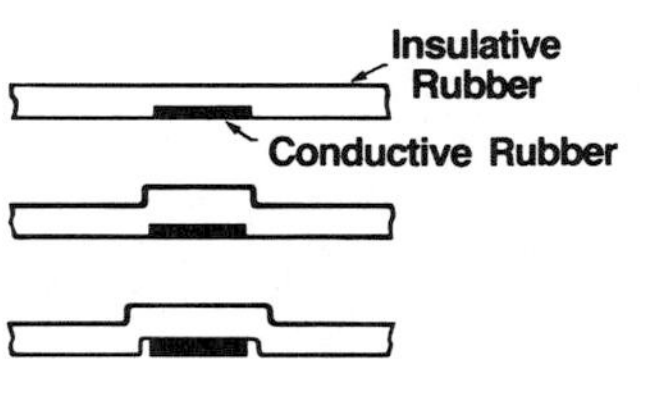

(b)

FIGURE 11.5 (*Continued*)

TABLE 11.2 Standardized Tolerances for Molded Silicone Switches

Inches	Millimeters
<0.250 ± 0.006	<6.36 ± 0.15
0.251–0.394 ± 0.008	6.37–10.02 ± 0.20
0.395–0.630 ± 0.008	10.03–16.00 ± 0.20
0.631–0.984 ± 0.010	16.01–25.00 ± 0.25
0.985–1.570 ± 0.014	25.01–39.89 ± 0.35
1.571–2.480 ± 0.016	39.90–63.00 ± 0.40
2.481–3.937 ± 0.020	63.01–99.99 ± 0.50
3.938–6.290 ± 0.028	100.00–159.77 ± 0.71
>6.290 ± 0.5%	>159.77 ± 0.5%

Source: Conductive Rubber Technology, Inc., Bothell, Wash.

classified in *standard, medium,* and *high-performance* grades. Most conductive rubber keypads are made with *standard-performance* material and, depending on various factors, have mechanical life expectancies, when molded, of 500,000 to 1,000,000 actuations. It should also be noted that most keypads are made with silicone rubber material that has a UL flammability rating of 94HB.

Table 11.3 depicts technical information on typical silicone rubber formulations. Particular attention should be directed to the *Appearance* entries in the table as it is very difficult for a conductive rubber keypad manufacturer to color match silicone rubber when dark gray or black base material is selected for a keypad because of its specific physical or flammability rating. Silicone rubber formulations that begin as translucent material or grayish-white material are normally used in most color matching applications because colored pigmentation can be added with relative ease. A unique design feature which adds versatility to the conductive rubber keypad is the capability to mix two different durometers of silicone rubber in the same keytop. Referred to as a *dual durometer* key in its final molded form, the example shown in Fig. 11.6 has the top portion of the key molded of 80 durometer rubber and the bottom portion, including the all-important membrane web, molded of 50 durometer rubber. The heat and pressure of the molding process inseparably joins the two durometers and, once molded, will not peel apart or separate. The advantages of this molding technique are twofold: first, the membrane web is molded from the softer, more flexible (and therefore longer life) durometer material, and second, the top portion of the key, in the harder durometer, retains the feel of a plastic key (which is generally more acceptable to keypad operators). Additionally, the harder durometer enhances the survivability of the key under harsh conditions of use. If the entire key was molded from the 80 durometer material, the tactile feel would be stiffer and mechanical life would decrease significantly. If the entire key was molded from the 50 durometer material, the softer feel of the key might be unsatisfactory to the keypad user and the key more susceptible to abuse. The combination of the two separate durometers produces a keypad which retains the best characteristics of both. The major design consideration here is that dual durometer keypads generally must be at least 0.236 in (6,0 mm) high to allow for the two-shot molding of the harder durometer keytop which is then molded to the base material of the keypad. If the specifying engineer is considering a dual durometer keypad, consult with the switch manufacturer to determine the best durometer characteristics for the keypad under design. Obtain samples from the manufacturer of the durometer pairs that most closely meet your requirements and collect comments from, ideally, the operator population or, if this is not feasible, other engineers or designers familiar with the operation of the final system. The determination of the durometer pairing is subjective, and no amount of calculation or analysis can replace the opinion of the ultimate user.

It should be noted that high-performance materials are more expensive than standard-performance materials and materials with flammability ratings of 94V-0 (or higher) are more expensive than those with ratings of 94HB. The following section discusses the flammability test procedures used to classify a material in either the 94V-0 (or higher) or 94HB category so the user can determine if the use of the more expensive material is necessary.[1]

Flammability Ratings. Flammability ratings for silicone rubber materials are derived in the same manner as similar ratings for plastic materials. They are in-

[1]Since procedures can change from time to time, the user should obtain the latest test procedures directly from the agency responsible for approval of the equipment under design.

TABLE 11.3 Silicone Rubber Formulations

Material	931	941	951	953	961	971	981	3801	5140	5150	5160	9511	9611
Typical properties before cure:													
Appearance	Trans.	Trans.	Trans.	Trans.	Grayish-white	Grayish-white	Grayish-white	Black	Trans.	Trans.	Trans.	Trans.	Trans.
Specific gravity, 25°C	1.10	1.12	1.15	1.15	1.22	1.30	1.42	1.20	1.09	1.11	1.12	1.14	1.14
Plasticity, Williams	130–210	200–260	230–290	230–330	270–340	330–400	420–500	600	160	170	175	200	205
After cure:													
Linear shrinkage, %	4.3	4.0	3.8	3.9	3.4	3.0	2.8	3.3	3.9	3.7	3.9	3.5	3.4
Hardness, JIS	30	40	50	50	60	70	80	73	40	50	60	50	59
Tensile strength, kg/cm^2	55	75	90	70	73	68	85	55	81	83	83	72	67
Elongation, %	530	450	340	400	280	200	130	180	550	480	410	290	290
Tear strength, kg/cm	8	11	14	18	15	15	10	10	14	19	15	8	10
Compression set, %	4.3	4.0	3.8	3.0	3.4	3.0	2.8	3.0	7.0	6.0	6.0	4.0	4.0
Flammability rating	94HB	94HB	94HB	94HB	94HB	94HB	94HB	94V-0	94HB	94HB	94HB	94HB	94HB
Fatigue test, cycle × 10,000	10	20–30	30–40	30–40	30–40	15–25	10–20	N/A	600–1000	400–800	200–400	200–300	150–200

Source: Glolite Sales, Ltd., Pauls Valley, Okla.

tended to serve as a *preliminary indication* of probable flammability in a particular application.

Flammability ratings are derived by using standard-size specimens and are intended to be used solely to measure and describe flammability properties of various materials in very closely controlled laboratory conditions. Actual response to flame and heat depends upon the size and form of the product and the end use of the product using the material. Other important characteristics identified by flammability testing include ease of ignition, burn rate, flame spread, intensity of burning, and products of combustion.

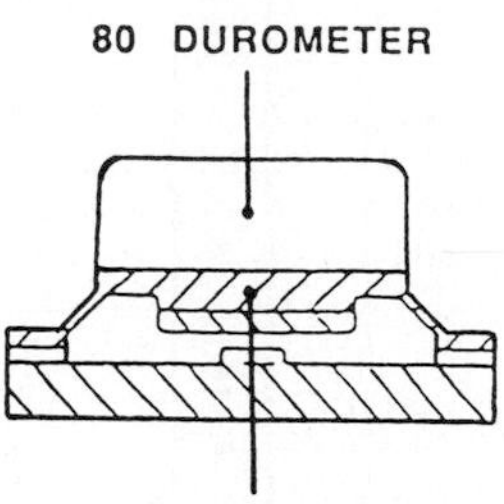

FIGURE 11.6 A dual durometer key. (*Courtesy of Glolite Sales, Ltd., Pauls Valley, Okla.*)

The final acceptance of the material itself is dependent upon its use in complete equipment which meets all applicable standards relating to the equipment.

The following describes the horizontal burning test for classifying material 94HB:

Materials shall be classified 94HB on the basis of test results obtained on small bar specimens when tested in very rigidly controlled laboratory conditions.

A material classified as *94HB* shall meet the following conditions:

A. Not have a burning rate exceeding 1.5 in (38,1 mm) per minute over a 3.0 in (76,2 mm) span for specimens having a thickness of 0.120–0.500 in (3,05-12,7 mm) or

B. Not have a burning rate exceeding 3.0 in (76,2 mm) per minute over a 3.0 in span for specimens having a thickness less than 0.120 in (3,05 mm) or

C. Cease to burn before the 4.0 in (102 mm) reference mark.

Additionally, the vertical burning test for classifying materials 94V-0/V-1 is as follows:

A material classified as *94V-0* shall meet the following conditions:

A. Not have any specimens that burn with flaming combustion for more than 10 seconds after either application of the test flame.

B. Not have a total flaming combustion time exceeding 50 seconds for the 10 flame applications for each set of five specimens.

C. Not have any specimens that burn with flaming or glowing combustion up to the holding clamp in the test fixture.

D. Not have any specimens that drip flaming particles that ignite the dry absorbent surgical cotton located 12 inches (305 mm) below the test specimen.

E. Not have any specimens with glowing combustion that persists for more than 30 seconds after the second removal of the test flame.

A material classified as *94V-1* shall meet the following conditions:

A. Not have any specimens that burn with flaming combustion for more than 30 seconds after either application of the test flame.

B. Not have a total flaming combustion time exceeding 250 seconds for the 10 flame applications for each set of five specimens.

C. Not have any specimens that burn with flaming or glowing combustion up to the holding clamp of the test fixture.

D. Not have any specimens that drip flaming particles that ignite the dry absorbent surgical cotton located 12 in (305 mm) below the test specimen.

E. Not have any specimens with glowing combustion that persists for more than 60 seconds after the second removal of the test flame. (1)

Table 11.4 depicts silicone rubber compounds developed specifically to meet 94V-0 of Flame Test UL 94. Material KE-5606U has excellent resistance to oil and tracking, has excellent compression set, and is widely used for electrical devices requiring exacting electrical properties, thermal resistance, and self-extinguishing properties. Material KE-5612GU has excellent resistance to heat, tracking, and arc, and good electrical insulation characteristics. (1)

Characteristics of Silicone Rubber

HEAT RESISTANCE: Compared with organic rubbers, most silicone rubbers have incomparably high resistance to heat. Silicone rubbers can be used for very long lengths of time at 302°F (150°C) with almost no change in material properties and, depending on the compound, can withstand more than 10,000 h of continuous service at 392°F (200°C) (Fig. 11.7). If necessary, silicone rubbers can operate in temperatures as high as 662°F (350°C) for short periods of time. Because of these characteristics, silicone rubbers are widely used in applications requiring high heat resistance.

When silicone rubber is heat aged in open air, the hardness of the material increases and elongation decreases. Conversely, the hardness of silicone rubbers tends to decrease when heated under a hermetically sealed condition and the heat-resistance serviceable life may sometimes be shorter if heated in open air. This phenomenon is caused by the breakage of the polysiloxane chains of the polymer. Several different grades of silicone rubbers are available which exhibit improved heat resistance under hermetically sealed conditions by combining special additives with a proper postcure agent. Special formulations and the selection of a proper curing agent operate to prevent softening of the silicone due to the breakage of siloxane chains.

CHEMICAL, OIL, AND SOLVENT RESISTANCE: At temperatures below 212°F (100°C), the oil resistance of silicone rubber is somewhat inferior to that of nitrile and

TABLE 11.4 Properties of Two Specific Silicone Rubber Materials

Material	KE-5606U	KE-5612GU
Appearance	Gray	Grayish-black
Specific gravity, 25°C	1.46	1.49
Plasticity, Williams	270–330	247
Hardness, JIS	53	59
Tensile strength, kg/cm^2	65	69
Elongation, %	300	340
Tear strength, kg/cm	17	15
Compression set	20	19
Volume resistivity, Ω-cm	1.3×10^{15}	2×10^{15}
Dielectric strength, kV/mm	27	29
Dielectric constant, 50 Hz	3.6	3.3
Dissipation factor, 50 Hz	53	51
Rebound resilience, %	55	51

Source: Glolite Sales, Ltd., Pauls Valley, Okla.

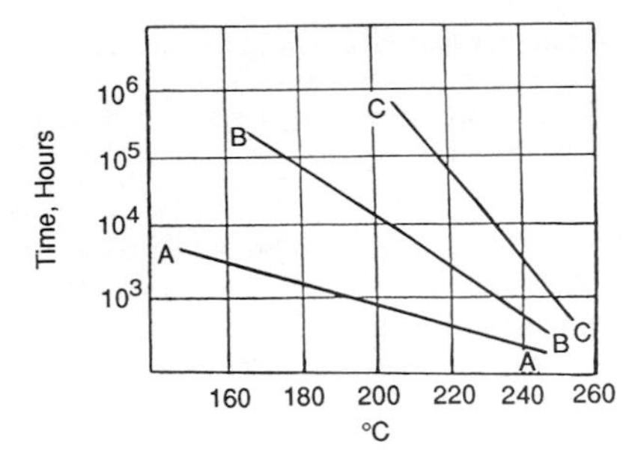

FIGURE 11.7 Serviceable life of silicone rubbers. (*Courtesy of Glolite Sales, Ltd., Pauls Valley, Okla.*)

TABLE 11.5 Oil Resistance of Silicone Rubber

			Changes in properties			
Oil type	Temperature	No. of hours	Hardness, points	Tensile strength, %	Elongation, %	Volume, %
ASTM 1	150/302	168	−10	−10	−10	−10
ASTM 3	150/302	168	−25	−20	−20	+40
GM hydramatic fluid	94/201	70	−35	−40	−5	+35
Ford brake fluid	150/302	72	−20	−60	−40	+15
Diesel fuel	50/122	168	−30	—	—	+105
Gasoline	23/73	168	−20	—	—	+165
Sydrol 500A fluid	70/158	168	−5	−10	+5	+10
Motor oil (SAE 30)	175/347	168	−8	−70	−65	−8

Source: Glolite Sales, Ltd., Pauls Valley, Okla.

chloroprene rubbers (Table 11.5), but at temperatures above 212°F (100°C), the oil resistance of silicone rubber is superior to all types of organic rubber. In addition to being resistant to oil, silicone rubbers are chemical and solvent resistant as well (Table 11.6). Polar organic compounds such as aniline, alcohol, and dilute alkaline and acid solutions only slightly affect the properties of silicone rubber, as can be evidenced by its volume swell of only 10 to 15 percent. On the other hand, silicone rubber swells when contacted with nonpolar solvents such as benzene, toluenes, and gasoline (Table 11.7), but the original properties are quickly restored when the solvents are removed.

Silicone rubber should not be used in the presence of highly concentrated acids and alkaline solutions because they permanently damage it. Other solvents affect the properties of silicone rubber in different ways, so it is important that silicone rubber be tested thoroughly before being used in areas where solvents are deployed.

Tactile Feel. Tactile feel is a term commonly used with rubber keypads to describe the way a key feels to the operator when it is actuated. While this defini-

TABLE 11.6 Chemical Resistance of Silicone Rubber

Chemical compound	Changes in properties, %				
	Hardness	Weight	Volume	Tensile strength	Ultimate elongation
Conc. nitric acid	−30	+10	+10	−80	+30
7% nitric acid	−2	1	1	−50	−30
Conc. sulfuric acid		Decomposed		Decomposed	Decomposed
10% sulfuric acid	−2	1	1	0	0
Acetic acid	+4	+3	+4	−20	+10
5% acetic acid	+8	+2	+2	−20	+10
Conc. hydrochloric acid	−6	+3	+4	−40	−20
10% hydrochloric acid	−4	+2	+2	−50	−50
20% sodium hydroxide	−2	−2	−1	−10	0
2% sodium hydroxide	−4	<1	<1	0	0
Conc. ammonia water	−4	+2	+1	−30	+10
10% ammonia water	−6	+2	+2	−20	0
Water	<1	<1	<1	0	0
Boiling water (70 h)	+2	<1	<1	−10	−10
Water at 70°C	<1	+1	<1	−10	+10
3% hydrogen peroxide	<1	<1	<1	0	+20

Source: Glolite Sales, Ltd., Pauls Valley, Okla.
Test condition: After dipping for 168 h at 25°C.

TABLE 11.7 Volume Swell after 168-h Dipping, 122°F (50°C)

Liquid	Nitrile rubber			Natural rubber	Butyl rubber	Silicone rubber
	28%	33%	38%			
Gasoline	15	10	6	250	240	260
ASTM 1 oil	−1	−1.5	−2	60	20	4
ASTM 3 oil	10	3	0.5	200	120	40
Diesel oil	20	12	5	250	250	150
Olive oil	−2	−2	−2	100	10	4
Lard	0.5	1	1.5	110	10	4
Formalin	10	10	10	6	0.5	1
Ethyl alcohol	20	20	18	3	2	15
Ethylene glycol	0.5	0.5	0.5	0.5	−0.2	1
Diethyl ether	50	30	20	170	90	270
Methyl ethyl ketone	250	250	250	85	15	150
Trichloroethylene	290	230	230	420	300	300
Carbon tetrachloride	110	75	65	420	275	300
Benzene	250	200	160	350	150	240
Aniline	360	380	420	15	10	7
Phenol	450	470	510	35	3	10
Cyclohexanol	50	40	25	55	7	25
Distilled water	10	11	12	10	5	2
Sea water	2	3	3	2	0.5	0.5

Source: Glolite Sales, Ltd., Pauls Valley, Okla.

tion for tactile feel is commonly used, it is misleading in that tactile feel is the result of a precise relationship between actuation force, contact force, return force, and stroke and travel. Keypads that demonstrate what is described by operators as "good" tactile feel do so because close attention has been given to the interaction of these four components.

In order for a switch or keypad to display good tactile feel it must have a snap ratio of at least 40 percent. The snap ratio of any keypad can be determined by utilizing the formula

$$\frac{F_1 - F_2}{F_1}$$

where F_1 = actuation force
F_2 = contact force

In order to have distinct tactile feeling (strong tactile feel) the snap ratio of the keypad should be 50 percent or higher. Table 11.8 illustrates the relationship of actuation force to subjective feel, snap ratio, and life. The higher the snap ratio the stronger the tactile feel. Conversely, if a keypad does not require distinct tactile feel the snap ratio should be less than 40 percent. The snap ratio range available for rubber keypads is 25 to 80 percent. Figure 11.8 compares force-deflection curves for "weak" and "strong" tactile feel. Figure 11.9 depicts a force-deflection curve that is typical for a rubber switch that has *overstroke,* such as is found in computer terminal keyboards and keypads. Rubber keypads that produce overstroke are always fitted with some type of plastic caps or covers for rigidity, as the molded rubber switch is hollow rather than solid in order to get the light tactile feel necessary for fast data entry type applications.

TABLE 11.8 Relationship of Actuation Force to Feel, Snap Ratio, and Life

Typical actuation force parameters*			
Actuation force, g	Tactile feel	Snap ratio, %	Typical life, actuations
40–60	Excellent	50	1,000,000
75–100	Excellent	50	1,000,000
100–150	Excellent	50	1,000,000
150–200	Good	40	500,000
200–300	Fair	25	300,000

Standardized actuation force tolerances†
50 g ± 15 g
75 g ± 20 g
100 g ± 25 g
125 g ± 30 g
150 g ± 35 g
175 g ± 40 g
200 g ± 50 g
250 g ± 50 g

Tactile feel is influenced by actuation force, stroke, and size of keytop. The data are based on similar size keytops with stroke of 1.0 mm.

**Source:* Glolite Sales, Ltd., Pauls Valley, Okla.

†*Source:* Conductive Rubber Technology, Inc., Bothell, Wash.

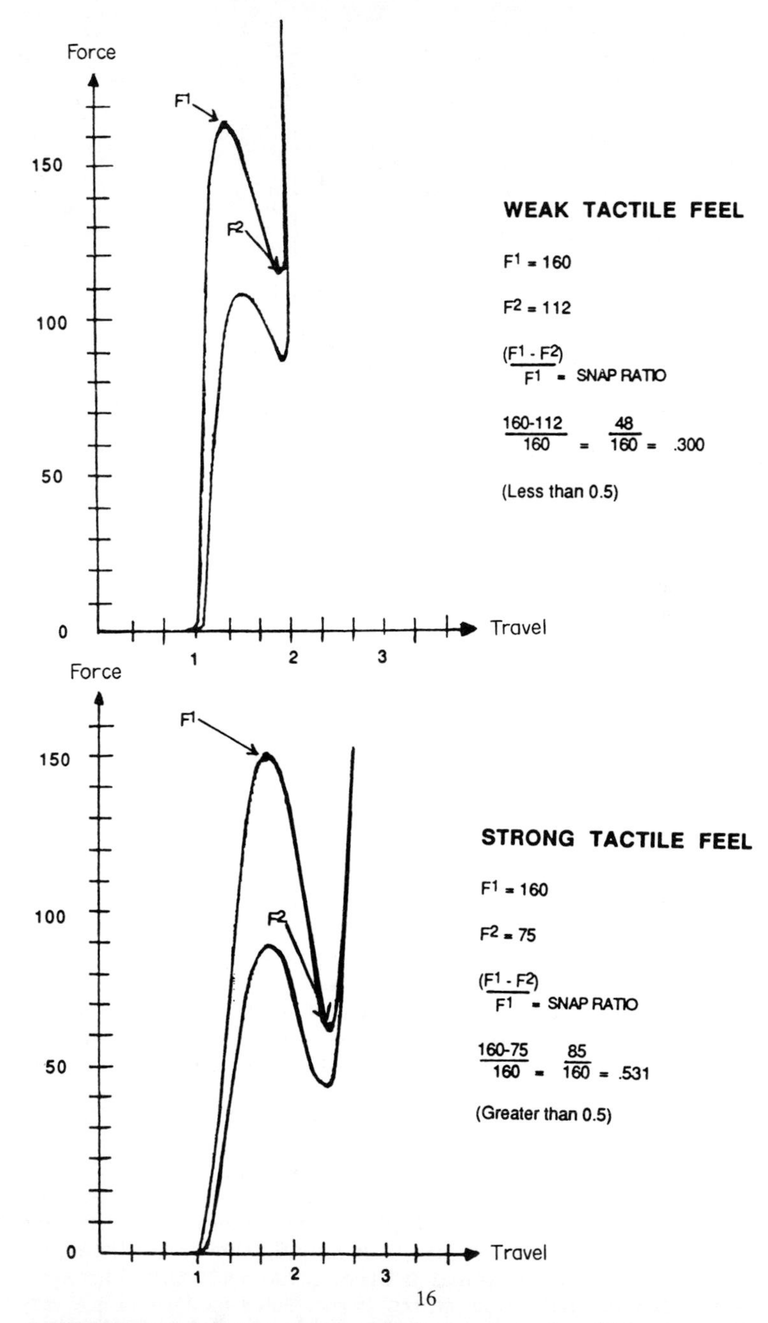

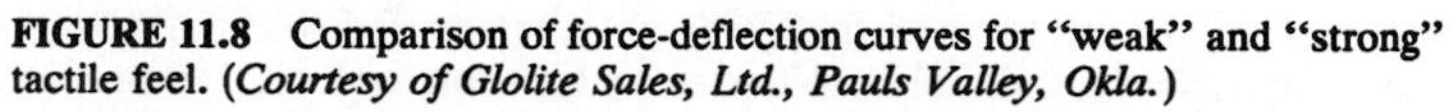

FIGURE 11.8 Comparison of force-deflection curves for "weak" and "strong" tactile feel. (*Courtesy of Glolite Sales, Ltd., Pauls Valley, Okla.*)

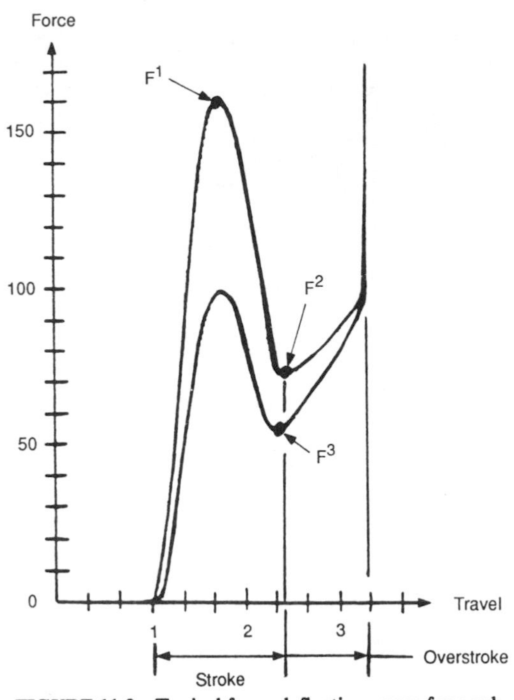

FIGURE 11.9 Typical force-deflection curve for a rubber switch with overstroke. (*Courtesy of Glolite Sales, Ltd., Pauls Valley, Okla.*)

It should be noted here that the snap ratio of a keypad also affects the *life* of the keypad. Keypads with high snap ratios exhibit shorter life than keypads with low snap ratios. Regardless, all keypads (high snap ratios as well as low snap ratios) should be able to realize a life of 250,000 to 500,000 actuations. It is not unusual for indirect contact switches (i.e., rubber keypads fitted with plastic keycaps) with very light actuation and return forces to yield life measurements in excess of 5,000,000 actuations.

It is difficult to recommend specific guidelines for creating the best tactile feel, but if actuation force and stroke are quantified it is possible to precisely design the key's membrane web in order to give the keypad a good tactile feel. It is imperative that the user understand that the design of the membrane web *must* be left to the switch manufacturer. The manufacturer must be allowed to shape the web as needed in order to achieve a certain stroke, actuation force, and life expectancy. High snap ratios create excellent tactile feel, but great care must be exercised to make certain the keypad also has a proper return force so that sticking keys are not encountered. Leave tactile feel to the switch manufacturer, but establish very strict parameters in the switch specification as to what type of tactile feel is desired. As demonstrated in Chap. 1, the best method for specifying tactile feel is the force-deflection (stroke) curve. Since tactile feel is a very subjective parameter, one method of finalizing a satisfactory force-deflection curve is to obtain samples of previously molded keypads from selected manufacturers, decide which sample "feels" the best for the application, and have the manufacturer measure the actuation force and stroke. Develop the force-deflection curve from these measurements and incorporate it into the switch specification. If the engineer wishes to make the tactile feel decision his or herself, it is recommended that a number of people (especially from the customer's operator population, if possible) actuate the samples and come to some sort of consensus about the right tactile feel. This can become a very frustrating exercise for the engineer when the tactile feel of several *different* force-deflection combinations are selected as the "best" and no particular combination stands out. People have their own ideas of what tactile feel works best for the individual, and if the engineer is unfortunate to be confronted with a number of strong but divergent opinions, the art of compromise should be practiced.

A general guideline that can be followed for establishment of an acceptable tactile feel is that large keys require higher actuation forces than small keys. This same rule also applies to key height. Tall keys require higher actuation force than short keys. A general rule for actuation force is that keys with heights of 0.394 to 0.590 in (10 to 15 mm) should have minimum actuation forces of 2.8 to 3.5 oz (80 to 100 g), while keys with heights of 0.590 to 0.985 in (15 to 25 mm) should have minimum actuation forces of 5.3 to 6.2 oz (150 to 175 g).

If the desired stroke and actuation force for a keypad are *not* known or fully defined, it is *imperative* that the following design rule be followed:

Always tool the keypad with relatively low actuation force and long stroke. Why? Because mold tooling can be easily modified to *increase* actuation force and *reduce* stroke (removing metal from the mold), but it is more difficult (and therefore more expensive in money and schedule) to modify tooling to *reduce* actuation force and *increase* stroke (adding metal to the mold).

It should also be noted that large and small keys with *identical actuation forces* will feel very different when they are actuated by the human finger. This difference of feeling is caused by the overall size of the switch and the fact that the force being applied is being dispersed or concentrated over a comparatively smaller or larger surface area. When keypads incorporate different-sized keys it is wise to use more than one actuation force in the keypad so that the tactile feel will remain fairly constant for all keys.

Determining Maximum Keystroke. Stroke and tactile feel in rubber switches are dependent upon the shape, thickness, and density (durometer) of the material used in the membrane web area of the keypad. Stroke typically ranges from 0.010 in (0,25 mm) to 0.200 in (5,0 mm) in rubber keypads and can be precisely controlled with proper design and accurate tooling.

Stroke, or travel, is depicted in Fig. 11.10 as the distance indicated by *d*, and is the distance between the PCB and the closest surface of the conductive pill on the individual switch.

The maximum stroke of any key and switch is *equal to or less than* the combined thickness of the base material and the height of the membrane area of the switch. When this dimension is determined it is important that the thickness of the conductive pill be subtracted from the total to accurately calculate the key's stroke. The normal thickness of a conductive pill is 0.015 in (0,4 mm) to 0.020 in (0,5 mm).

In order to increase the stroke and travel of any key, the base thickness must be increased or the membrane web area must be made steeper than originally designed.

Sticking Keys. One problem associated with poor design practices relating to conductive rubber is sticking keys. It is important to note that rubber keypads should always have a *minimum* return force of 1 oz (30 g) to prevent sticking keys

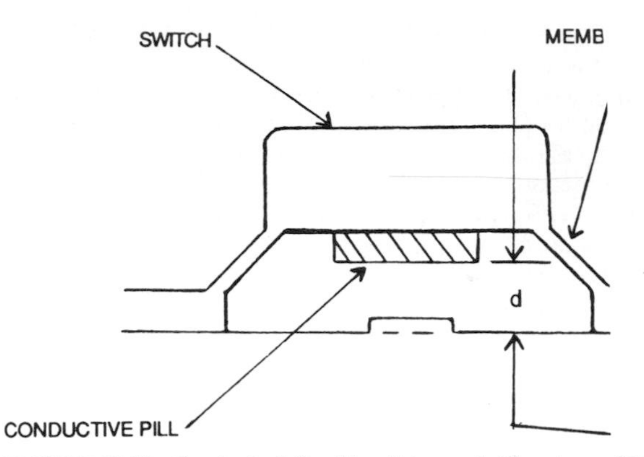

FIGURE 11.10 Stroke is defined by distance *d*. (*Courtesy of Glolite Sales, Ltd., Pauls Valley, Okla.*)

and other related problems, and the following considerations should be part of a designer's checklist to minimize the occurrence of sticking keys:

1. Obstruction of switch membrane web: If the rubber keypad is covered with a faceplate (bezel), it is imperative that close attention be paid to the design of this faceplate. The faceplate (bezel) should incorporate a 45° chamfer at least 0.040 in (1,0 mm) high on the bottom side around each key to prevent interference with the operation of the membrane web of the switch. This feature is typically required when the bezel will fit very close to the membrane web of the switch and the clearance between the two is very small.

2. Key opening clearance: In addition to being careful not to obstruct the operation of the key's membrane web, it is important that enough clearance is designed into the faceplate's key opening. A general rule of thumb for this clearance dimension is that the opening be 0.012 in (0,3 mm) (minimum) larger than the key itself on all sides of the key.

3. Switch return force: Another common cause for sticking keys is that the key's return force is not high enough to overcome the friction created by the bezel. A key's return force is typically 25 to 30 percent of its actuation force. It is suggested that keys with printed legends (direct contact switches, i.e., the human finger operates the key directly) should never have an actuation force of less than 2.8 oz (80 g) or a return force of less than 1 oz (30 g). Close attention should also be paid to the relationship between a key's actuation force and desired travel when tactile feel is being finalized. Always consult with the potential switch manufacturer's engineering department for assistance in this somewhat complex, but quantifiable, area.

4. Flash in the key opening: Make certain that all key openings are free of flash. To guarantee that all flash has been removed it is recommended that the key openings be cleaned with emery paper.

5. Absence of air channels or obstruction of air channels: In order for keys to be able to return to their normal rest position, it is critical that air channels be incorporated into the keypad design. Keys may be vented to the outside of the keypad or, if the keypad needs to be waterproof, vented to each other. Air channels are typically at least 0.008 in (2,0 mm) wide and 0.012 in (3,0 mm) deep.

Make certain that this important feature is *not* overlooked when the keypad design is finalized.

6. Make certain that the edge radius of the rubber key is *always less* than the radius of the key opening in the faceplate (bezel). Overlooking this detail can cause keys to hang up on the bezel even if all other design guidelines have been followed closely.

7. Incorrect switch and key height: In order to make certain that a key's stroke does not create sticking problems with the bezel, very close attention must be given to the key's height. The key's height *must be equal* to the combined thickness of the bezel and the base (apron) of the keypad *plus* the stroke of the individual key. When this value is calculated, a minimum clearance dimension of 0.020 in (0,5 mm) *must be added* to the figure to guarantee that the key or switch will not "hang up" on the bezel when actuated. Figure 11.11 illustrates this important relationship. (1)

Switch Reliability and Switch Life

Switch reliability and switch life depend on the membrane web style chosen, the density of the material in the keypad, and the quality of the material itself. Some switch manufacturers, such as Glolite Sales, Ltd., Pauls Valley, Okla., have developed several grades of silicone rubber compounds which are several times more resistant to fatigue by bending than conventional grades of silicone rubber. Actuation force, snap ratio, and stroke also influence switch life, as does proper curing of the keypad, so great care should be exercised when these parameters are identified and the keypad is manufactured.

Generally speaking, square keys demonstrate shorter life than circular keys because of the stress that is applied to the four corners of the key's membrane web. Circular keys have equal pressure exerted over the entire surface of the membrane web; hence these keys usually demonstrate considerably longer life than their square counterparts. Careful attention should be paid to this phenomenon when the keypad is being designed. All other things being equal, switch life is reduced as higher durometer (hardness) material is used, and switch life is also reduced as actuation force is increased.

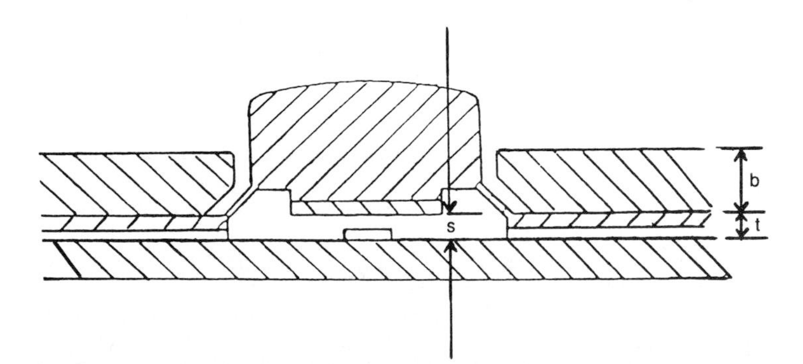

FIGURE 11.11 The minimum height of any switch or key is determined by adding the following values: base thickness (t) of keypad + bezel thickness (b) + stroke (s) + 0.02 in (0,5 mm). (*Courtesy of Glolite Sales, Ltd., Pauls Valley, Okla.*)

A commonly accepted definition of *mechanical life* is that the switch is considered functional until its initial actuation force degrades by 20 percent. The test condition and parameter associated with this definition of life is that the switch is continuously activated at the rate of 2 times per second at a force of 5.3 oz (150 g). By contrast, a commonly accepted definition of *electrical life* is that the contact resistance of the switch cannot exceed 200 Ω. The test condition and parameter for this definition of life is that an actuation force of 5.3 oz (150 g) is applied against a gold-plated contact on a PCB. As noted elsewhere in this chapter, several different types of contacts are available for use in rubber keypads, and each exhibits unique electrical characteristics.

Additionally, it should be noted that contact resistance for rubber switches varies depending on the plating used on the PCB. Conductive rubber switches can be used with gold-plated, nickel-plated, and screened carbon PCBs with high reliability, but the contact resistance will be different for each board. Gold-plated PCBs demonstrate the lowest contact resistance (less than 100 Ω), while screened carbon PCBs consistently measure much higher contact resistances (typically 1000 Ω). The use of tin-lead solder-plated PCBs is not recommended with conductive rubber switches because the carbon-filled conductive pills are not abrasive enough to keep the contact on the PCB sufficiently clean for reliable contact.

Key Rocking

Key rocking is a condition of conductive rubber switches wherein a key, when pressed by a human finger, has a tendency to tip, or rock, to one side or the other during its stroke. In order to minimize this undesirable feature, the following design suggestions should be given careful consideration:

A. Minimize keystroke: Specify travel within 0.03 to 0.04 in (0,8 to 1,0 mm) for switches with tactile feel.

B. Design keys to incorporate slightly concave tops: Concave keytops give a more solid feel than flat keytops.

  1. If a concave keytop is used, be certain that the radius of the concavity is not overly severe if it is to be printed with some type of graphics.
  2. If the radius of the keytop is too severe the printing will become fuzzy because the ink will have a tendency, during application, to run to the center of the key. In addition, it will be very difficult for the manufacturer to control the consistency of the graphics because of the severity of the key slope.

C. Limit height of key: Keytop should not be higher than 0.100 in (2,5 mm) above the faceplate (bezel) surface.

D. Add antirocking pins (stabilizing pads) to the bottoms of all keys:

  1. Antirocking pins are frequently added to the inner surfaces of keys to negate key rocking. The pins are molded as part of the base silicone rubber material of the keypad.
  2. Make certain that the height of the antirocking pins does not interfere with the switch contact.
  3. Incorporate antirocking pins for switches that have a stroke of 0.04 in (1,0 mm) or more.

E. Design the conductive rubber contact to stabilize the key:

  1. Try to design the contact so that it conforms to the key shape. Multiple contacts can be used for large keys, and the availability of different-shaped contacts (i.e., circular, oval, square) from the potential manufacturer should be investigated.

2. As a general rule, design the contact under the key to cover approximately 50 percent of the key area. This will promote switch stability and help ensure reliable switch contact. If this general rule cannot be followed because of space limitations on the PCB and the switch contact area is less than 50 percent of the key area, it is even more important to incorporate antirocking pins into the keypad so that reliable contact can be made despite the use of undersize contacts.

Figure 11.12 depicts the incorporation of antirocking pins on a typical conductive rubber key. It is important to note that a minimum clearance of 0.020 in (0,5 mm) must be maintained from the bottom inside edge of the membrane web of any key to any other feature on the inner surface of that same key. This rule applies whether that feature is a conductive contact or antirocking pins. This clearance is critical to ensure that the conductive rubber "puck" comes into contact with the PCB in the correct manner.

Typically, antirocking pins are approximately 0.04 in (1,0 mm) in diameter. The antirocking pin must have a height that is smaller than the height of the conductive contact itself so that the pins do not prevent the contact from reaching the PCB.

Printed Circuit Board Design. Conductive rubber keypads are reliable and versatile; however, the environment in which the keypad will be used must be considered in determining the type of plating applied to the PCB which will enhance the long life and trouble-free operation inherent in this technology.

As previously discussed, the type of PCB plating that offers the lowest contact resistance is gold over nickel. However, other viable options exist depending on the application.

The type of plating used for switch contacts on the PCB should be carefully considered to maximize switch life. Most manufacturers recommend that only gold plating be used on the switch contacts. Nickel, tin-lead, and bare copper are

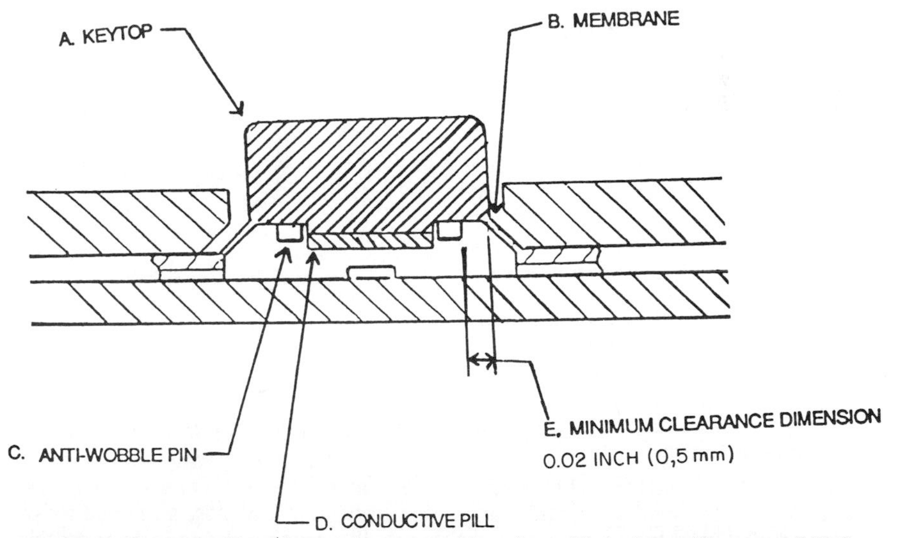

FIGURE 11.12 Antirocking pins are incorporated into the molded rubber. (*Courtesy of Glolite Sales, Ltd., Pauls Valley, Okla.*)

not recommended due to the inherently poor wiping action of the silicone rubber carbon contacts. Surface oxides can quickly build up on these metals, causing a sharp increase in electrical contact resistance. This can lead to intermittent functioning of the switch and eventual switch failure.

As an alternative to expensive gold plating, some suppliers of conductive rubber switches offer conductive carbon ink which can be applied directly on a hard PCB substrate using a simple screen printing and heat curing process. One manufacturer, Advanced Connector Technology, Ventura, Calif., offers a product called Carbon Ink #2000 to be used in conjunction with silicone rubber switch PCB's to increase switch life. The manufacturer recommends that the entire PCB switch contact be composed of conductive carbon ink only (no etched metal) and that the carbon ink overlap only the metal land areas (usually copper) connecting the PCB switch trace patterns. The carbon ink will not oxidize, is etch resistant, and can withstand immersion in a 500°F (260°C) wave solder bath with no switch contact deterioration. According to the company, replacement of gold-plated PCB switch contacts with conductive carbon ink can result in PCB cost savings of as much as 50 percent! In addition, switch life is usually increased since the PCB carbon contacts remove fewer carbon particles from the conductive silicone rubber switch contacts than do PCB metallic contacts such as gold. One caution should be observed, however; because of their abrasive action, edgeboard connectors and metal dome switches (Chap. 9) are *not* recommended for use with carbon contacts on PCBs.

It should be noted that conductive carbon contacts used in conductive rubber switches demonstrate very stable contact resistance and long life (at least 50,000,000 cycles). Contact problems with conductive rubber keypads are often related to selection, or application, of the plating on the PCB, and extra care should therefore be taken in PCB design. Consultation with a reliable conductive rubber switch manufacturer, as early in the system design stage as feasible, concerning selection of materials and plating for the PCB will provide benefits by way of long-term dependable performance of the keypad.

Design of PCB Contact Pads for Conductive Rubber Switches. Contact pads on PCBs should be equal to or up to 1.25 times larger than the conductive contact on the rubber key. The minimum distance between contact fingers on the pad and the minimum size of the pad traces should be 0.010 in (0,25 mm). In most cases, the recommended maximum spacing between the fingers on the pad is 0.015 in (0,38 mm). Figure 11.13 depicts some common pad patterns for conductive rubber keypad PCBs. Although no particular pad pattern is recommended by switch manufacturers, the pad pattern selected for a particular application should offer as many *shorting paths* as possible to guarantee switch operation.

FIGURE 11.13 Common pad patterns for conductive rubber switches. (*Courtesy of Shin-Etsu Polymer America, Inc, Union City, Calif.*)

Keypad Colors and Graphics

Keypad Colors. As previously indicated, silicone rubber formulations that begin as translucent or grayish-white material can be easily pigmented for color matching of the basic keypad to a selected color standard. The Pantone color system is commonly used in the industry as one of these standards, and conductive rubber manufacturers can do a remarkable job of matching keypad colors to this criterion. Another color standard sometimes used in the rubber keypad industry is the FED-STD-595 system, a government specification developed to standardize color selection for military products, although this is less widely used than the Pantone system. Most conductive rubber switch manufacturers can also match to special color chips provided by the user.

Although it is obvious that the keys of the keypad are usually molded in the same color as the basic keypad, a unique characteristic of the molding process allows individual keys or groups of keys to be color-coded to contrast with or complement the base color or other keys. This feature, when combined with the nearly unlimited graphics capabilities of printed keypads, permits production of keypads which are attractive and easy to operate. This, along with the flexibility of design, color matching capabilities, and cost effectiveness of conductive rubber switches, make them a favored switch technology for industrial designers, particularly for products destined for the consumer marketplace.

Keypad Graphics. Almost all keypad legend and graphics information applied to rubber keypads is *surface printed* (silk-screened) using a special silicone ink that, after curing, becomes permanently bonded to the base silicone rubber material during the manufacturing process. A specially formulated silicone ink is used in the printing process and is a true rubber ink, rather than a paint. Silk screening provides sharply defined graphics, and the process is ideal for keypads containing multiple colors, as each color is applied separately with a different screen.

Once the keypad has been screened with this special ink the keypad is put into a high-temperature oven for curing and bonding of the ink to the keypad. After curing, the ink and keypad are, in effect, "one material," and it is very difficult to remove the graphics from the keypad.

Silk-screen application of graphics successfully meets the needs of most keypad applications, but additional means of extending the durability of graphics life have been developed by conductive rubber manufacturers.

One method, called *overcoating,* involves coating the graphics layer with a translucent silicone ink (providing either a matte or a shiny finish) and curing the overcoat in the same manner as the graphics layer, during which time it becomes, as the graphics layer did, an integral part of the keypad. Overcoating typically doubles the life of silk-screen graphics and is especially useful in applications where the keypad is exposed to very harsh conditions, such as when an operator wears gloves during operation of the keypad or even when the keypad is expected to endure an excessive number of operations (i.e., the ENTER key of a numeric entry keypad). In some cases, two overcoatings may be applied to even further protect the graphics from abrasion. Because of the additional labor involved in applying the overcoating, a keypad with this feature will be more expensive than one without, but the user may decide that the extra cost is worth the security of knowing the graphics are well protected.

A second method of protecting graphics from abrasion is called the *diffusion* process in which the graphics are sublimated *below* the surface of the silicone rubber switch material itself. The depth of this special diffusion printing is typically 0.004 to 0.006 in (0,10 to 0,15 mm) below the top surface of the keypad and typically lasts three to four times longer than silk-screen printing. The abrasion resistance of diffuse printing can be further enhanced by the application of an overcoat similar to that applied over silk-screened graphics layers. However, diffusion printing can be done only on flat keytops and on keys without radii. Additionally, diffusion printing will also require that the keytops be a minimum height, typically 0.275 in (7,0 mm), to allow for the special tooling necessary to perform the diffusion process. Because of this special tooling and the details of the process itself, the user can expect keypad unit price and tooling charges to be higher than the standard silk-screen–overcoat process. If the specifying engineer decides the operating environment of the proposed keypad is harsh enough to justify consideration of diffusion printing, potential switch manufacturers should be questioned about their ability to process diffusion printing (many can't) and their experience with the process. Securing samples of the switch manufacturer's previous *production* processing of diffusion printing can assist the user in determining the quality capabilities of the manufacturer. (1)

To satisfy the demands for an immortal legend for products in constant contact with cloth, gloves, pencil erasers, sharp instruments, dirt, and grit, Conductive Rubber Technology, Bothell, Wash., offers the *engraved printed legend.* The legend, engraved 0.012 in (0,3 mm) below the surface of the key and filled with a contrasting silicone ink, has 100 times the abrasion resistance of the surface printed legend, with a minimum line width of 0.015 in (0,4 mm). Immortality doesn't come cheap, however, as this process costs approximately 50 percent more than surface printing. (5)

Once the graphics process has been decided, it is recommended that the specifying engineer and the switch manufacturer identify an acceptable abrasion test, for inclusion in the switch specification, that the graphics on the keypad can be expected to meet. One abrasion test common in the industry states the number of times the graphic layer will be subjected to back-and-forth rubbing by a rubber eraser and at what force the eraser is applied to the graphics during the test. The acceptance criterion might be that more than 50 percent of the graphics must be visible after the test to be considered successful or that the graphics can still be read under certain lighting conditions, even though faded, after test completion. The point to be emphasized is that the results of an abrasion test can be interpreted in a number of ways and the best time for the supplier and the user to decide on acceptance criteria is when the specification is written, not after the keypad is delivered to the user.

Some examples of abrasion tests included in specifications include:

1. Resistance to abrasion:
   a. Test: Rub the printed surface with a plastic eraser under a 17.6-oz (500-g) force.
   b. Failure criteria: Graphics cannot be read, from a distance of 18 in (457 mm) under normal office lighting, after 1500 cycles.
2. Resistance to key striking:
   a. Test: Strike the keytop with a device approximating the texture and density of the human finger with 7-oz (200-g) force at a rate of 3.5 times per second.
   b. Failure criteria: Graphics cannot be read, from a distance of 1 ft (304,8 mm) under normal office lighting, after 500,000 cycles.
3. Resistance to peeling:
   a. Test: Apply adhesive tape to printed surface and peel off.
   b. Failure criteria: Graphics display any type of peeling prior to 100 applications of the tape test.
4. Resistance to fingerprint liquid:
   a. Test: Liberally apply synthetic fingerprint liquid on printed surface and leave for 240 h at 104°F (40°C), 95 percent RH.
   b. Failure criteria: Graphics display easily detectable changes in printing quality.

5. Resistance to ultraviolet rays:
   a. Test: Apply 20-W sterilizing ultraviolet ray to printed surface from distance of 6 in (152 mm) for 72 h.
   b. Failure criteria: Graphics demonstrate changes from original color when compared with color chip (or with graphics of a nonexposed keypad).

Some basic design suggestions to follow for conductive rubber keypad graphics artwork are:

1. Supply positive image artwork to the switch manufacturer:
   a. If the specifying engineer chooses to supply artwork for keypad graphics to the switch manufacturer, the format for the artwork should be agreed upon beforehand. Normally, switch manufacturers request positive image artwork, but it is always best to check with the selected manufacturer.
   b. If the specifying engineer prefers that the switch manufacturer supply final artwork for the graphics, the engineer should specify preferences for font style, print style, and print size, and should *always* insist on artwork review and sign-off prior to any production parts being supplied.
2. Finalize artwork for the keypad before production tooling has been completed:
   a. Artwork charges (if done by the manufacturer) and silk-screen charges are always part of the original tooling charges quoted.
   b. Any and all changes in artwork will require new and additional silk screens, translating to additional tooling charges. (1)

Figure 11.14 depicts the various limitations, tolerances, and spacings required to silk-screen graphics on rubber keypads.

Recommended Design Dimensions

Figure 11.15 depicts recommended design dimensions standard for the conductive rubber industry. Deviations from these dimensions and the tolerances shown in Table 11.2 should be discussed with potential switch suppliers as early in the design phase as possible.

Backlighting. Conductive rubber keypads can be backlighted in a number of ways, the most common being that depicted in Fig. 11.16*a, b* and *c.* LEDs (radial lead, formed and surface mounted) are frequently used to achieve backlighting because of their low operating temperature and relatively small size.

Diffused backlighting can be achieved only by using translucent silicone rubber (refer to Table 11.3) and a light pipe or by using the base material of the keypad itself as a light pipe (Fig. 11.16*c*). Regardless, it must be remembered that conductive rubber keys cannot be screened or printed on the sides, so light will escape from the side of the key as well as be emitted from the top surface of the switch. A small number of conductive rubber manufacturers have developed methods of spraying opaque silicone inks on the sides of keys to block light in this manner, but the additional cost is significant.

Both positive image and negative image legends can be backlit when translucent material is used for the keypad's construction. Attention to the following notes will help the specifying engineer achieve the best results:

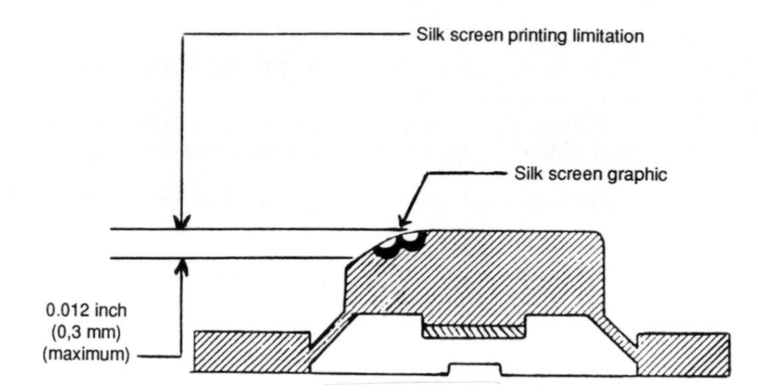

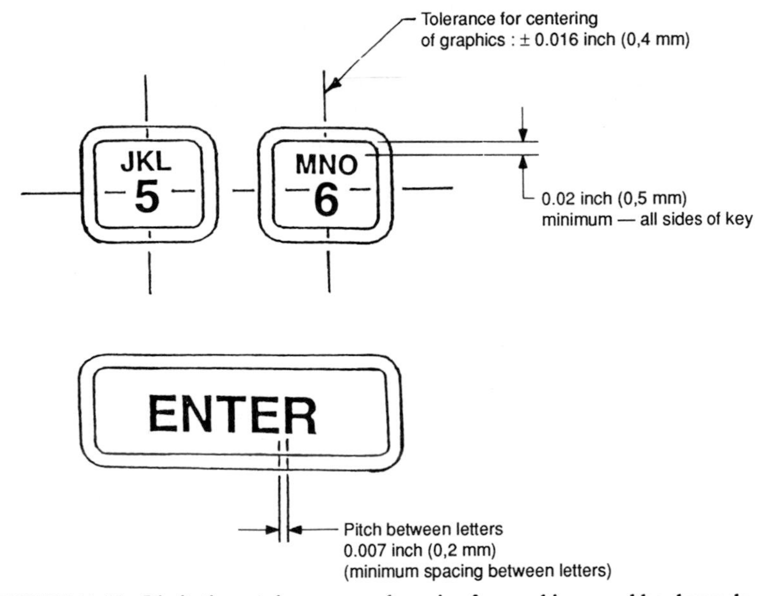

FIGURE 11.14 Limitations, tolerances, and spacing for graphics on rubber keypads. (*Courtesy of Glolite Sales, Ltd., Pauls Valley, Okla.*)

- The bottom of the base material of the keypad can be screened white to enhance backlighting capabilities. If the bottom of the keypad is screened white, no light will be lost to the PCB.
- Discuss with the manufacturer any additional costs in having the conductive pill silk-screened white prior to molding. This is done so that no dark spots or shadows appear during backlighting.

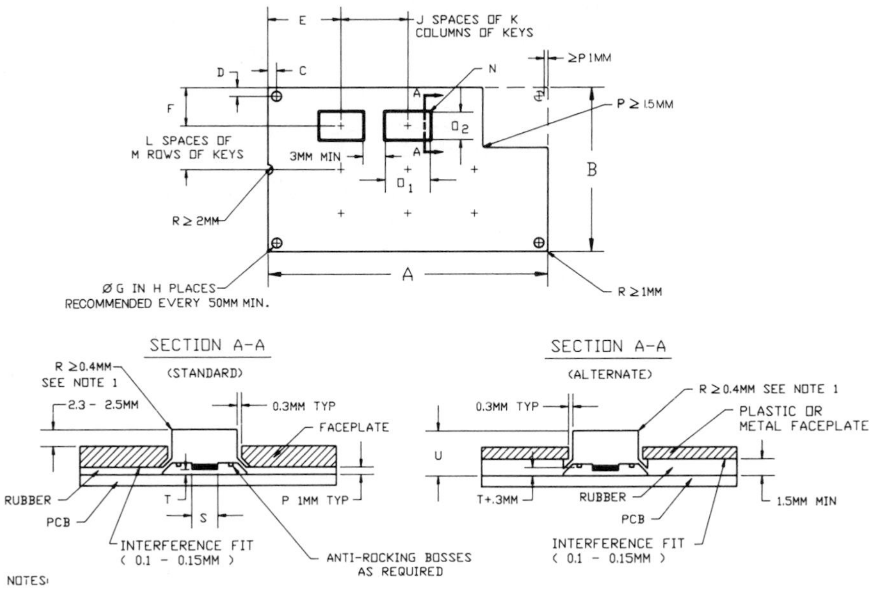

FIGURE 11.15 Standardized design dimensions for conductive rubber switches. (*Courtesy of Conductive Rubber Technology, Inc., Bothell, Wash.*)

- Discuss adding a small amount of white diffuser material to the base translucent material to prevent hot spots during backlighting, particularly if using individual lamps as the light source. (1)

Additional methods of backlighting include EL (electroluminescent) lamps and woven fiber-optic lamps. These methods are discussed further in Chap. 10.

GLOSSARY

Actuation force. The minimum force required to compress the membrane web of the conductive rubber switch so that contact is made with the PCB. Generally, larger keys require higher actuation forces than small keys.

Air channel. Air path from key to key or to the outside of the keypad which allows the keys to "breathe" and return to normal position after actuation. Keys should be vented to each other and not to the outside if the keypad needs to be watertight.

Alignment hole. A hole in the rubber keypad that is used for aligning keypads that are typically more than 3 in (76,2 mm) in length. The alignment holes mate with pins protruding from the rear of the bezel.

Base. The silicone sheet material that forms the apron of the keypad which joins all keys together.

Bezel. The faceplate, typically made of plastic or metal, that covers a keypad.

Conductive contact (conductive puck, conductive pill). The current-carrying contact, manufactured from silicone rubber impregnated with amorphous carbon, permanently

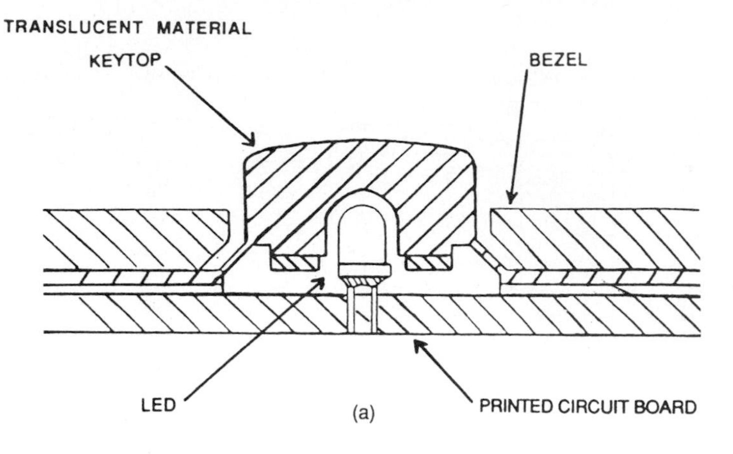

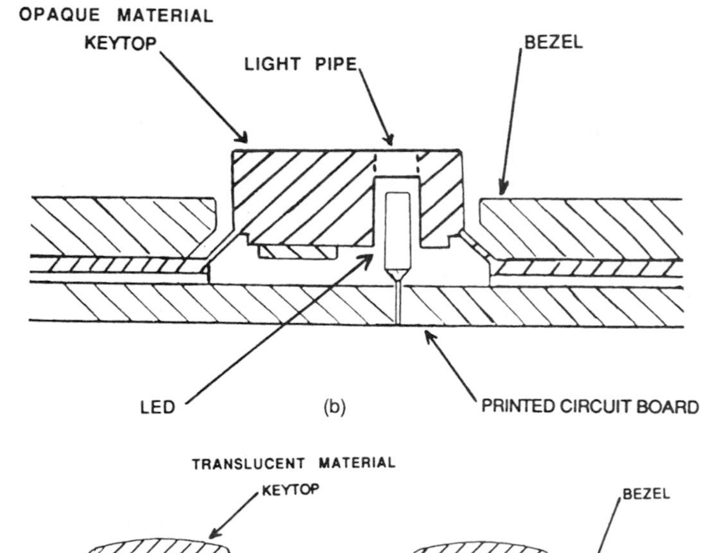

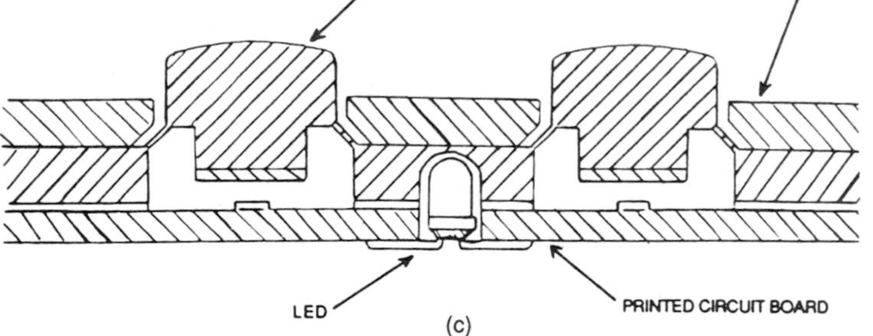

FIGURE 11.16 Typical conductive rubber backlighting techniques: (*a*) Centralized lamp, (*b*) lamp under light pipe, and (*c*) indirect lighting. (*Courtesy of Glolite Sales, Ltd., Pauls Valley, Okla.*)

molded into the basic keypad rubber at each key location. Usually produced in a cylindrical shape, the conductive contact completes electrical connection with the PCB when the keytop is actuated.

Contact force. The force a switch typically realizes when contact is made with the PCB.

Contact resistance. The resistance realized between the surface of the conductive contact of the switch and the contact pad on the PCB when the switch is actuated.

Durometer. A measurement of the hardness of rubber materials. The higher the durometer reading, the harder the rubber.

Membrane web. The nonconductive skirt attached between the key and the apron of a conductive rubber switch which provides the tactile feel and stroke characteristics of the switch. The membrane web shape is always designed by the manufacturer to provide the stroke, actuation force, and return force required by the switch specification.

Mounting boss. An alignment feature that is molded on the bottom of a keypad to help align the keypad with its PCB at assembly.

Return force. The force a switch typically realizes after contact has been made with the PCB.

Snap ratio. The difference between the actuation force (F_1) of a switch and the contact force (F_2) of the same switch divided by the actuation force (F_1):

$$\frac{F_1 - F_2}{F_1}$$

Snap ratio is important since it has direct influence on tactile feel and the life of the keypad.
Tactile feel
A subjective term used to describe the perceived response of a conductive rubber switch during actuation travel and at switch contact. In conductive rubber switches, tactile feel is controlled by the shape and thickness of the membrane web of each switch position and is the most critical process in manufacturing rubber keypads.

REFERENCES

1. "Conductive Rubber Switch Design Guide—Long Form," Glolite Sales, Ltd., Pauls Valley, Okla., 1990.
2. "Conductive Rubber Switch Design Guide—Short Form," Glolite Sales, Ltd., Pauls Valley, Okla., 1990.
3. "Successful Product Development Depends on Knowing a Good Idea When You See One," Shin-Etsu Polymer America Inc., Union City, Calif., 1990.
4. "Rubber Keyboards Product Information," Memtron Technologies, Inc., Frankenmuth, Mich., 1991.
5. "Application Notes—Silicone Rubber Printing Methods," Conductive Rubber Technology, Inc., Bothell, Wash., 1991.

CHAPTER 12
DIP SWITCHES*

A DIP switch is a *DIP* (*dual in-line package*) configuration containing a number of switches. Depending on the manufacturer and style chosen, the number of settable switches in a DIP switch package typically range from 1 to 12. A dual in-line package is an industry standard that was created by the manufacturers of integrated circuits. This standard package outline allows many different types of integrated circuits to be plugged into the same socket or the same pattern on a PCB. The principal dimensional standards are 0.10 in (2,54 mm) between adjacent terminals and 0.30 in (7,62 mm) between terminals across the package (Fig. 12.1). (1)

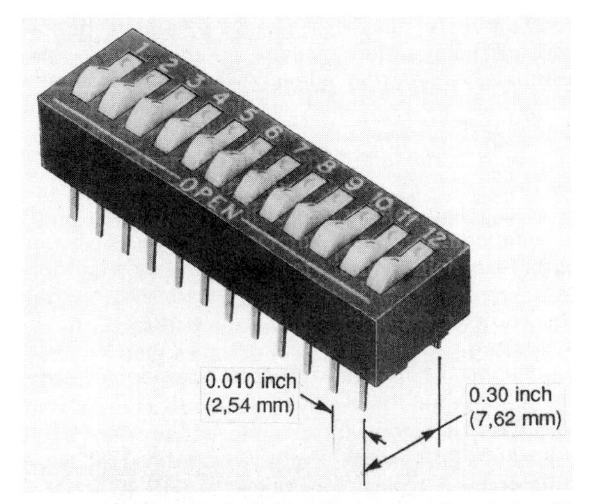

FIGURE 12.1 A DIP (dual in-line package) switch with rocker actuation. Principal dimensional standards are *between* terminals and *across* terminals. (*Courtesy of Grayhill, Inc., La Grange, Ill.*)

*Numbers in parentheses indicate items in the References at the end of this chapter.

How DIP Switches Function

Since the switch condition (open or closed) determines the circuitry input, two types of switch actuation can easily provide the required function. These two types are *rocker actuation* and *slide actuation* (not to be confused with slide *switches*). Some manufacturers provide either rocker actuation or slide actuation. One major manufacturer of DIP switches, Grayhill, Inc., La Grange, Ill., provides both rocker and slide types.

The selection of rocker or slide actuation in a DIP switch is a matter of user preference, often dictated by how easily the user can determine switch condition (open or closed) by looking at the actuator position.

Some initial design criteria for DIP switches include wiping action and a high contact force since a back panel switch operates at low or logic level currents and voltages; immunity to shock and vibration since the DIP switch may be used in an industrial process control application; and positive positioning so the switch would snap into the open or closed position without "hanging up" between positions. The following descriptions of both rocker- and slide-actuated DIP switches illustrate the approach one manufacturer took in meeting the design criteria.

Figure 12.2 depicts a cross section of a typical rocker-actuated SPST DIP

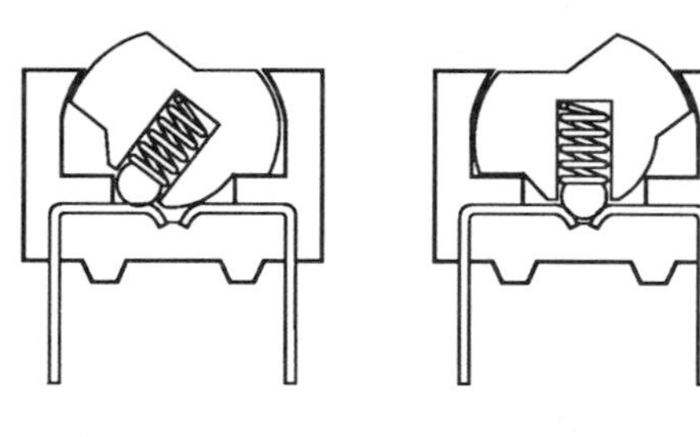

Open position Closed position

FIGURE 12.2 Cross section of a rocker-actuated DIP switch in the open and closed positions. Note that the moving contact (ball) fits securely into detent contact in the closed position to provide immunity to shock and vibration, and positive positioning. (*Courtesy of Grayhill, Inc., La Grange, Ill.*)

switch in the open and closed positions. As the ball slides from the open to the closed position and vice versa, wiping action takes place at the formed contact. The spring-loaded contact (sliding ball) provides positive positioning and resistance to shock and vibration. The ball and the contacts are usually lubricated for long life. Figure 12.3 depicts typical mechanical dimensions and the circuitry for a rocker-actuated SPST DIP switch, similar to the device illustrated in Fig. 12.1.

Figure 12.4 illustrates a typical slide-operated DIP switch utilizing a ball and detent contact style similar to the rocker-actuated type depicted in Fig. 12.2. A spring-loaded sliding ball is the shorting member which provides all the advantages of high contact pressure, wiping action, optimum immunity to shock and vibration, positive positioning, and long life. Note the depression in which the ball rests in the open position. It should also be noted that the contact area is off-center to permit additional circuitry options. (1)

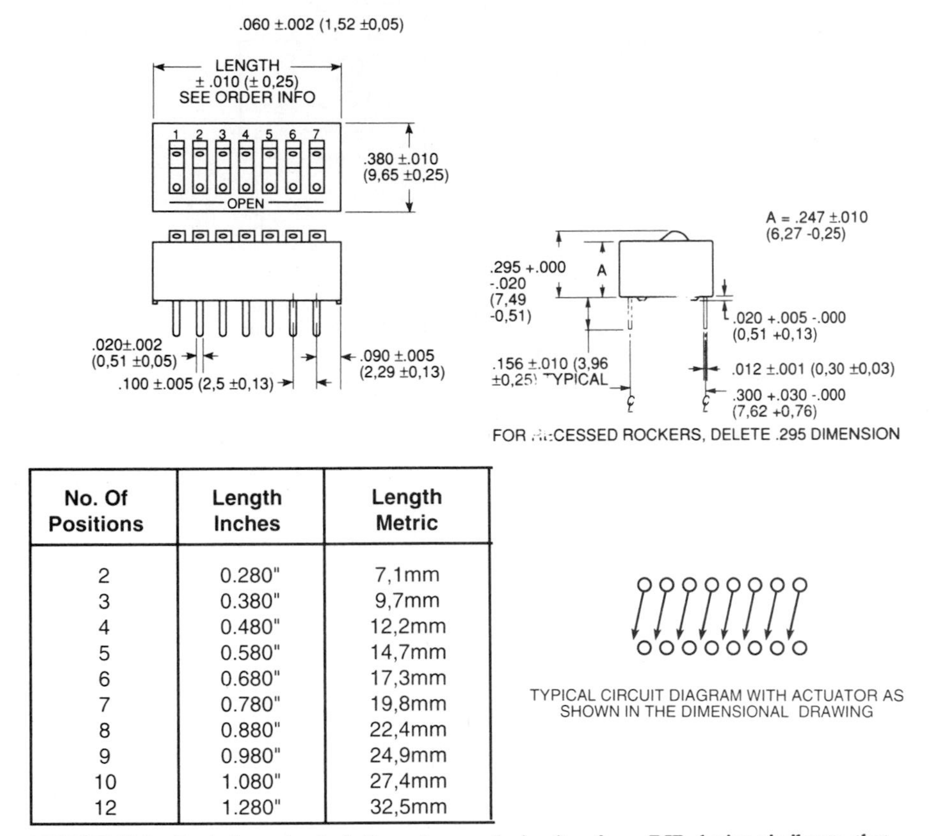

No. Of Positions	Length Inches	Length Metric
2	0.280"	7,1mm
3	0.380"	9,7mm
4	0.480"	12,2mm
5	0.580"	14,7mm
6	0.680"	17,3mm
7	0.780"	19,8mm
8	0.880"	22,4mm
9	0.980"	24,9mm
10	1.080"	27,4mm
12	1.280"	32,5mm

FIGURE 12.3 Typical mechanical dimensions and circuitry for a DIP device similar to that shown in Fig. 12.1. (*Courtesy of Grayhill, Inc., La Grange, Ill.*)

Sealed DIP Switches

Since DIP switches are assembled to PCBs, they must be able to withstand the rigors of fluxing, soldering, and cleaning operations. Cleaning solutions, contaminated with flux and washed into the contact system, combine with the grease to leave a residue on the contacts when the cleaning solution evaporates. This residue can cause high contact resistance or even an open circuit. To protect the switch contact mechanism from contamination, DIP switches are available with epoxy sealing. This effectively seals the terminals from contamination. The top of the switch around the actuator must also be sealed. This is accomplished by a *tape seal* across the actuators. A projecting tab from the tape seal allows the tape to be removed after the cleaning operation. An even more effective seal can be accomplished if the rocker or slide actuators are recessed below the top surface of the package. A recessed slide actuator version is shown in Fig. 12.5*a* and *b*. Another obvious advantage to the recessed actuator versions is the prevention of accidental actuation. (1)

The surface-mount (SMT) DIP switch shown in Fig. 12.6, manufactured by Grayhill, La Grange, Ill., features terminals on standard 0.10-in (2,54-mm) cen-

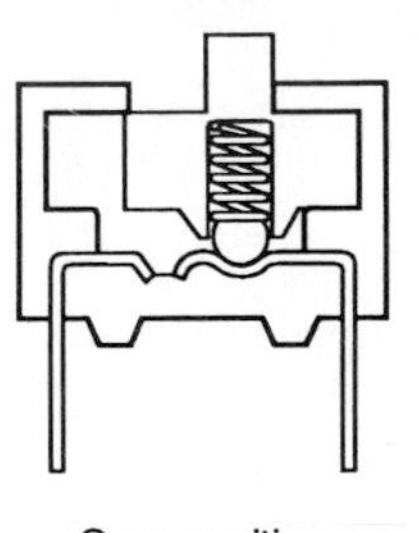

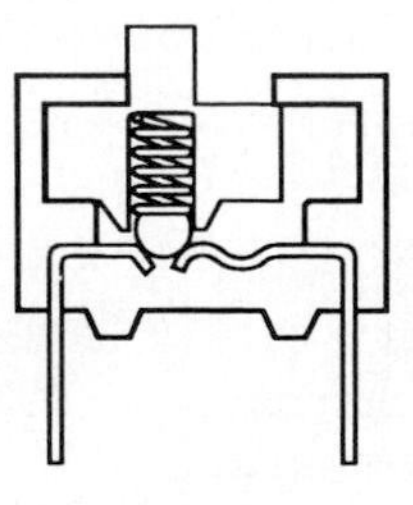

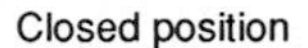

FIGURE 12.4 Cross section of a slide actuator DIP switch in the open and closed positions. Note the double detent system, one detent to hold the switch *open* and the second to hold the switch *closed,* under shock and vibration conditions. (*Courtesy of Grayhill, Inc., La Grange, Ill.*)

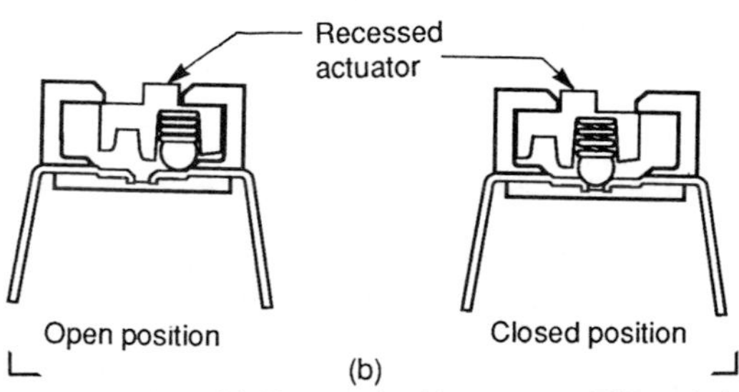

FIGURE 12.5 (*a*) Recessed slide actuator DIP switch with top sealing tape removed; (*b*) cross section showing recessed slide design which permits use of sealing tape and also prevents accidental actuation. (*Courtesy of Grayhill, Inc., La Grange, Ill.*)

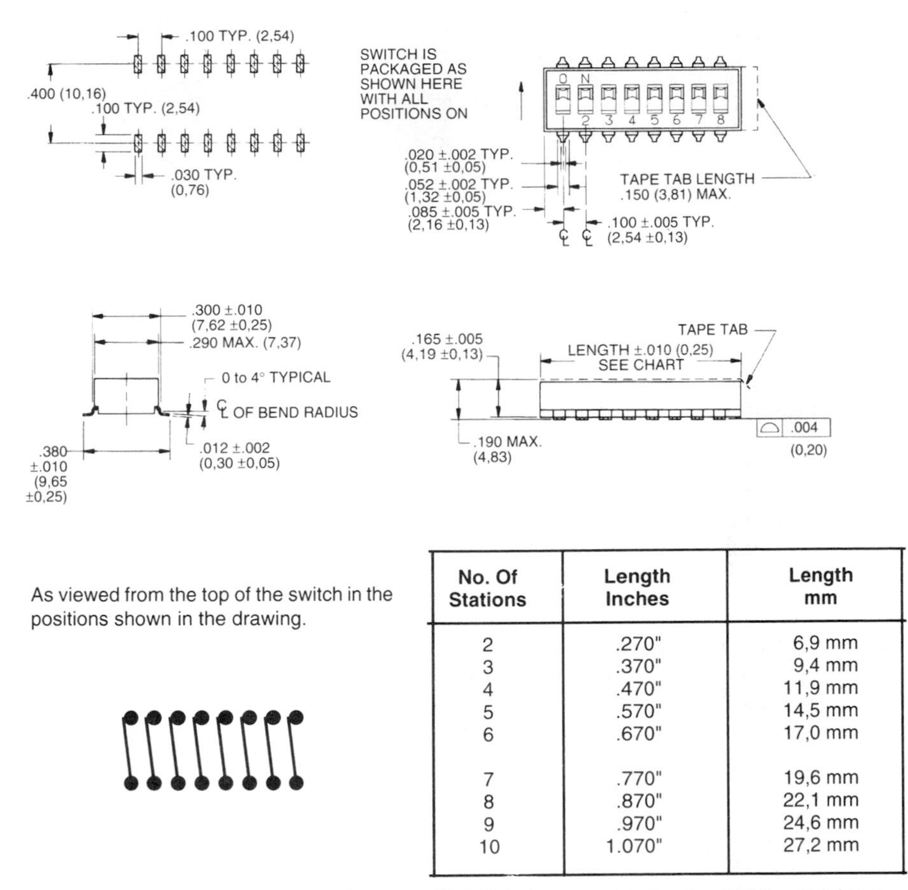

No. Of Stations	Length Inches	Length mm
2	.270"	6,9 mm
3	.370"	9,4 mm
4	.470"	11,9 mm
5	.570"	14,5 mm
6	.670"	17,0 mm
7	.770"	19,6 mm
8	.870"	22,1 mm
9	.970"	24,6 mm
10	1.070"	27,2 mm

FIGURE 12.6 A typical configuration for an SMT (surface-mount technology) DIP switch incorporating recessed slide actuators and top seal tape. (*Courtesy of Grayhill, Inc., La Grange, Ill.*)

ters, flush slide actuators, and a top tape seal which is easily removable. Figure 12.7 depicts typical mechanical, electrical, and environmental ratings the user can expect in a DIP switch of this design. (2)

A sealed[1] *ultraminiature* SMT DIP switch manufactured by C&K Components, Inc., Newton, Mass., is illustrated in Fig. 12.8. Ultraminiature refers to the 0.050-in (1,27-mm) terminal spacing. Typical mechanical, electrical, and environmental specifications for this device are shown in Fig. 12.9. The package incorporates recessed slide actuators and removable top tape seal.

Another design approach to the cleaning problem is shown in Fig. 12.10*a* and *b*. These *"washable"* DIP switches from C&K Components, Inc., Newton, Mass., incorporate an open-base design which allows cleaners to flush through

[1]Process sealed construction withstands IR and vapor phase reflow soldering and cleaning processes.

Ratings	
Mechanical life: Operations per switch position	5000
Make and break current rating: Operations per switch position at these resistive loads	
1 mA, 5 V dc; 50 mA, 30 V dc; or 150 mA, 30 V dc	
10 mA, 30 V dc; or 10 mA, 50 mV dc	
10 mA, 50 mV dc; or 25 mA, 24 V dc; or 100 mA, 6 V dc	2000
Contact resistance: Initially	≤20 mΩ
After life, at 10 mA, 50 V dc, open circuit	≤100 mΩ
Insulation resistance:	
Minimum, at 100 V dc between adjacent closed contacts and also across open switch contacts:	
Initially	5000 MΩ
After life	1000 MΩ
Dielectric strength: Minimum voltage (ac, rms) measured between adjacent closed contacts and also across open switch contacts:	
Initially	1000 V
After life	500 V
Current carry rating: Maximum rise of 20°C	3 A
Switch capacitance: At 1 MHz	2 pF
Operating temperature	−25 to +85°C
Storage temperature	−40 to +85°C
SMT processing temperature	260°C max
Surface Mount Switches	
Additional Information	
Processing temperature: 260°C maximum	
Processing position: Switch is to be processed with all actuators in the closed (ON) position as shipped	

FIGURE 12.7 Typical mechanical, electrical, and environmental ratings for the DIP switch depicted in Fig. 12.6. (*Courtesy of Grayhill, Inc., La Grange, Ill.*)

the package—no boots, caps, or tape is required to seal the unit. An additional feature of Fig. 12.10*b* is the *right-angle mounting* style provided by the terminal configuration. This permits greater access to the DIP switch where PCB-to-PCB spacing is minimal. (3)

How DIP Switches Are Used

A DIP switch is used as a back panel switch or logic card selector switch. The DIP switch can be used to program a multifunction PCB circuit to perform one specific function. Most programming or selection functions use SPST circuitry, but slide actuation DIP switches are also available in 2PST, 3PST and 4PST. In addition, some DIP switch manufacturers offer SPDT (Fig. 12.11) and DPDT versions. The closed or open condition of each switch provides the input.

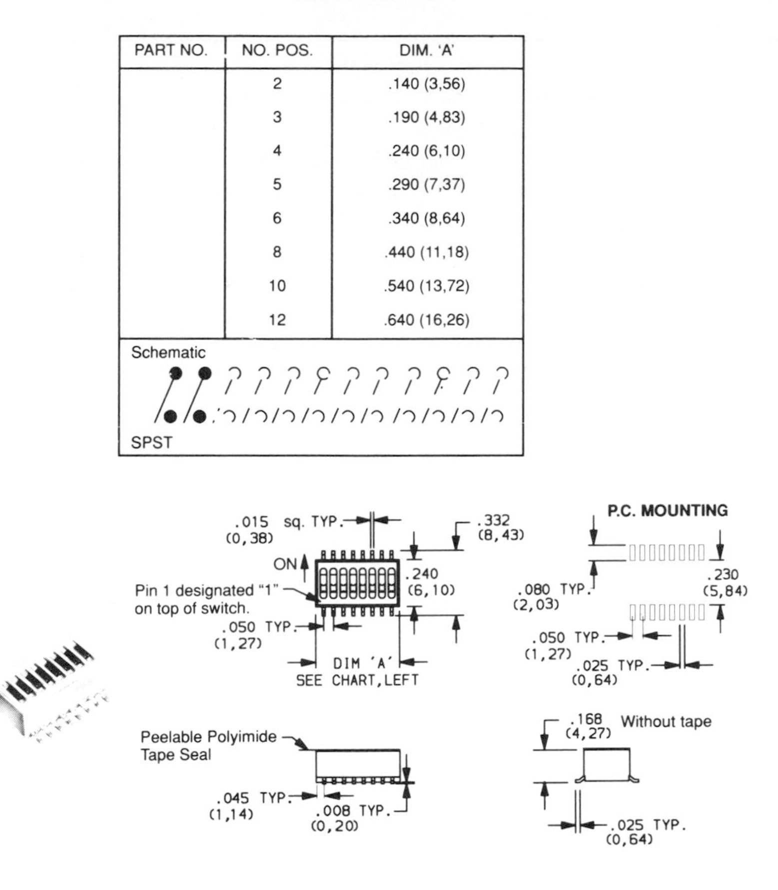

PART NO.	NO. POS.	DIM. 'A'
	2	.140 (3,56)
	3	.190 (4,83)
	4	.240 (6,10)
	5	.290 (7,37)
	6	.340 (8,64)
	8	.440 (11,18)
	10	.540 (13,72)
	12	.640 (16,26)

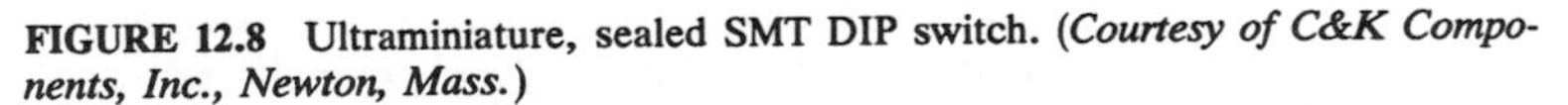

CAUTION: PC mounting layouts and pads as shown are designed to be compatible with the latest equipment and reflow techniques. Care should be exercised, however, in the design and location of PC lands to suit individual needs. Orientation relative to reflow direction may significantly impact solder joint integrity.

FIGURE 12.8 Ultraminiature, sealed SMT DIP switch. (*Courtesy of C&K Components, Inc., Newton, Mass.*)

A DIP switch can be used to address a specific portion of a computer memory. It can be used as a test switch to determine circuit operation. In peripheral equipment, designed to connect to a variety of computers, a DIP switch can be used to set baud rates (data transfer speeds) or to match impedances. If system manufacturers require a change in operation of their equipment, this can be done by adjustment of a DIP switch. Manufacturers of computer peripherals, computer input terminals, test equipment, telephone and telecommunications equipment, and process control equipment are all users of DIP switches.

A simple application example is the use of a DIP switch to program the change device in a vending machine. The proper combination of switches in the closed and open positions provides the proper change return for one unit price. The combina-

Specifications

Switch function: SPST 2, 3, 4, 5, 6, 8, 10, 12 pos.

Contact rating: 0.4 VA max at 20 V ac or dc max

Mechanical and electrical life: 2000 make-and-break cycles at full load, each circuit

Contact resistance: Below 30 mΩ typical initial at 2–4 V dc, 100 mA

Insulation resistance: 10^9 Ω min

Dielectric strength: 500 V rms min at sea level between adjacent terminals

Storage and operating temperature: −40 to 85°C

Solderability: per MIL-STD-202F method 208D, or EIA RS-186E method 9 (1 h steam aging)

Materials

Case and cover: glass-filled LCP (UL 94V-0)

Actuator: glass-filled PES (UL 94V-0)

Contacts: copper alloy, with gold plate over nickel plate

Terminals: copper alloy, with tin-lead alloy over nickel plate. Tin-lead coats all sides and ends of terminals. All terminals insert molded

Tape seal: polyimide

FIGURE 12.9 Typical mechanical, electrical, and environmental ratings for the ultraminiature DIP switch depicted in Fig. 12.8. (*Courtesy of C&K Components, Inc., Newton, Mass.*)

tion can be easily changed for new unit prices. An even more familiar example is in the ordinary garage door opener where a DIP switch can be used to set the same user-specified frequency in both the receiver and transmitter units. (1)

Binary-Coded Rotary DIP Switches

Yet another version of the popular DIP package, the *binary-coded rotary DIP switch* is available in a number of styles, one family of which is shown in Fig. 12.12. This style, manufactured by Grayhill, Inc., La Grange, Ill., can be configured with octal, binary-coded decimal, or hexadecimal codes, with either standard or complement output (Fig. 12.13). The device is available in either SMT or through-hole (perpendicular and right-angle) styles (Fig. 12.14) and with flush or extended actuators. (2)

A binary-coded rotary DIP switch from C&K Components, Inc., Newton, Mass. (Fig. 12.15), functions similarly to the unit shown in Fig. 12.12 but features unique actuator configurations (Fig. 12.16) inspired by human engineering considerations. The "wheel" and "knob" actuators of Fig. 12.16 are effective when unambiguous reading of the coded value is essential. [Adapted from (3).]

The specifying engineer should be aware that many DIP switches can be procured from the manufacturer packaged in tape, reels, and tubes to operate in component insertion machinery.

Because of the rapid progress in DIP switch technology, always contact the switch manufacturer concerning up-to-date ratings of DIP switches, as even published specifications can sometimes be incorrect or obsolete.

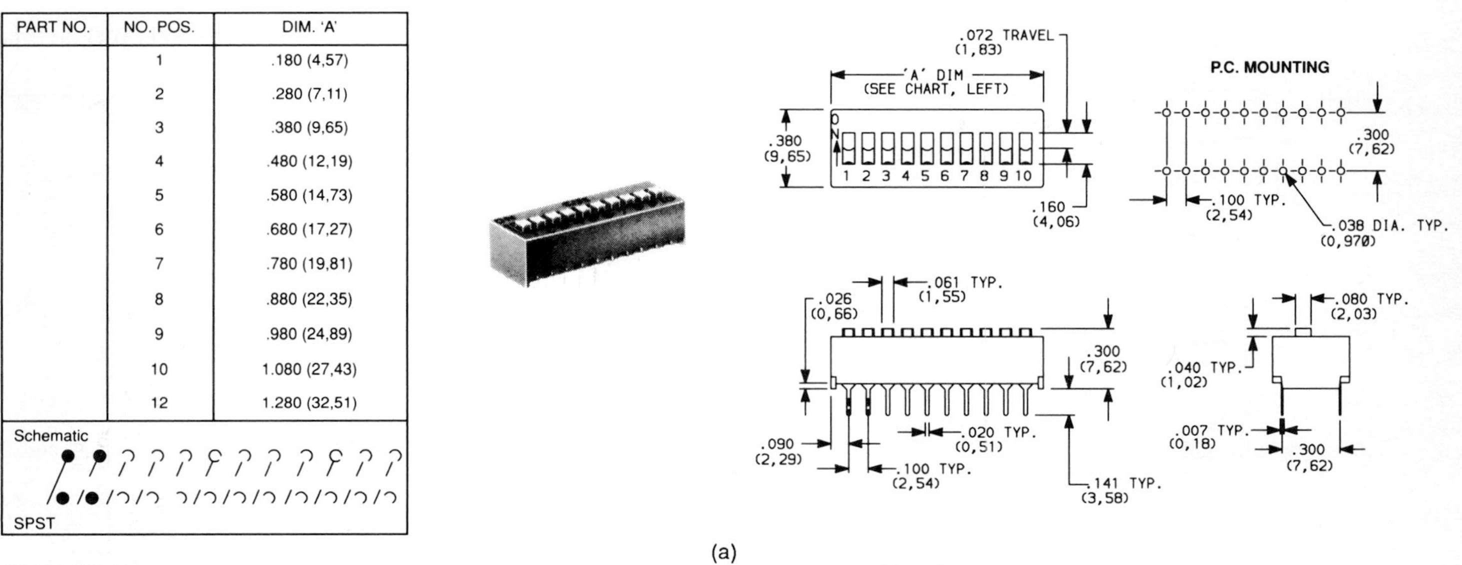

PART NO.	NO. POS.	DIM. 'A'
	1	.180 (4,57)
	2	.280 (7,11)
	3	.380 (9,65)
	4	.480 (12,19)
	5	.580 (14,73)
	6	.680 (17,27)
	7	.780 (19,81)
	8	.880 (22,35)
	9	.980 (24,89)
	10	1.080 (27,43)
	12	1.280 (32,51)

Schematic

SPST

(a)

FIGURE 12.10 "Washable" DIP switches incorporate an open-base design to flush out cleaners. Note the right-angle design in (*b*) which enables free access to the DIP switch from the side of densely packaged PCBs. (*Courtesy of C&K Components Inc., Newton, Mass.*)

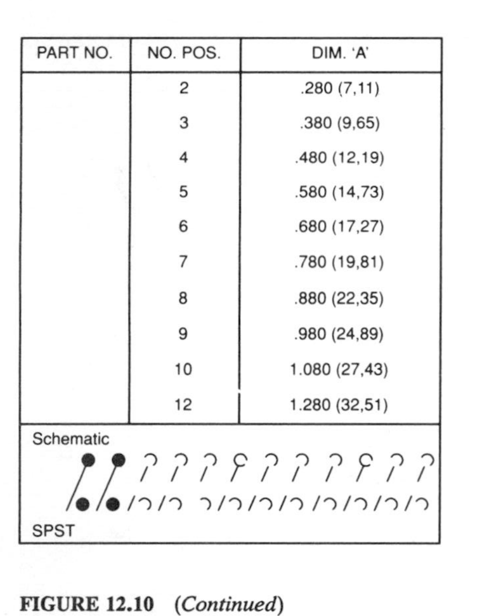

PART NO.	NO. POS.	DIM. 'A'
	2	.280 (7,11)
	3	.380 (9,65)
	4	.480 (12,19)
	5	.580 (14,73)
	6	.680 (17,27)
	7	.780 (19,81)
	8	.880 (22,35)
	9	.980 (24,89)
	10	1.080 (27,43)
	12	1.280 (32,51)

Schematic

SPST

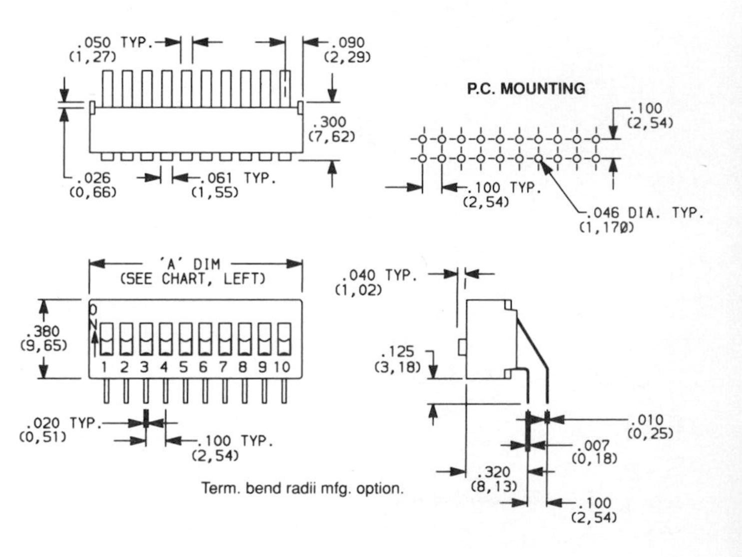

(b)

FIGURE 12.10 *(Continued)*

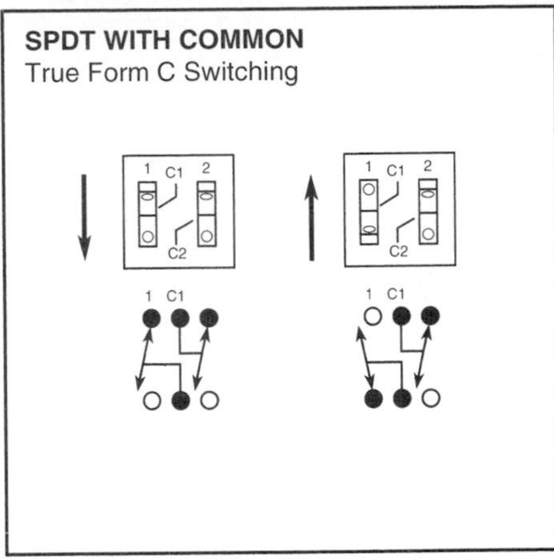

FIGURE 12.11 A SPDT DIP switch with rocker actuators. (*Courtesy of Grayhill, Inc., La Grange, Ill.*)

FIGURE 12.12 A family of binary-coded rotary DIP switches. (*Courtesy of Grayhill, Inc., La Grange, Ill.*)

CODE & TRUTH TABLES

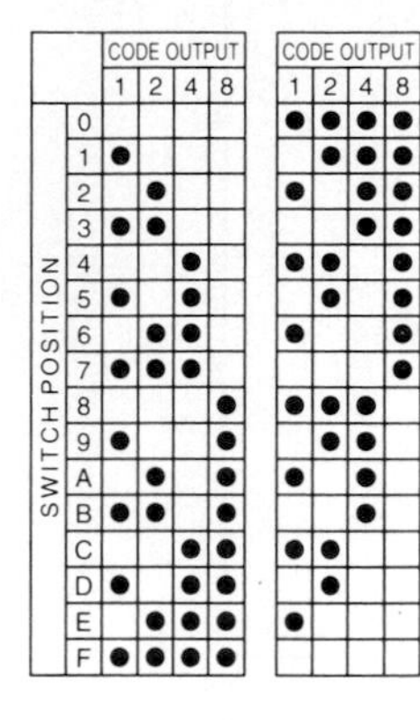

Switch Position	Standard Output: Code Output 1	2	4	8	Complement Output: Code Output 1	2	4	8
0					●	●	●	●
1	●					●	●	●
2		●			●		●	●
3	●	●					●	●
4			●		●	●		●
5	●		●			●		●
6		●	●		●			●
7	●	●	●					●
8				●	●	●	●	
9	●			●		●	●	
A		●		●	●		●	
B	●	●		●			●	
C			●	●	●	●		
D	●		●	●		●		
E		●	●	●	●			
F	●	●	●	●				

Dot indicates terminal to common connection. All switches are continuous rotation.

Octal and Octal Complement outputs are 0 thru 7 positions.

BCD and BCD Complement outputs are 0 thru 9 positions.

Hexadecimal and Hexadecimal Complement outputs are 0 thru F positions.

Standard codes have natural color rotors; complements have red.

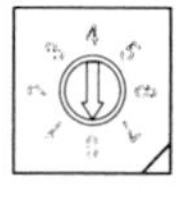

Octal - 8 position

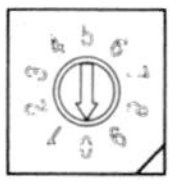

BCD - 10 position

Hex - 16 position

All actuation types are available in octal (8), binary coded decimal (10), or hexadecimal (16) codes; with either standard or complement output.

FIGURE 12.13 **Binary-coded rotary DIP switches are available with octal, decimal, and hexadecimal codes, with either standard or complement output. (*Courtesy of Grayhill, Inc., La Grange, Ill.*)**

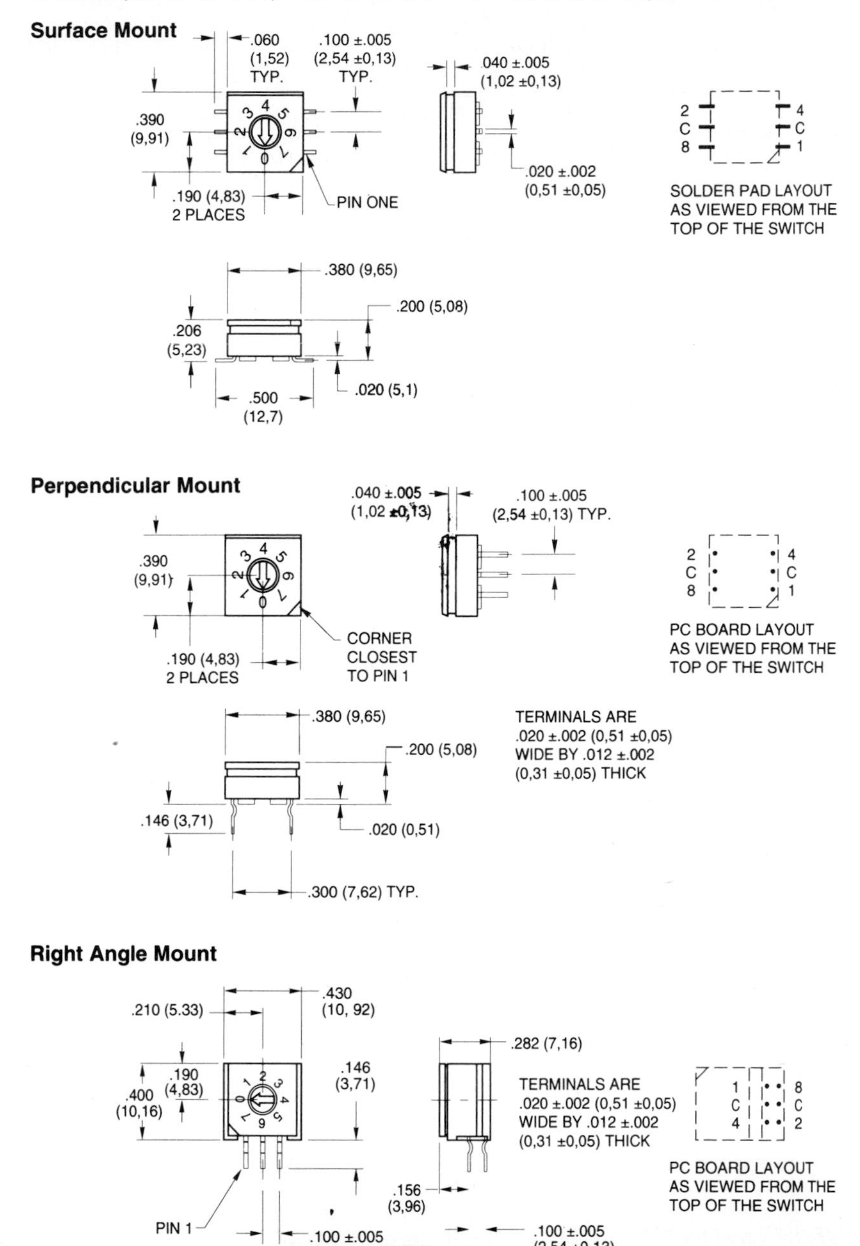

FIGURE 12.14 Surface-mount and two through-hole style (perpendicular and right-angle) mounts are available for the binary-coded rotary DIP switch family shown in Fig. 12.12. (*Courtesy of Grayhill, Inc., La Grange, Ill.*)

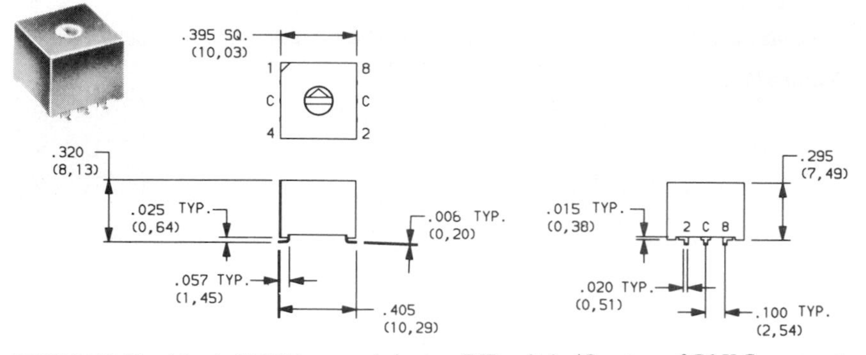

FIGURE 12.15 A basic SMT binary-coded rotary DIP switch. (*Courtesy of C&K Components, Inc., Newton, Mass.*)

GLOSSARY

Binary-coded rotary DIP switch. Basically a rotary switch in a DIP package, the device produces coded outputs in octal, decimal, and hexadecimal, with either standard or complement output.

DIP (dual in-line package). An electronics industry standard created by integrated circuit manufacturers. This standard package outline allows many different types of ICs to be plugged into the same socket or the same pattern on a PCB.

Right-angle mounting. A DIP switch configured with terminals at right angles to the top of the switch to permit the device to be accessed from the side of the PCB on which it is mounted. This configuration allows easy switch settings in electronics packages with minimal PCB-to-PCB spacing or where access to the front or back side of the PCB is difficult or impossible.

Rocker actuation. DIP switch actuator style in which a "teeter-totter" motion is used to open and close each individual switch, similar to the motion used in rocker switches.

Slide actuation. DIP switch actuator style in which a back-and-forth sliding motion is used to open and close each individual switch, similar to the motion used in slide switches.

Tape seal. A polyester or polyimide film used to seal the top of a DIP switch, over the rocker or slide actuator, to prevent contamination from entering the switch during soldering and cleaning operations. The tape is removable after cleaning is completed.

Ultraminiature. As related to DIP switches, ultraminiature describes smaller than standard DIP packages, based on a principal dimensional standard of 0.050 in (1,27 mm) between terminals, compared with a standard DIP dimensional standard of 0.100 in (2,54 mm) between terminals.

Washable. A washable DIP switch is one which incorporates an open-base design to allow cleaners to flush through the package; therefore, no tape seal is required.

REFERENCES

1. "SPST Circuitry DIP Switches," Grayhill, Inc., La Grange, Ill., 1987.
2. "Engineering Catalog No.1," Grayhill, Inc., La Grange, Ill., 1991.
3. "Catalog No. 9105—Switches, Newton Division," C&K Components, Inc., Newton, Mass., 1991.

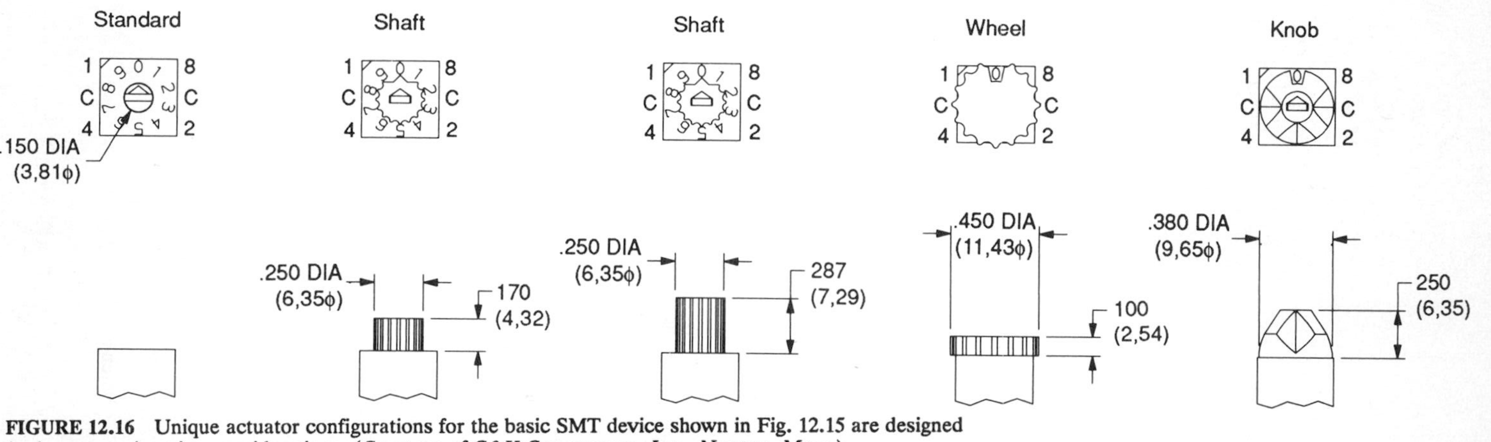

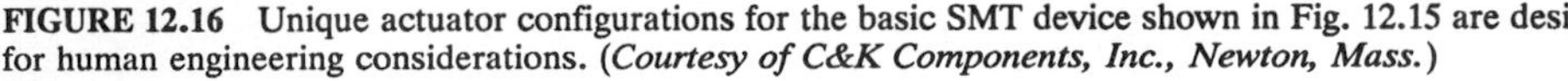

FIGURE 12.16 Unique actuator configurations for the basic SMT device shown in Fig. 12.15 are designed for human engineering considerations. (*Courtesy of C&K Components, Inc., Newton, Mass.*)

CHAPTER 13
TOUCH SCREENS, TOUCH SWITCHES, AND LIGHT PENS*

A *touch screen* is a touch-sensitive switch array mounted in front of a CRT or flat panel display (Fig. 13.1). It is activated by the touch of a finger or pointing stylus. The switching means can be (1) *digital or analog resistive membrane,* (2) intersecting *infrared (IR) light beams,* (3) *capacitive,* (4) *surface acoustic wave technology,* (5) *piezoelectric sensor,* or (6) *light pens.* An application designed specifically for touch input can create the most direct interaction between a computer and a human being. Large databases can be accessed quickly and effortlessly by people who have never before interacted with a database program or a computer. Complicated processes can be made understandable to the factory

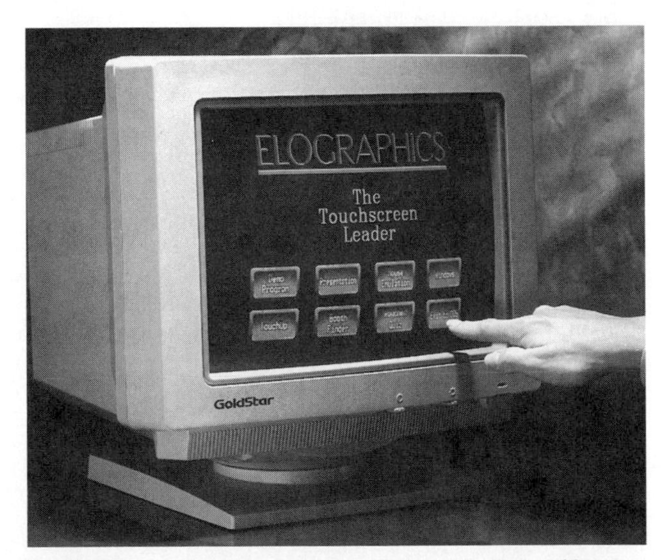

FIGURE 13.1 A touch screen mounts in front of a display. (*Courtesy of Elographics, Oak Ridge, Tenn.*)

*Numbers in parentheses indicate items in the References at the end of this chapter.

worker, whose touch sets a process in motion. As each process is activated, animated graphics displayed on the screen show the worker that the process is taking place and has been executed correctly. In touch screen–based video training systems, nothing comes between the student and the subject matter, and pointing to the correct answer is a process as old as teaching itself. Portable instruments and laptop computers, designed to operate in situations where no "desktop" space is available, can provide complete mouse functionality through a touch screen. New book-sized portable units, consisting of a CD player and a touch screen, are providing convenient access to encyclopedic volumes of information. The current expansion in the development of these types of systems is largely due to the new touch screen tools available to the systems integrator. (1)

A *touch switch* is a discrete switch operated by the operator's finger or a stylus interrupting a light beam. Although referred to as a "touch" switch, no physical contact with the mechanism is needed to complete the switching function. As discussed later in the chapter, this noncontact technique has the potential to minimize carpal tunnel syndrome (CTS) injuries in the workplace. (2)

A *light pen* is a hand-held electro-optical computer data entry device, used by an operator to indicate a specific location on the face of a CRT to the host computer. The light pen is held much like a pen or pencil, with the tip of the light pen being touched on the CRT face or activating a finger-operated switch on the light pen, at the point of interest to the operator. Light pens can be used in place of keyboards, mice, digitizing tablets, and touch screens. (3)

TRANSPARENT MEMBRANE TOUCH SCREENS

How Membrane Touch Screens Function

Membrane touch screens (also referred to as resistive touch screens) are transparent switch assemblies designed to be installed over various types of displays including CRT, vacuum fluorescent (VF), electroluminescent (EL), plasma, and liquid crystal displays (LCDs). The touch screen provides direct operator-to-screen interaction where a switching function is needed to augment the display symbology. Figure 13.2 shows a typical application for a membrane touch screen. The membrane assembly is often attached to a transparent acrylic or polycarbonate base for rigidity.

Two mechanically similar, but electrically unique, membrane touch screen technologies are available, the matrix (or digital) style and the analog style.

Digital resistive touch screens utilize two separate layers of transparent conductive film with parallel separator lines etched in their coatings (Fig. 13.3) to form isolated conductive lanes or stripes, one stripe for each row or column required in the final switch assembly. The conductive films are oriented so that the conductive stripes are perpendicular to each other (Fig. 13.4). With minimal finger pressure, the films are brought into physical contact, permitting current to flow at the point of contact. In effect, the conductive surfaces make contact, in an *X-Y* fashion, and create a switch closure.

Analog resistive touch screens are similar to digital resistive touch screens in construction, but the entire active area is a single switch (Fig. 13.5); no isolation lines are incorporated. When both layers make contact, one layer measures the voltage in the *X* direction, the other in the *Y* direction (Fig. 13.6). The host computer interprets the voltages as *X-Y* coordinates.

FIGURE 13.2 A typical application for a digital or analog resistive membrane touch screen. (*Courtesy of Memtron Technologies, Inc., Frankenmuth, Mich.*)

CRT-Membrane Touch Screen Combinations. Membrane touch screens are used extensively in conjunction with CRTs in applications such as point of sale (POS), medical, telecommunications, and public information systems. Standard touch screens are available in sizes from 5 in (127 mm) to 25 in (635 mm) diagonal. Installation of the touch screen is simply a matter of attaching the touch screen between the CRT and the touch screen bezel. An O-ring or water sealing gasket is recommended for industrial and other harsh environment applications. Contrast-enhancement filters, ESD shields, and multicolor silk-screen overlays are available. (4) Curvature of the touch screen to match that of the CRT is accomplished at the factory (Fig. 13.7).

VF-Membrane Touch Screen Combinations. VF (vacuum fluorescent) displays utilize membrane touch screens in applications ranging from automated test equipment (ATE) to machine tool control systems. Since the VF display is flat, membrane touch screen manufacturers recommend using the touch screen without the backing plate in order to minimize parallax. (4)

EL and Plasma Display–Membrane Touch Screen Combinations. Membrane touch screens are used in combination with EL (electroluminescent) and plasma displays for both military and aerospace applications and process control equipment. The inherent flatness of the display allows the use of touch screens without

FIGURE 13.3 A typical digital resistive membrane touch screen with its encode electronics board attached to the touch screen via flex tails. (*Courtesy of Transparent Devices, Inc., Newbury Park, Calif.*)

the acrylic or polycarbonate backing plate. Circular polarizers are recommended for membrane touch screens used with EL displays to enhance the visual characteristics of the display. (4)

LCD (Liquid Crystal Display)–Membrane Touch Screen Combinations. The use of membrane touch screens in front of LCDs is preferred for indoor applications where the LCD is backlighted. Outdoor applications using the LCD–membrane touch screen combination are not recommended because the internal reflections caused by (1), the two different layers of indium tin oxide (ITO)–sputtered material, (2) the air gap between the two layers, and (3) the ambient light differential between the outside and inside glass layers of the LCD limit the legibility of the LCD–touch screen combination in outdoor applications. Because of the construction of an LCD wherein the glass is unsupported across the front of the display, it is recommended that the touch screen be mounted on the LCD support frame. In applications such as portable, hand-held terminals and battery-operated equipment, the LCD–membrane touch screen combination has a distinct advantage by virtue of its lower power consumption and the elimination of permanent switches surrounding the LCD assembly. Additionally, the membrane touch screen adds virtually unlimited "soft" key capability. (4)

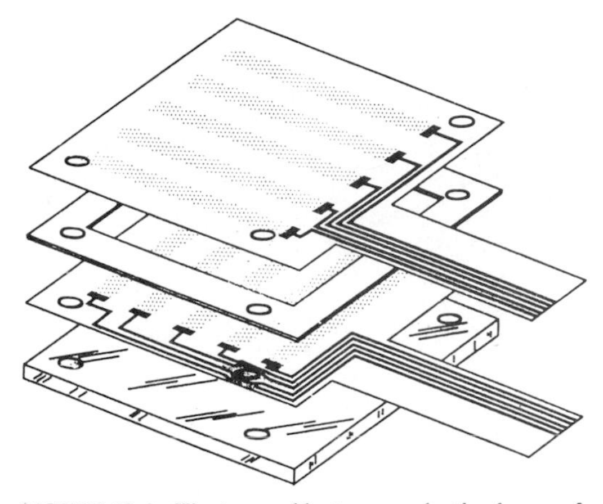

FIGURE 13.4 The top and bottom conductive layers of a digital membrane touch screen are oriented perpendicular to each other. (*Courtesy of Memtron Technologies, Inc., Frankenmuth, Mich.*)

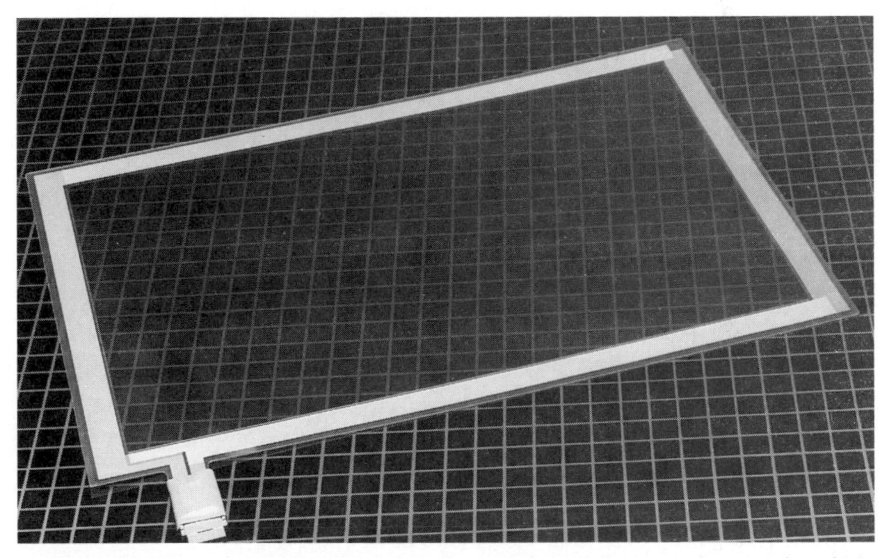

FIGURE 13.5 The entire active area of an analog resistive membrane touch screen is a single switch. (*Courtesy of Elographics, Oak Ridge, Tenn.*)

How Membrane Touch Screens Are Constructed

A membrane touch screen is a transparent, normally open, momentary-contact touch panel nearly identical in construction to membrane switches (Chap. 8). The basic switch is made of two layers of 0.005- to 0.007-in (0,127- to 0,177-mm) optical grade polyester film with a 0.0005- to 0.002-in (0,0127- to 0,051-mm) separation be-

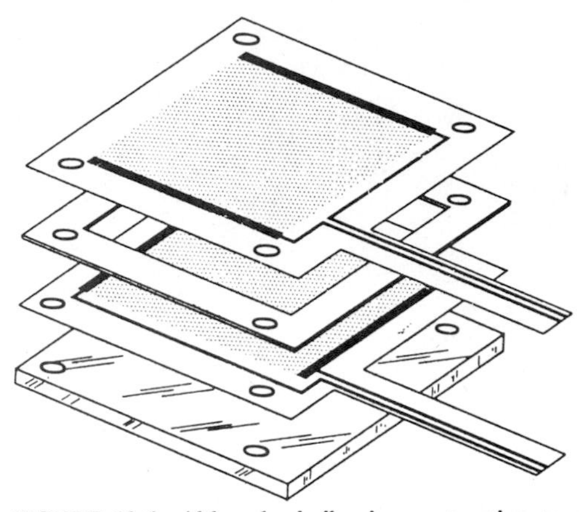

FIGURE 13.6 Although similar in construction to a digital resistive touch screen, the analog touch screen does not require etched conductive stripes. (*Courtesy of Memtron Technologies, Inc., Frankenmuth, Mich.*)

tween layers. The inside, or facing, surfaces of both film layers are sputter-coated with a transparent electrically conductive thin film coating, usually indium tin oxide (ITO), although other conductive materials such as gold are being used successfully.

In the digital resistive touch screen, one film layer has horizontal traces (rows) and the other film layer has vertical traces (columns). The traces are usually configured by etching the circuit pattern into the already sputtered film material, much like etching the circuit pattern into the copper layer of a printed wiring board (Fig. 13.8). An alternative method is to mask the film material with the circuit pattern prior to the sputtering operation, followed by removal of the masking material to reveal the desired pattern in the ITO conductive layer. In the analog resistive-style touch screen, no conductor patterns are etched in the conductive layer (Fig. 13.9).

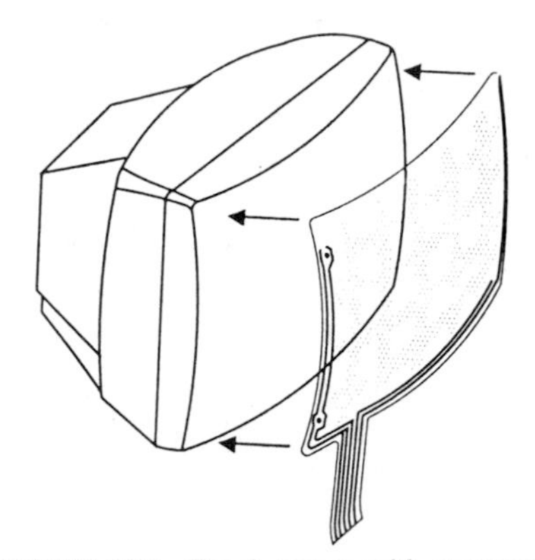

FIGURE 13.7 Touch screen with curvatures to match the face of CRT displays are available. (*Courtesy of Memtron Technologies, Inc., Frankenmuth, Mich.*)

Conductive epoxy conductors are then silk-screened onto the conductive side of the film, outside of the transparent viewing area, as a means of extending the ITO conductors (either etched or nonetched) out to the I/O connector. The conductive epoxy conductors act as a more robust continuation of the ITO pattern itself while providing a means of connection to the touch screen. Typically, the interconnection to the touch screen is made through a membrane switch-style clincher connector installed at the end of the flex tail carrying the conductive epoxy conductors.

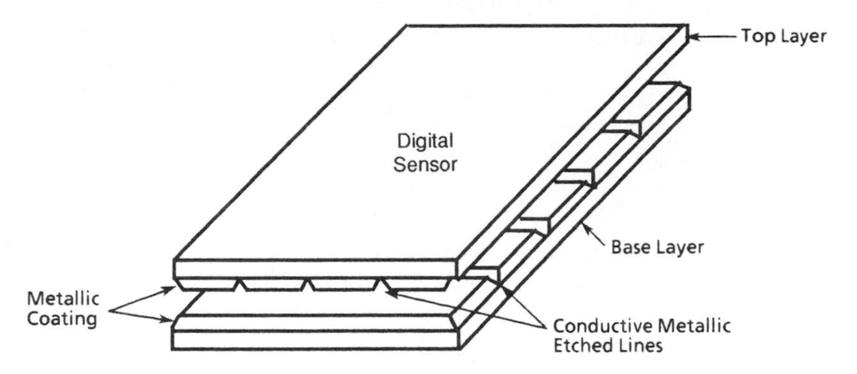

FIGURE 13.8 The circuit pattern for digital membrane touch screens is etched into the sputtered conductive material on the substrate. (*Courtesy of Carroll Touch, Round Rock, Tex.*)

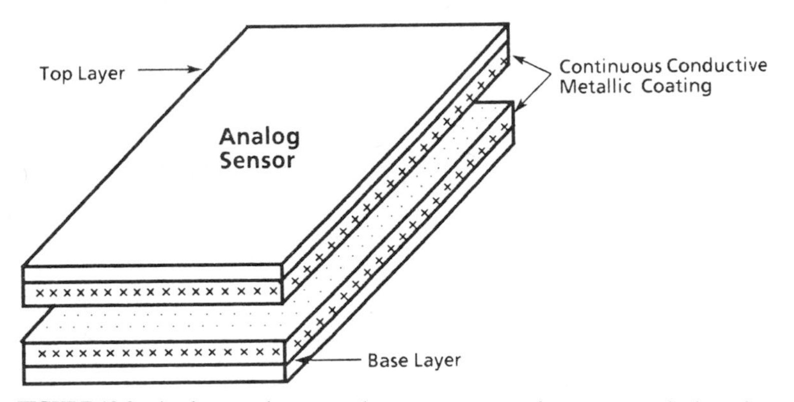

FIGURE 13.9 Analog membrane touch screens use a continuous, nonetched conductive layer. (*Courtesy of Carroll Touch, Round Rock, Tex.*)

In both the digital and analog resistive touch screens, the accurate separation of the two conductive layers when the two halves are assembled is accomplished (Fig. 13.10) either by applying precisely controlled dots of transparent nonconductive material (insulators) silk-screened to the inner side of one of the ITO coated films or by adhering glass microspheres to the same inner surface. The dots, or spheres, maintain the precise separation between the two conductive films, preserving the consistency of actuation force across the entire active area of the touch screen. A double-sided frame adhesive is used to permanently bond the two film halves together. Touch screens that are designed for use in front of CRTs are laminated to an acrylic or polycarbonate backing plate to provide rigidity for the touch screen while enhancing user protection by acting as an implosion shield.

When the digital resistive touch screen is assembled, the intersection of rows on one layer and columns on the other layer designates discrete switch positions which, when pressed into contact by the operator's finger or stylus, are scanned

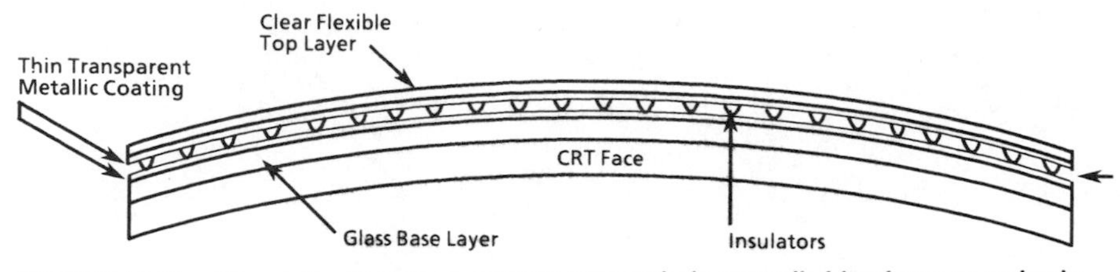

FIGURE 13.10 All resistive-type touch screens use precisely controlled insulators to maintain separation between top and bottom layers. (*Courtesy of Carroll Touch, Round Rock, Tex.*)

electronically to decode the *X-Y* position of the switch contact. In fact, digital touch screens are scanned in a manner similar to matrix keyboards.

In the analog resistive touch screen, light finger pressure causes internal electrical contact at the point of touch. This contact supplies the controller with the analog voltage needed for digitization. A companion serial or bus interfacing controller impresses a voltage across the conductive coating on the glass substrate. This voltage is alternated between the *X* and *Y* directions, allowing separate voltage readings for both the *X* and *Y* axis. When touched, the cover sheet conductive coating relays a voltage to the analog-to-digital circuitry of the controller, and digitized coordinate pairs are transmitted to the computer.

As is true with all touch screen technologies, a number of companies who manufacture resistive touch screens also provide the encoder interface, usually a single printed wiring board assembly containing all the scanning electronics necessary to encode the touch screen switch closures (refer back to Fig. 13.3). Off-the-shelf encoder boards allow the engineer to save time and are a relatively inexpensive way to produce working prototypes of touch screen–display assemblies.

Typical membrane touch screen specifications are depicted in Table 13.1. Always check with the selected touch screen manufacturer for more detailed parameters.

INFRARED (IR) BEAM TOUCH SCREEN TECHNOLOGY

How Infrared Beam Touch Screens Function

Infrared (IR) beam touch screens differ from membrane touch screens in that intersecting beams of infrared light take the place of the membrane itself as the switching mechanism. To produce the grid of light, the optomatrix frame or sensor unit surrounds the display. The optomatrix frame is composed of infrared light-emitting diodes (LEDs) and the phototransistors or light-detecting devices. The LEDs are arranged along two adjacent sides of the frame, while the phototransistors are placed along the other two sides of the frame, opposite the LEDs. The frame is attached to the face of the display and concealed behind a bezel. The controller sequentially pulses the LEDs, creating an invisible grid of infrared light beams just in front of the display surface (Fig. 13.11). When the screen is touched, the light beams from certain LEDs are obstructed and do not reach their corresponding phototransistors. The touch controller constantly monitors the

TABLE 13.1 Typical Membrane Switch Touch Screen Specifications

	Commercial	Medical Industrial	Military
Mechanical:			
Actuation force*	20–40 grams	40–80 grams	100–500 grams
Actuation travel	0.015 mm	0.03 mm	0.05 mm
Thickness (w/o sup.)	0.5 mm	0.5 mm	0.5 mm
Thickness (w/ sup.)	2–4 mm	2–4 mm	2–4 mm
Mechanical life	6 million	6 million	6 million
Electrical:			
Contact resistance:			
Open	20 megohm	20 megohm	20 megohm
Close	10–30 kohm	10–30 kohm	10–30 kohm
Operating voltage	5 volts	5 volts	5 volts
Operating current	5 microamp	5 microamp	5 microamp
ESD shield	1 kV	15 kV	35 kV
Optical:			
Light transmissivity @ 550 nanometers	73–80%	65–73%	50–65%
Anti-Glare (gloss)	55–95	55–95	55–95
Environmental:			
Temperature:			
Operating	0 to 35°C	−10 to 55°C	−40 to 85°C
Nonoperating	0 to 35°C	−10 to 55°C	−40 to 85°C
Altitude:			
Operating	3,000 meters	3,000 meters	3,000 meters
Nonoperating	12,000 meters	12,000 meters	12,000 meters
Vibration	0.7 G @ 15 Hz	0.7 G @ 15 Hz	0.7 G @ 15 Hz
	1.3 G @ 25 Hz	1.3 G @ 25 Hz	1.3 G @ 25 Hz
	3 G @ 55 Hz	3 G @ 55 Hz	3 G @ 55 Hz
Shock	30 G for 11 sec.	30 G for 11 sec.	30 G for 11 sec.

Source: Transparent Devices, Inc., Newbury Park, Calif.
*Using a 0.375 in diameter RTV tip.

presence or, in this case, the absence of the infrared light beams received by the phototransistors and thus determines the X and Y coordinates of the touch activation. The controller transmits the X and Y coordinates to the host computer where an *application program* causes the action selected to occur.

The touch controller includes the electronics which are used to process the information received from the sensor unit and to interface with the host computer. (5) Figure 13.12 depicts a method one manufacturer has selected for packaging an IR touch screen assembly. Figure 13.13 shows the ease of installation of a clip-on style IR touch screen used in process control systems.

Breaking the beam with the operator's finger or a stylus is the only motion required to produce a switch closure; no force is necessary to complete the switch closure. Since no force is required, the operator's hand, arm, and fingers do not need to be held at unusual angles in order to operate the touch screen. IR beam touch switches are commonly referred to as optoelectric or optoelectronic switches.

Industrial Electronic Engineers, Inc. (IEE), Van Nuys, Calif., manufactures a line of touch-input optoelectric switch matrices for installation over the company's

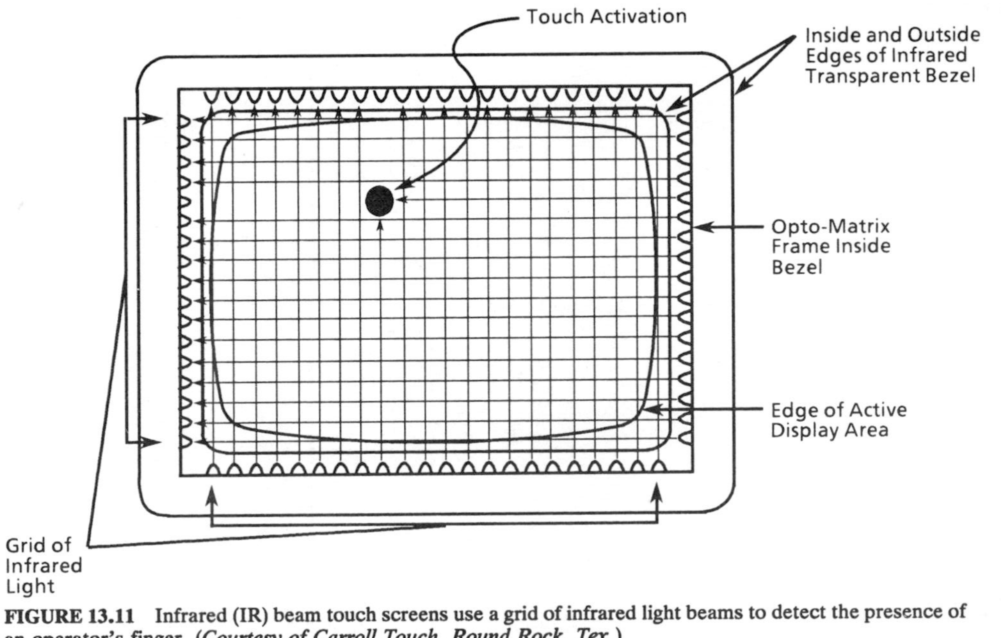

FIGURE 13.11 Infrared (IR) beam touch screens use a grid of infrared light beams to detect the presence of an operator's finger. (*Courtesy of Carroll Touch, Round Rock, Tex.*)

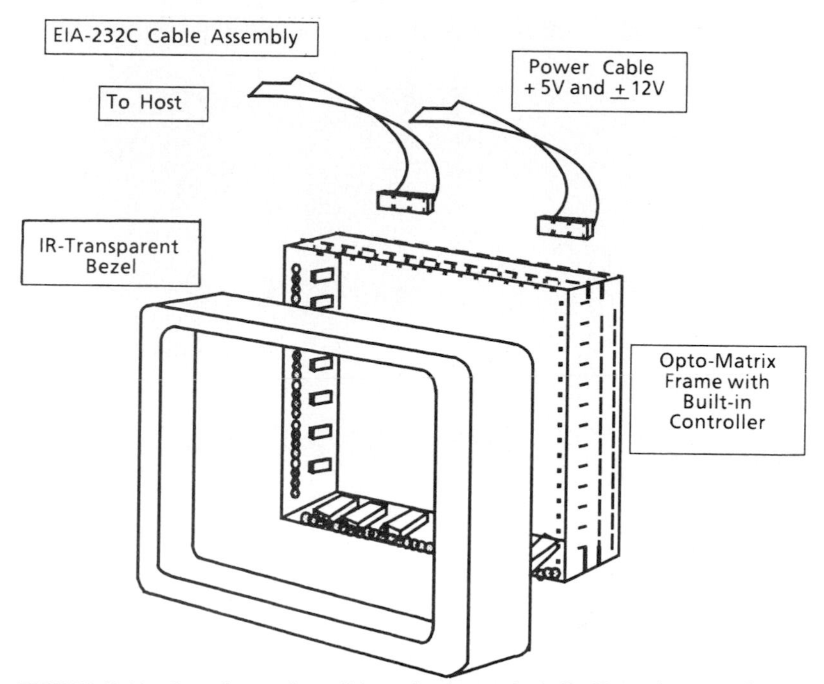

FIGURE 13.12 One of several possible packaging methods for IR touch screen electronics. (*Courtesy of Carroll Touch, Round Rock, Tex.*)

plasma and VF (vacuum fluorescent) displays. The company's 4284 series PEP (peripheral entry panel) will be examined to illustrate how an optical touch screen system is utilized.

The 4284 (Fig. 13.14) is an intelligent display module integrating a 12-line by 40-character dc gas plasma display with an optoelectric switch matrix. The optoelectric switch is formed by horizontal and vertical light beams generated by IR LEDs. This array yields 240 active switch locations of approximately 0.20 in (5,08 mm) by 0.20 in (5,08 mm). Interaction is achieved when the operator reads the display message and responds by finger or pointer contact to the touch-sensitive switch location on the viewing screen. When an intersecting set of beams is broken by the touch, the 4284 outputs the switch position to the host system. According to the manufacturer, the switch matrix will not respond to ambient light, and error-checking algorithms in the microprocessor software ensure switch output reliability.

FIGURE 13.13 Clip-on style IR touch screens provide ease of assembly to display systems. (*Carroll Touch, Round Rock, Tex.*)

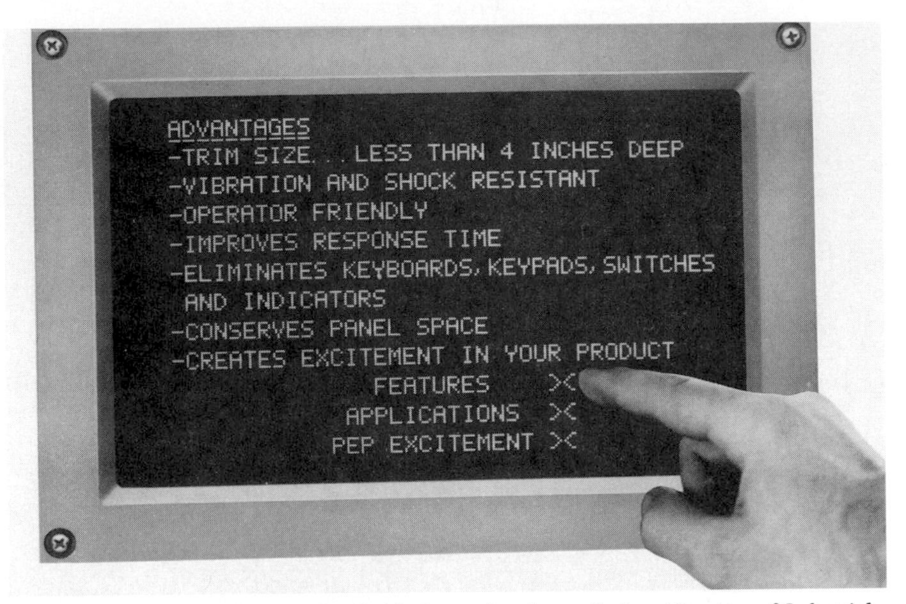

FIGURE 13.14 IR touch screen installed in front of a plasma display. (*Courtesy of Industrial Electronic Engineers, Inc., Van Nuys, Calif.*)

The 4284 provides bidirectional communications with the host system and a microprocessor controls the switch overlay and serial communications (which include display inputs and switch outputs). The entire module, including the optoelectric matrix, can be manufactured with a dripproof option. Additionally, the display and optoelectric touch system self-tests. (6)

Lucas Deeco Corporation, Hayward, Calif., packages an IR touch screen over an EL (electroluminescent) display panel. The company's infrared array employs an 80 × 25 array that is capable of operating in varying light conditions (even direct sunlight). Two Lucas Deeco models, the M4ST and the ST2200, feature full sealing, with the ST2200 capable of meeting NEMA 4 and 12 requirements. The model M4ST features what the company calls programmable touch sensitivity which allows variable touch sizes and touch time for applications requiring "fail-safe" touch operation. Using this feature, the M4ST can be programmed to look for a specific size of touch area and/or an area to be touched for a time duration before an operation is invoked. Button draw routines provide the programmer with simple "define and draw" of buttons and labels. Up to 120 buttons can be displayed on a display page and up to 120 sets of these buttons and their respective responses can be defined. [Adapted from (7) and (8).]

Typical environmental specifications for IR touch screens are shown in Table 13.2.

CAPACITIVE TOUCH SCREEN TECHNOLOGY

Capacitive touch screen technology uses a glass substrate (overlay) over the surface of the display screen. The user must touch the overlay substrate with a fin-

TABLE 13.2 Typical IR Touch Screen Specifications

Operating temperature	0–50°C
Storage temperature	−20 to +70°C
Relative humidity	0–95% (noncondensing)
Vibration:	
Operating	2.5 g at 10–100 Hz
Nonoperating	2.5 g at 10–100 Hz
Mechanical shock	15 g, 11 ms pulse width

Source: Industrial Electronic Engineers, Inc., Van Nuys, Calif.

ger to activate the system. The body capacitance of the user's finger is added to the capacitance already present on the substrate. The increase in capacitance is measured by the touch system and the location of the touch is detected.

Capacitive touch systems are similar to resistive systems in that they utilize a touch sensor which is a glass overlay. However, the capacitive sensor has only one layer. This layer is a glass substrate with a thin metallic coating, identical to the bottom layer of a resistive sensor (Fig. 13.15). Typically, capacitive systems

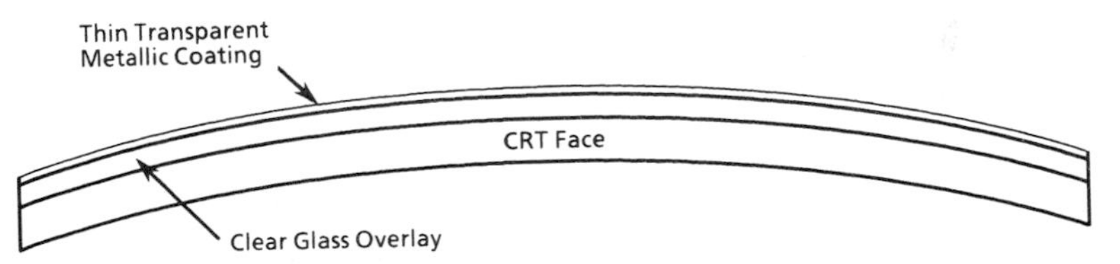

FIGURE 13.15 Capacitive touch screens feature a transparent conductive coating on a glass substrate. (*Courtesy of Carroll Touch, Round Rock, Tex.*)

have four timing circuits attached to the four corners of the glass sensor. These circuits are connected to the sensor using four conductive wires. Each of the timing circuits oscillates at a specific frequency which is stored in memory.

To determine the touch activation, the effect of the user's body capacitance on the timing circuits is measured. Whenever the conductive surface of the sensor is touched, body capacitance is added to each of the four timing circuits. When the extra body capacitance is added to the capacitance already present in the timing circuit, the frequency at which the timing circuit oscillates decreases.

The controller measures the different effect on each of the timing circuits to determine the X and Y coordinates of the touch activation. The X and Y coordinates are then transmitted to the host computer where an application program causes a predetermined action to occur. (5)

SURFACE ACOUSTIC WAVE (SAW) TOUCH SCREEN TECHNOLOGY

Surface acoustic wave (also referred to as SAW) technology takes advantage of the ability of inaudible, high-frequency acoustic waves to travel over the surface of glass at very precise speeds in very straight lines. The wavelength of the signal used in these touch screens is about 0.02 in (0,51 mm). The technology has a high signal-to-noise ratio, which lets the controller determine the center of each touch within ±0.5 wavelength, regardless of the screen's dimensions. Surface acoustic screens (Fig. 13.16) have the second highest resolution of the touch-input alternatives, with up to 100 touch points per in (25.4 mm), which is usually more than sufficient for most applications.

In SAW operation, a pair of piezoelectric transducers convert a 5.53-MHz electrical signal generated by the controller into surface acoustic waves. The *X*-axis transducer is located in the upper left-hand corner of the screen, the *Y*-axis transducer in the lower right-hand corner. Each transducer-generated signal travels along the edge of the screen, encountering a reflective array printed directly along the edge of a glass panel. The array consists of a series of 0.002 in (0,050 mm) thick by 0.50 in (12,7 mm) wide diagonal parallel lines made of powdered glass. Each array element reflects a small part of the acoustic signal (0.2 percent) over the screen surface. The remainder of the signal travels on to the next array element, where it is similarly reflected. By the time the original signal reaches the opposite corner of the touch screen, almost all its energy has been reflected out over the display surface. To compensate for the reduction of the original signal as it travels over—and is reflected away from—the edge of the screen, the reflective array elements are placed progressively closer together, ensuring consistency in the strength of the waves traveling across the screen.

The reflected portions of each signal are met by mirror-image reflective arrays on the bottom and left side of the touch screen. These second arrays reflect the signals to two receiving transducers, one each for the *X* and *Y* signals. When a touch occurs, the pointing device absorbs a portion of the energy flowing in both the *X* and *Y* directions, attenuating the signal. By comparing the speed of the re-

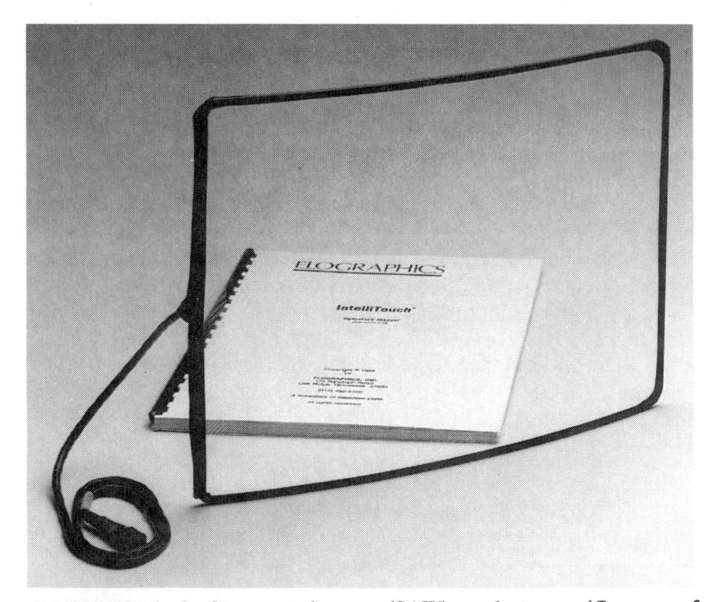

FIGURE 13.16 Surface acoustic wave (SAW) touch screen. (*Courtesy of Elographics, Oak Ridge, Tenn.*)

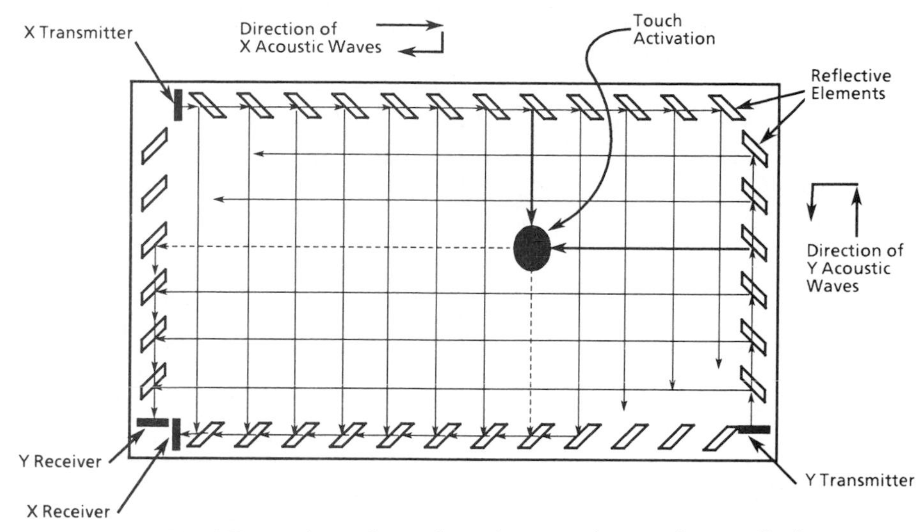

FIGURE 13.17 *X* and *Y* transmitters alternately send an acoustic wave along a reflective array on each axis. The wave is partially reflected across the screen to a complementary reflective array that redirects the wave to a receiver. A finger touching the screen weakens the wave signal, indicating a touch has occurred. (*Courtesy of Carroll Touch, Round Rock, Tex.*)

ceived signal and the known speed of sound waves in glass, the controller calculates the touch location (Fig. 13.17).

Surface acoustic wave takes advantage of the fact that a pliable pointing device such as a finger can deliver various degrees of contact area and pressure, absorbing more or less energy. SAW's ability to sense pressure as well as location adds a Z-axis dimension to touch input, offering some interesting application possibilities (Fig. 13.18). Pressure sensing lets SAW touch emulate other input devices, such as a mouse. A light touch can be used to "drag" the cursor across active touch zones without activation, for example, while a heavy touch can be used to "click," or activate, a command.

Pressure sensing can also be used with pull-down menus, with a light touch to pull the menu down and a heavy touch to activate a menu choice. With other two-axis methods of handling pull-down menus, finger liftoff activates the menu choice. The drawback of this method, of course, is that users can't lift their finger from the display without activating a touch zone. The *Z* axis can also be used to control scrolling speed, robotic movement, or material flow. Several levels of Z-axis differentiation are possible and current screens support 16 levels, but most applications are best served with no more than two or three.

Like the other touch-input alternatives, SAW hasn't been free of problems. Before the technology could be used in actual applications, for example, developers had to deal with the fact that foreign matter on the screen, such as heavy grease or water drops, can absorb a portion of the acoustic signal and register as a touch. However, anything smaller than a single acoustic wavelength, such as normal environmental residue, won't register.

The solution to this problem was to digitize the base amplitude level and continually compare it with touch-screen signals. If the touch-screen signal ampli-

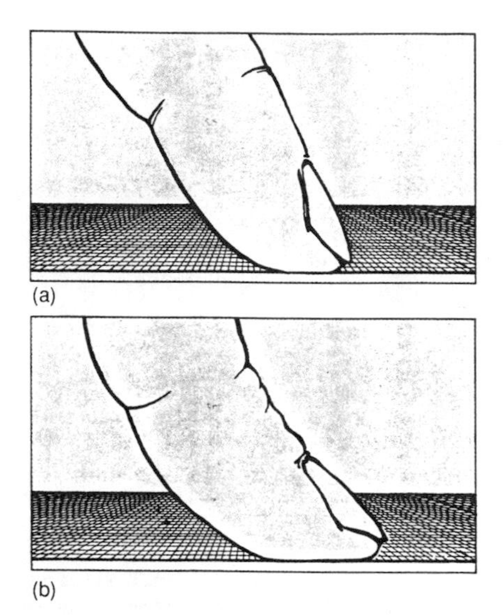

FIGURE 13.18 SAW touch screen technology lets users work with pliable pointers that can deliver varying amounts of pressure and contact area to the screen. A light touch (*a*) causes less deformation, thus producing a smaller contact area and absorbing less area than a heavy touch (*b*). (*Reprinted with permission from the March 15, 1988, issue of* Computer Design. *Copyright 1988, PennWell Publishing Company, Advanced Technology Group.*)

tude is lower than the base trace, a touch has occurred. If the signal continues to be attenuated for a longer time than a normal touch would take, the controller assumes that the touch is an "artifact" and ignores it. (9)

PIEZOELECTRIC SENSOR TOUCH SCREEN TECHNOLOGY

Piezoelectric, or pressure-sensitive, touch technology uses pressure transducers to determine the location of a touch activation. The piezoelectric sensor is a clear glass plate, similar to the overlay technology sensors. Pressure transducers are attached to the four corners of the sensor. The sensor is mounted in front of the display screen with the transducers in contact with the surface of the display screen (Fig. 13.19).

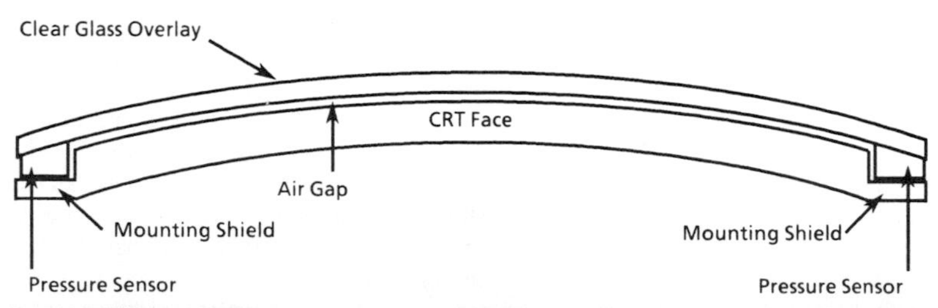

FIGURE 13.19 When a user touches the piezoelectric sensor touch screen, varied pressure readings are recorded at each corner sensor. (*Courtesy of Carroll Touch, Round Rock, Tex.*)

When a user touches the piezoelectric sensor, pressure is exerted on each of the transducers and a pressure reading is recorded. The reading at each of the pressure transducers varies depending on the location of the touch. The controller measures the different readings on the four transducers and a firmware algorithm converts the values detected into *X* and *Y* coordinates. (5)

TOUCH SCREEN INTERFACE

Touch Screen Controllers

Although a touch screen represents the most simple interactive device to the user, it has not been a simple input device from the system developer's standpoint. Unlike a keyboard, a touch screen requires a separate controller and driver. Controllers for touch screens have been nonstandard and drivers were unsupported by programming languages. Additionally, there was until recently a shortage of general-purpose, easily utilized development tools for creating touch screen applications.

However, the recently increased popularity and acceptance of touch screens, in an ever-widening marketplace, has rapidly altered this situation. New software development packages now include sophisticated drivers, development utilities, and complete authoring systems, most of which make efficient use of a developer's other resources. New mouse and keyboard emulation software speeds the process of converting existing programs to accept touch input. On the hardware side, controllers (Fig. 13.20) are becoming less proprietary and more easily assimilated into an existing system. It has finally become a fairly simple matter to add touch input capability to any application program.

The main components in a touch screen–based system are the touch screen, a controller, driver software, and the application program. Choosing the appropriate type of touch screen for a given application involves interlocking decisions based on light transmission, resolution, price, speed, durability, resistance to vandalism, the type of stylus which can be used, NRE costs, and aesthetics. Information on these and other touch screen parameters is readily available from a wide range of touch screen manufacturers.

Unfortunately, it is more difficult to find written material on specific controllers, drivers, and development programs. These components are seldom discussed in trade journals, yet it is these very items which most affect the construction of a touch screen–based system.

All touch screens require a controller; most are microprocessor-based (the Intel 8031 and Motorola 6811 are typically used for this purpose). The controller drives the screen, processes the incoming data (including filtering and debouncing), and transfers the processed touch coordinates to the host.

All touch screens share another common characteristic: each is designed to work with a specific touch screen. This lack of standardization can be a problem when a developer tries to switch from one manufacturer's touch screen to another.

There are two commonly used types of controllers: serial and bus. On a serial controller, data are transferred to the host system as a serial stream via an RS232 port. Bus controllers plug directly into the PC bus and are accessed via I/O ports.

Touch screen controllers may be either configured with switches and jumpers or down-loaded parameters from the host. Several common modes and communication techniques are possible. Options can include the data rate and format,

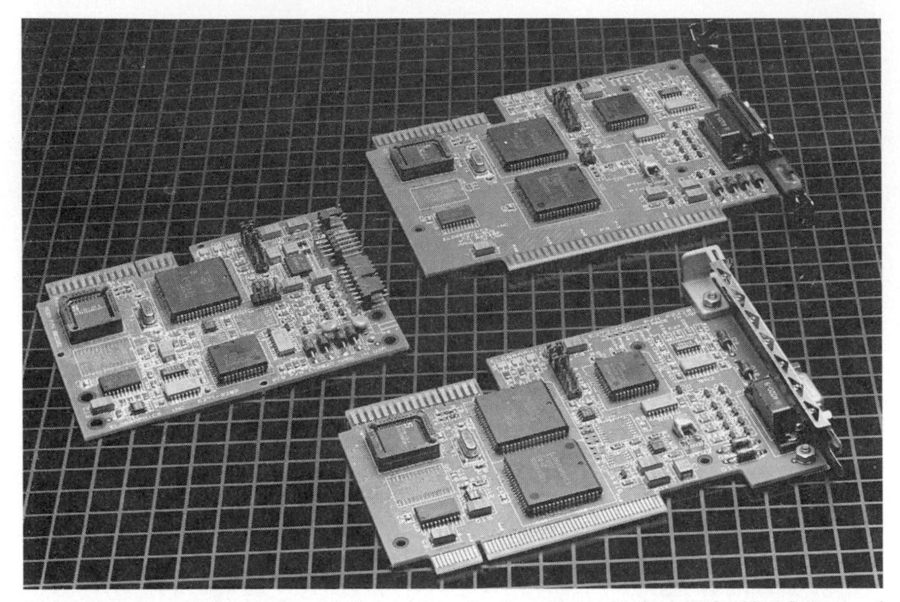

FIGURE 13.20 Off-the-shelf controllers such as the SmartSet, with Serial RS-232, PC-Bus, and Micro Channel® interfaces, are available for a wide variety of touch screen applications. (*Courtesy of Elographics, Oak Ridge, Tenn.*)

axis inversion, filtering options, conversion enable conditions, and operating modes. The operating mode chosen by the developer should be dictated by the *type* of application being designed and the *users* of the application.

Operating modes include:

- *Single-point or enter-point:* a single coordinate pair is transmitted whenever a touch occurs. No further coordinates are communicated until the finger is lifted (untouch) and the touch screen is retouched.
- *Continuous or stream:* The controller continuously sends data as long as the screen is being touched.
- *Exit-point or untouch:* Coordinates are reported by the controller only when the touch is released. This mode can be combined with the other modes and is particularly useful when the finger is sliding over a menu, then lifts off to activate a given position.

Polled or Interrupt

Controllers can be operated in either a polled or an interrupt mode. In the polled mode, the host polls a status port. When data are available, the X and Y values are read from the data ports. In bus controllers, this port is on the card. In serial implementations, the port resides on the UART (universal asynchronous receiver-transmitter). Polling software is easily implemented and may be sufficient for many applications.

The interrupt mode, however, is more efficient because it does not require constant monitoring of the status port. In this mode, the host is interrupted when touch data are available. The host processor then executes an interrupt routine which retrieves the touch data and places them in a buffer. The interrupt routine returns to the application without affecting it. The application program can check the buffer and process the touch at its convenience, unconcerned with how the data are placed in the buffer.

The interrupt mode should be used if the application has other duties besides polling. If polled mode is used and the application is occupied when the controller sends a "data ready" signal, that signal and subsequent data will be lost until the application resumes polling the port.

Serial Controllers

Serial controllers are the most universal, since RS232 ports are a standard fixture on almost every computer. These controllers can be mounted within the display enclosure if there is sufficient space. Additionally, sufficient power may be available from the monitor to run the controller. This is often true for resistive-membrane controllers; many run on a single low-voltage power supply. On larger displays, 19 in (483 mm) and larger, there is usually room for a separate power supply if one is required. Serial controllers are also placed in separate enclosures outside the display. Serial controllers transmit *X/Y* information in ASCII or binary data packets. Data packets have an identifiable beginning and/or end so the driver can stay synchronized with the controller. ASCII packets are usually separated by a carriage return <CR> and line feed <LF>. For example, 105,063 <CR> <LF> is a nine-character string (<CR> and <LF> are used for sequencing). In binary form, each packet consists of *X* and *Y* bytes and a sequence byte or sequence bits within the data bytes. These packets are at least three characters.

Obviously, the ASCII form requires more transmission time and host overhead. In either binary or ASCII form, serial controllers are less efficient than bus controllers, because they are so character-intense and polling or interrupt service is required for each character.

Bus Controllers

Bus controllers interface most directly with the host. They communicate complete *X* and *Y* data after a single polling sequence or interrupt. No synchronization is necessary. A bus controller operating in the interrupt mode is the most efficient controller-mode combination. These attributes, and the fact that the cost of serial and bus controllers is equivalent, have prompted developers to choose a bus controller whenever possible.

A properly designed bus card should be capable of working in a polled or interrupt mode and should be selectable for different port addresses and interrupts. This flexibility allows the controller to be easily assimilated into a system and work with peripherals which would otherwise compete for interrupts or ports, such as network cards or multiple I/O cards.

Because of the design of the PC bus, there is a limit to the number of interrupts available for the bus controller. Table 13.3 lists the devices assigned to each interrupt in a PC/AT and PC/XT.

TABLE 13.3 Device Assignments to Each Interrupt in a PC/AT and PC/XT

Interrupt	XT	AT
2	IBM EGA, IBM network	Interrupts 8–15 (shared)
3	COM2	COM2
4	COM1	COM1
5	Hard disk controller	LPT2
6	Floppy disk controller	Floppy disk controller
7	LPT1	LPT1

Source: Elographics, Oak Ridge, Tenn.

Unassigned interrupts are seldom available. On an AT, interrupt 5 may be free. It is also possible for the driver software to disable the interrupt line drivers of contending devices, such as serial port controllers, while the touch screen is in operation.

When confronted with this limitation, some developers opt for a serial controller with a multiport card, because the multiport card uses only on interrupt. Unfortunately there are no standards in this environment; it is possible that the driver software will not work when a multiport serial card is used.

Custom Controllers

Developers creating applications for large-volume distribution or incorporating a touch screen into a portable instrument may wish to design their own controller. A controller's necessary electronic components can be integrated into a system's main printed circuit board and use its microprocessor. Since this reduces the number of boards needed, power consumption, and the cost of the controller, it is a popular alternative for those working with portable instruments, dedicated instruments, or consumer electronics.

Incorporating a touch screen controller into the unit's electronics is most easily done when using a resistive-membrane touch screen. It is possible to do this with capacitive and surface-acoustic-wave screens, but they are more demanding of the controller and require more computing power.

Calibration

The need for calibration is unique to the touch screen. Unlike mouse or keyboard applications where the cursor is part of the image, a touch screen is a physical overlay with an independent coordinate system. Only by knowing the position of the image can the host software convert touch screen coordinates into screen coordinates.

Calibration is especially necessary when using touch screens over CRTs because the video image varies among monitors. The image is affected by horizontal and vertical adjustments on the monitor and by physical mounting of the touch screen.

Additional calibration complications include image blooming, where bright-colored images expand. Another phenomenon, called the "pincushion" effect, causes the corners of the display to be stretched. Poor linearity can cause similarly sized boxes to be larger at the edges of the screen than they are in the middle of the screen or vice versa. Even changing video modes can affect the screen display. In a 19-in (483-mm) screen, for example, switching from EGA to CGA can cause the image to shrink by as much as 3 in (76,2 mm)! The more one learns about these display characteristics, the more obvious it becomes that calibration must be a major consideration when designing a touch screen–based application.

There are two ways to calibrate a touch screen: using hardware-based or software-based adjustments. If hardware-based calibration is used, each time recalibration is needed the computer must be opened up and adjustments made to the potentiometers on the touch screen's controller. This can be avoided if the calibration is accomplished with software.

There are a variety of software-based calibration techniques, from simple to complex. It is a mistake to assume that the more complex techniques, such as those using rotation algorithms or least-squares algorithms, result in better calibration. The most sophisticated calibration formulas cannot overcome the parallax problems encountered by users of different sizes or even one user changing positions as the touch screen is operated.

It is possible to overcome all these difficulties by proper application program design. If this is done, even the simplest calibration technique, which uses two points displayed at opposite, diagonal corners, is sufficient for most applications.

Figure 13.21 shows a touch screen and its calibration point coordinates. The inner rectangle indicates the position of the image. Anything outside this area is referred to as the "overscan" area and is inaccessible to a program. The * points, at the extremes of the image, are given two names. One name represents raw

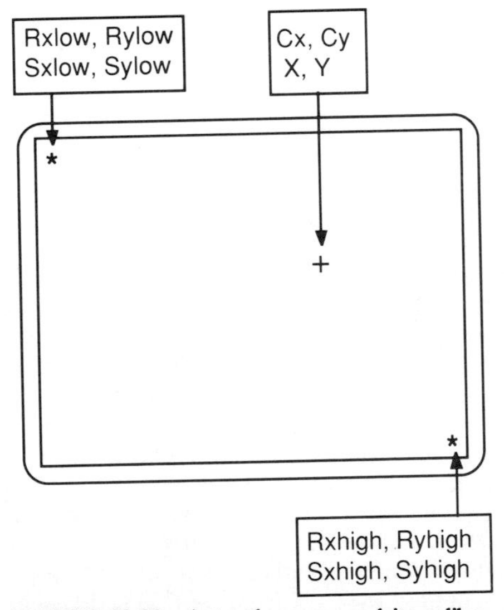

FIGURE 13.21 A touch screen and its calibration point coordinates. (*Courtesy of Elographics, Oak Ridge, Tenn.*)

coordinates (denoted by R); the other represents screen coordinates (denoted by S). The point of touch is at the +. This point is also given two names: *Cx, Cy* for the raw coordinates and *X, Y* for the screen coordinates.

The * coordinates are obtained through a calibration program written by the system developer. This program simply displays a target at one extreme point, lets the user touch it, and then repeats the process for the opposite extreme. The calibration program stores the raw coordinates for each target in a file or nonvolatile memory. The driver or application program will later load these points and use them in the conversion formula.

Consider an 80 × 25 screen coordinate matrix. The conversion process must be performed for both *X* and *Y*. The *X* formula is

$$X = [(^\wedge Sx/^\wedge Rx)(Cx - Rx\text{low})] + Sx\text{low}$$

where Cx = raw coordinate at + in the X axis
X = translated coordinates at + in screen coordinates
$^\wedge Rx$ = Rxhigh − Rxlow (range of calibration coordinates)
$^\wedge Sx$ = Sxhigh − Sxlow (range of screen coordinates, e.g., 79 = 80 − 1)

The screen coordinates in this example are 1 to 80 in *X* and 1 to 25 in *Y*. Therefore, *Sx*high is 80. In fact, any coordinate scaling may be used, such as 0 to 999 or −10 to +10.

A similar formula, substituting *Y* for *X* values, accomplishes the *Y* conversion process. This formula allows any origin and scaling, independent of the touch screen and controller. The calibration conversion and scaling functions can be performed by either the controller, the driver, or the application program.

Drivers

In most touch screen systems, the details of hardware interrupt handling, buffering, handshaking, and coordinate conversion are performed by the driver software. The driver can be easily accessed from assembly language or any high-level language. It typically becomes part of the operating system or is simply one part of a touch library which is incorporated into the touch application. Once the touch driver is installed, the application program receives scaled *X* and *Y* coordinates from the driver.

Drivers fall into several categories. Some allow the application, once written, to be ported easily from one type of touch screen to another. Others are based on keyboard or mouse emulation, which let the developer easily add touch to existing applications (see Emulators, later in this chapter).

Drivers that make the application portable do so by making the touch screen controller-independent, but only within the product line of a given manufacturer. Thus the application can be moved from one type of touch screen to another (resistive-membrane to surface-acoustic-wave, for example), from one screen size to another, and from a serial port to a bus port, all without changing code or recompiling. A variety of function calls are available for this type of driver. Language bindings are provided for C, Pascal, and BASIC, which turn every driver function into a subroutine call. Other languages can be accommodated similarly, by translating the source code provided.

Application Development Tools

New tools automate portions of the development process. One module, a screen generator for text-based applications, can import screens or help the developer create them from scratch. This type of module does not generate program code for screens but instead creates a portable, object-oriented data table which describes each screen. Although used extensively in PC-based applications, the module is suitable for non-PC applications because the portable data code it generates can be used on custom hardware.

The machine-independent binary table produced by this module is converted back into a screen image by a run-time library. The library repaints the screens out of the table, paints and removes overlays, and translates raw touches into zones.

To simplify the process of creating graphics-based applications, another new software program makes it possible to take snapshots of screen images already created with other software, port the snapshots into the developmental tool, and then develop touch zones simply by touching the zone boundaries on the screen, eliminating the tedious, error-prone process of typing in zone coordinates. Programs like this one represent a breakthrough for touch screen application development, because they allow the use of powerful and familiar off-the-shelf graphics software. Any package using text modes, or CGA, EGA, VGA, and Hercules graphics can be used.

Once touch zones have been defined, they must be described and labeled. The program then generates these zone identification tables automatically. The tables can be in a language-independent binary file form or in source code (C, Pascal, or assembly). The zone identification tables can be built into the application at compile time, linked directly to the application, or loaded at run time.

Emulators

Emulators help the developer convert existing applications to accept touch input. Keyboard emulators take advantage of the fact that every language can write to the screen and accept input from the keyboard; a touch can generate the same keystrokes as those which would be typed manually. A language's normal keyboard input command is used to receive these keystrokes. Applications developed with programs, a variety of BASICs, or even batch files can be easily modified using a keyboard emulator. One program requires only two lines of code per screen to set up the interaction between the application and the touch screen. This keyboard emulator can also be used to display a captured screen image, which is particularly significant for developers working with text-only applications such as dBase. Using this emulator, a graphics screen can be incorporated into any application, including a dBase application or batch file, using one command.

Mouse emulators have recently become available which are compatible with Microsoft® Windows and MOUSE.COM (DOS). The challenge in creating these emulators was to determine the most logical way for the touch screen to emulate mouse button functions. A variety of techniques have been used, including clicking on first touch or clicking on untouch (finger lift-off). The right technique for a given application should be dictated by the application itself and the sophistication of its users.

Both hardware-based and software-based mouse emulators can use off-the-shelf software packages without modification. However, while hardware-based emulators are the easiest to install (the touch screen controller is simply plugged into the serial mouse port), software-based emulators offer more options to the developer who wants to use the best interaction method for a particular application. Like the hardware-based emulator, the basic clicking of the left and right buttons, dragging, and double clicks can be performed. Additional options available with software emulators include the ability to turn off the cursor (when a finger is the pointer, a cursor is often redundant). Audible signals can provide the feedback normally given by clicking mouse buttons. In addition, if a software-based emulator is used, the mouse can be running concurrently with the touch screen or it can be used alone.

Using keyboard and mouse emulators to develop touch screen applications has its drawbacks. An application program designed for use with one type of input device may not work well with another input device. The inherent "look and feel" of a properly designed touch application program will differ markedly from a program written specifically for a mouse or keyboard. Today's new touch development tools have made it a fairly simple task to build touch into an existing computer environment without the aid of an emulator, taking advantage of the unique attributes of the touch interface. (1)

LIGHT BEAM TOUCH SWITCHES

How Light Beam Touch Switches Function

Light beam touch switches operate in a manner similar to light beam touch screens where the activation means is a physical interrupt of the intersecting light beam. A touch switch (Fig. 13.22) manufactured by Banner Engineering, Minneapolis, Minn., features a built-in SPDT electromechanical output relay (one NO contact, one NC contact) activated whenever a finger, introduced into the "touch area" of the switch, interrupts the device's infrared sensing beam. Called the OPTO-TOUCH this series of optical touch buttons are designed to replace mechanical pushbuttons and toggle switches. Both the OTB (momentary action) and LTB (alternate action) switches are engineered to fit into the same package (Fig. 13.23) which is designed to fit standard 1.19-in (30,0-mm) mounting holes for oiltight pushbutton switches (Chap. 6). Two LEDs are located on the top of the device, one LED indicating when the unit is powered on and the other LED indicating activation of the output relay. It should be noted that the alternate action output of the LTB series changes state only when the "touch area" is cleared by removing the finger and then touched again, whereas the momentary action output of the OTB series activates the output relay only when a finger is present in the "touch area." The relay is activated for as long as the OTB's sensing

FIGURE 13.22 A light beam touch switch. (*Courtesy of Banner Engineering Corp., Minneapolis, Minn.*)

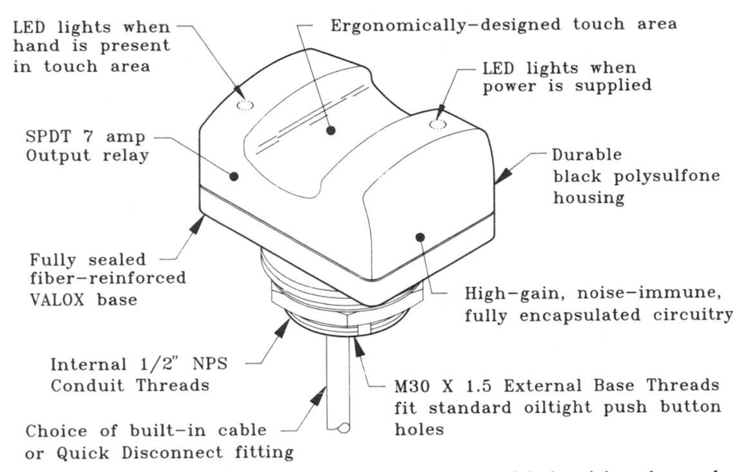

FIGURE 13.23 Features of the rugged packaging of industrial-grade touch switch. (*Courtesy of Banner Engineering Corp., Minneapolis, Minn.*)

beam is blocked and deactivates the moment the block is removed. Response time of output for both models is 80 ms.

The OPTO-TOUCH switch is highly immune to both the single and mixed EMI and RFI noise sources found in typical industrial settings. Because modern industrial environments contain *multiple* interference sources of mixed types, Banner incorporated the concept of "simultaneous interference sources" to optimize the design and testing of the OPTO-TOUCH switch. (2)

The OPTO-TOUCH wiring diagram shown in Fig. 13.24 is valid for both LTB and OTB series switches.

Ambient light immunity of the LTB and OTB switches is 120,000 lux (direct sunlight), but note the following two cautionary notes:

1. Prolonged exposure to direct outdoor sunlight will cause embrittlement of the polysulfone housing. However, ordinary window glass effectively filters longer-wavelength ultraviolet and provides excellent protection from sunlight.
2. The user should note the following warning in Banner's OPTO-TOUCH brochure: *WARNING!* Optical touch buttons are intended as general-purpose initiators, and are *NOT* safety devices. Like most solid-state devices, they are equally as likely to fail in the conducting ("ON") state as in the nonconducting ("OFF") state. If OPTO-TOUCH optical touch buttons are used to initiate a machine or operation in which false operation could be dangerous, *point-of-operation guarding devices* must be installed and maintained to meet all appropriate OSHA and ANSI B11.1 regulations and standards.

Since OPTO-TOUCH switches are activated by light beam interruption, no physical pressure is required to operate the switch. The ergonomic design of the switch is intended to minimize the hand, wrist, and arm stresses that can result from repetitive operation of mechanical pushbuttons. These stresses, over time, can cause tendon, nerve, and neurovascular disorders. These disorders, called CTDs (cumulative trauma disorders), can lead to decreased productivity and employee morale, and increased fatigue and health problems.

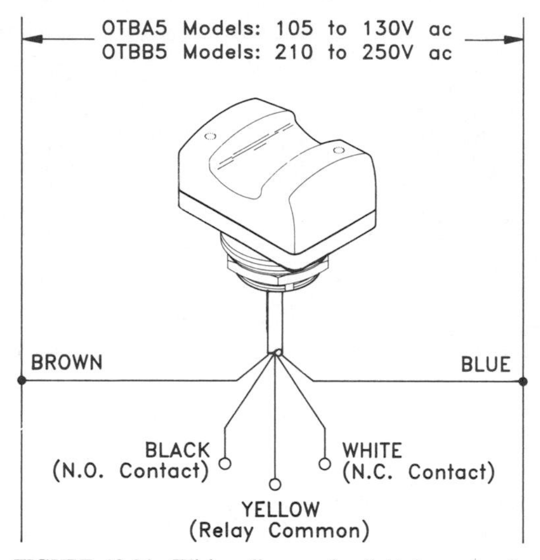

FIGURE 13.24 Wiring diagram for light beam touch switch. (*Courtesy of Banner Engineering Corp., Minneapolis, Minn.*)

Switches which require little or no pressure to actuate or which do not require the wrist to be in a flexed or extended (as opposed to "neutral") position while using the fingers can help minimize or even eliminate CTDs. (2)

Additional discussion concerning CTDs can be found at the end of this chapter.

How Light Beam Touch Switches Are Constructed

Once again referring to Fig. 13.23, the rugged construction of the OPTO-SWITCH begins with the cover, injection molded of black polysulfone, which attaches to the black fiber reinforced VALOX® base. All electronics are contained within the totally sealed, nonmetallic enclosure formed by the cover and base assembly and are fully epoxy encapsulated, resulting in the OPTO-SWITCH carrying ratings of NEMA 1,3,4,4X,12, and 13 (see Appendix 3). The threaded base of the switch has M30 × 1,5 external threads and ½-in NPSM internal threads. The switch requires a 1.19 in (30,0 mm) diameter mounting hole (fits standard automotive-size "jumbo" legend plates and oiltight pushbutton holes).

A summary of OPTO-TOUCH switch environmental and performance characteristics is shown in Table 13.4. (2)

LIGHT PENS

How Light Pens Function

A light pen is part of an overall system composed of the light pen (Fig. 13.25), the host system to which it is connected, and the software that controls its function.

TABLE 13.4 Summary of OPTO-TOUCH Environmental and Performance Characteristics

Characteristic	Models LTB and OTB specifications
Temperature range (operating)	−4 to +122°F (−20 to +50°C)
Response time of output	80 ms
Ambient light immunity	120,000 lux (direct sunlight)
Output configuration	SPDT electromechanical relay (one NO contact, one NC contact)
Maximum voltage	250 V ac (or 30 V dc)
Maximum current	7 A (resistive load)
Minimum load	100 mA at 24 V
Relay life (mechanical)	50,000,000 operations (minimum)
Relay life (electrical) at full resistive load	100,000 operations (minimum)

Source: Banner Engineering Co. (2).

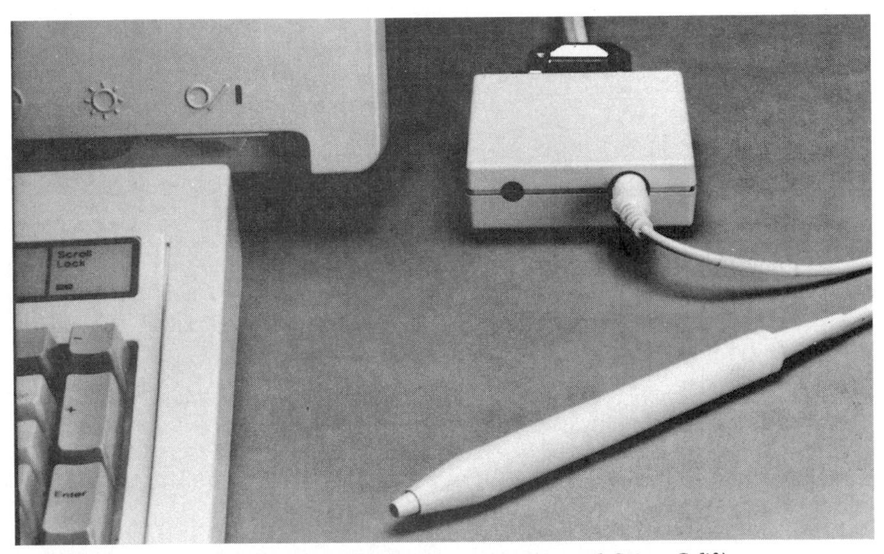

FIGURE 13.25 A light pen. (*Courtesy of Design Technology, El Cajon, Calif.*)

The system hardware itself determines the resolution and accuracy which the system is capable of attaining and varies according to phosphor and monitor type and the light pen's resolution. The software determines the sophistication level and speed at which the light pen may perform. Light pen location information is generated by using a combination of lenses, an aperture, and a photodiode, all contained within the pen itself. The light pen acquires and magnifies the signal produced by the phosphor dots (pixels) as they are being struck by the electron beam on the CRT face (Fig. 13.26). In order to better understand how the light pen functions, let us first examine how a simple black-and-white CRT operates.

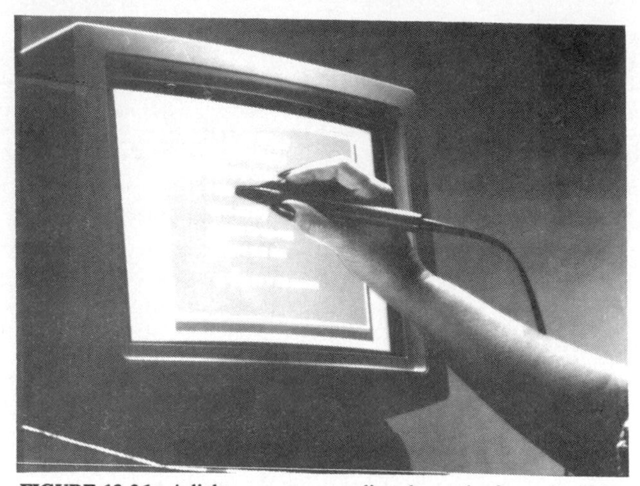

FIGURE 13.26 A light pen operates directly on the face of a CRT display. (*Courtesy of Design Technology, El Cajon, Calif.*)

An electron beam sweeps from left to right on the backside of the CRT face, "painting" the phosphor dots (pixels) that make up the CRT's imaging capability. Each horizontal sweep is called a line, and the total number of lines from top to bottom of the CRT is called a frame. The ultimate resolution of the CRT is determined by the number of pixels per line and the number of lines per frame. Both are controlled by a video generation circuit, made up of two counters, horizontal and vertical, and driven by an oscillator called the dot clock.

The horizontal and vertical counters address the screen random access memory (RAM) and, depending upon the software, determine whether each pixel will be "on" (white) or "off" (black) as the electron beam travels across the screen. The white and black condition on the screen may be varied to "shades of gray" if the system RAM is more than one bit per pixel.

When the electron beam strikes a pixel within the light pen's field of view, the light emitted by the tiny dot of phosphor illuminates the photodiode in the pen, causing an increase in signal current. The output of the photodiode is amplified within the pen by a high-gain, high-frequency amplifier. The amplified signal is compared with a threshold voltage; when the threshold is exceeded, the light pen circuit generates an output pulse.

This output pulse, when gated by the computer, will latch the address information of the screen's horizontal and vertical counters into a buffer so that it may be read by the computer. In this way, the computer determines the location of the pen on the screen.

The type of light pen needed for each CRT is determined by the type of phosphor on the CRT (color, black and white, monochrome) and what function the light pen is required to perform (e.g., single pixel, pick work, character block, menu selection). Electronics for light pens are designed to be suitable for color, monochrome, and long-persistence phosphor (e.g., IBM) CRTs.

Having the shape of a familiar-looking "tool," the light pen is often used as intuitively as a pencil and paper, even by untrained operators. No special hand-eye coordination or computer skills are required in order for an operator to become proficient in the use of the light pen. When the point of interest is located

FIGURE 13.27 This "long-range" light pen operates at a distance from the CRT display. (*Courtesy of Design Technology, El Cajon, Calif.*)

on the CRT face, the operator simply designates that point by pushing the light pen against the CRT face to actuate a push-type switch on the nose tip or actuate a miniature switch on the side of the light pen. Another advantage of the light pen is that no additional desk space is required, as the pen, when not in use, can be attached to a holder on the side of the CRT and removed only when needed. (3)

One unique use for light pen technology is shown in Fig. 13.27. This "long-range" wand uses special optics to permit physically impaired operators to interact with CRT displays from a distance. Table 13.5 shows general electrical and optical characteristics of typical light pens.

How Light Pens Are Constructed

Design Technology, El Cajon, Calif., manufactures a line of light pens designed for use in such high-abuse environments as the CAD/CAM, process control, menu selection, data entry, medical, and gaming industries. Design Technology

TABLE 13.5 General Electro-optical Parameters for Light Pens

Signal output	TTL compatible, norm high, active low
Spectral response	400–1000 nm
Ambient rejection	100 fL
ESD protection	Up to 25 kV

Source: Design Technology (3).

uses SMT techniques to mount the amplifier electronics onto miniature printed wiring boards which are installed inside an injection-molded ABS housing simulating the appearance of a standard fountain pen. Stainless-steel and aluminum housings are also available. The photodiode is securely shock-mounted behind the optical-grade lens in the nose of the pen. Some models feature interchangeable nose tips, ideal for field replacement or multiple applications where changing the nose tip allows one pen to work with both color and monochrome CRTs.

A coiled, straight conductive, armored PVC shielded or shielded straight cable exits from the end of the pen opposite the nose tip. A choice of connectors attached to the interface cable is available, such as modular or d-subminiature. RFI, EMI, and ESD protection is afforded through both case *insulated* and case *isolated* design techniques. (3)

Typical Environmental Specifications

Table 13.6 shows typical environmental specifications for light pens.

TABLE 13.6 Typical Environmental Specifications for Light Pens

Parameter	Specification
Mean time between failures	275,000 h (MIL-HDBK-217D)
Temperature (operating)	+30 to +176°F (0 to +80°C)
Temperature (storage)	−40 to +176°F (−40 to +80°C)
Relative humidity	0 to 95% noncondensing
Shock	MIL-STD-810C
Vibration	MIL-STD-810C
Altitude	0 to 20,000 ft (0 to 6096 m)
Radiated susceptibility	MIL-STD-461A

Source: Design Technology (3).

CUMULATIVE TRAUMA DISORDER (CTD)

Cumulative trauma disorders (CTDs) are injuries caused over a period of time by repeated exposure to stress on particular body parts. The most common CTD to affect machine operators is carpal tunnel syndrome (CTS). Carpal tunnel syndrome is caused by using the fingers while the wrist is in a flexed or extended (other than "neutral") position. The farther the wrist is bent from the neutral position, the more muscle pressure and tendon tension is required to do a given amount of work. (2)

In other times this pain in the hand was called anything from "writer's cramp" to "washerwoman's thumb." Deriving its name from the Greek *karpos,* or wrist, the carpal tunnel is the passageway, composed of bone and ligament, through which a major nerve system of the forearm passes into the hand. These nerves control the muscles in this area, as well as the nine tendons that allow a person's fingers to flex. The wear and tear of repeated movement thickens the lubricating membrane of the tendons and presses the nerves up against the hard bone.[1]. Tendons in the underside of the wrist become compressed and inflamed, resulting in pressure on the median nerve, which supplies feeling in the thumb, index, middle, and ring fingers. Health effects can range from a slight numbness and tingling to severe pain and muscle atrophy. The more tendon tension exerted while the wrist is bent, the more potentially severe are the effects.

Repeated pressing of mechanical pushbuttons has been identified as a contributor to CTS. Splints or gloves designed to prevent CTS by restraining the wrist at a neutral or near-neutral position are sometimes worn by machine operators. An engineer involved in the design of systems requiring an operator to continually actuate pushbuttons, especially those with high actuation force, would be wise to consider touch-type switches to minimize the occurrence of CTS. [Adapted from (2).]

BUILDING A TOUCH APPLICATIONS PROGRAM

The user interface for a given software program can make or break that program in the marketplace. This is particularly true in touch screen–based systems, where the interface and the software appear as one entity to the user. There are some tricks to creating a functional, and logical, touch-based application.

Consider the User

Touch screens are not static input devices which can be used only one way. Because they are software-based, they are as flexible as software itself. Although users of touch screens can span from a three-year-old to an adult, it is possible to separate all users into two categories: trained and untrained. A trained user is anyone who has been taught—even for only a few seconds—how the touch screen interface works. An untrained user is any person who walks up to the touch screen for the first time and must immediately understand how to interact with the system. For this untrained user, such as a person using a touch screen in a public-information kiosk, active touch zones should be obvious and well separated by dead zones. To overcome parallax and calibration problems, which occur because users are all sizes (and can even be caused by one user shifting positions), the touch zones should be large. Placing a visual target in each zone, even if it is just a word inside a box, will help the user touch the correct spot on the touch screen. The *point mode* (which activates immediately upon touch) is best for this type of application, since it is the most obvious and natural way to interact with a touch screen.

Trained users, such as workers in a factory, become proficient with the touch screen interface in a very short period of time. This knowledge is reinforced constantly as they continue to operate familiar processes with the touch screen. Their requirements thus differ significantly from those of the untrained user. More sophisticated, mouselike techniques can be used. It is not necessary to separate touch zones with dead zones; in fact, when the finger is sliding over menu selections, the

[1]Excerpted from the University of California, Berkeley Wellness Letter, © Health Letter Associates, 1991.

active zones should be right next to each other. Just like a mouse-based application, the finger can be used to touch a selection at the top of a screen, which will cause a pull-down menu to be displayed; then the finger can slide to the menu choice. As soon as the finger lifts off the menu choice, the command is activated. Obviously, in the case of the trained user, the *stream mode* (continuous tracking of the finger's position) is more suitable than the point mode.

Tracking the finger's position with highlighting or reverse video minimizes calibration problems because the mind focuses on the highlighted item rather than the position of the finger, just as it does when a mouse is being used.

Provide Feedback

Both the trained and untrained user *must* receive feedback, either visually or audibly, immediately upon touch. Selections can be highlighted with reverse video or a color change. Beeps can indicate that a valid or invalid choice has been made. Feedback is essential because touch screens lack the tactile response typical of keyboards or mice.

If no feedback occurs when the touch is activated, the user will assume the touch has not "taken" and will continue to press on the screen. Feedback of some kind assures the user that the touch has registered and gives the system the time it needs to present the next screen.

Consider the Environment

Is the touch screen kiosk going to be located in a noisy factory (Fig. 13.28) or in a quiet office? Audible beeps built into the system as a feedback mechanism will be heard in the quiet environment but possibly not in the noisy environment.

Is the kiosk going to be located in a public place? Is vandalism a concern? When considering which type of touch screen technology to use, remember that

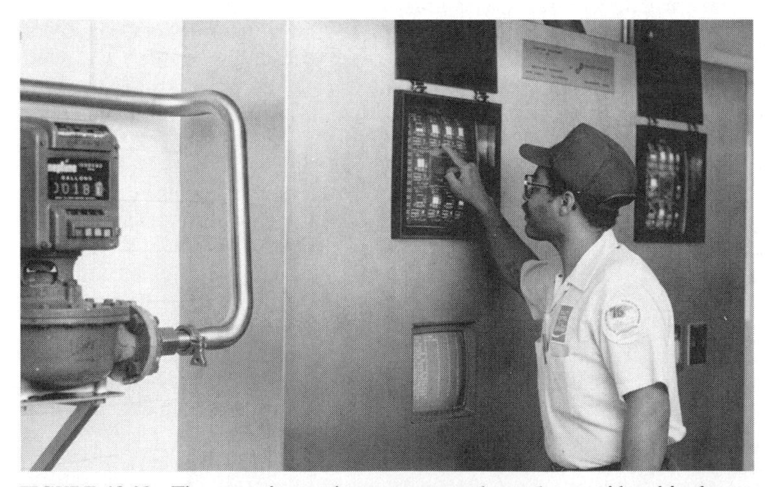

FIGURE 13.28 The operating environment must always be considered in the selection of touch screen technology. (*Courtesy of Elographics, Oak Ridge, Tenn.*)

an uncoated glass surface provides the best defense against vandalism. Surface acoustic wave and infrared touch screens meet this criterion.

Make It Bright and Colorful

Bright, varied colors make any application more attractive. However, in a touch screen–based system, they also provide some additional benefits.

All monitors have a reflection; it is most noticeable when the monitor is turned off. Some touch screens increase the reflection of the monitor's surface. Fortunately, the human eye focuses on the image displayed rather than the reflection in the screen, just as it does when looking out a window. By keeping the images bright and interesting (Fig. 13.29), the designer can help the user see the application without the distraction of reflections. Bright colors also make fingerprints "disappear." (1) The system designer should exercise some degree of caution in the selection of colors for touch-screen applications, particularly if the color is to highlight a critical touch selection by the operator. The population, in general, has preconceived notions, established by convention, previous usage, or observation, that certain colors are reserved for specific applications. For example, red is a color usually perceived as an alert "flag" to stop an operation or signal that a portion of the system is inoperative. Yellow is often perceived as a cautionary alert or advice to the operator that a marginal condition exists. The importance of understanding the "audience" who will use the touch screen and display should not be minimized, particularly if the system is meant for use by military operators. In fact, MIL-STD-1472, the Department of Defense document

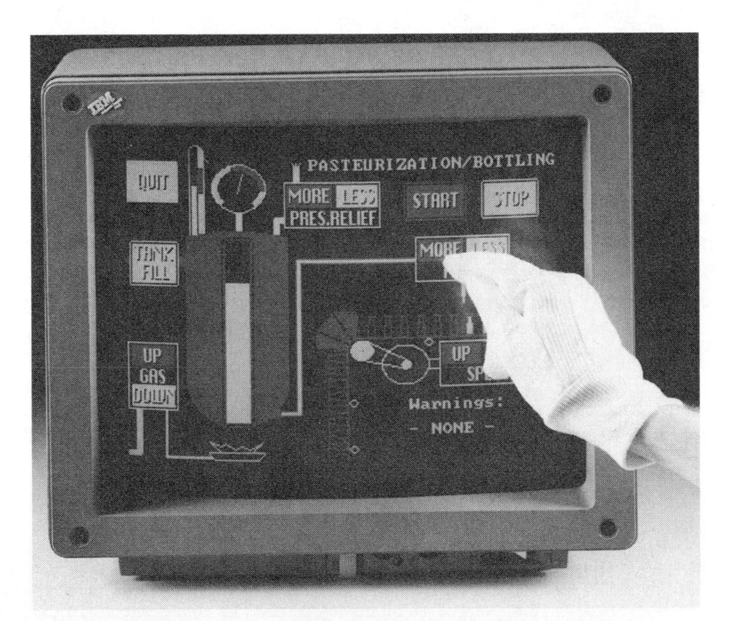

FIGURE 13.29 Bright, colorful images on the display enhance touch screen "friendliness." (*Courtesy of Elographics, Oak Ridge, Tenn.*)

concerning human engineering design criteria for military systems, specifically defines color coding of displays as follows:

a. RED shall be used to alert an operator that the system or any portion of the system is inoperative, or that a successful mission is not possible until appropriate corrective or override action is taken ("error," "no-go," "failure," "malfunction," etc.).
b. FLASHING RED shall be used only to denote emergency conditions which require operator action to be taken without undue delay, or to avert impending personnel injury, equipment damage, or both.
c. YELLOW shall be used to advise an operator that a condition exists which is marginal. YELLOW shall also be used to alert the operator to situations where caution, recheck, or unexpected delay is necessary.
d. GREEN shall be used to indicate that the monitored equipment is in tolerance or a condition is satisfactory and that it is all right to proceed (e.g., "ready," "go ahead," "in tolerance").
e. WHITE shall be used to indicate system conditions that do not have "right" or "wrong" implications, such as alternative functions or transitory conditions, provided such indication does not imply success or failure of operations.[2] (10)

This is not to imply that the system designer should select colors from MIL-STD-1472 or even try to design a commercial touch screen and display application using military standards. But it *is* prudent to consider the potential operators of the system, whether the system will ever be operated in a high-stress environment (in which an operator might make a touch choice not by the alphanumeric information on the display but by color choice only!), or even the potential for color blindness in the operator population (*especially* if the touch choices are keyed to color perception only).

GLOSSARY

Ambient light. The level of visible and invisible light in the area surrounding the touch system.

Analog resistive touch screen. One of two types of resistive membrane touch screen. Analog touch screens measure resistance levels in both *X* and *Y* axis to determine accurate position information on touch location.

Application program. A program which provides the interface between the user and the computer system. The program typically presents displays, accepts input, and processes data. The touch system is treated as an additional source of input for the application program.

Bezel. The front panel surrounding the display screen. In standard infrared (IR) systems, the bezel is made of plastic or a plastic-based material which allows transmission of infrared beams.

[2] MIL-STD-1472, "Human Engineering Design Criteria for Military Systems, Equipment and Facilities," published by the Department of Defense, is an excellent reference source for the system or hardware designer who faces design decisions based on the integration of a human operator into the system, subsystem, or equipment. This source enumerates, in detail, the capabilities and limitations of the human part of the "loop." Another excellent resource for human factors information is Woodson and Conover's "Human Engineering Guide for Equipment Designers" (see bibliography).

Capacitance. The property of a capacitor which determines how much charge can be stored in the capacitor. Equal to the quantity of stored charge divided by the voltage across the device when the charge was stored.

Capacitive overlay. An overlay system that uses human body capacitance (actual electrical potential) to trigger a touch. The capacitive touch screen has timing circuits on a glass surface connected to conductive leads with a clear metal coating. When a finger touches the screen, the additional body capacitance causes the frequency at which the timing circuit oscillates to change. The difference between these frequencies is measured and the location of touch is determined.

Continuous mode. An operating mode in which coordinates are transmitted as long as a stylus is in the touch-active area. The stylus does not have to be moving to be detected. Also referred to as *Stream.*

Controller. The component of the touch system which includes the electronics to interface with the sensor unit and the firmware to process the touch location. It also handles the communications protocol to interface to the host.

Coordinate. A set of numbers which represent a position on the *X* and *Y* axes. These numbers identify the location of a stylus within the touch-active area.

Coordinate report. A report from the touch controller to the host system indicating the location of a stylus within the touch-active area.

Dead zone (see *Guard band*). Also referred to as dead band. An area around a touch target which the touch system does not report to the touch.

Detector (see *Phototransistor*). A device which receives light signals from emitters.

Digital resistive touch screen. One of two types of resistive membrane touch screen. Digital touch screens use perpendicular conductive layers to determine an *X-Y* coordinate at the location of the touch.

Display. The surface of the screen, or the entire unit itself. For example, monitors and terminals are referred to as displays.

Driver module. A program which handles communication between the touch system and the application program.

EIA-232-C. The industry standard serial interface between data terminal equipment and data communications equipment employing serial binary data interchange.

Emitter (see *LED*). Device which radiates light when a current is passed through it.

Enter-point mode. An operating mode in which coordinates are reported only when a stylus enters the touch-active area. No additional reports are transmitted until an empty touch-active area is detected, indicating that the stylus has left the touch-active area. Also called *single point mode.*

Exit-point mode. Also called exit or *untouch* mode. An operating mode in which touch coordinates are transmitted only when the stylus leaves the target area. May also be used in conjunction with other operating modes.

Firmware. A program or instruction set that has been stored in a read only memory (ROM) device.

Guard band. See *Dead zone.*

Header code. An ASCII character that directly precedes touch data to differentiate them from terminal data when touch and terminal data are transmitted over the same EIA-232-C line.

Infrared (IR). A part of the light spectrum that cannot be seen by the naked eye. Designates those wavelengths just beyond the red end of the visible spectrum.

Input device. Method used to enter data into a computer such as a keyboard, touch system, light pen, mouse, or joy stick.

Interface. The mechanical or electrical boundary where two pieces of hardware meet. Also, the way in which two pieces of equipment communicate with each other electrically, including signal levels and connector pin assignments as well as software protocols.

LED (see *Emitter*). Light-emitting diode. A solid-state, semiconductor device that emits light when current passes through it.

Microprocessor. A large-scale integrated device that performs central processing unit functions. Typically packaged in a single chip.

Operating mode. A set of conditions for the transmission of touch coordinates. For example, in one mode, a set of coordinates is sent only when a stylus enters the touch-active area; in another mode, coordinates are sent continuously as long as a stylus is present.

Optoelectric (optoelectronic). Refers to an IR touch screen or the entire touch screen system.

Optomatrix. The grid of light beams produced by the touch frame assembly.

Overlay. A touch input system which mounts directly over a display system. Composed of sheets of polyester film or film and glass that senses the presence of a stylus within the touch-active area, using pressure either by a switch closure or by changes in resistance or capacitance.

Parameter. A characteristic of an operating condition that is determined by a set value.

Phototransistor (see *Detector*). Semiconductor device which converts light energy into electrical energy.

Piezoelectric. Piezoelectric touch technology uses a glass plate over the surface of the display. Pressure-sensitive crystals are embedded under the four corners of the glass. The location of the touch is determined by measuring and comparing the different pressures at the four corners.

Pixel. Picture element. The smallest area or dot on a video display that can be accessed individually.

Point mode. The touch mode in which the touch screen activates immediately upon touch.

Point-of-operation guarding device. Electronic or mechanical devices designed to prevent dangerous operation of a machine.

Resistive overlay. A resistive overlay system consists of a bottom layer of glass (or plastic) and a top layer of plastic, separated by tiny plastic insulators. Both layers are coated with a metallic compound. When the two layers are brought into contact, a switch closure results.

Resolution. Touch system resolution refers to the number of touch-active points on each axis of the touch-active matrix. Display resolution is the number of individually addressable picture elements (pixels) on each axis of the active display area.

Response. The time required for the touch system to transmit the touch location to the host such that the next display appears in an acceptable time frame to the user.

Scan rate. The rate at which the optical devices of an infrared touch system are pulsed.

Sensor unit. The part of the touch system that "senses" the touch target position. The sensor unit can be resistive overlay, capacitive overlay, surface acoustic wave overlay, piezoelectric, or scanning infrared.

Single-point mode. See *Enter-point mode.*

Stream. See *Continuous mode.*

Stylus. Any object or instrument used to activate a touch system, such as finger, pen, or conductive rod.

Surface acoustic wave (SAW). A surface acoustic wave transmits surface acoustic waves through a glass substrate. When the substrate is touched, the user's finger causes a change in the energy measured by the system, and the location of the touch is detected.

Target. An area of the display screen which is designated as a touch area in order to accomplish a designated task. The coordinates of this area are then recognized by the application program.

Touch-active area. Positions on the touch screen where touches are recognized by the application software.

Touch controller. See *Controller.*

Touch screen. Any touch-activated switch array designed to be installed in front of a display.

Trailer codes. An ASCII character that directly follows touch data to differentiate them from terminal data sent when touch and terminal data are transmitted over the same EIA-232-C line. See *Header code.*

Untouch. See *Exit-point mode.*

***X* coordinate.** The position along the horizontal axis of the display where a touch occurs.

***Y* coordinate.** The position along the vertical axis of the display where a touch occurs.

REFERENCES

1. "New Tools Make It Easier to Build Touch into a System," Elographics, Oak Ridge, Tenn., 1990.
2. "OPTO-TOUCH Optical Touch Buttons, Brochure No. 28211/28437," Banner Engineering Co., Minneapolis, Minn., 1990.
3. "Design Technology Takes You Back to the Future," Design Technology, El Cajon, Calif., 1991.
4. "Digital Transparent Touch Systems," Transparent Devices, Inc., Newbury Park, Calif., 1988.
5. "Touch Handbook," Carroll Touch, Round Rock, Tex., 1989.
6. "Touch Screen Display, Series 4284," Industrial Electronic Engineers, Inc., Van Nuys, Calif., 1991.
7. "M4ST Sealtouch Graphic Display Module," Lucas Deeco Corporation, Hayward, Calif., 1991.
8. "Model ST2200 Sealed VT Emulation with Touch Screen," Lucas Deeco Corporation, Hayward, Calif., 1991.
9. Platshon, Mark, "Acoustic Touch Technology Adds a New Input Dimension," *Computer Design,* March 1988.
10. "MIL-STD-1472D, Human Engineering Design Criteria for Military Systems, Equipment and Facilities," U.S. Department of Defense, 1989.

CHAPTER 14

PHOTOELECTRIC SENSORS AND PROXIMITY SWITCHES*

A *photoelectric sensor* is an electrical device which detects a visible or invisible beam of light and responds to a change in light intensity received by the sensor. The optical system of a photoelectric sensor is designed for one of four basic *sensing modes: opposed, retroreflective, diffuse,* or *convergent.* Most sensing situations may be solved by applying a choice of these modes; however, there is usually a "best" mode for each set of sensing variables.

In the opposed mode sensing mode, the emitter and the receiver are positioned opposite to each other so that the light from the emitter shines directly at the receiver. An object then breaks the light beam which is established between the two. The retroreflective sensing mode is also called the "reflex" or "retro" mode.

A retroreflective sensor contains both the emitter and the receiver. A light beam is established between the sensor and a special retroreflective target, called a *retroreflector.* As in opposed sensing, an object is sensed when it interrupts this beam.

Like retroreflective, diffuse sensors contain both the emitter and the receiver. Diffuse sensing is synonymous with photoelectric "proximity" sensing. In the diffuse sensing mode the emitted light strikes the object at some arbitrary angle and the light is then diffused from the surface at all angles. The receiver captures some small percentage of the diffused light. In diffuse sensing the object to be detected, itself, becomes the reflective surface. The object actually "makes" the beam, instead of breaking the beam.

Convergent-beam sensing is a special variation of the diffuse mode. Using additional optics, a very small, intense, and well-defined image is produced at a fixed distance from the front surface of the sensor. Like diffuse sensors, a convergent-beam sensor receives light which is reflected directly back from an object.

A sensor operating in the *proximity sensing mode* senses the presence of an object in close proximity to the device. Inductive proximity switches sense metallic objects and are normally used to perform logic functions in machine control applications. Capacitive proximity switches can sense both metallic and nonmetallic targets but are most often used for product detection, sensing whether a particular product or level of a product is present.

*Numbers in parentheses indicate items in the References at the end of this chapter.

PHOTOELECTRIC SENSORS

How Photoelectric Sensors Function

Sensing with light beams has been popular since the 1950s. For many years, a typical photoelectric sensing system consisted, in part, of an incandescent lamp and a light-sensitive resistive device called a *photocell.* The filament of the bulb was carefully placed exactly at the focus of a lens so that its light could be projected across the sensing area to the photocell (Fig. 14.1). The photocell was positioned carefully at the focus of its own lens. It was important that the photocell receive as much light as possible from its incandescent light source so that it could differentiate its beam from all the other light which its lens gathered in.

Early photoelectric designs had some basic shortcomings. One was bulb burnout. Incandescent bulbs lost their intensity and eventually failed. Filaments would sag and fall away from the lens focus. Incandescent lamps could not survive temperature extremes or high vibration. Another problem was critical sensor alignment. Lenses with very long focal lengths and narrow beam angles were needed in order to get enough light energy from the light bulb to its photocell. All these problems quickly disappeared with the availability of efficient LEDs in the 1970s.

An LED is a solid-state semiconductor which emits a small amount of light when current flows through it in the forward direction. LEDs can be manufactured to emit green, yellow, red, or infrared light. Infrared light is invisible to the human eye, as shown in Fig. 14.2. Because LEDs are solid-state, they will last for the entire useful life of the sensor. Since the LED will not need replacement for the life of a sensor, the circuitry of sensors may be totally encapsulated, making them far more rugged and reliable than their former incandescent equivalents. LEDs are not sensitive to vibration and shock and can handle a wide temperature

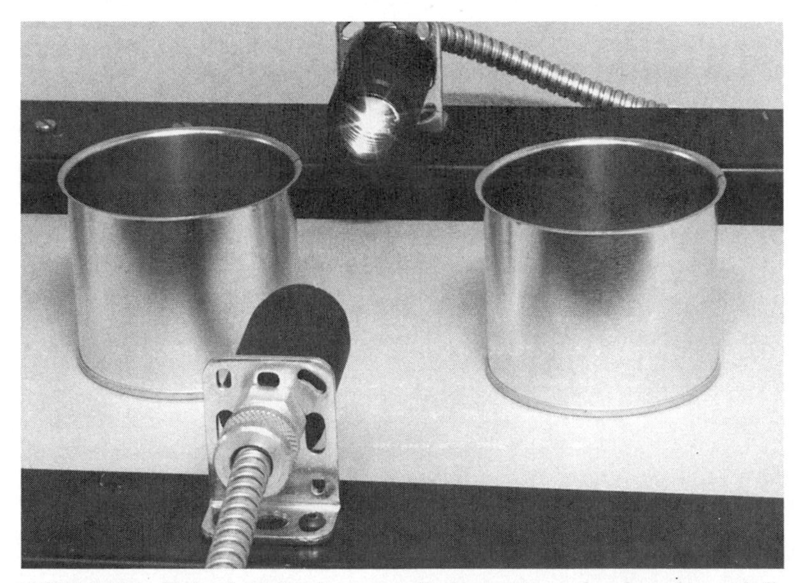

FIGURE 14.1 Early photoelectric sensors. Early sensors used an incandescent bulb as the light source. *(Courtesy of Banner Engineering, Minneapolis, Minn.)*

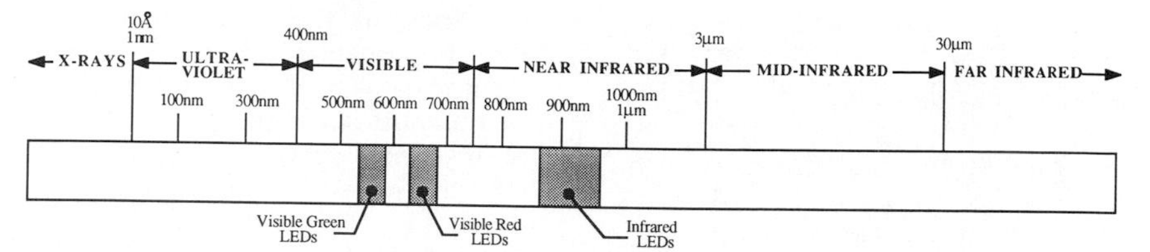

FIGURE 14.2 The light spectrum. (*Courtesy of Banner Engineering, Minneapolis, Minn.*)

range. There is, however, a trade-off in the area of light intensity; LEDs produce only a small percentage of the light generated by an incandescent bulb of the same size. Infrared types are the most efficient LED light generators and were the only type of LED offered in photoelectric sensors until 1975. However, this *invisible* infrared light, though ideal for security and film processing applications, was initially not well received by those accustomed to *visually* aligning and checking incandescent emitters.

By 1970, photoelectric sensor designers had recognized that LEDs had a benefit much more profound than long life. Unlike their incandescent equivalents, LEDs can be turned "on" and "off" (or modulated) at a high rate of speed, typically at a frequency of several kilohertz (Fig. 14.3). This modulating of the LED means that the amplifier of the *phototransistor* receiver can be "tuned" to the frequency of modulation and will amplify *only* light signals pulsing at that frequency. This is analogous to the transmission and reception of a radio wave of a particular frequency. The modulated LED light source of a photoelectric sensor is usually called the transmitter (or emitter), and the tuned photodevice is called the receiver. Phototransistors have prevailed, over *photodiodes* and photo-

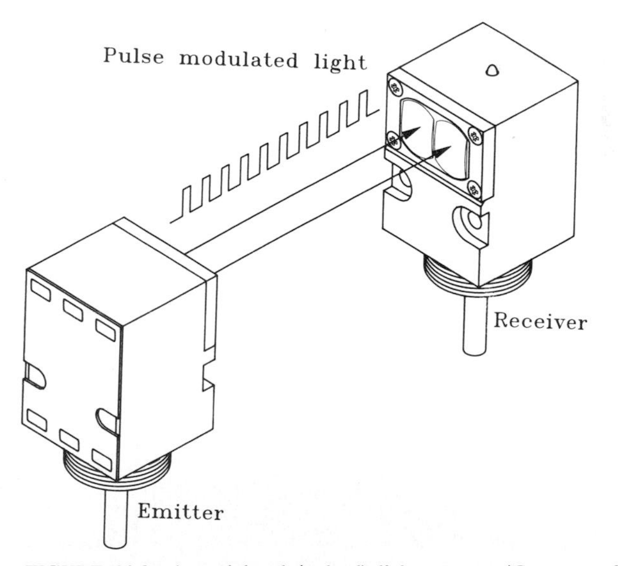

FIGURE 14.3 A modulated (pulsed) light source. (*Courtesy of Banner Engineering, Minneapolis, Minn.*)

FIGURE 14.4 Typical photocell (left) and phototransistor (right). (*Courtesy of Banner Engineering, Minneapolis, Minn.*)

darlingtons, as the most widely used receiver optoelement in industrial photoelectric sensor design, offering the best trade-off between light sensitivity and response speed compared with photoresistive and other photojunction devices (Fig. 14.4). Photocells are used whenever greater sensitivity to visible wavelengths is required, as in some color registration and ambient light detection applications. Photodiodes are generally reserved for applications requiring either extremely fast response time or linear response over several magnitudes of light-level change.

There is a common misconception that because an infrared LED system is invisible, it must therefore be powerful. The apparent high level of optical energy in a modulated photoelectric sensing system has, in itself, little to do with the wavelength of the LED. Remember that an LED emits only a fraction of the light energy of an incandescent bulb of the same size. It is the modulation of an LED sensing system that accounts for its power.

The gain of a nonmodulated amplifier is limited to the point at which the receiver recognizes ambient light. A nonmodulated sensor may be powerful only if its receiver can be made to "see" only the light from its emitter. A nonmodulated sensor may be powerful only if its receiver can be made to "see" only the light from its emitter. This requires the use of lenses with very long focal length and/or mechanical shielding of the receiver lens from ambient light. In contrast, a modulated receiver ignores ambient light and responds only to its modulated light source. As a result, the gain of a modulated receiver may be turned up to a very high level (Fig. 14.5).

There is, however, a limit to a modulated sensor's immunity to ambient light. Extremely bright ambient light sources may sometimes present problems. No modulated photoelectric receiver will function normally if it is pointed directly into the sunlight. If you have ever focused sunlight through a magnifying glass onto a piece of paper, you know that you can easily focus enough energy to start the paper on fire. Replace the magnifying glass with a sensor lens, and the paper with a phototransistor, and it becomes easy to understand why the receiver shuts down when the sensor is pointed directly at the sun. This is called ambient light saturation.

Infrared LEDs were found to be the most efficient types and were also the best spectral match to phototransistors (Fig. 14.6). However, photoelectric sensors used to detect *color differences* (as in color registration sensing applications)

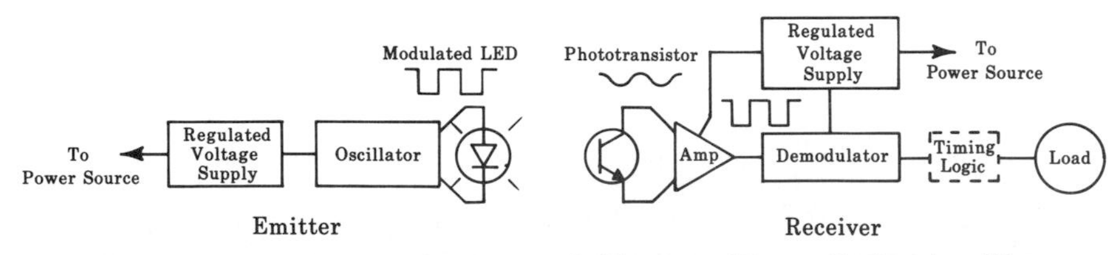

FIGURE 14.5 A modulated photoelectric control. (*Courtesy of Banner Engineering, Minneapolis, Minn.*)

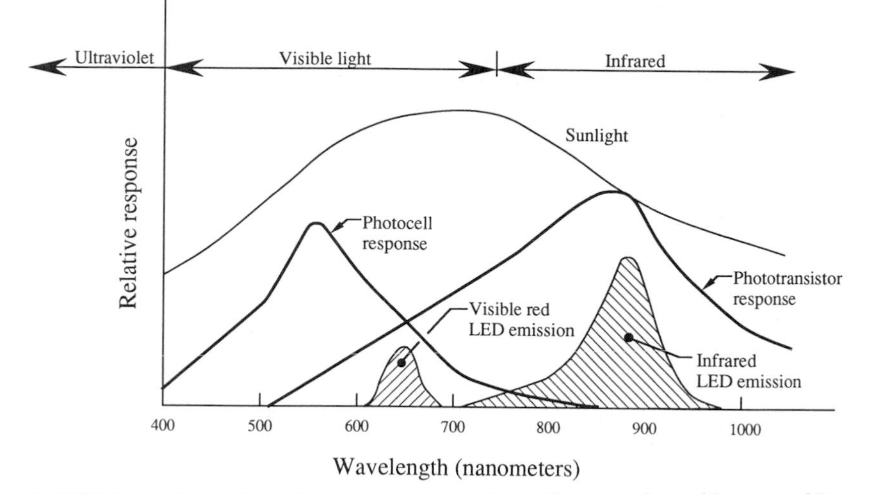

FIGURE 14.6 Comparison of spectral response: photocell vs. transistor. (*Courtesy of Banner Engineering, Minneapolis, Minn.*)

require a visible light source. As a result, color sensors continued to use photocell receivers and incandescent lamps while sensor designers awaited the development of more efficient visible LEDs. Today, with the advent of improved visible LEDs, most color registration sensors are modulated and utilize colored LEDs as emitters.

Sensing Modes. The optical system of any photoelectric sensor is designed for one of three basic sensing modes: opposed, retroreflective, or proximity. The photoelectric *proximity sensing mode* is further divided into four submodes: diffuse proximity, divergent-beam proximity, convergent-beam proximity, and *background suppression* proximity.

Opposed Mode. Opposed mode scanning is often referred to as "direct scanning" and is sometimes called the "beam-break" mode. In the opposed mode, the emitter and receiver are positioned opposite each other so that the sensing energy from the emitter is aimed directly at the receiver. An object is detected when it interrupts the sensing path established between the two sensing components (Fig. 14.7). Opposed mode sensing was historically the first photoelectric sensing mode. In the early days of nonmodulated photoelectrics, problems of difficult emitter-receiver alignment gave the opposed mode a bad reputation that still exists to some extent. With today's high-powered modulated designs, however, it is very easy to align most opposed mode photoelectric sensors. Alignment of a sensor means positioning the sensor(s) so that the maximum amount of emitted energy reaches the receiver sensing element. In opposed sensing, this means that the emitter and the receiver are positioned relative to each other so that the radiated energy from the emitter is centered on the field of view of the receiver.

Sensing range is specified for all sensors. For opposed sensors, range is the maximum operating distance between the emitter and the receiver. A sensor's *effective beam* is the working part of the beam: it is the portion of the beam that

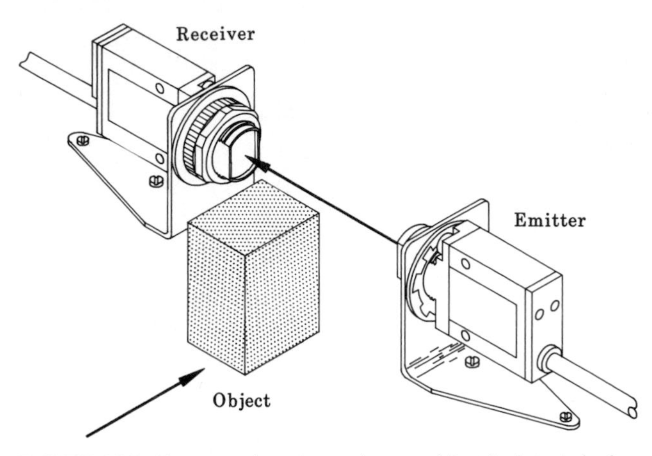

FIGURE 14.7 In opposed mode sensing, an object is detected when it interrupts the sensing path established between the two sensing components. (*Courtesy of Banner Engineering, Minneapolis, Minn.*)

must be completely interrupted in order for an object to be reliably sensed. As shown in Fig. 14.8, the effective beam of an opposed mode sensor pair may be pictured as a rod that connects the emitter lens to the receiver lens. This rod will be tapered if the two lenses are of different sizes. The effective beam should not be confused with the actual radiation pattern of the emitter or with the field of view of the receiver.

The effective beam size of a standard opposed mode photoelectric sensor pair may be too large to detect small parts or inspect small profiles, or for a very accurate position sensing. In such cases, opposed mode photoelectric sensor lenses can usually be *apertured* to reduce the size of the effective beam. Creating an aperture can be as easy as drilling a hole or milling a slot in a thin metal plate and locating the plate directly in front of the lens, with the opening on the lens centerline. When selecting an aperture material, it is important to remember that the powerful beam of many modulated opposed mode photoelectric sensors can actually penetrate many nonmetallic materials to varying degrees.

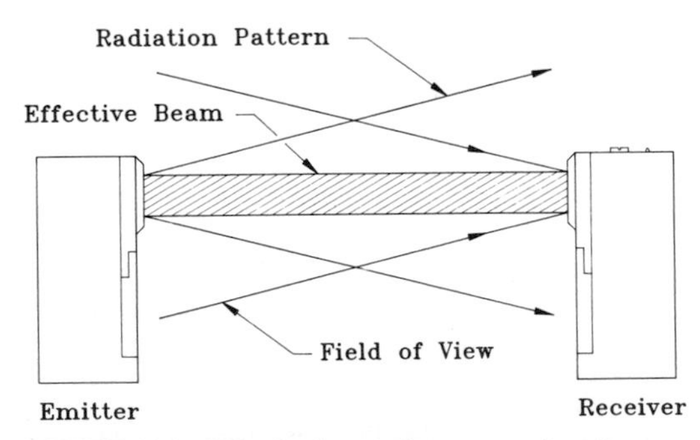

FIGURE 14.8 Effective beam of an opposed mode sensor pair. (*Courtesy of Banner Engineering, Minneapolis, Minn.*)

Apertures reduce the amount of light energy that can pass through a lens by an amount equal to the lens area reduction. For example, if a 1-in (25,4-mm) beam is apertured down to ¼ in (6,3 mm), the amount of optical energy passing through the apertured lens is equal to $(1/4)^2$ = 1/16th the amount of energy through the 1-in lens. *This energy loss is doubled if apertures are used on both the emitter and the receiver.* If the object to be detected will always pass very close to either the emitter or the receiver, an aperture may be required on only one side of the process. In this case, the size of the effective beam is equal to the size of the aperture on the apertured side and uniformly expands to the size of the lens on the unapertured side. The effective beam is therefore cone-shaped.

The goal in any application requiring the detection of small parts in an opposed beam is to size and shape the effective beam to be smaller than the smallest profile that will ever need to be detected, while retaining as much lens area as possible. Often the easiest way to size and shape an effective beam to match a part profile is to use a glass fiber-optic assembly that has its sensing end terminated in the desired shape. Fiber-optic assemblies for sensor applications are discussed in detail later in this chapter.

The very high power of some modulated LED opposed sensor pairs (especially when used at close range) can create a "flooding" effect of light energy around an object that is equal to or even slightly larger than the effective beam. This is another reason to ensure that the size of the effective beam is always smaller than the profile of the object to be detected.

Retroreflective Mode. The photoelectric retroreflective sensing mode is also called the "reflex" mode, or simply the "retro" mode (Fig. 14.9).

A retroreflective sensor contains both the emitter and receiver circuitry. A light beam is established between the emitter, the retroreflective target, and the receiver. Just as in opposed mode sensing, an object is sensed when it interrupts the beam.

Retroreflective range is defined as the distance from the sensor to its retroreflective target. The effective beam is usually cone-shaped and connects the periphery of the retrosensor lens (or lens pair) to that of the retroreflective

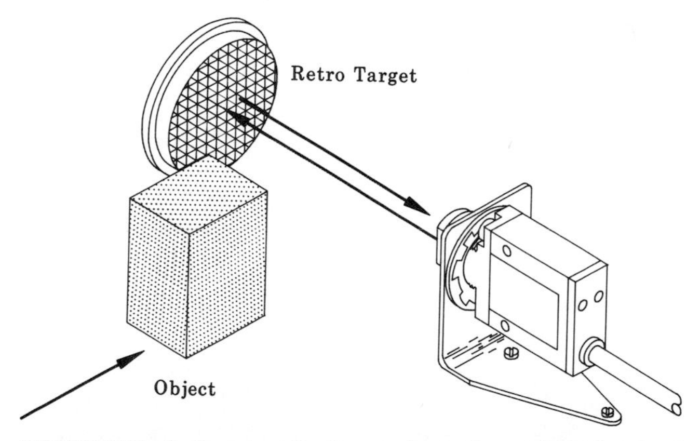

FIGURE 14.9 In the retroreflective sensing mode, an object is sensed when it interrupts a light beam established between the emitter, the retroreflective target, and the receiver. (*Courtesy of Banner Engineering, Minneapolis, Minn.*)

target. The exception to this is at close range, where the size of the retrobeam has not expanded enough to at least fill the target area.

Retroreflective targets are also called "retroreflectors" or "retrotargets." Most retroreflective targets are made up of many small corner-cube prisms, each of which has three mutually perpendicular surfaces and a hypotenuse face. A light beam that enters a *corner-cube prism* through its hypotenuse face is reflected by the three surfaces and emerges back through the hypotenuse face parallel to the entering beam. In this way, the retroreflective target returns the light beam to its source. Because of this feature, corner-cube plastic retroreflectors are commonly used for highway markers and vehicle safety reflectors. A clear glass sphere also has the ability to return a light beam back to its source, but a coating of glass beads is not as efficient a reflector as is a molded array of corner cubes.

A single mirrored surface may also be used with a retroreflective sensor. Unlike the corner cube, however, light striking a flat mirror surface is reflected at an angle that is equal and opposite to the angle of incidence (Fig. 14.10). This is called specular reflection. In order for a retroreflective sensor to "see" its light reflected from a flat mirrored surface, it must be positioned so that its emitted beam strikes the mirror exactly perpendicular to its surface. A retroreflector, on the other hand, is very forgiving and returns incident light to its source at angles up to about 20° from the perpendicular. This property makes the retroreflective sensors much easier to align to their retrotargets.

A good retroreflector returns about 3000 times as much light to its sensor as does a piece of white typing paper. This is why it is easy for a retroreflective sensor to recognize only the light returned from its retroreflector. If the object that is to interrupt a retroreflective beam is *itself* highly reflective, however, it is

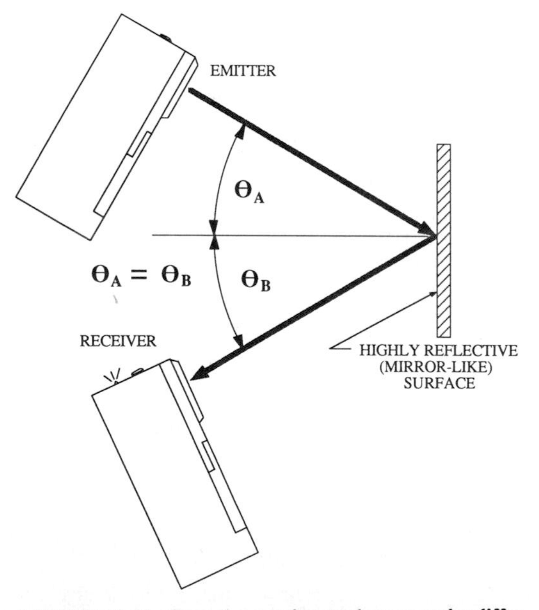

FIGURE 14.10 Specular sensing mode senses the difference between shiny and dull surfaces. (*Courtesy of Banner Engineering, Minneapolis, Minn.*)

possible for the object to slip through the retroflective beam without being detected. This retroreflective sensing problem is called *proxing,* and relatively simple methods exist to deal with it.

If a shiny object has flat sides and passes through a retroreflective beam with a predictable orientation, the cure for proxing is to orient the beam so that the object's specular surface reflects the beam away from the sensor. This is called scanning at a skew angle to the object's surface. The skew angle only needs to be 10 to 15° (or more) to be effective. This solution to proxing may, however, be complicated if the shiny object has a rounded (radiused) surface or if the object presents itself to the beam at an unpredictable angle. In these cases, the best mounting scheme, although less convenient, has the beam striking the object at *both* a vertical and horizontal skew angle.

With recent improvements in LED technology, the use of visible light LEDs as photoelectric emitters has increased. When equipped with a visible emitter, a retrosensor may be aimed like a flashlight at a retroreflective target. When the reflection of the beam is seen on the retroreflector, correct alignment is assured. This principle is also of benefit when a visible emitter is used in an opposed mode photoelectric system. A retrotarget is placed directly in front of the lens of the receiver, and the emitter is aligned by sighting the visible beam on the target. The retrotarget is then removed, and the emitter and receiver orientations are "fine-tuned" for optimum alignment.

Polarizing filters, when used on *visible* retroreflective sensors, can significantly reduce the potential for proxing. A polarizing filter is placed in front of both the emitter and receiver lenses, with the filters oriented so that the planes of polarization are at 90° to one another. When the light is emitted, it is polarized "vertically." When the light reflects from a corner-cube retrotarget, its plane of polarization is rotated 90°, and only the polarized target–reflected light is allowed to pass through the polarized receiver filter and into the receiver. When the polarized emitted light strikes the shiny surface of the object being detected, its plane of polarization is *not* rotated, and the returned nonpolarized beam is blocked from entering the receiver. This scheme is very effective for eliminating proxing. One word of caution, however: like a good pair of sunglasses, polarizing filters reduce the amount of optical power available in a retrobeam by more than 50 percent. This is an especially important consideration whenever the environment is very dirty or where sensing range is long. In addition, polarized retrosensors work *only* with corner-cube retroreflective materials.

Proximity Mode. Proximity mode sensing involves detecting an object that is directly in front of a sensor by detecting the sensor's own transmitted energy back from the object's surface. For example, an object is sensed when its surface reflects a sound wave back to an ultrasonic proximity sensor. Both the emitter and receiver are on the same side of the object, usually together in the same housing. In proximity sensing modes, an object, when present, actually "makes" (establishes) a beam rather than interrupts the beam. Photoelectric proximity sensors have several different optical arrangements, and each has its own sensing mode designation: diffuse, divergent-beam, convergent-beam, and background-suppression.

Diffuse mode sensors are the most commonly used type of photoelectric proximity sensor. In the diffuse sensing mode, the emitted light strikes the surface of an object at some arbitrary angle. The light is then diffused from that surface at many angles (Fig. 14.11). The receiver can be at some other arbitrary angle, and some small portion of the diffused light will reach it. Generally speaking, the diffuse sensing mode is an inefficient mode, since the receiver looks for a relatively

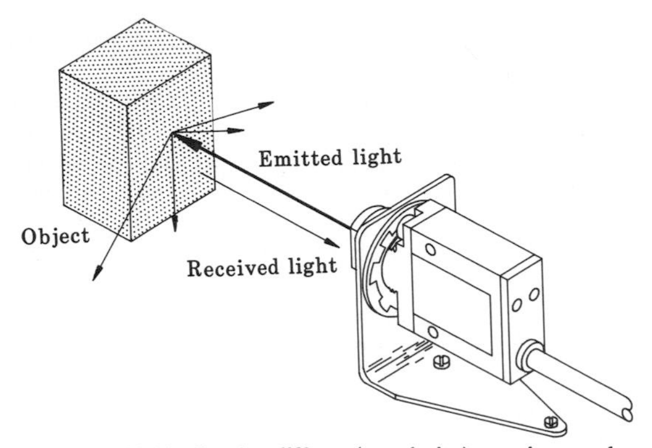

FIGURE 14.11 In the diffuse (proximity) sensing mode, emitted light strikes an object at an arbitrary angle, is diffused, and a small portion of the diffused light reaches the receiver. (*Courtesy of Banner Engineering, Minneapolis, Minn.*)

small amount of light that is bounced back from a surface. Also, the diffuse mode, like the other proximity sensing modes, is dramatically influenced by the reflectivity of the surface being sensed. A bright white surface will be sensed at a greater range than a dull black surface. The specified sensing range of a diffuse sensor is referenced to a material which has about the reflectivity of white paper. The actual material to be sensed may be either more or less reflective than white paper and, as a result, may be sensed at either a longer or a shorter range (see Table 14.1, p. 14.18).

Most diffuse mode sensors use lenses to collimate emitted light rays and to gather in more received light. While lenses help a great deal to extend the range of diffuse sensors, they also increase the criticality of the sensing angle to a shiny or glossy surface. Because all such surfaces are mirrorlike to some degree, the reflection is more specular than diffuse. Most diffuse sensors can guarantee a return light signal only if the shiny surface of the material presents itself perfectly parallel to the sensor lens. This is usually not possible with radiused parts like bottles or shiny cans. It is also a concern when detecting webs of metal foil or poly film where there is any amount of web "flutter."

In addition to assessing the reflectivity of the object to be detected by a diffuse sensor, it is just as important to consider the reflectivity of any background object which lies in the scanning path of the sensor. Locating a background object beyond the specified range of the sensor does not guarantee that it will not be detected, especially if the background object is more reflective than white paper. The worst situation involves a nearby background material with a reflectivity which approaches (or exceeds) the reflectivity of the object to be sensed. A good rule of thumb is to locate the sensor so that the nearest background object is at least *3 times* the distance from the diffuse sensor to the object to be sensed. Diffuse sensing can be a highly reliable sensing mode for presence sensing if the details of the sensing conditions are carefully evaluated.

Divergent mode sensors are devices in which the collimating lenses have been eliminated, making the sensor ideal for short-range applications. The lensless sensor, of course, has a shortened sensing range, but it is much less dependent upon the angle of incidence of its light to a shiny surface that does fall within its reduced range (Fig. 14.12).

The range of any proximity mode sensor also may be affected by the size and profile of the object to be detected. A large object that fills the sensor's beam area will return more energy to the receiver than a small object that only partially fills the beam. A divergent sensor responds better to objects within about 1 in (25,4 mm) of its sensing elements than does a diffuse mode sensor. As a result, divergent mode sensors can successfully sense objects with very small profiles, like yarn or wire.

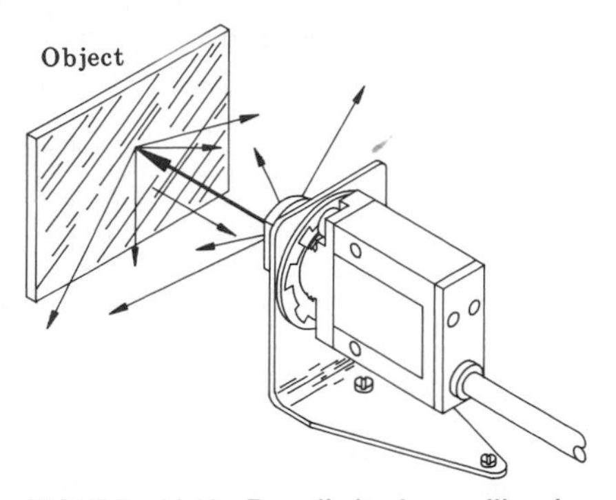

FIGURE 14.12 By eliminating collimating lenses in divergent proximity mode sensors, the sensing range is shortened, but the sensor is less dependent on the angle of incidence of its reflected light. (*Courtesy of Banner Engineering, Minneapolis, Minn.*)

The *convergent-beam mode* is another proximity mode that is effective for sensing small objects. Most convergent-beam sensors use a lens system that focuses the emitted light to an exact point in front of the sensor and focuses the receiver element at the same point. This design produces a small, intense, and well-defined sensing area at a fixed distance from the sensor lens (Fig. 14.13). This is a very efficient use of reflective sensing energy. Objects with small profiles are reliably sensed. Also, materials of very low reflectivity that cannot be sensed with diffuse or divergent mode sensors can often be sensed reliably using the convergent-beam mode.

The range of a convergent-beam sensor is defined as its focus point, which is fixed. This means that the distance from a convergent-beam device to the surface to be sensed must be more or less closely controlled. Every convergent-beam sensor will detect an object of a given reflectivity at its focus point, plus and minus some distance. This sensing area, centered on the focus point, is called the

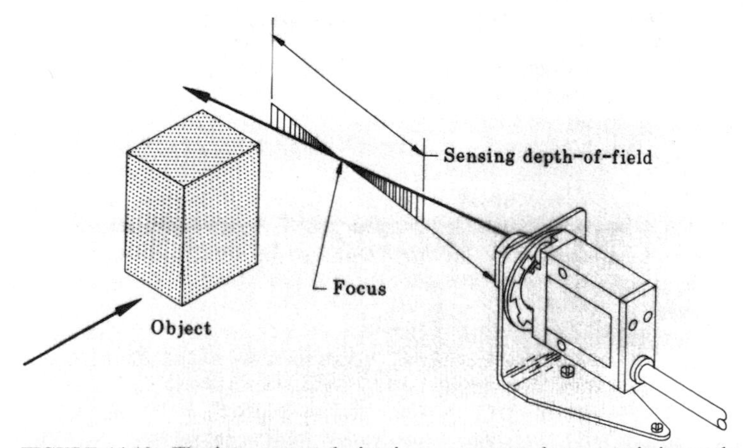

FIGURE 14.13 The lens system design in a convergent-beam proximity mode sensor produces a small, intense, well-defined sensing area. (*Courtesy of Banner Engineering, Minneapolis, Minn.*)

sensor's depth of field. The size of the depth of field depends upon the sensor design and the reflectivity of the object to be sensed. The depth of field of precise-focus convergent-beam sensors is very small. Such sensors may be used for precise position sensing or profile inspection. The depth of field of *mechanical convergent-beam* sensors is relatively large. As the name suggests, mechanical convergent-beam sensors direct a lensed emitter and a separate lensed receiver toward a common point ahead of the sensor. With the proper bracketing, any opposed mode sensor pair may be configured for the mechanical convergent-beam mode.

One specialized use of mechanical convergence is for the sensing of specular reflections (Fig. 14.10). This involves positioning a lensed emitter and receiver at equal and opposite angles (from perpendicular) to a glossy or mirrorlike surface. *The distance from the shiny surface to the sensors must remain constant.* Specular reflection is useful for sensing the difference between a shiny and a dull surface. It is particularly useful for detecting the presence of materials that do not offer enough height differential from their background to be recognized by a convergent-beam sensor. For example, the specular mode may be used to sense the presence of cloth material of any color (a "diffuse material") on a steel sewing machine work surface (a "shiny surface").

It is often necessary to detect objects that pass the sensor within a specified range, while ignoring other stationary or moving objects in the background. One advantage of convergent-beam sensors is that objects beyond the far limit of the depth of field are ignored. It is important to remember, however, that the near and far limits of a convergent-beam sensor's depth of field are dependent upon the reflectivity of the object in the scan path. Background objects of high reflectivity will be sensed at a greater distance than objects of low reflectivity.

Background-suppression sensors have a definite limit to their sensing range and ignore objects that lie beyond this range, regardless of object surface reflectivity (Fig. 14.14). Background-suppression sensors compare the amount of reflected light that is received by *two* optoelements. A target is recognized as long as the amount of light reaching receiver R2 is equal to or greater than the amount "seen" by R1. The sensor's output is canceled as soon as the amount of light at R1 becomes greater than the amount of light at R2.

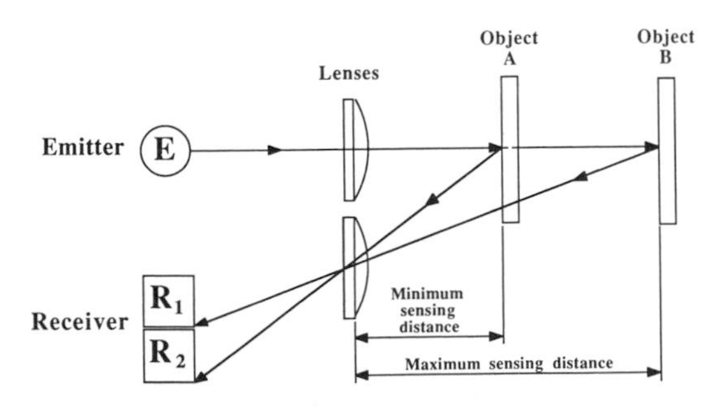

FIGURE 14.14 Background suppression proximity mode sensors, using *two* optoelements, ignore objects that lie beyond their sensing range, regardless of object surface reflectivity. (*Courtesy of Banner Engineering, Minneapolis, Minn.*)

Beam Patterns. A *beam pattern* is included as part of the description for each photoelectric sensor. It includes information that may be useful for predicting the performance of the sensor. All beam patterns are drawn in two dimensions; symmetry around the optical axis is assumed, and the shape of the pattern is assumed to be the same in all sensing planes (note, however, that this is not always an accurate assumption). Beam patterns are drawn for perfectly clean sensing conditions, optimum angular sensor alignment, and the proper sensor sensitivity (gain) setting for the specified range. Maximum light energy occurs along the sensor's optical axis, and light energy decreases with movement toward the beam pattern boundaries. Beam pattern dimensions are typical for the sensor being described and should not be considered exact. Also, beam pattern information is different for each sensing mode.

Opposed Mode Beam Patterns. Beam patterns for opposed sensors represent the area within which the receiver will effectively "see" the emitted light beam. The horizontal scale is the separation distance between the emitter and receiver. The vertical scale is the width of the active beam, measured on either side of the optical axis of the emitter or receiver lens. It is assumed that there is no angular misalignment between the emitter and the receiver. In other words, the optical axis of the emitter lens is kept exactly parallel to the optical axis of the receiver lens while plotting the pattern. Even small amounts of angular misalignment will significantly affect the size of the sensing area of most opposed sensor pairs, except at close range.

Opposed beam patterns predict how closely adjacent to one another parallel opposed sensor pairs may be placed without generating optical crosstalk from one pair to the next. A typical beam pattern for an opposed mode sensor pair is shown in Fig. 14.15. This pattern predicts that, at an opposed sensing distance of 4 ft (1,22 m), a receiver that is kept perfectly parallel to its emitter will "see" enough light for operation at up to just over 8 in (203,2 mm) in any direction from the optical axis of the emitter. This means that adjacent emitter-receiver pairs may be safely placed parallel to each other as close as about 10 in (254 mm) apart (i.e., safely more than 8 in apart) without optical crosstalk from an emitter to the wrong receiver (Fig. 14.16).

Figure 14.17 shows how parallel-beam spacing may be cut in half by *alternat-*

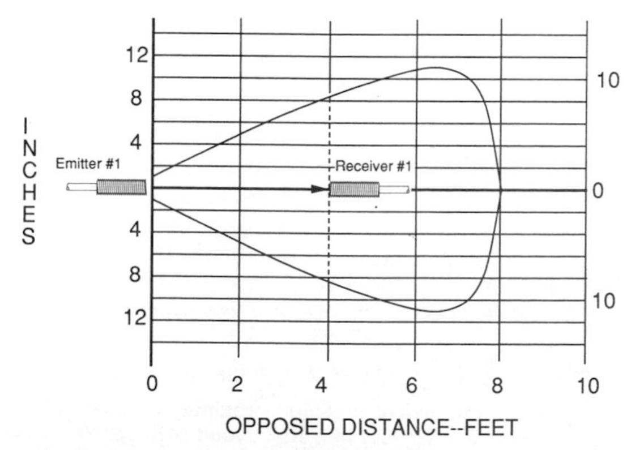

FIGURE 14.15 Typical opposed mode beam pattern. (*Courtesy of Banner Engineering, Minneapolis, Minn.*)

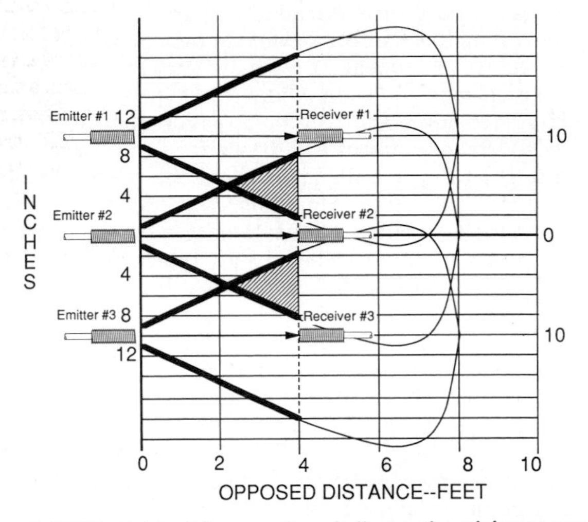

FIGURE 14.16 A beam pattern indicates the minimum separation required to avoid crosstalk between adjacent opposed mode sensor pairs. The spacing for three opposed pairs is shown. (*Courtesy of Banner Engineering, Minneapolis, Minn.*)

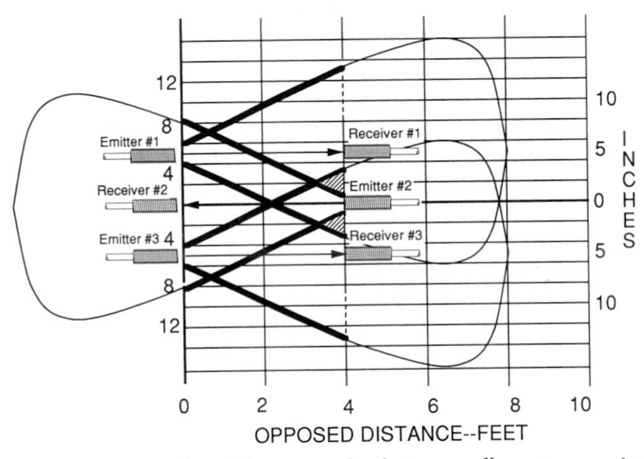

FIGURE 14.17 The minimum spacing between adjacent opposed sensor pairs is cut in half if emitters and receivers are alternated on each side. Spacing for three opposed sensor pairs (staggered) is shown. (*Courtesy of Banner Engineering, Minneapolis, Minn.*)

ing emitter-receiver-emitter-receiver, etc., on each side of the sensing area. Whenever *only two* opposed beams are involved in the sensing scheme, they may be placed in this manner as closely together as the dimensions of the sensors permit without causing direct optical crosstalk.

However, whenever emitters and receivers that are on the same side of the sensing area get very close together, typically 2 in (50,8 mm) or less, the potential for reflective crosstalk (i.e., "proxing") increases. Since the receivers in opposed mode are "looking" for dark (i.e., beam blocked) for object detection, the light detected by a receiver due to reflective crosstalk may cause an object in the sensing area to slip past undetected.

Another common way to minimize optical crosstalk between adjacent opposed sensor pairs is to include a slight angle in the emitter or receiver mounting to intentionally misalign the outermost beams of the array. For example, in Fig. 14.16, emitter 1 could be rotated to direct its beam slightly "up" and away from the view of receiver 2. Similarly, emitter 3 could be rotated slightly "down" and away from receiver 2.

Yet another way to minimize optical crosstalk is to separate adjacent emitter-receiver pairs both horizontally *and* vertically. The diagonal separation between adjacent beams is determined by the beam pattern. In this way, adjacent beams may be placed on closer centers in one dimension. This is possible whenever the object that is to be sensed is large in cross section and when available space permits this approach to sensor mounting. When adjacent opposed beams are placed on very close centers, optical crosstalk can be eliminated by *multiplexing* the sensors in the array. True photoelectric multiplexing enables ("turns on") each modulated emitter only during the time that it samples the output of its associated receiver. As a result, the chance of false response of any receiver to the wrong light source is eliminated.

Opposed mode beam pattern information is also useful for predicting the area within which an emitter and receiver will align when one is moving relative to the other, as with automatic vehicle guidance systems. The beam pattern represents the largest typical sensing area when sensor sensitivity is adjusted to match range specifications. The boundary of the beam pattern will shrink with decreased sensitivity setting and may expand with increased sensitivity.

Retroreflective Mode Beam Patterns. Beam patterns for retroreflective sensors are typically plotted using a 3 in (76,2 mm) diameter plastic corner-cube-type retroreflector, the beam pattern representing the boundary within which the sensor will respond to this retrotarget (Fig. 14.18). The retroreflective target is kept perpendicular to the sensor's optical axis when plotting the pattern. The horizontal scale is the distance from the retrosensor to the retrotarget. The vertical scale is the farthest distance on either side of the sensor's optical axis where a retrotarget can establish a retroreflective beam with the sensor. A "retro" beam pattern indicates how one retrotarget will interact with multiple parallel retroreflective sensors that are mounted on close centers. The beam pattern also predicts whether a 3-in (76,2-mm) reflector will be detected if it is traveling past the sensor parallel to the sensor face or vice versa. Most important, a retroreflective beam pattern is an accurate depiction of the size of the active beam area at distances of a few feet or more from the sensor. It is always good practice, if possible, to capture the entire emitted beam within the retroreflective target area. The beam pattern indicates how much reflector area is needed at any distance where the beam size is greater than 3 in (76,2 mm) wide.

Proximity Mode Beam Patterns. The beam pattern for any proximity mode photoelectric sensor represents the boundary within which the edge of a light-

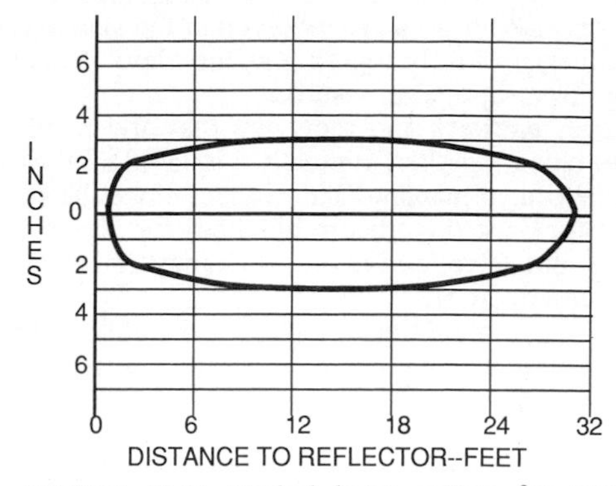

FIGURE 14.18 Typical beam pattern for retroreflective sensors. (*Courtesy of Banner Engineering, Minneapolis, Minn.*)

colored diffuse surface will be detected as it moves past the sensor. Beam patterns for diffuse, convergent, divergent, and background suppression mode sensors are developed using a Kodak 90 percent reflectance white test card, which is about 10 percent more reflective than most white typing paper. The beam pattern will be smaller for materials that are less reflective and may be larger for surfaces of greater reflectivity. The test card used to plot the pattern measures 8 by 10 in (203,2 by 254 mm). Objects that are substantially smaller may decrease the size of the beam pattern at long ranges. Also, the angle of incidence of the beam to a *shiny* surface has a pronounced effect on the size and the shape of a diffuse mode beam pattern.

The horizontal scale is the distance from the sensor to the reflective surface. The vertical scale is the width of the active beam measured on either side of the optical axis (Fig. 14.19). *The beam pattern for any diffuse, convergent, divergent, background suppression sensor is equivalent to the sensor's effective beam.*

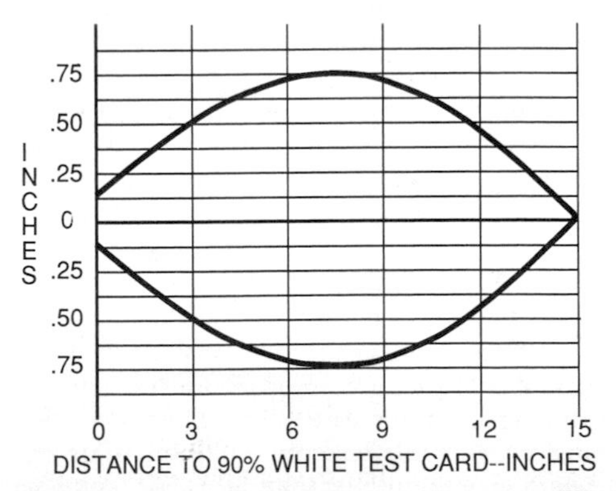

FIGURE 14.19 Typical beam pattern for diffuse proximity mode sensors. (*Courtesy of Banner Engineering, Minneapolis, Minn.*)

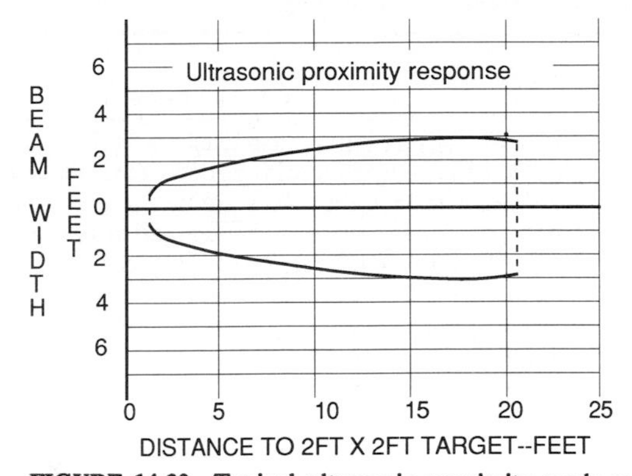

FIGURE 14.20 Typical ultrasonic proximity mode response pattern. (*Courtesy of Banner Engineering, Minneapolis, Minn.*)

The beam pattern (more commonly called the *response pattern*) for an ultrasonic proximity sensor is drawn for a square, solid, flat surface (Fig. 14.20). The size of the target is specified by the manufacturer for each type of sensor. The size of an ultrasonic proximity response pattern is affected by the size, shape, texture, and density of the material being sensed.

Excess Gain. *Excess gain* is a measurement that may be used to predict the reliability of any sensing system. As its name suggests, excess gain is a measurement of the sensing energy falling on the receiver element of a sensing system *over and above* the minimum amount required to just operate the sensor's amplifier.

Once a signal is established between the emitter and the receiver of any sensor or sensing system, there may be attenuation (reduction) of that signal resulting from dirt, smoke, moisture, or other contaminants in the sensing environment. The excess gain of a sensing system may be seen as the extra sensing energy that is available to overcome this attenuation.

Excess gain is usually clearly specified for photoelectric sensors. In equation form:

$$\text{Excess gain (EG)} = \frac{\text{light energy falling on receiver element}}{\text{sensor's amplifier threshold}}$$

The *threshold* is the level of sensing energy required by the sensor's amplifier to cause its output to change state (i.e., to switch "on" or "off"). In a modulated photoelectric system, excess gain is measured as a voltage (typically at millivolt levels), usually at the first stage of receiver amplification. This measured voltage is compared with the amplifier's threshold voltage level to determine the excess gain. There is an excess gain of one (usually expressed as "1×" or "one times") when the measured voltage is at the amplifier threshold level. If 50 percent of the original light energy becomes attenuated, a minimum of 2× excess gain is required to overcome light loss. Similarly, if 80 percent of a sensor's light is lost to attenuation (i.e., only 20 percent left), an available excess gain of at least 5× is required.

TABLE 14.1 Guidelines for Excess Gain Values

Minimum excess gain required	Operating environment
1.5×	Clean air: no dirt buildup on lenses or reflectors
5×	Slightly dirty: slight buildup of dust, dirt, oil, moisture, etc., on lenses or reflectors; lenses are cleaned on a regular schedule
10×	Moderately dirty: obvious contamination of lenses or reflectors, but not obscured; lenses cleaned occasionally or when necessary
50×	Very dirty: heavy contamination of lenses; heavy fog, mist, dust, smoke, or oil film; minimal cleaning of lenses

Source: Banner Engineering, Minneapolis, Minn.

If the general conditions in the sensing area are known, the excess gain levels listed in Table 14.1 may be used as guidelines for assuring that the sensor's light energy will not be entirely lost to attenuation. Table 14.1 lists an excess gain of 1.5× (i.e., 50 percent more energy than the minimum for operation) for a perfectly clean environment. This amount includes a safety factor for subtle sensing variables such as gradual sensor misalignment and small changes in the sensing environment. At excess gains above 50×, sensors will begin to *burn through* (i.e., "see" through) paper and other materials with similar optical density.

The excess gain that is available from any sensor or sensing system may be plotted as a function of distance (Fig. 14.21). Excess gain curves are plotted for conditions of perfectly clean air and maximum receiver gain and are an important part of every photoelectric sensor specification. For example, the excess gain curves shown in product catalogs published by Banner Engineering, Minneapolis, Minn., represent the lowest *guaranteed* excess gain available for the particular model highlighted. Most sensors are factory calibrated to the excess gain curve. Sensors that have a gain adjustment (also called "sensitivity control") can usually be field-adjusted to exceed the excess gain specifications; however, this is never guaranteed by sensor manufacturers.

The excess gain curve in Fig. 14.21 suggests that operation of this opposed mode sensor pair is possible in a perfectly clean environment (excess gain ≥1.5×) at distances up to 10 ft (3,04 m) and in a moderately dirty area (excess gain ≥10×) up to 4 ft (1,22 m) apart. At distances of less than 1 ft (304 mm), these sensors will operate in nearly any environment.

Excess Gain–Opposed Mode Sensing. The relationship between excess gain and sensing distance is different for each photoelectric sensing mode. For example, the excess gain of an opposed mode sensor pair is directly related to sensing distance by the inverse square law. If the sensing distance is doubled, the excess gain is reduced by a factor of $(1/2)^2$ = one-fourth. Similarly, if the sensing distance is tripled, the excess gain curve is reduced by a factor of $(1/3)^2$ = one-ninth, and so on. As a result, the excess gain curve for opposed mode sensors is always a straight line when plotted on a loglog scale.

Since the light from the emitter goes directly to the receiver, opposed mode sensing makes the most efficient use of sensing energy. Therefore, the excess gain that is available from opposed mode sensors is much greater than from any other photoelectric sensing mode.

Excess Gain–Retroreflective Mode Sensing. The shape of excess gain curves for the other sensing modes is not as predictable. Retroreflective excess gain

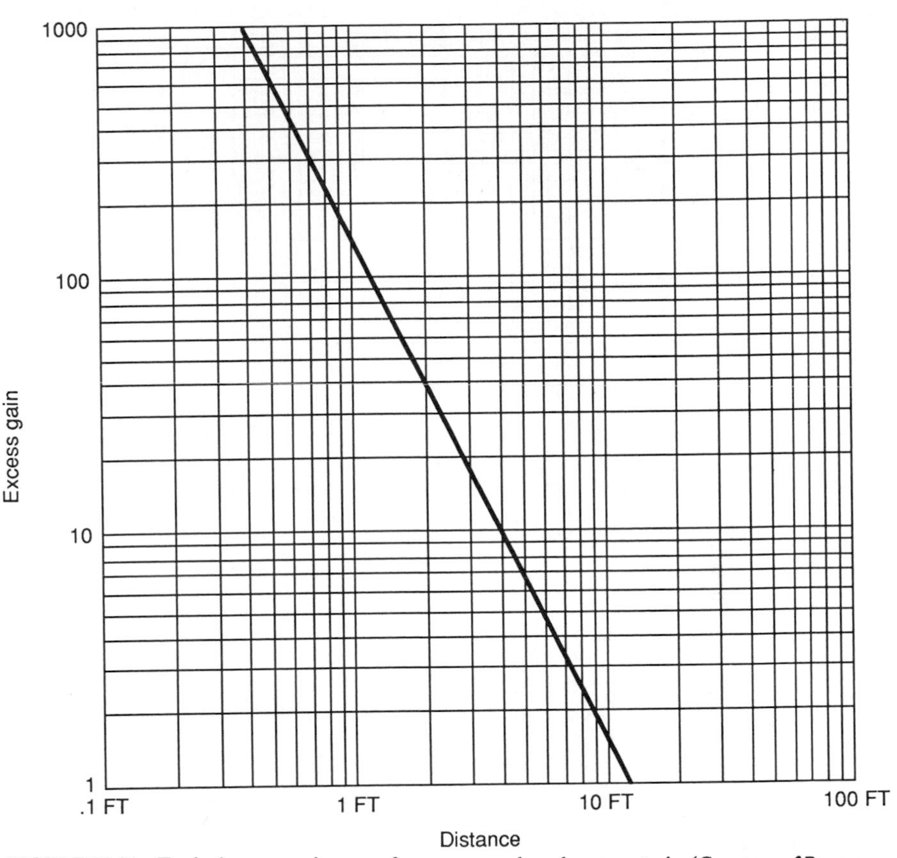

FIGURE 14.21 Typical excess gain curve for an opposed mode sensor pair. (*Courtesy of Banner Engineering, Minneapolis, Minn.*)

curves are usually plotted using a 3-in (76,2-mm) diameter retroreflector (Fig. 14.22), except where otherwise noted in the sensor manufacturer's literature. The shape of retroreflective excess gain curves is affected by the size of the retroreflective target. Several retrotargets, used together in a cluster, will usually result in longer sensing range and a higher maximum excess gain. A smaller corner-cube reflector yields a smaller curve.

Most retroreflective sensors are designed for long-range performance and use separate lenses for the emitter and the receiver. A good retroreflector has the property of returning most of the incoming light directly back to the sensor. At close ranges, the retroreflector sends most of the incoming light directly back into the emitter lens. As a result, many two-lens retroreflective sensors suffer a blind spot at close ranges, which is evident on excess gain curves.

Excess Gain–Proximity Mode Sensing. Generally speaking, photoelectric proximity modes are inefficient sensing modes. The receiver must "look" for a relatively small amount of light that is bounced back directly from the surface of an object. As a result, the excess gain available from a proximity mode sensor is usually lower than that of the other photoelectric sensing modes.

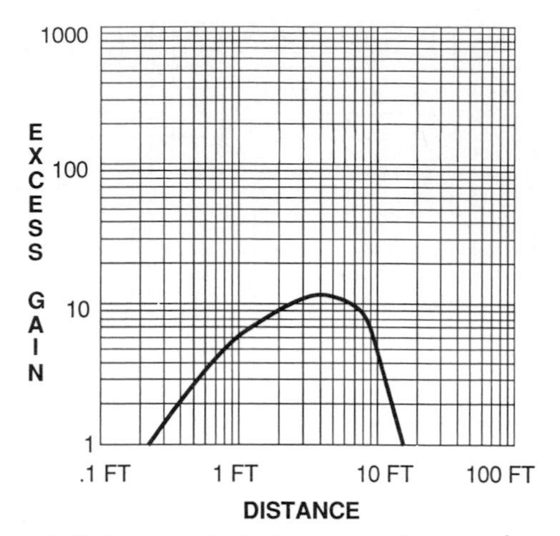

FIGURE 14.22 Typical excess gain curve for a retroreflective mode sensor. (*Courtesy of Banner Engineering, Minneapolis, Minn.*)

The curves for diffuse, convergent, divergent, and background-suppression sensors are plotted using a Kodak 90 percent reflectance white test card as the reference material. The excess gain of diffuse sensors is dramatically influenced by the reflectivity of the surface to be sensed. Any material surface may be ranked for its reflectivity compared with the Kodak 90 percent reflectance white reference card (Table 14.2).

In Table 14.2, the numbers in the "Excess Gain Required" column indicate the *minimum* excess gain that is required to sense the material. For example, if the material to be sensed is opaque black plastic (excess gain required = 6.4), then the diffuse sensor with the excess gain curve of Fig. 14.23 will "see" the material from 0 (zero) to 10 in (254 mm). This, of course, assumes perfect sensing conditions.

To get the actual required excess gain for diffuse sensing of any material, multiply the material's *reflectivity factor* by the excess gain level that is required for the sensing conditions (from Table 14.1). For example, to sense black opaque plastic in a slightly dirty environment, the minimum required excess gain is

$$\text{Excess gain required} = 6.4 \text{ (reflectivity factor)} \times 5 \text{ (min EG required)} = 32$$

Under these conditions, the diffuse sensor of Fig. 14.23 will reliably sense the black plastic from ½ in (12,7 mm) to 4 in (101,6 mm), even after there is a slight buildup of dirt on the lens.

The excess gain of diffuse mode sensors is also affected by the size and the profile of the object to be detected. The excess gain curves assume a white test card that fills the entire area of the diffuse sensor's effective beam. If the object to be detected fills only a portion of the sensor's effective beam, proportionately less light energy will be returned to the receiver. Like the diffuse mode, the excess gain of divergent mode sensors is affected by the reflectivity and the size of the object to be sensed. However, the effect of these variables is less noticeable in divergent sensing, simply because divergent mode sensors lose their sensing

TABLE 14.2 Relative Reflectivity Chart

Material	Reflectivity, %	Excess gain required
Kodak white test card	90	1
White paper	80	1.1
Newspaper (with print)	55	1.6
Tissue paper: 2 ply	47	1.9
1 ply	35	2.6
Masking tape	75	1.2
Kraft paper, cardboard	70	1.3
Dimension lumber (pine, dry, clean)	75	1.2
Rough wood pallet (clean)	20	4.5
Beer foam	70	1.3
Clear plastic bottle*	40	2.3
Translucent brown plastic bottle*	60	1.5
Opaque white plastic*	87	1.0
Opaque black plastic (nylon)*	14	6.4
Black neoprene	4	22.5
Black foam carpet backing	2	45
Black rubber tire wall	1.5	60
Natural aluminum, unfinished*	140	0.6
Natural aluminum, straightlined*	105	0.9
Black anodized aluminum, unfinished*	115	0.8
Black anodized aluminum, straightlined*	50	1.8
Stainless steel, microfinish*	400	0.2
Stainless steel, brushed*	120	0.8

Source: Banner Engineering, Minneapolis, Minn.

*For materials with shiny or glossy surfaces, the reflectivity figure represents the maximum light return, with the sensor beam exactly perpendicular to the material surface.

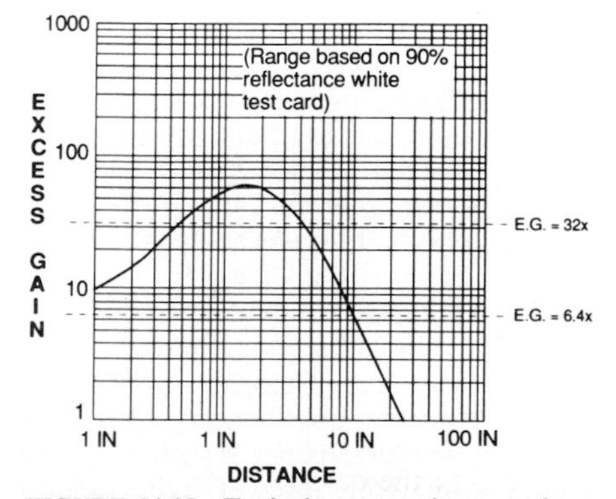

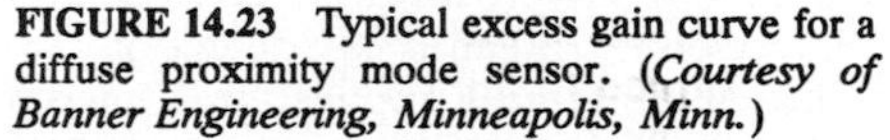

FIGURE 14.23 Typical excess gain curve for a diffuse proximity mode sensor. (*Courtesy of Banner Engineering, Minneapolis, Minn.*)

ability within such a short range. Since most of the energy of a convergent-beam sensor is concentrated at its focus, the maximum available excess gain is much higher than for any of the other proximity modes. This relatively high excess gain allows the detection of materials of very low reflectivity, where diffuse, divergent, and background suppression mode sensors would fail. The effect of an object's relative reflectivity is most noticeable in the size of the resultant depth of field. Also, because the effective beam of a convergent-beam sensor is so small, even objects with narrow profiles can return a relatively high percentage of the incident light.

The concept of excess gain is not intended to be an exact science but rather is a guideline for the sensor selection process. Knowing values from an excess gain curve can be valuable information for predicting the success of a particular sensor in a given sensing environment. *In most sensing situations, high excess gain relates directly to sensing reliability.*

Contrast. All photoelectric sensing applications involve differentiating between two received light levels. Contrast is the ratio of the amount of light falling on the receiver in the "light" state compared with the "dark" state. Contrast is also referred to as the "light-to-dark-ratio," as represented by the following equation:

$$\text{Contrast} = \frac{\text{light level at receiver in light condition}}{\text{light level at receiver in dark condition}}$$

It is always important to choose the sensor or lensing option that will optimize contrast in any photoelectric sensing situation. Many situations, like a cardboard box breaking a retroreflective beam, are applications with infinitely high contrast ratios. In this type of high-contrast application, sensor selection simply involves verifying that there will be enough available excess gain for reliable operation in the sensing environment.

Many of today's industrial photoelectric sensing applications are not so straightforward. Most problems with contrast in opposed and retroreflective applications occur when (1) the beam must be blocked by a material that is not opaque, or (2) less than 100 percent of the effective beam is blocked.

When proximity mode sensors are used, most low-contrast problems occur where there is a close background object directly in the scanning path. This problem is compounded when the background object's reflectivity is greater than the reflectivity of the object to be detected. Background-suppression or ultrasonic proximity mode sensors can often deal successfully with this problem. *As a general rule, a contrast of 3 is the minimum for any sensing situation.* This is usually enough to overcome the effect of subtle variables that cause light-level changes, like small amounts of dirt buildup on the lenses or inconsistencies in the product being sensed. Table 14.3 gives suggested guidelines for contrast values.

Close Differential Sensing. Some applications offer contrast of *less* than 3, regardless of the sensing method used. These low-contrast situations fall into the category of *close differential sensing* applications. Most color registration applications qualify as close differential sensing. Another common close differential situation involves breaking a relatively large effective beam with a small part, as in ejected small part detection or thread break detection.

Whenever a close differential sensing application is encountered, use of an *ac-coupled amplifier* should be considered. Most sensing systems, including all self-contained sensors, use dc-coupled amplifiers. A dc-coupled amplifier is one that amplifies *all* received signal levels. Ac-coupled amplifiers may sometimes be

TABLE 14.3 Contrast Values and Corresponding Guidelines

Contrast	Recommendation
1.2 or less	*Unreliable:* evaluate alternative sensing schemes
1.2–2	*Poor contrast:* consider sensors with ac-coupled amplification
2–3	*Low contrast:* sensing environment must remain clean and all other sensing variables must remain stable
3–10	*Good contrast:* minor sensing system variables will not affect sensing reliability
10 or greater	*Excellent contrast:* sensing should remain reliable as long as the sensing system has enough excess gain for operation

used more reliably in close differential sensing, since they *amplify only ac (changing) signals* while completely ignoring dc (steady) signals. This means that very small changes in received light level can be highly amplified. As useful as they are, however, ac-coupled amplifiers should be avoided whenever the contrast is high enough for a dc-coupled device. Because they are so sensitive to very small signal changes, ac-coupled amplifiers may respond to conditions like electrical "noise" or sensor vibration. Additionally, ac-coupled amplifiers require a sensing event to occur at a minimum rate of change. As a general guideline, a target must move into the sensing beam at a speed of 1 in (25,4 mm) per second.

In the contrast range of 2 to 3, consider a dc-coupled device as first choice. However, in order for a dc-coupled sensor to be reliable in this low-contrast range, sensing variables like dirt buildup on lenses, reflectivity, or translucency of the part being sensed, and the mechanics of the sensing system *must* remain constant. If it is known that these variables might gradually change over time, ac-coupled amplification should be considered.

Measuring Contrast. Contrast may be calculated if excess gain values are known for both light and dark conditions:

$$\text{Contrast} = \frac{\text{excess gain (light condition)}}{\text{excess gain (dark condition)}}$$

Contrast should always be considered when choosing a sensor and should always be maximized by alignment and gain adjustment during sensor installation. Optimizing the difference in the amount of received light between light and dark conditions of any photoelectric sensing application *will always increase* the reliability of the sensing system.

Sensor Outputs. The output of a self-contained sensor or of the remote amplifier of a component sensing system is either digital or analog. A *digital output* (Fig. 14.24) is more commonly called a switched output and has only two states: "on" and "off." "On" and "off," in this case, refer to the status of the load that the sensor output is controlling. The load might be an indicator light, an audible alarm, a clutch or brake mechanism, a solenoid valve or actuator, or a switching relay. The load might also be the input circuit to a timer, counter, programmable logic controller, or computer.

In photoelectrics, the sensing event (input) and the switched output state are characterized together by one of two sensing terms. *Light operate* describes a sensing system that will energize its output when the receiver "sees" more than a set

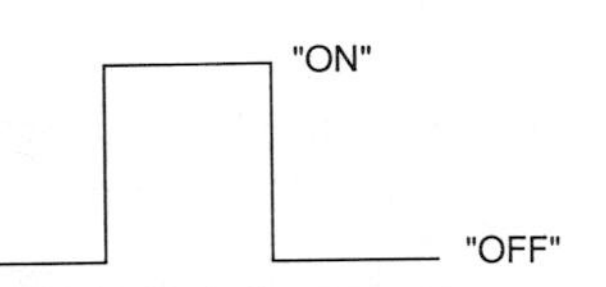

FIGURE 14.24 Digital sensor output: output is either "on" or "off." (*Courtesy of Banner Engineering, Minneapolis, Minn.*)

amount of light. *Dark operate* means that the sensor's output will energize when its receiver is sufficiently dark.

In an opposed mode sensing system (Fig. 14.25), "dark operate" means that the output energizes its load when an object is present (breaking the beam). The light condition occurs when the object is absent.

In a retroreflective sensing system (Fig. 14.26), the conditions are the same as in the opposed mode. The dark condition occurs when the object is present, and the receiver sees light when the object is absent.

These conditions are reversed in all proximity sensing modes (Fig. 14.27). The light condition occurs when the object is present, "making" (establishing) the beam. When the object is absent, no light is returned to the receiver.

An *analog output* (Fig. 14.28) is one that varies over a range of voltage (or current) and is proportional to some sensing parameter. The output of an analog photoelectric sensor is proportional to the strength of the received light signal.

The output of an analog ultrasonic proximity sensor is proportional to the distance between the sensor and the object that is returning the sound echo. The output is proportional to the time required for the echo to return to the sensor.

Sensors with analog outputs are useful in many process control applications where it is necessary to monitor an object's position or size or translucency and to provide a continuously variable control signal for another analog device, like a motor speed control.

Response Time. Every sensor is specified for its response time. The response time of a sensor or sensing system is the maximum amount of time required to respond to a change in the input signal (e.g., a sensing event). It is time between the leading edge (or trailing edge) of a sensing event and the change in the sensor's output. With a switched output, the response time is the time required for the output to switch from "off" to "on" or from "on" to "off." These two times are not always equal. With an analog output, the response time is the maximum

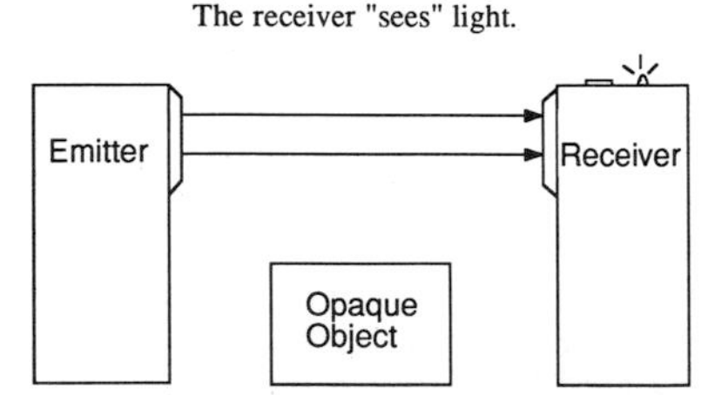

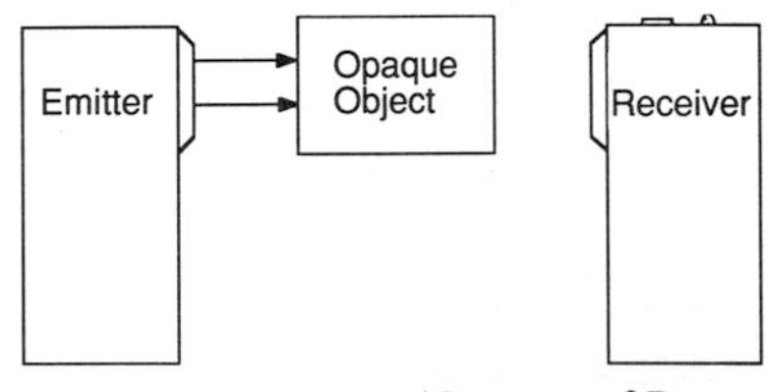

FIGURE 14.25 Light operate vs. dark operate for an opposed mode system. (*Courtesy of Banner Engineering, Minneapolis, Minn.*)

LIGHT OPERATE

The output is energized when the beam is unblocked.

The sensor "sees" light.

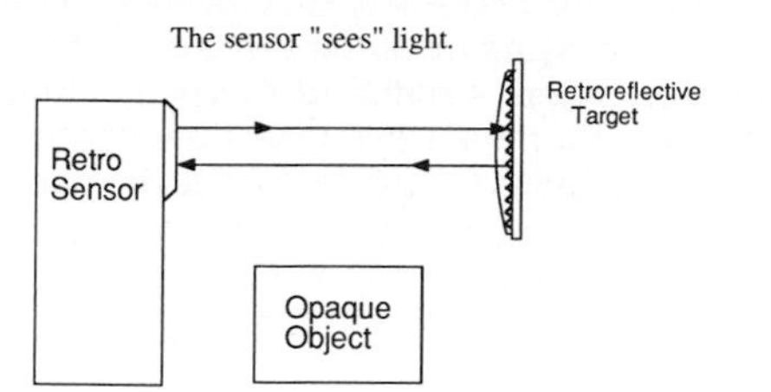

DARK OPERATE

The output is energized when an object blocks the light from reaching the retroreflective target.
The sensor "sees" dark.

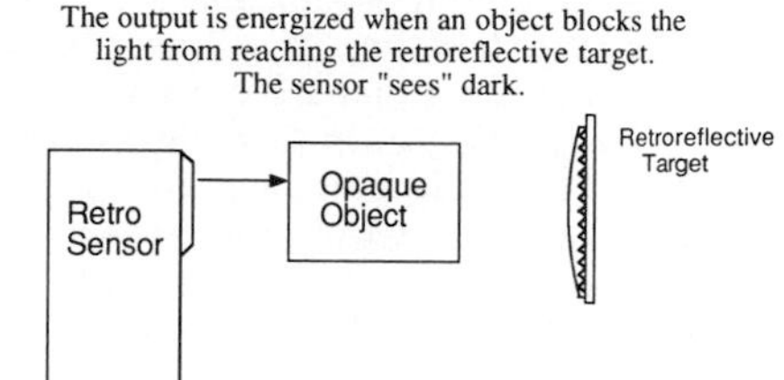

FIGURE 14.26 **Light operate vs. dark operate for a retroreflective mode sensor. (*Courtesy of Banner Engineering, Minneapolis, Minn.*)**

LIGHT OPERATE

The output is energized when light is reflected directly from an object surface.
The sensor "sees" light.

Proximity Sensor

Reflective Object

DARK OPERATE

The output is energized when no object is present in front of the sensor to return the emitted light.
The sensor "sees" dark.

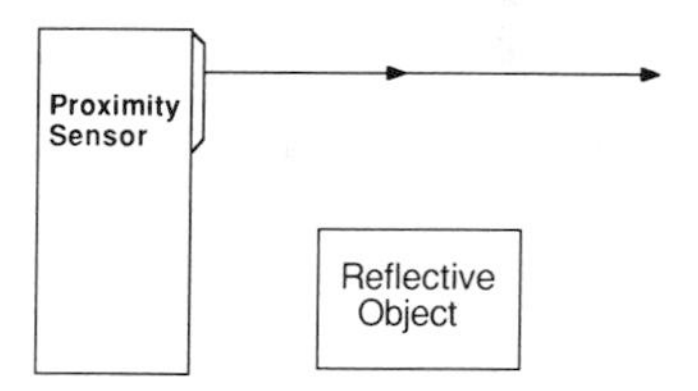

FIGURE 14.27 **Light operate vs. dark operate for a proximity mode sensor (diffuse, divergent, convergent, and background-suppression modes). (*Courtesy of Banner Engineering, Minneapolis, Minn.*)**

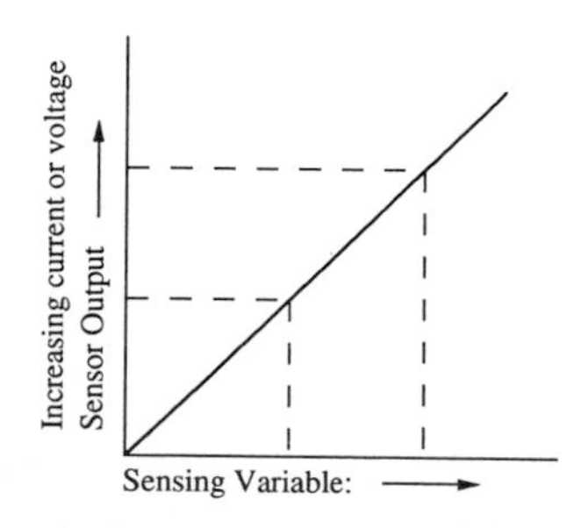

Photoelectric: Increase (or decrease) in received light level
-or-
Ultrasonic: Movement of object toward (or away from) sensor

FIGURE 14.28 **Analog sensor output: output varies over a range of voltage or current and is proportional to a sensing parameter. (*Courtesy of Banner Engineering, Minneapolis, Minn.*)**

time required for the output to swing from minimum to maximum or from maximum to minimum. Again, these two times are not necessarily equal.

The response time of a sensor is not *always* an important specification. For example, sensors that are used to detect boxes passing on a conveyor do not require fast response. In fact, time delays are sometimes added to extend sensing response to avoid nuisance trips or to add simple timing logic for flow control applications. Response time does become important when detecting high-speed events and becomes quite critical when detecting small objects moving at high speed. Narrow gaps between objects or short times between sensing events must also be considered when verifying that a sensor's response is fast enough for the application.

Required Sensor Response Time. The required sensor response time may be calculated for a particular sensing application when the size, speed, and spacing of the objects to be detected are known:

$$\text{Required sensor response time} = \frac{\text{apparent width of object passing sensor}}{\text{speed (velocity) of object passing sensor}}$$

As an example, consider an application in which O-ring packets on a conveyor are counted by a convergent-beam sensor (Fig. 14.29). The following information is known:

1. The O-ring packets are processed at a rate of 600 per minute.
2. The packets are 3 in (76,2 mm) wide.
3. The packets are equally spaced with about 1 in (25,4 mm) separation between adjacent packets.

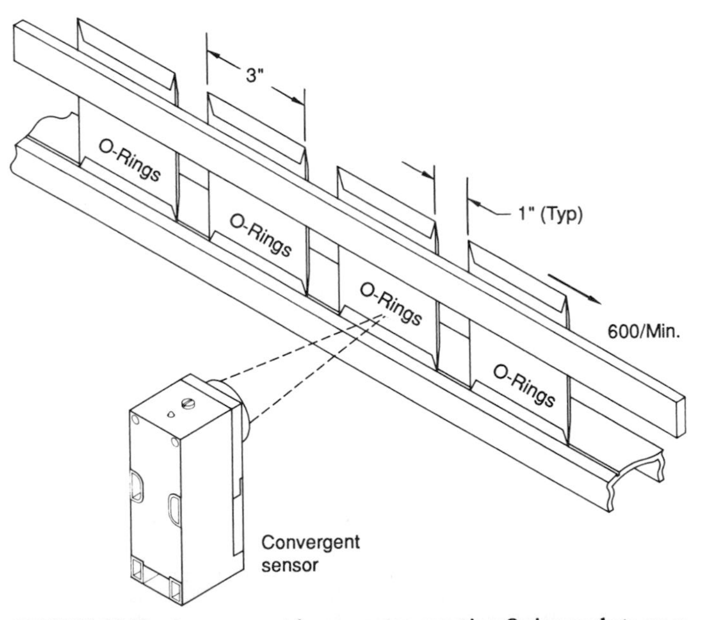

FIGURE 14.29 A convergent-beam sensor counting O-ring packets on a conveyor. (*Courtesy of Banner Engineering, Minneapolis, Minn.*)

To compute the required sensor response time, the processing rate is first converted to packet speed:

1. 600 packets per minute = 10 packets per second.
2. Each packet accounts for a 3-in (76,2-mm) packet width + 1 in (25,4 mm) space = 4 in (101,6 mm) of linear travel.
3. Speed of the packets = 4 in (101,6 mm) per packet × 10 packets per second = 40 in (1016 mm) per second.

The time during which a packet is "seen" by the convergent-beam sensor is

Time of light condition = object width = 3 in (76,2 mm) = 0.075 s
or 75 ms as total time of packet passing sensor

In this application, a sensor with a specified response time of less than 25 ms will work in this counting application. It is always wise to include a safety factor and to choose a sensor with a response time faster than required.

Response Requirements for Rotating Objects. When sensing a rotating object, the calculation for the required sensor response time is the same. The only additional calculation is conversion of rotational speed to linear speed. For example, calculate the required sensor response time for sensing a retroreflective target on a rotating shaft (Fig. 14.30), given the following information:

1. The target is a 1-in (25,4-mm) square piece of retroreflective tape on a 3.25-in (82,5-mm) diameter shaft.
2. Maximum shaft speed is 600 revolutions/minute = 10 revolutions/second.

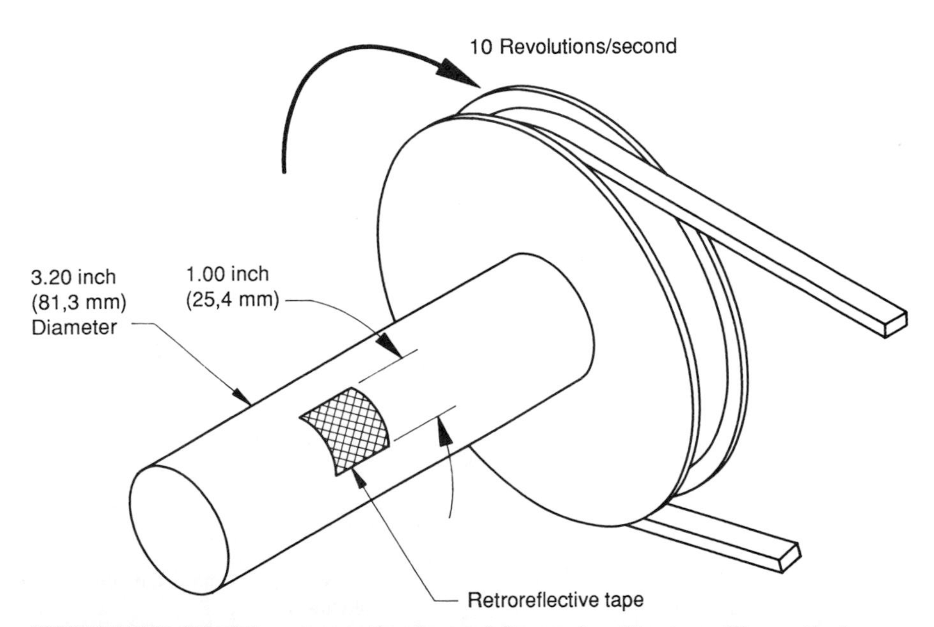

FIGURE 14.30 Calculating response time for a rotating member. (*Courtesy of Banner Engineering, Minneapolis, Minn.*)

To convert rotational speed to linear velocity:

1. Circumference of the shaft = π × diameter = π × 3.25 in (82,5 mm) = 10 in (254 mm).
2. Linear velocity on shaft circumference = 10 in (254 mm) per revolution × 10 revolutions/second = 100 in (2540 mm)/s.

The required sensor response time is:

$$\text{Time of light condition} = \frac{\text{target length}}{\text{linear speed}} = \frac{1\text{ in (25,4 mm)}}{100\text{ in (2540 mm)/s}} = 0.01\text{ s}$$

or 10 ms as total time sensor "sees" retrotape

Ten milliseconds is the fastest response requirement in this application, since the untaped portion of the circumference is 9 times longer. A retroreflective sensor with a small effective beam and a response time faster than 10 ms will reliably sense the tape at the maximum shaft speed. To ease the response time requirement in applications like these that simply require one pulse per revolution (or per cycle), a target should cover 50 percent of the shaft circumference so that half of the revolution is light time and the other half is dark time.

Response Time Requirements for Small Objects. A safe assumption to make when calculating the response time requirement for an object with a small cross section is that the object must fill 100 percent of the sensor's effective beam to be detected. Whenever the size of a small object begins to approach the size of the effective beam, the apparent size of the object as "seen" by the sensor becomes less than the actual width of the object. A safe assumption in these situations is to reduce the apparent size of the object by an amount equal to the diameter of the effective beam at the sensing location. As a result, the required response time decreases:

$$\text{Required response time} = \frac{\text{width of object} - \text{diameter of effective beam}}{\text{speed of object through beam}}$$

To illustrate the effect of small objects on response time requirements, consider the following example of a small pin that breaks the beam of an opposed sensor pair (Fig. 14.31):

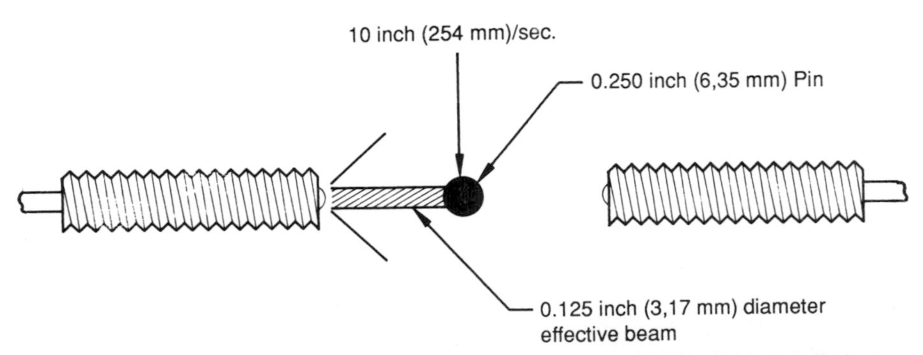

FIGURE 14.31 A 0.250-in (6,35-mm) diameter pin is sensed in a 0.125-in (3,17-mm) diameter effective beam. (*Courtesy of Banner Engineering, Minneapolis, Minn.*)

1. 0.25-in (6,35-mm) diameter pins pass through the beam of an opposed emitter-receiver pair that has a 0.12-in (3,18-mm) diameter effective beam.
2. The maximum speed of the pins is 10 in (254 mm) per second.

Computing the required sensor response time:

$$\text{Time of dark condition} = \frac{\text{pin diameter} - \text{effective beam diameter}}{\text{speed of pin through beam}}$$

$$= \frac{0.25 \text{ in } (6{,}35 \text{ mm}) - 0.12 \text{ in } (3{,}18 \text{ mm})}{10 \text{ in } (254 \text{ mm})/\text{s}}$$

$$= \frac{0.12 \text{ in } (3{,}18 \text{ mm})}{10 \text{ in } (254 \text{ mm})/\text{s}}$$

$$= 0.012 \text{ s } (21 \text{ ms})$$

The addition of apertures on the emitter-receiver pair (Fig. 14.32) will ease the response time requirement because the pin will block the smaller effective beam for a longer time. If 0.040-in (1,01-mm) diameter apertures are used:

$$\text{Time of dark condition} = \frac{0.25 \text{ in } (6{,}35 \text{ mm}) - 0.04 \text{ in } (1{,}01 \text{ mm})}{10 \text{ in } (254 \text{ mm})/\text{s}}$$

$$= \frac{0.21 \text{ in } (5{,}33 \text{ mm})}{10 \text{ in } (254 \text{ mm})/\text{s}}$$

$$= 0.021 \text{ s } (21 \text{ ms})$$

Because of resulting low excess gain, it is usually impractical to aperture an opposed beam to less than about 0.02 in (0,51 mm). Objects with cross sections smaller than about 0.03 in (0,76 mm) are usually sensed most reliably using one of the proximity sensing modes. The wider the proximity beam, the longer a small part will be sensed. This eases the sensor response requirement. A divergent-

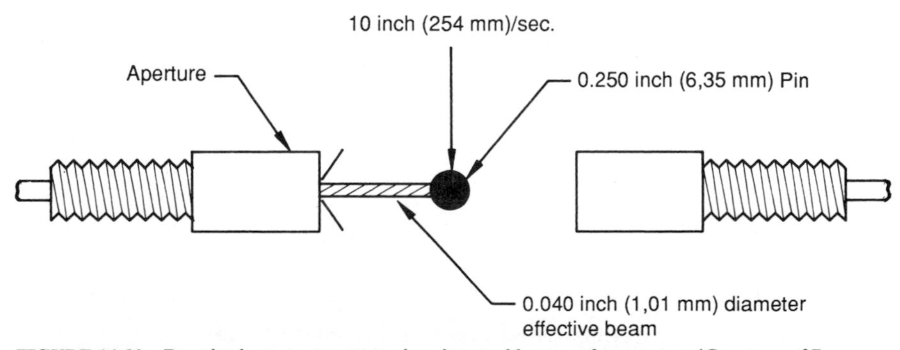

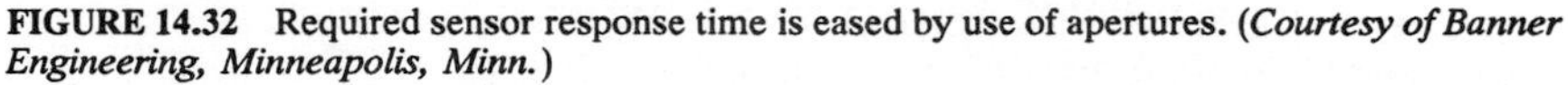

FIGURE 14.32 Required sensor response time is eased by use of apertures. (*Courtesy of Banner Engineering, Minneapolis, Minn.*)

beam sensor or a bifurcated fiber-optic sensor are preferred for sensing very small profiles.

When sensing narrow gaps, opposed mode sensors should have a wide beam so that light is seen through the gap for a long time. Use of individual fiber optics with a rectangular termination is one way to shape the effective beam and ease sensor response requirements. When sensing narrow gaps with a proximity sensor, the small effective beam of a convergent mode sensor is usually preferred.

Response Time of a Load. The *response time* of a load is the maximum time required to energize and/or deenergize a particular load and is included in the load's specifications (Fig. 14.33). In general, solid-state loads like counters and solid-state relays have faster response times than electromechanical devices like solenoids and contactors. The response speed characteristics of any load to be controlled by a sensor's output should be checked to be sure that the duration of the output signal from the sensor and the time between adjacent outputs are both long enough to allow the load to react properly. In situations where the load is too slow to react, a delay timer may be required between the sensor output and the load to extend the duration of the sensor's output signal. An even better solution involves changing the sensing geometry, if possible, to equalize the durations of the light and dark ("on" and "off") times.

Photoelectric Sensor Selection Process: Evaluating Photoelectric Sensing Mode Options

An important part of any sensor selection process involves determination of the best sensing mode for the application. The best sensing mode is the mode that yields the greatest amount of sensing signal differential between the conditions of target present and target absent (i.e., the most sensing *contrast*), while maintaining enough sensing signal (i.e., enough *excess gain*) to comfortably overcome any attenuation caused by conditions in the sensing environment.

To determine which mode will yield the highest sensing contrast, you must evaluate the properties of the target to be sensed, including:

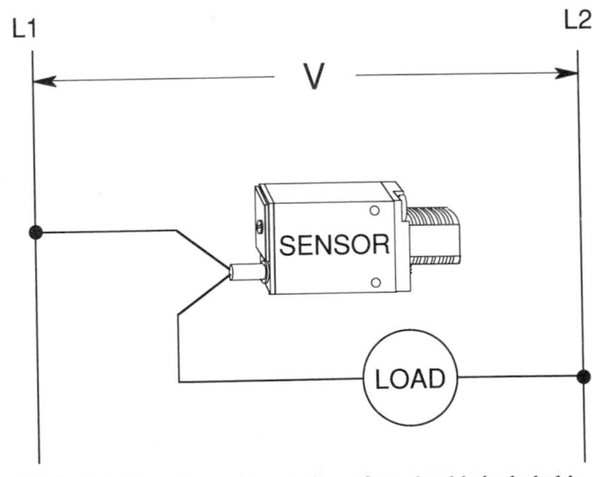

FIGURE 14.33 The response time of any load is included in its specifications. (*Courtesy of Banner Engineering, Minneapolis, Minn.*)

- Part size and/or profile
- Optical opacity
- Optical or acoustical (ultrasonic) surface reflectivity

The geometry of the application should be analyzed to determine whether sensing of the target will be at a repeatable point or if the target, instead, will pass at random distances and/or with random orientations to the sensor. It may also be necessary to evaluate the properties of other objects (if present) in the sensing path to ensure that they do not interfere with sensing of the target.

Opposed Mode Sensing—Uses and Advantages

General Application Rule. Use opposed mode photoelectric sensors wherever possible as this type of sensor will always provide the most reliable sensing system, as long as the object to be detected is opaque to light (i.e., if the object *completely* blocks the opposed light beam). Exception: an inductive proximity sensor becomes a first choice for sensing of *metal* objects that pass close enough to the sensor for reliable detection.

Excess Gain. Opposed sensors offer the highest levels of excess gain. Types of applications requiring high levels of excess gain include:

1. Sensing through heavy dirt, dust, mist, condensation, oil film, etc.
2. Long-range scanning
3. Precise position sensing or small part detection using small apertures.
4. Detection of opaque solids or liquids inside closed thin-walled boxes. Opposed mode sensors can sometimes be used to "burn through" thin-walled boxes or containers to detect the presence, absence, or level of the product inside.

Parts Counting. Opposed sensors are usually the most reliable for accurate parts counting, largely owing to their well-defined effective beam.

Object Reflectivity. Use of opposed mode sensors eliminates the variable of surface reflectivity or color.

Mechanical Convergence. A pair of opposed mode sensors may be positioned to converge mechanically at a point ahead of the sensor pair. This type of configuration usually results in more depth of field compared with convergent-beam proximity sensors.

Specular Reflection. One specialized use of a mechanically converged emitter and receiver pair is to detect the difference between a shiny and a dull surface. A shiny surface will return emitted light to a receiver if the two units are mounted at equal and opposite angles to the perpendicular to the shiny surface (Fig. 14.10). The light will be diffused by any nonreflective surface that covers or replaces the shiny surface.

Opposed Mode Sensing—Application Cautions

Clear Materials. Opposed mode photoelectric sensors should be avoided for detection of translucent or transparent materials, with the following exceptions:

1. Many of the above materials can be reliably sensed by using special-purpose opposed mode sensor pairs, such as Banner Engineering's Mini-Beam clear plastic detection sensors, which take advantage of the polarizing properties of many plastics.
2. Most glass containers have a thick bottom section of glass that may usually be

used to reliably block an opposed beam that has been properly shaped with rectangular apertures.

Very Small Parts. Avoid trying to detect objects that interrupt less than 100 percent of an opposed effective beam area. Use apertures, lenses, or fiber optics to shape the effective beam to match the profile of a small part.

Too Much Excess Gain. Some opposed mode pairs have so much excess gain when they are used at close range that they tend to "burn through" thin opaque materials like paper, cloth, and plastics. In situations where it becomes difficult to set a sensitivity control operating point because of too much excess gain, the signal may need to be mechanically attenuated by addition of apertures over the lenses or by intentional sensor misalignment.

Retroreflective Mode Sensing—Uses and Advantages

General Application Rule. Use a retroreflective sensor in lieu of the opposed mode where sensing is possible from only one side.

Applications. Retroreflective mode sensing is the most popular sensing mode in conveyor applications where objects are large (e.g., boxes, cartons), where the environment is relatively clean, and where scanning ranges are from about 2 to 10 ft (610 mm to 3,05 m).

Retroreflective Mode Sensing—Application Cautions

Excess Gain. Avoid using retroreflective sensors on the basis of convenience only, especially where it is important to have high excess gain. Retrosensors offer much less available excess gain compared with an equivalent opposed sensor pair at the same range. Additionally, retrosensors lose excess gain *twice as fast* as opposed mode sensors, because of dirt buildup on both the retrotarget and the sensor lenses.

Effective Beam. Avoid using retrosensors for detecting small objects or for precise positioning control as it is difficult to create a small effective beam with this device.

Clear Materials. Avoid using retroreflective sensors for detecting translucent or transparent materials. It is true that a retrolight beam must pass through a translucent material two times before reaching the receiver, but it is usually difficult to optimize excess gain without sacrificing optical contrast (or vice versa).

Shiny Materials. Use the retroreflective mode with caution when sensing materials with shiny surfaces. The optics of a good-quality retroreflective sensor are designed and assembled with great care to minimize "proxing." Yet a shiny surface that presents itself perfectly parallel to a retroreflective sensor lens may return enough light to cause the object to pass by the sensor undetected.

Target Size. Except at close range, the size of the retrotarget becomes important. Use as large a target area as is possible or practical. A "cluster" of several standard targets is often most convenient. Be aware, also, that the efficiency of different retrotarget material types varies widely.

Short Range. Most retroreflective sensors are designed for long-range sensing and suffer a "blind spot" at close range. Excess gain curves and beam patterns warn of this problem.

Proximity (Diffuse Mode) Sensing—Uses and Advantages

General Application Rule. Diffuse mode sensors are used for straightforward product presence sensing applications when neither opposed nor retroreflective sensing is practical, and where the sensor-to-product distance is from a few inches (50 mm) to a few feet (600 mm). Diffuse mode sensors, being sensitive to differences in surface reflectivity, are useful for applications that require monitoring of surface conditions that relate to differences in optical reflectivity.

Convenience. Diffuse mode (and indeed all proximity mode) sensors require mounting of only one item: the sensor itself. However, in order to avoid a marginal sensing situation, this attractive convenience should not take precedence over an analysis of the sensing conditions.

Proximity (Diffuse Mode) Sensing—Application Cautions

Reflectivity. The response of a diffuse sensor is *dramatically* influenced by the surface reflectivity of the object to be sensed. The performance of diffuse mode (and all proximity mode) sensors is referenced to a 90 percent reflectance Kodak white test card; the material to be sensed should be ranked for its relative reflectivity to the same standard (see Table 14.2).

Shiny Surfaces. Diffuse sensors use collimating lenses for maximizing sensing range. As a result, response to a specular surface is sensitive to scanning angle. Divergent and convergent mode sensors are much more forgiving to orientation of the sensor to shiny surfaces.

Background Objects. As a general rule, verify that the distance from a diffuse sensor to the nearest background object is at *least four times* the distance from the sensor to the surface to be detected. This rule assumes that the reflectivity of the background surface is less than or equal to the reflectivity of the surface to be detected. Attempts to "dial out" the background objects by simply reducing amplifier gain can do *nothing* to improve the existing amount of optical contrast.

Small Parts Detection. Diffuse sensors have less sensing range when used to sense objects with small reflective area than when used to sense objects with larger reflective area.

Excess Gain. Most diffuse mode sensors lose their gain very rapidly as dirt and moisture accumulate on their lenses. In addition, buildup of dirt on the lenses of high-gain diffuse sensors may couple enough light from the emitter to the receiver to "lock on" the sensor in the light condition.

Count Accuracy. Diffuse mode sensors are usually a poor choice for applications that require accurate counting of parts. Diffuse sensors are particularly unreliable for counting glass or shiny objects, small parts, objects with irregular surfaces, or parts that will pass by the sensor at varying distances.

Proximity (Divergent Mode) Sensing—Uses and Advantages

General Application Rule. Divergent mode sensors reliably sense shiny radiused objects and are tolerant of shiny surfaces that vibrate, such as metal foil webs. Divergent mode sensors are particularly useful for sensing clear materials.

Clear Materials. Divergent mode sensors reliably sense clear plastic films or bags that bounce or "flutter." However, sensing range is limited to a few inches (50 mm) or less. Beyond this range, ultrasonic sensors should be considered for sensing of clear materials.

Small Objects. Divergent mode sensors do not exhibit the "blind spot" that diffuse sensors have for small objects at close range.

Shiny Surfaces. Divergent mode sensors are not sensitive to the angle of view to a specular surface.

Background Rejection. Divergent mode sensors run out of excess gain very rapidly with increasing range. They often may be used successfully in areas where there is a background object that lies just beyond the sensor's range. Note, however, that highly reflective objects will be recognized at greater distances than objects of low reflectivity.

Proximity (Divergent Mode) Sensing—Application Cautions

Side Sensitivity. The field of view of a divergent mode sensor is extremely wide. Objects that are off to any side of the sensor may be sensed. *Do not* recess divergent mode optics into a mounting hole.

Excess Gain. The divergent sensing mode is very inefficient. Divergent sensors offer only low levels of excess gain at sensing distances beyond 1 in (25,4 mm). They should be used only in clean to slightly dirty environments.

Proximity (Convergent Mode) Sensing—Uses and Advantages

General Application Rule. The lens system of most convergent-beam sensors focuses the emitted light to an exact point in front of the sensor. The receiver element is focused at the same point. This very efficient use of reflective sensing energy enables convergent-beam sensors to reliably sense objects with small profiles.

Excess Gain. Convergent-beam sensors make the most efficient use of reflective sensing energy. As a result, it becomes possible to detect some objects with low optical reflectivity when opposed or retroreflective sensors cannot be used. The high excess gain at the focus of a convergent-beam sensor makes the angle of view to a shiny surface forgiving, compared with the diffuse mode.

Counting Radiused Objects. Convergent-beam sensing is a good choice for counting bottles, jars, or cans, where there is no space between adjacent products (i.e., where opposed sensors cannot be used). Usually, convergent-beam sensors with infrared light sources provide the most reliable count.

Accurate Positioning. The effective beam of most convergent sensors is well defined, especially at the focus point. It is a good second choice, after opposed, for accurate position sensing of edges that travel through the focus point at right angles to the scan direction. Convergent-beam sensing becomes a first choice for accurate position sensing of clear material, such as plate glass.

Fill-Level Applications. Convergent-beam sensors may be used in some applications for detecting the fill level of materials in an open container, where the opening is too small or the surface to be sensed is too unstable to allow use of an ultrasonic proximity detector.

Color Sensing. Convergent-beam sensors with visible green LED light sources are used for color-registration sensing. Convergent-beam sensors with visible red LED light sources may also be used for sensing large color differences like black on white.

Height Differential. Convergent-beam sensors can sometimes be used for sensing height differences or for detecting the presence of an object ahead of an immediate background (e.g., parts riding on a conveyor), when opposed, retroreflective, background-suppression, or ultrasonic proximity sensors cannot be used.

Proximity (Convergent Mode) Sensing—Application Cautions

Depth of Field. Convergent-beam sensors require that the surface to be detected pass at (or close to) the focus distance from the sensor lens. Avoid using convergent-beam sensors for detection of objects that pass at an unpredictable distance from the sensor.

Effect of Relative Surface Reflectivity. Consider the reflectivity of the surface to be detected. The distance within which a convergent-beam sensor will detect an object (i.e., the sensor's "depth of field") is relative to that object's optical reflectivity. If a shiny background object returns unwanted light, tilt or rotate the sensor to move the sensing beam away from perpendicular to the shiny surface.

Proximity (Background-Suppression Mode) Sensing—Uses and Advantages

General Application Rule. Background suppression sensors ignore objects that lie beyond their sensing range, regardless of object surface reflectivity. Because of this feature, background-suppression sensors may be used to verify the presence of a part or feature of an assembly that is directly ahead of another reflective surface.

Definite Range Limit. Background-suppression sensors have a defined cutoff point at the far end of their range. Even highly reflective background objects will be ignored.

Proximity (Background-Suppression Mode) Sensing—Application Cautions

Shiny Surfaces. The beam angle to a specular surface may affect the location of a background-suppression sensor's cutoff point.

Blind Spot. The optical design of most background-suppression sensors creates a *minimum* sensing distance for small surfaces and for objects of low reflectivity.

Fiber-Optic (Glass and Plastic) Mode Sensing—Uses and Advantages

General Application Rule. Fiber-optic assemblies are used *in addition to* a photoelectric sensor to fill a variety of sensing requirements. The configuration of the fiber-optic assembly determines the sensing mode:

1. Opposed mode fiber optics: Opposed mode fiber-optic sensing calls for two individual fiber-optic assemblies. They usually plug into the same fiber-optic sensor, and the fibers are routed to opposite sides of the process (Fig. 14.34).
2. Proximity mode fiber optics: A bifurcated fiber optic plugs into a fiber-optic sensor to become a wide-beam diffuse (divergent-beam) proximity mode sensor.
3. Retroreflective mode fiber optics: The retroreflective sensing mode is configured using a bifurcated glass fiber-optic assembly with a threaded end tip and a small bundle size plus a lens threaded onto the sensing end.

Tight Sensing Locations. The small size and flexibility of fiber-optic assemblies allow positioning and mounting in tight spaces. Fiber-optic assemblies are routinely made with sensing tips as small as a hypodermic needle (Fig. 14.35).

Inherent Noise Immunity. A fiber-optic assembly is a mechanical part and is completely immune to electrical noise (RFI and EMI).

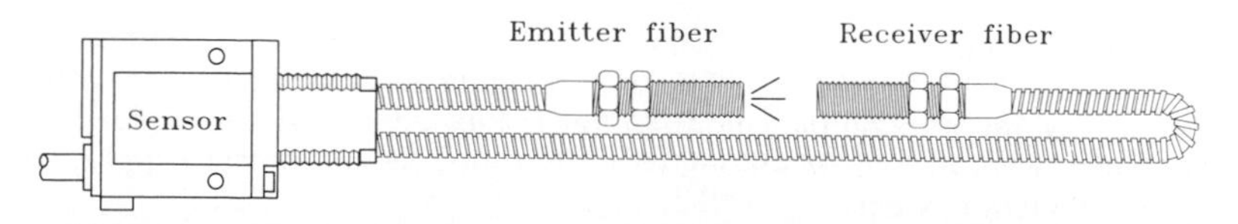

FIGURE 14.34 Two individual fiber-optic assemblies plug into a fiber-optic sensor for opposed mode sensing. (*Courtesy of Banner Engineering, Minneapolis, Minn.*)

FIGURE 14.35 Fiber-optic assemblies may be terminated in needlelike probes. (*Courtesy of Banner Engineering, Minneapolis, Minn.*)

Explosionproof Design. Fibers can safely pipe light into and back out of hazardous locations. However, the photoelectric sensor itself must be kept outside of the explosive environment, unless it is explicitly designed to safely remain in that type of environment.

Vibration and Shock. Optical fibers are very low in mass, enabling fiber-optic assemblies to withstand high levels of vibration and/or mechanical shock.

Custom Sensor Design. It is relatively easy, fast, and inexpensive to make a special fiber-optic assembly to fit a specific sensing or mounting requirement.

High-Temperature Applications (Glass Only). Most glass fiber-optic assemblies are constructed to withstand continuous duty at 480°F (248°C) and even higher with special bonding agents.

Extreme Sensing Environments (Glass Only). A glass fiber-optic assembly can be constructed to survive mechanically in areas of corrosive materials and/or extreme moisture.

Shaping of Effective Beam (Glass Only). The bundle of glass fibers often can be terminated on the sensing end to match the profile of a small object.

Fiber-Optic (Glass and Plastic) Mode Sensing—Application Cautions

Sensing System Cost. Fiber optics always add cost to a system, since a fiber-optic assembly is always a part *in addition* to a basic photoelectric sensor.

Excess Gain. A large percentage of the sensing light energy is lost when coupling light to and from a fiber. As a result, sensing ranges are relatively short and excess gain levels are generally low.

Fiber Breakage (Glass Only). Glass fiber strands fracture if bent too sharply or if repeatedly flexed.

Radiation. Glass fibers tend to darken and lose their light transmission properties in the presence of heavy x-radiation.

Continuing the Sensor Selection Process: Evaluating Sensor Package Options

Sensors may be grouped as either self-contained or *remote* types. Self-contained photoelectric sensors are those types that contain the element(s), amplifier, power supply, and output switch all in a single package (Fig. 14.36). Remote sensors, on the other hand, contain only the sensing element(s). The other circuitry is contained within an amplifier module that is located elsewhere, typically in a control panel. Remote sensors, along with their module and power supply, comprise a component system (Fig. 14.37).

A third alternative in photoelectric sensor "packaging" is *fiber optics.* Fiber-optic "light pipes," used *along with* either self-contained or remote sensors, are

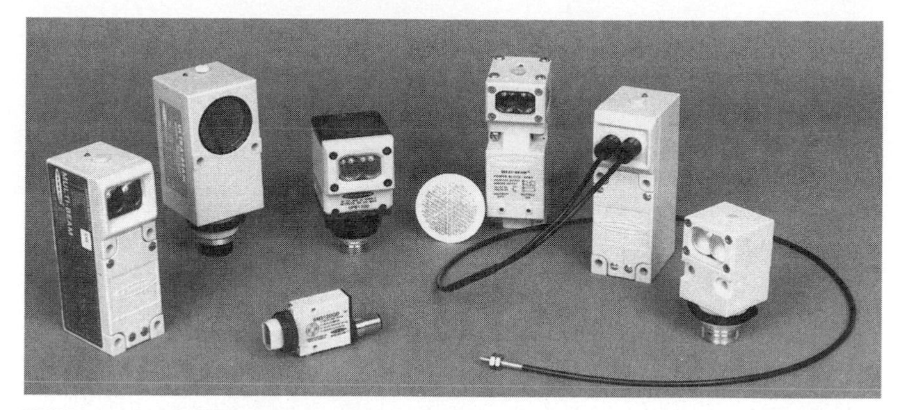

FIGURE 14.36 Self-contained sensors. (*Courtesy of Banner Engineering, Minneapolis, Minn.*)

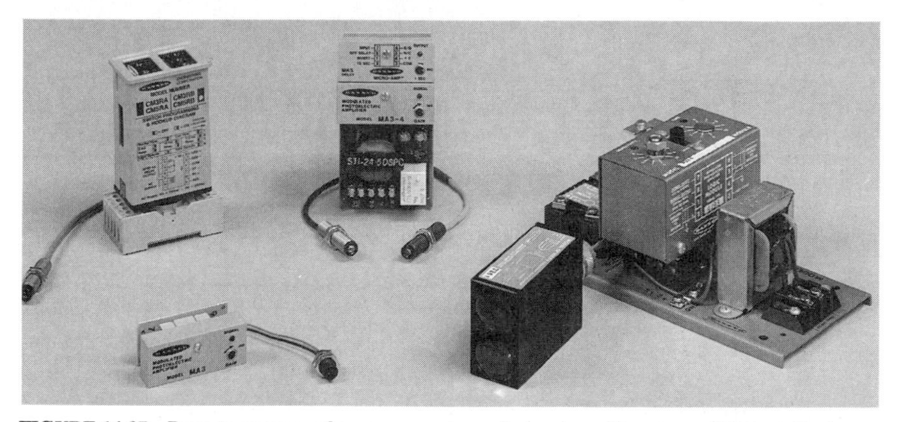

FIGURE 14.37 Remote sensors of a component sensing system. (*Courtesy of Banner Engineering, Minneapolis, Minn.*)

purely passive, mechanical components of the sensing system (Fig. 14.38). Fiber optics are used in the many sensing situations where space is too restricted or the environment too hostile even for remote sensors (component systems). Since fiber optics contain no electrical circuitry and have no moving parts, they can safely pipe light into and out of hazardous sensing locations and withstand hostile environmental conditions. Moreover, fiber optics are completely immune to all forms of electrical "noise" and may be used to isolate the electronics of a sensing system from known sources of electrical interference.

An optical fiber consists of a glass or plastic core surrounded by a layer of cladding material. The cladding material is less dense than the core material and consequently has a lower index of refraction. The optical principle of *total internal reflection* states that any ray of light that hits the boundary between two materials with different densities (in this case, the core and the cladding) will be totally reflected, provided that the angle of incidence is less than a certain critical value (ϕ).

FIGURE 14.38 Fiber-optic "light pipes." *(Courtesy of Banner Engineering, Minneapolis, Minn.)*

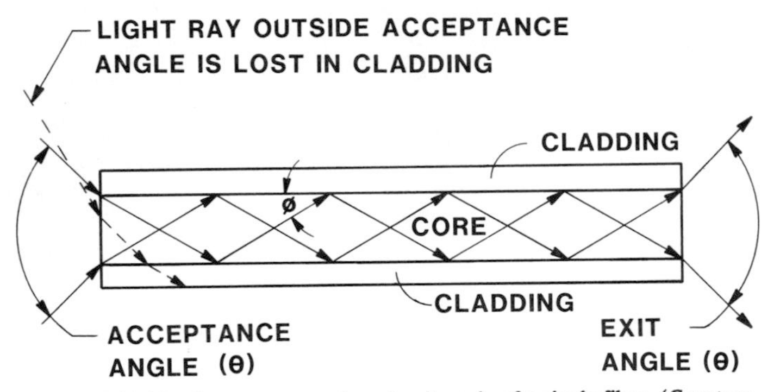

FIGURE 14.39 Acceptance angle and exit angle of a single fiber. *(Courtesy of Banner Engineering, Minneapolis, Minn.)*

Figure 14.39 illustrates two light rays (inside the angle of acceptance) that are repeatedly reflected along the length of the fiber. The light rays exit the opposite end at approximately the entry angle. Another light ray (outside the angle of acceptance) is lost into the cladding. Note that the acceptance angle is slightly larger than twice ϕ. This is because the rays are bent slightly as they pass from the air into the more dense fiber material.

The principle of total internal reflection works regardless of whether the fiber is straight or bent (within a defined minimum bend radius). Most fiber-optic assemblies are flexible and allow easy routing through tight areas to the sensing location.

Glass Fiber Optics. Glass fiber optics used for photoelectric light pipes are made up of a bundle of very small [usually about 0.002 in (0,051 mm) in diameter] glass fiber strands. A typical glass fiber-optic assembly consists of several thousand cladded glass fibers protected by a sheathing material, usually a flexible armored cable. The cable terminates in an end tip that is partially filled with rigid clear epoxy. The sensing face is optically polished so that the end of each fiber is

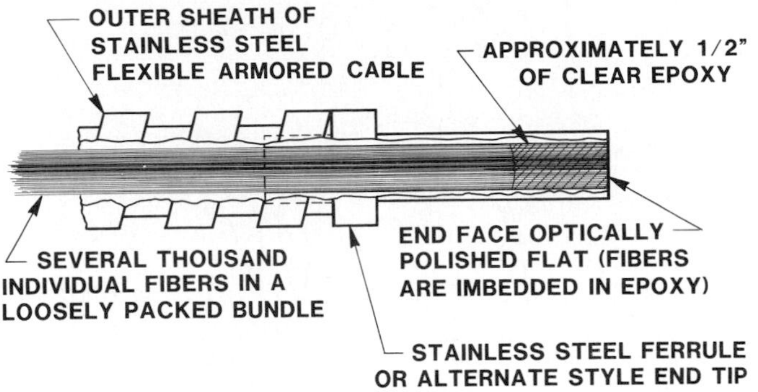

FIGURE 14.40 Construction of a typical glass fiber bundle. (*Courtesy of Banner Engineering, Minneapolis, Minn.*)

perfectly flat. The degree of care taken in the polishing process dramatically affects the light coupling efficiency of the fiber bundle (see Fig. 14.40).

There are two types of fiber-optic bundles. One type, the *coherent bundle,* is used in medical instruments and in borescopes. Coherent fiber-optic assemblies have each fiber carefully lined up from one end to the other, in such a way that an image at one end may be viewed at the opposite end. Coherent bundles are expensive to manufacture. Because the production of a clear image is irrelevant in most fiber-optic sensing applications, almost all glass fiber-optic assemblies use the much less costly *randomized bundles,* in which no special care is taken to line up corresponding fiber ends.

It is relatively easy, fast, and inexpensive to create a glass fiber-optic assembly to fit a specific space or sensing environment. These are called *special fiber-optic assemblies.* The bundle may even be shaped at the sensing end to create a beam to "match" the profile of the object to be sensed.

The outer sheath of a glass fiber-optic assembly is usually stainless-steel flexible conduit but may be PVC or some other type of flexible plastic tubing. Even when a nonarmored outer covering is used, a protective steel coil is usually retained beneath the sheath to protect the fiber bundle.

Most glass fiber-optic assemblies are very rugged and perform reliably in extreme temperatures. The most common problem experienced with glass fibers is breakage of the individual strands resulting from sharp bending or continued flexing, as occurs on reciprocating mechanisms.

Plastic Fiber Optics. Plastic fiber optics are *single* strands of fiber-optic material, typically 0.01 to 0.06 in (0,25 to 1,52 mm) in diameter. They can be routed into extremely tight areas. Most plastic fiber-optic assemblies are terminated on the sensing end with a probe and/or a threaded mounting tip. The control (sensor) end of a plastic fiber-optic assembly is left unterminated so that it may be easily cut by the user to the proper length.

Unlike glass fiber optics, plastic fibers survive well under repeated flexing. In fact, precoiled plastic fiber optics are available for sensing applications on reciprocating mechanisms. Plastic, however, does absorb certain bands of light wavelengths, including the light from most infrared LEDs (see Fig. 14.41). Consequently, plastic fiber optics require a *visible* light source, like a visible red LED, for effective sensing. Plastic fiber optics also are less tolerant of temperature extremes and are sensitive to many chemicals and solvents.

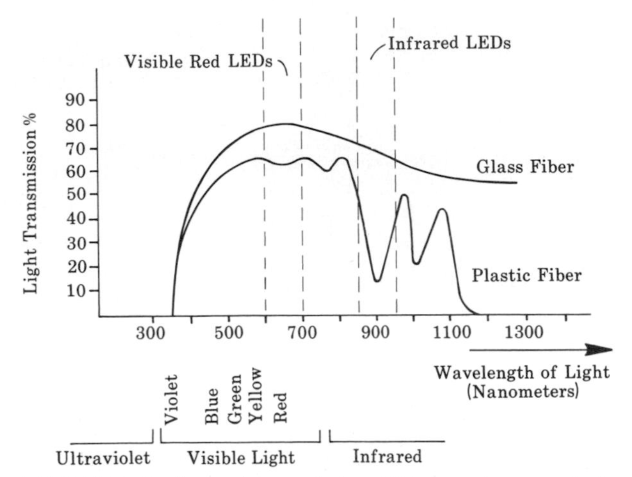

FIGURE 14.41 Spectral transmission efficiency: glass vs. plastic fiber optics. (*Courtesy of Banner Engineering, Minneapolis, Minn.*)

Self-Contained Sensors—Uses and Advantages

Simplicity of Wiring. Self-contained sensors require only a source of voltage to power them, and they can interface directly to a load. Placing the power supply, amplifier, output circuitry, and even, occasionally, timing logic right in the sensor can considerably reduce required cabinet space and associated wiring. Without a need to house control relays and logic modules in a control cabinet, self-contained sensors are usually the most cost-effective choices. Additionally, the rules governing cable runs for self-contained sensors are much less strict compared with remote sensors. Cable lengths well in excess of 100 ft (30,5 m) are usually possible.

Ease of Alignment. Self-contained sensors contain the amplifier circuitry, so they also have alignment indicators as an integral part of the sensor.

Self-Contained Sensors—Application Cautions

Accessibility of Controls. Self-contained sensors should be avoided where it is known that the sensitivity or timing logic controls must be adjusted frequently. The adjustment controls of self-contained sensors are sometimes difficult to access. The need for sensitivity adjustment should always be eliminated by optimizing sensing contrast. The need for timing adjustment should always be engineered out of a sensing system through use of mechanical references. However, there are times when either sensitivity or timing must be changed for different setups. In these situations, it is often much more convenient to adjust the controls of component amplifiers and/or logic modules that are located in a control panel.

Temperature Limitations. Avoid using self-contained sensors in temperatures exceeding 158°F (70°C). The temperature limit specifications for some self-contained sensors are even lower. *Always* check the manufacturer's recommendations before specifying a particular model, even if you have used that particular model in the past.

Self-Contained Sensor Types and Features. Self-contained sensors are either *modular* or *one-piece.* Modular sensors (Fig. 14.42) offer the benefit of flexibility in

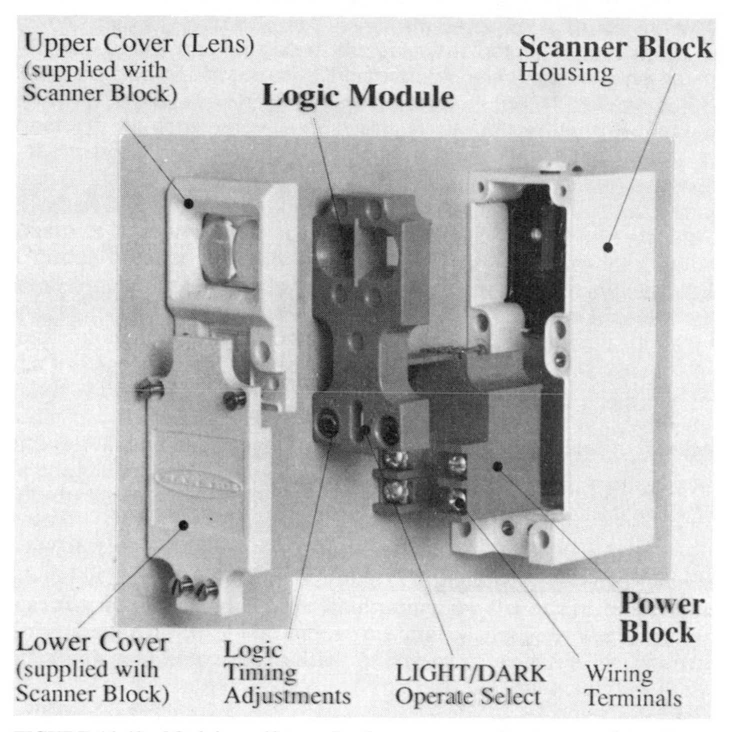

FIGURE 14.42 Modular self-contained sensors permit a large variety of sensor configurations, resulting in the correct sensor for any photoelectric application. (*Courtesy of Banner Engineering, Minneapolis, Minn.*)

sensing system design and revision. From the standpoint of system design, the modular approach permits tailoring exactly the right sensor for the application from a relatively small selection of sensor components. In addition, as sensing requirements change, a modular sensor may be modified by a simple swap of a sensor block, power block, or logic module. One-piece self-contained sensors are not as versatile as modular designs. The sensing mode, input voltage, and output configuration are dictated by the particular model. There is also no provision for the addition of timing logic within the sensor. However, some self-contained sensors are designed so that the lens may be replaced. As a result, the sensing mode of an installed sensor may sometimes be converted with a simple lens change.

An important consideration to keep in mind when selecting a sensor is *access to sensor controls* once the sensor has been installed.

Remote Sensors—Uses and Advantages

Small Sensor Size. The remote sensors of a component system contain only the sensing element(s). As a result, they may be used where small sensor size is required.

Accessibility of Controls. The amplifiers of a component system typically have easily accessible sensitivity and timing controls. This is a consideration when it is known that adjustments will have to be made frequently.

High Temperature. Some remote sensors may be placed in locations with temperatures up to 212°F (100°C). However, the amplifier and timing modules of any component system must be kept relatively cool, typically below 120°F (50°C).

Price Advantage. Multiple remote sensors may sometimes be wired into a *single* amplifier to reduce the overall cost of a sensing system.

Remote Sensors—Application Cautions

Alignment Indicator. Most remote sensors do not have an amplifier, and so they cannot have an integral alignment indicator. Instead, an alignment indicator is housed with the amplifier module, back in a control cabinet.

Wiring Precautions. When using component systems, it is very important to follow the rules for connecting the remote sensors to their amplifiers. This is especially true for modulated remote sensors:

1. Avoid running remote sensor cables in wireways together with power-carrying conductors.
2. *Always* use shielded cables, and connect the shield at the amplifier.
3. Avoid running remote sensor cable lengths longer than specified for the amplifier to be used.
4. Avoid running remote sensor cables through areas of known extreme electrical interference (e.g., areas of inductive welding or arc welding).
5. When splicing additional cable length to remote sensor leads, never combine emitter and receiver wires in one common cable. The result in a modulated system will be electrical "crosstalk" within the cable, causing a "lockup" condition in the amplifier.

Sensor Size. The size, shape, and mounting configuration of any sensor are important criteria for most sensing applications. Since a large variety of sensors are available, *always* carefully check the sensor manufacturer's published literature concerning these important attributes, and if in doubt, contact the manufacturer directly or through the manufacturer's representative.

Sensor Housing and Lens Material. Not all sensing locations are high and dry and protected from threatening elements. For some applications, sensor housing materials may be of major importance when choosing a sensor. For example, in areas with very high moisture levels it is usually best to select a noncorrosive thermoplastic housing like Valox®. In areas where industrial solvents are used, a metal housing may be needed. Attention to the material used for a lens or for a transducer cover may also be necessary in some environments. For example, a glass lens may be necessary in areas where there is acid or solvent splash. On the other hand, glass lenses may not be allowed in food processing applications.

Continuing the Sensor Selection Process: Electrical Considerations

Sensor Supply Voltage. Every sensor selection process requires an examination of the system to determine what voltage is available to power the sensor(s). Low-voltage dc sensors are usually specified whenever the interface will be to a low-voltage dc circuit or load. Low-voltage sensors also provide a relative degree of electrical safety. High-voltage ac sensors are also selected based on the sensor interface (to an ac load). However, ac sensors are often selected simply for the convenience of using readily available ac "line" voltage for sensor power. High voltage also provides a relative degree of electrical "noise" immunity.

Sensor Interface. A key step in any sensor selection process is to analyze the load to which the sensor or sensing system will interface. Here the term "load" is used in a general sense to describe an electromechanical device (e.g., a solenoid, clutch, brake, contactor), and resistive loads (lamps, heaters, etc.), or it may be an input to a circuit (e.g., a counter, programmable logic controller, electronic speed control).

The first step in analyzing a sensor-to-load interface is to determine whether the load requires a switched signal or an analog signal from the sensor. In other words, is the sensing system to switch a load "on" and "off" with a digital signal, or will the sensor provide an analog signal to a meter or control circuit? Switched outputs are typical of most sensing situations, including presence or absence, go or no-go, limit control, and counting applications. Examples of devices that require an analog signal from a sensor include meters, data recorders, speed controls, and analog inputs to programmable logic controllers. The output of most sensors and sensing systems is digital.

All sensors with digital outputs have an output switch of one type or another, which interfaces the sensor to the load. An understanding of the configuration and the capabilities of each type of switch is important in the sensor selection process. The major division in switch types is between electromechanical and solid-state.

Automated factories have adopted the programmable logic controller (PLC) to direct process control. Today's sensor is more likely to be used to supply data to a computer or controller than to perform an actual control function. For applications such as these, the benefits of solid-state relays, such as reliability and speed, become important. Because of this, the output device of most self-contained photoelectric controls is a solid-state relay.

ULTRASONIC SENSORS

How Ultrasonic Sensors Function

Ultrasonic sensors (Fig. 14.43) emit and receive sound energy at frequencies above the range of human hearing (above about 20 kHz). Ultrasonic transducers vibrate with the application of ac voltage, alternately compressing and expanding air molecules to send "waves" of ultrasonic sound outward from the face of the transducer. Ultrasonic sensors are categorized by transducer type: either electrostatic or piezoelectric. Electrostatic types can sense objects up to several feet away by reflection of ultrasound waves from the object's surface, while piezoelectric types are generally used for sensing at shorter ranges. Ultrasonic sensors are designed for either opposed or proximity mode sensing.

The transducer of an ultrasonic proximity sensor also receives "echoes" of ultrasonic waves that are located within its response pattern. Requirements for very long range (up to 20 ft is common) proximity sensing are satisfied by electrostatic sensors. Piezoelectric sensors usually have a much shorter proximity range, typically up to 3 ft (914 mm), but can be sealed for protection against harsher operating conditions.

It is possible to shape an opposed ultrasonic beam by using waveguides. Waveguides attach to the transducer of the ultrasonic receiver (and sometimes to the emitter). With waveguides attached, the receiver is less likely to respond to sound echoes that approach from the side. This makes for more reliable detection of small objects that interrupt the ultrasonic beam.

Selection of an ultrasonic sensor is based upon application requirements that are best handled only by that mode. For example, linear analog reflective positioning sensing or very long range (several feet) proximity presence sensing are application requirements that are reliably filled only by ultrasonic proximity sensors.

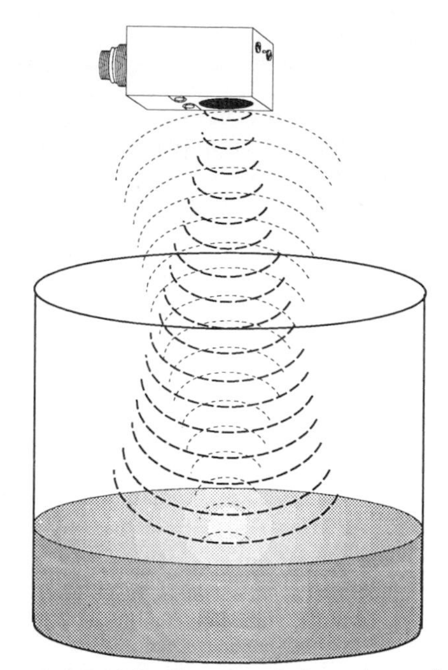

FIGURE 14.43 Detection of ultrasonic sound energy. (*Courtesy of Banner Engineering, Minneapolis, Minn.*)

Ultrasonic Proximity Sensing Mode—Uses and Advantages

Long-Range Proximity Sensing. Electrostatic-type ultrasonic proximity detectors are able to sense large targets at up to 20 ft (6,1 m) away. By comparison, the maximum range of high-powered diffuse mode photoelectric proximity detectors is about 10 ft (3,0 m).

Sensing Not Dependent on Surface Color or Optical Reflectivity. An ultrasonic proximity detector's response to a clear glass plate is exactly the same as to a shiny steel plate.

Sensing Repeatability. Ultrasonic proximity mode sensors with digital (switched) outputs have excellent repeat sensing accuracy along the scan direction. Switching hysteresis is relatively low, making it possible to ignore immediate background objects, even at long sensing distances.

Analog Response. The response of analog ultrasonic proximity sensors is highly linear with sensing distance. This makes ultrasonic sensors ideal for many level-monitoring applications. Analog response makes it possible for an ultrasonic sensor to sense within a window where a near and a far sensing limit are selected. This type of sensing is called *windowing,* or is sometimes referred to as *ranging.*

Ultrasonic Proximity Sensing Mode—Application Cautions

Sensitivity to Sensing Angle. Ultrasonic proximity sensors must view a surface (especially a hard, flat surface) squarely (perpendicularly) in order to receive enough sound echo. This is especially true of short-range piezoelectric types. Also, reliable sensing requires a minimum target surface area, which is specified in the manufacturer's literature for each sensor type.

Minimum Sensing Range. Most ultrasonic proximity sensors have a minimum specified sensing distance. At closer range, it is possible for a double-bounce echo (sensor-to-target-to-sensor-to-target-to-sensor) to be falsely sensed.

Effects of Ambient Changes. Changes in the sensing area such as temperature, pressure, humidity, air turbulence, and airborne particles affect ultrasonic response. However, these variables are usually noticed only in distance measurement application or in ranging applications where the sensing window is small.

False Response to Background Noise. Ultrasonic sensors evaluate a series of ultrasonic impulses instead of a single pulse. This gives them good immunity to the harmonics of background noise. Even so, any ultrasonic sensor is likely to respond falsely to some loud noises. The "hissing" sound produced by air hoses and relief valves is particularly troublesome.

Slow Response. Ultrasonic proximity sensors require time to "listen" for their echo and to allow the transducer to stop ringing after each transmission burst. As a result, sensor response times are typically slow, at about 0.1 s. This is usually not a disadvantage in most level-sensing and distance-measurement applications, and in fact, extended response times are advantageous in some applications.

Effects of Target Density and Surface Texture. Targets of low density, like foam and cloth, tend to absorb sound energy. Some materials may not be sensed at all at longer range. Smooth surfaces reflect sound energy more efficiently than rough surfaces; however, the sensing angle to a smooth surface is generally more critical than to a rough surface.

Obviously, the selection of photoelectric or ultrasonic sensors requires that attention be given to the details of each sensing situation. System variables or constraints often make the sensor selection process seem complex, but by contacting the potential sensor manufacturer(s) early in the system design process, the task can be greatly simplified. [Adapted from (1) and (2).]

For further, more detailed discussions concerning photoelectric sensing, the reader is referred to an exceptional treatise on the subject entitled "The Handbook of Photoelectric Sensing," available from Banner Engineering, Minneapolis, Minn.

GLOSSARY

AC-coupled amplifier. Ac-coupled amplifiers are most often used in photoelectric sensing to amplify the analog signal from a nonmodulated remote sensor. Because ac-coupled amplifiers are sensitive to very small signal changes, they may respond to unwanted conditions like sensor vibration or electrical "noise" and hence should be avoided except when they are the only solution to a close differential sensing situation. See also *DC-coupled amplifier.*

Alignment. Positioning of a sensor so that the maximum amount of the emitted energy reaches the receiver sensing element.

Ambient light receiver. A nonmodulated photoelectric receiver that is used to detect differences in ambient light level (using sunlight or incandescent, fluorescent, infrared, or laser light sources).

Ambient light saturation. A condition wherein extremely bright ambient light sources, i.e., sunlight, cause abnormal functioning of modulated photoelectric receivers.

Analog output. A sensor output that varies over a range of voltage (or current) and is *proportional* to some sensing parameter (as opposed to a digital output).

Aperture (apertured). The size of a lens opening. A mechanical part attached to a lens used to restrict the size of a lens opening. Apertures are used to limit the amount of light reaching a photoelectric receiver.

Attenuation. Lessening of sensing energy caused by environmental elements such as dirt, dust, moisture, or other contaminants in the sensing area.

Background suppression. A photoelectric proximity sensing mode with response that is similar to a diffuse sensor, but with a defined range limit.

Beam-break. See *Opposed sensing mode.*

Beam pattern. A two-dimensional graph of a sensor's response. Beam patterns are helpful in predicting the performance of the sensor.

Burn-through. The ability of high-powered modulated opposed mode sensors to "see" through paper, thin cardboard, opaque plastics, and materials of similar optical density.

Capacitive sensor. Capacitive proximity sensors are triggered by a change in the surrounding dielectric. The transducer of a capacitive sensor is configured to act as the plate of a capacitor. The dielectric property of any object present in the sensing field increases the capacitance of the transducer circuit and, in turn, changes the frequency of an oscillator circuit. A detector circuit senses this change in frequency and signals the output to change state.

Close differential sensing. Sensing situations with low optical contrast (less than 3 to 1). Includes most color registration sensing applications. Close differential sensing often requires the use of an ac-coupled amplifier.

Collimation. Optical collimation is the process by which a lens converts a divergent beam into a parallel beam of light.

Contrast (optical). The ratio of the amount of light falling on the receiver in the "light" state compared with the "dark" state. Contrast is also referred to as the "light-to-dark-ratio" as expressed by the equation

$$\text{Contrast} = \frac{\text{light level at receiver (light condition)}}{\text{light level at receiver (dark condition)}} = \frac{\text{excess gain (light condition)}}{\text{excess gain (dark condition)}}$$

Optimizing the contrast in any sensing situation will increase the reliability of the sensing system.

Convergent-beam sensing mode. A special variation of diffuse mode photoelectric proximity sensing which uses additional optics to create a small, intense, and well-defined image at a fixed distance from the front surface of the sensor lens. Also called "fixed-focus proximity mode."

Corner-cube reflector (corner-cube prism). A prism having three mutually perpendicular surfaces and a hypotenuse face. Light entering through the hypotenuse face is reflected by each of the three surfaces and emerges back through the hypotenuse face parallel to the entering beam. The light beam is returned to its source. Corner-cube geometry is used for retroreflective materials. See *Retroreflector.*

Dark operate mode. The initiation of a photoelectric sensor's output (or of timing logic) when the receiver goes sufficiently dark.

DC-coupled amplifier. An amplifier in which all signal changes, slow or fast, are amplified. See also *AC-coupled amplifier.*

Depth of field. The range of distance within which a sensor has response. See *Convergent beam sensing mode.*

Diffuse sensing mode. A photoelectric proximity sensing mode in which light from the emitter strikes a surface of an object at some arbitrary angle and is diffused from the surface at all angles. The object is detected when the receiver captures some small percentage of the diffused light. Also called the "direct reflection mode" or simply the photoelectric "proximity mode."

Digital output. A sensor output that exists in only one of two states: "on" or "off." The output of most sensors and sensing systems is digital.

Divergent sensing mode. A variation of the diffuse photoelectric sensing mode in which the emitted beam and the receiver's field of view are both very wide. Divergent sensors are particularly useful for sensing transparent or translucent materials or for sensing objects with irregular surfaces (e.g., webs with "flutter").

Effective beam. The "working" part of a photoelectric beam. The portion of a beam that must be completely interrupted in order for an object to be reliably sensed. Not to be confused with the actual radiation pattern of the emitter or with the field of view of the receiver.

Emitter (photoelectric)

1. The sensor containing the light source in an opposed mode photoelectric sensing pair (see *Opposed sensing mode*).
2. The light-emitting device within any photoelectric sensor (LED, incandescent bulb, laser diode, etc.).

Excess gain. A measurement of the amount of light energy falling on the receiver of a sensing system over and above the minimum amount of light required to just operate the sensor's amplifier. In equation form:

$$\text{Excess gain} = \frac{\text{light energy falling on receiver}}{\text{amplifier threshold}}$$

Excess gain values are used in the sensor selection process to predict the reliability of a photoelectric sensor in a known sensing environment.

Fiber optics. Transparent fibers of glass or plastic used for conducting and guiding light energy. Used in photoelectrics as "light pipes" to conduct sensing light into and out of a sensing area.

Field of view. Refers to the area of response of a photoelectric sensor.

Gain adjustment. See *Sensitivity adjustment.*

Inductive proximity sensor. Sensors with an oscillator and coil which radiate an electromagnetic field that induces eddy currents on the surface of metallic objects approaching the sensor face. The eddy currents dampen the oscillator energy. This energy loss is sensed as a voltage drop, which causes a change in the sensor's output state.

Light curtain (light screen). An array of photoelectric sensing beams configured to sense objects passing anywhere through an area (sensing plane).

Light operate mode. The program mode for a photoelectric sensor in which the output energizes (or the timing logic begins) when the receiver becomes sufficiently light.

Logic module. A sensing system accessory that interprets one or more input signals (e.g., from sensors, limit switches, or other logic modules) and modifies and/or combines those input signals for control of a process.

Mechanical convergence. A less precise form of convergent sensing compared with the optical convergent-beam sensing mode. In mechanical convergence, an emitter and a receiver are simply angled toward a common point, ahead of the sensor(s). Depth of field is controlled by adjusting the angle between the emitter and the receiver.

Multiplexing. A scheme in which an electronic control circuit interrogates each sensor of an array in sequence. "True" photoelectric multiplexing enables each modulated emitter only during the time that it samples the output of the associated receiver.

Null. This term is used in analog sensing and control to describe the minimum voltage (or current) in an analog output range. Analog sensors have an adjustment for setting the null value.

Opaque. A term used to describe a material that blocks the passage of light energy. "Opacity" is the relative ability of a material to obstruct the passage of light.

Opposed sensing mode. A photoelectric sensing mode in which the emitter and receiver are positioned opposite each other so that the light from the emitter shines directly at the receiver. An object then breaks the light beam that is established between the two. Opposed sensing is the most efficient photoelectric sensing mode and offers the highest level of optical energy to overcome lens contamination, sensor misalignment, or long scanning ranges. Also often referred to as "direct scanning" and sometimes called the "beam-break" mode.

Photocell. A resistive photosensitive device in which the resistance varies in inverse proportion to the amount of incident light. Such devices are characterized by resistances of from 1000 Ω to 1 MΩ, response speed of several milliseconds, and color response roughly equivalent to that of the human eye.

Photodiode. A semiconductor diode in which the reverse current varies with illumination. Characterized by linearity of its output over several magnitudes of light intensity, very fast response times, and wide range of color response.

Photoelectric sensor. A device which detects a visible or invisible beam of light and responds to a change in received light intensity.

Phototransistor. A photojunction device in which current flow is directly proportional to the amount of incident light. Phototransistors are well matched spectrally to infrared LEDs.

Polarizing filter. A filter that polarizes light passing through it. Also called "antiglare filters" when used on retroreflective mode sensors.

Proximity sensing mode. Direct sensing of an object by its presence in front of a sensor. For example, an object is sensed when its surface reflects a sound wave back to an ultrasonic proximity sensor. Also see *Diffuse sensing mode.*

Proxing. In retroreflective sensing, "proxing" describes undesirable reflection of the sensing beam directly back from an object that is supposed to *break* the beam. When sufficient light is reflected from the object back to the sensor, the sensor thinks it is seeing the retroreflective target, and the object may pass undetected. This is a common problem encountered when attempting to retroreflectively sense highly reflective objects.

Radiation pattern. See *Effective beam.*

Range (sensing range). The specified maximum operating distance of a sensor or sensing system:

- *Opposed sensing mode:* the distance from the emitter to the receiver
- *Retroreflective sensing mode:* the distance from the sensor to the retrotarget
- *Proximity sensing modes:* the distance from the sensor to the object being sensed

Ranging. See *Sensing window.*

Reflectivity (relative). A measure of the efficiency of any material surface as a reflector of light, compared with a Kodak white test card which is arbitrarily rated at 90 percent reflectivity. Relative reflectivity is of great importance in photoelectric proximity modes (diffuse, divergent, and convergent), where the more reflective an object is, the easier it is to sense.

Registration mark. A mark printed on a material using a color that provides optical contrast to the material color. The mark is sensed as it moves past a photoelectric color sensor.

Remote sensor. The remote photoelectric sensor of a *component system* contains only the optical elements. As a result, remote sensors are generally the smallest and the most tolerant of hostile sensing environments.

Rep rate (scan rate). In photoelectrics, this term is used to describe the time taken to interrogate each optoelement in a scanned array of receivers (e.g., for a line-scan camera, a multiplexed array of emitters and receivers, or a light curtain). It is the time from the start of one complete scan until the start of the next scan. The rep rate is one parameter that determines the system response speed.

Response pattern. See *Beam pattern.*

Response time (response speed). The time required for the output of a sensor or sensing system to respond to a change of the input signal (e.g., a sensing event). Response time of a sensor becomes extremely important when detecting small objects moving at high speed.

Retroreflective sensing mode. Also called the "reflex" mode or simply the "retro" mode. A retroreflective photoelectric sensor contains both the emitter and receiver. A light beam is established between the sensor and a special retroreflective target. As in opposed sensing, an object is sensed when it interrupts this beam.

Retroreflector. A reflector used in retroreflective sensing to return the emitted light directly back to the sensor. The most efficient type have corner-cube geometry (see *Corner-cube reflector.*

Sensing end. Term used to describe the end of any fiber-optic cable at which sensing takes place (i.e., the end at which objects to be sensed are located).

Sensing mode. The particular method a sensing system utilizes to sense the presence or absence of an object. The optical system of any photoelectric sensor is designed for one of three basic sensing modes: opposed, retroreflective, or proximity.

Sensing window. The range of distance over which the sensor will recognize an object and produce an output. The near and far limits of the sensing window are set by the null and span controls, respectively. Sensing within a window is also called "ranging."

Sensitivity adjustment. An adjustment made to a sensor's amplifier that determines the sensor's ability to discriminate between different levels of received energy (e.g., between two light levels reaching a photoelectric receiver). Sometimes called the "gain adjustment."

Specular sensing mode (specular reflection). A photoelectric sensing mode where an emitter and a receiver are mounted at equal and opposite angles from the perpendicular to a highly reflective (mirrorlike) surface.

Threshold (of a photoelectric amplifier). The value of voltage in a dc-coupled photoelectric amplifier that causes the output of the sensor or sensing system to change state. This voltage level is directly related to the amount of light that is incident on the photoelectric receiver.

Tracer beam. A visible red beam used as an alignment aid. The tracer beam is inactive, in that it does not supply any of the sensing energy.

Translucent. Term used to describe materials that have the property of *reflecting* a part and *transmitting* a part of incident radiation.

Ultrasonic. Sound energy at frequencies just above the range of human hearing, above about 20 kHz. The use of ultrasound is of advantage in many sensing applications because of its ability to detect objects without regard to their reflectivity or translucency to light.

Window. See *Sensing window.*

REFERENCES

1. "Handbook of Photoelectric Sensing." Banner Engineering Corporation, Minneapolis, Minn., 1991.
2. "Photoelectric Controls, 1991–1992." Banner Engineering Corporation, Minneapolis, Minn., 1991.

CHAPTER 15

OVERCURRENT PROTECTION DEVICES*

The design function of electronic and electromechanical devices is inevitably complicated by the presence of two conflicting design considerations—normal operation and abnormal operation. Significant design effort is devoted to normal operating conditions since this is the benchmark by which customer satisfaction is usually measured. However, whether it is simply a remote possibility or a routine condition, the abnormal operating situation must also be addressed at some point in the design development.

No matter how carefully designed, electric circuits are occasionally subject to fault or overload operating conditions. This is true whether those circuits carry power loads or support microamp communications. It is in those abnormal operating situations that circuit breakers become a design engineer's best friend. Circuit protection is generally the last thing considered in the design scheme. Despite this, it needs to be *carefully* selected and specified. (1)

Electrical current flows in a closed loop and transmits energy from one point to another. The current path is sized to accommodate the amount of flow and to withstand the circuit voltage. Any flow in excess of the design value of the system will result in loss, damage, or both. In an electric circuit this excess flow of current is referred to as *overcurrent.* Overcurrents, since they exceed design or rated values, are always abnormal—whether they are caused by an abnormal condition, such as a stalled electric motor, or by an abnormal current path, such as a short circuit. The severity of an overcurrent, or its potential for damage, depends on two principal factors: the size of the overcurrent relative to the rated current of the circuit, and the total duration of the overcurrent flow. Any action which limits the size of an overcurrent for the duration of its flow will lessen the harmful impact of the overcurrent. Any device which accomplishes this protective action is an *overcurrent protection device.* (2)

There are many examples of overcurrent protection devices such as fuses, electromechanical circuit breakers, and solid-state power switches. However, for the purposes of this book and its orientation to switches and switchlike devices, this chapter concerns itself exclusively with *circuit breakers* and *solid-state power switches.*

A circuit breaker is an automatic switching device that opens a circuit and interrupts the flow of current when an overload condition occurs in the circuit which the circuit breaker is protecting. The three main types of circuit breakers

*Numbers in parentheses indicate items in the References at the end of this chapter.

are (1) thermal, (2) magnetic, and (3) thermal-magnetic. Thermal circuit breakers use a current-sensing element which responds to the heat generated by an overload current. *Thermal breakers* are available with either a bimetal or "hot wire" sensor. *Magnetic breakers* are further broken down into two categories: (1) the instantaneous and (2) the delayed-dampened or hydraulic. Each uses a magnetic solenoid to trigger the tripping mechanism.

A *thermal-magnetic breaker* reacts in much the same manner as a thermal breaker. However, it contains a magnetic circuit which assists the tripping mechanism in getting off the line in high-current short-circuit situations (several hundred times the rating).

Solid-state power switches are power semiconductor devices used as pure on/off switching devices in overcurrent protection applications. They are not operated in their linear or amplifying mode; therefore, the devices are rated by their on-state current-carrying capabilities and their off-state voltage-blocking capabilities. Power semiconductor switching devices are classified into two broad categories: continuous drive devices and latching devices. Continuous drive devices require a continuous control signal, either a gate terminal voltage or a gate terminal current, to maintain forward or on-state conduction. Latching power devices need only a gating pulse signal to switch states. [Adapted from (2).]

OVERVIEW OF OVERCURRENT PROTECTION

Overcurrents and protective devices are not new subjects. Soon after Volta constructed his first electrochemical cell or Faraday spun his first disk generator, someone else graciously supplied these inventors with their first short circuit loads. Patents on mechanical circuit-breaking devices, in fact, go back to the late 1800s.

In a practical sense, we can say that no advance in electrical science can proceed without a corresponding advance in protection science. A design engineer should never design a new electronic power supply that does not automatically protect its solid-state power components in case of a shorted circuit. Protection from overcurrent damage must be inherent to any new development in electrical devices. Anything less leaves the device or circuit susceptible to damage or total destruction within a relatively short time.

Overcurrent protection devices are utilized in every conceivable electrical system where there is the possibility of overcurrent damage (Fig. 15.1). As an example, consider the typical industrial laboratory electrical system shown in Fig. 15.2. The figure shows a one-line diagram of the radial distribution of electrical energy, starting from the utility distribution substation, going through the industrial plant, and ending in a small laboratory personal computer. The system is said to be radial since all *branch circuits,* including the utility branch circuits, radiate from central tie points. There is only a single feed line for each circuit. There are other network-type distribution systems for utilities, but the radial system is the most common and the simplest to protect.

Overcurrent protection is seen to be a series connection of *cascading* current-interrupting devices. Starting from the load end, the circuit has a dual-element, or slow-blow fuse at the input of the power supply to the personal computer. This fuse will open the 120-V circuit for any large fault within the computer. The large inrush current, occurring for a very short time when the computer is first turned on, is masked by the slow element within the fuse. Very large fault currents are

FIGURE 15.1 A wide variety of circuit protection devices are available to the engineer. (*Courtesy of Philips Technologies, Airpax Protector Group, Cambridge, Md.*)

detected and cleared by the fast element within the fuse. Protection against excess load at the plug strip is provided by the thermal circuit breaker within the plug strip. The thermal circuit breaker depends on differential expansion of dissimilar metals, which forces the *mechanical* opening of circuit contacts.

The 120-V single-phase branch circuit, within the laboratory which supplies the plug strip, has its own branch breaker in the laboratory's main breaker box or panel board. This branch breaker is a combination thermal and magnetic, or thermal-magnetic, breaker. It has a *bimetal* element which, when heated by an overcurrent, will trip the device. It also has a magnetic-assist winding which, by a solenoid-type effect, speeds the response under heavy fault currents.

All the branch circuits on a given phase of the laboratory's three-phase system join within the main breaker box and pass through the main circuit breaker of that phase, which is also a thermal-magnetic unit. This main breaker is purely for backup protection. If, for any reason, a branch circuit breaker fails to interrupt overcurrents on that particular phase within the laboratory wiring, the main breaker will open a short time after the branch breaker should have opened.

Backup is an important function in overload protection. In a purely radial system, such as the laboratory of Fig. 15.2, it is easy to see the cascade action in which each overcurrent protection device backs up the devices downstream from it. If the computer power supply fuse fails to function properly, the plug strip thermal breaker will respond, after a certain *coordination delay.* If it should also fail, the branch breaker should back them both up, again after a certain coordination delay. This coordination delay is needed by any backup device to give the primary protection device—the device which is electrically closest to the overload or fault—a chance to respond first. The coordination delay is the principal means by which a backup system is selective in its protection.

Selectivity is the property of a protection system by which only the minimum amount of system functions are disconnected in order to alleviate an overcurrent situation. A power delivery system which is selectively protected will be far more reliable than one which is not. For example, in our laboratory system of Fig.

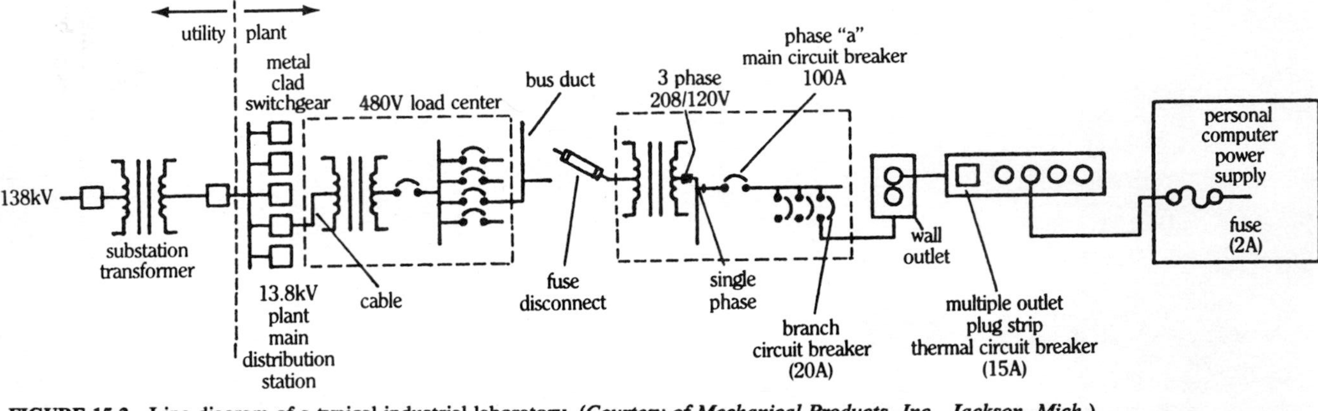

FIGURE 15.2 Line diagram of a typical industrial laboratory. (*Courtesy of Mechanical Products, Inc., Jackson, Mich.*)

15.2, a short within the computer power cord should be attended to only by the thermal breaker in the plug strip. All other loads on the branch circuit, as well as the remaining loads within the laboratory, should continue to be served. Even if the breaker within the plug strip fails to respond to the fault within the computer power cord, and the branch breaker in the main breaker box is forced into interruptive action, only that particular branch circuit is deenergized. Loads on the other branch circuits within the laboratory still continue to be served. In order for a fault within the computer cord to cause a total blackout within the laboratory, two series-connected breakers would have to fail simultaneously—the probability of which is extremely small.

The ability of a particular overcurrent protection device to interrupt a given level of overcurrent depends on the device *sensitivity*. In general, all overcurrent protection devices, no matter the type or principles of operation, respond faster when the levels of overcurrent are higher.

Coordination of overcurrent protection requires that the user-engineer have detailed knowledge of the total range of the response for particular protection devices. This information is contained in the "trip time vs. current curves," commonly referred to as the *trip curves*. A trip time–current curve displays the range of, and the times of response for, the currents for which the device will interrupt current flow at a given level of circuit voltage. For example, the time-current curves for the protection devices in our laboratory example are shown superimposed in Fig. 15.3.

The *rated current* for a device is the highest steady-state current level at which the device will not trip for a given ambient temperature. The steady-state trip current is referred to as the *ultimate trip current*.[1] The ratings for the dual-element fuse in the computer power supply, the plug strip thermal breaker, the branch circuit thermal-magnetic breaker, and the main circuit thermal-magnetic breaker are 2, 15, 20, and 100 A, respectively. Note that, except for the fuse curve, each time-current curve is shown as a shaded area, representing the range of response for each device. Manufacturing tolerances and material property inconsistencies are responsible for these banded sets of responses. Even with a finite width to the time-current curves, it is easy to see the selectivity and coordination between the different protection devices. For any given steady-state level of overcurrent, simply read up the trip time–current plot, at that level of current, to determine the order of response.

Consider the following example from our laboratory system: Assume a frayed line cord finally shorts during some mechanical movement. Assume also that there is enough resistance within the circuit, plug strip, and line cord system to limit the resulting fault current to 300 A.

The level of current is 2000 percent (20 times) of the rated current of the plug strip thermal breaker and is beyond the normal range of the published trip time specifications for thermal breakers (100 to 1000 percent of the rated current). Thus the exact trip time range of the thermal unit is indeterminate.

At high levels of fault current, greater than 150 A in this case, we can see the inherent speed advantage of magnetic detection of overcurrents. This is evidenced by the fact that the response curve for the thermal-magnetic branch circuit breaker "knees" downward sharply at current levels between 150 and 200 A. At these and higher currents, the magnetic detection mechanism within the thermal-magnetic unit is dominant. The response curve for the unit crosses over

[1]In some protection device specifications this ultimate trip current is higher (i.e., 115 percent minimum to 138 percent maximum) than the rated current of the device.

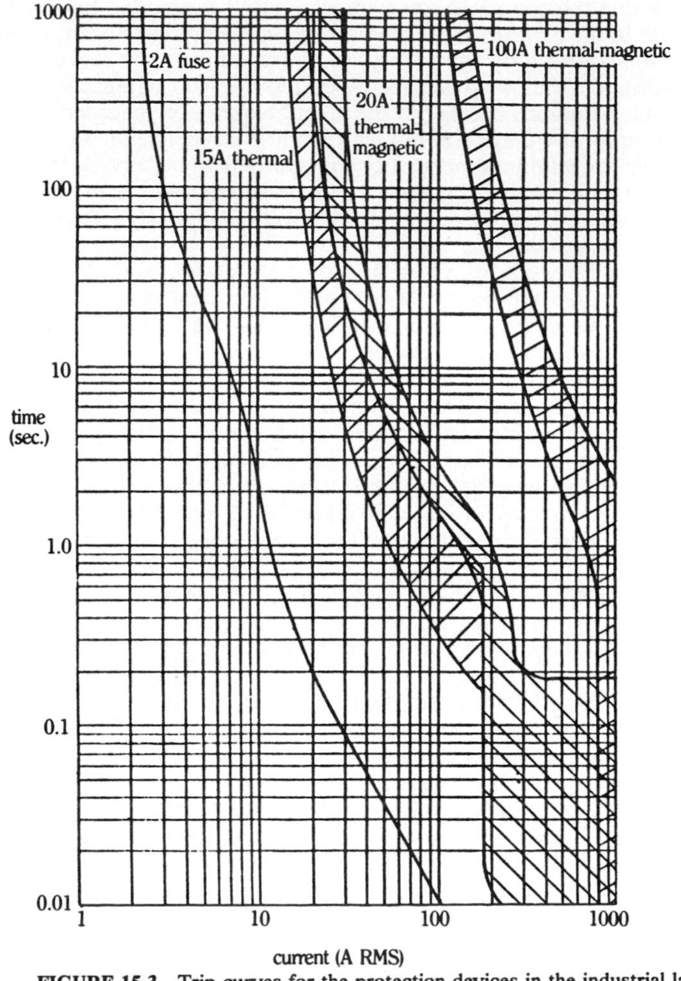

FIGURE 15.3 Trip curves for the protection devices in the industrial laboratory shown in Fig. 15.2. (*Courtesy of Mechanical Products, Inc., Jackson, Mich.*)

the plug strip thermal breaker response curve (assuming that it extends past its 1000 percent limit), and coordination between the two interrupters is lost. The range of response for the thermal-magnetic breaker at 100 A is 8 to 185 ms. Should both the plug strip breaker and the branch circuit breaker fail to operate, the main laboratory breaker should clear the fault within 11 to 40 s. (2)

Primary Hazards of Overcurrent

In the generic application for circuit protection within the confines of appliance or equipment design, there are three significant categories to consider:

1. Avoidance of *hazardous conditions* for operating personnel and others indirectly involved
2. Any *reduction* in size of current-carrying conductors
3. *Isolating a faulted function*—leaving the normally operating functions able to continue

The safety issue is undoubtedly the most important of these three. The primary safety focus is to eliminate two types of hazards: (1) electrical shock and (2) smoke and fire. The safety consideration manifests itself most often in the primary-input circuit protector. In this application, power enters the equipment and is immediately met by a protector ready to take the system off-line when required.

The matter of *current reduction* is really a traditional definition for circuit protection application. Anywhere there is a reduction in the size of a current-carrying medium, the potential exists for a fault to exceed the limits of that medium. For example, a 12-gauge wire is designed to carry ≤20 A and is protected accordingly. If a branch circuit is taken from this 20-A circuit utilizing 18-gauge wire (which, depending upon insulation, is rated to carry 3 A), then a dangerous overload in the 3-A circuit (say, 15 A or 500 percent overload) would *not* be noticed by the 20-A circuit. The result is a hazardous condition in the low-ampere circuits. This situation exists in many electrical outlet situations; for instance, when a 14-gauge–15-A line cord is plugged into a 12-gauge–20-A circuit within the building.

The third issue involves the design of fault-tolerant equipment and relates to the case illustrated in Fig. 15.4 involving distributed power through branch circuits. If the low-ampere circuits are independent functions, a fault in one circuit should not take the whole system down. By inserting a circuit breaker in each of the branches, the fault is isolated to the least affected circuit leaving the others to function normally.

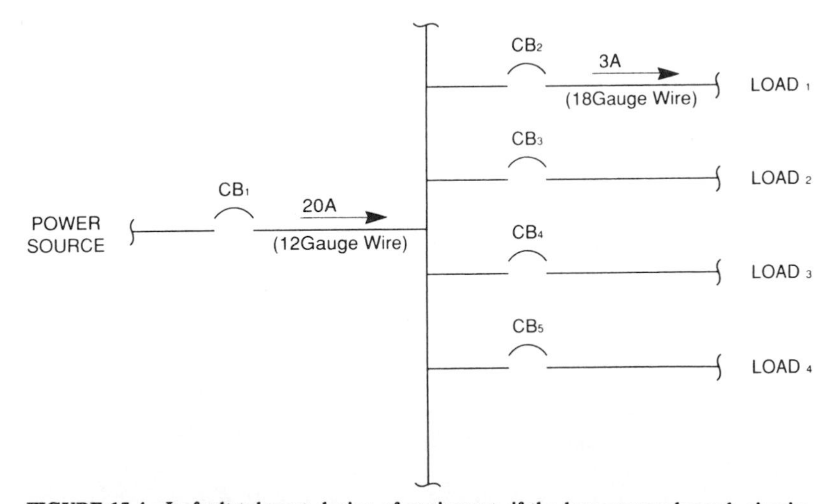

FIGURE 15.4 In fault-tolerant design of equipment, if the low-ampere branch circuits are independent functions, a fault in one circuit should not take the whole system down. By inserting a circuit breaker in each of the branches, the fault is isolated to the least affected circuit leaving the others to function normally. (*Courtesy of Mechanical Products, Inc., Jackson, Mich.*)

Types of Abnormal Conditions

In any discussion of circuit protection, it is convenient to separate fault conditions into two types. They are distinguished not only by virtue of their magnitude but also by their cause-and-effect relationship. The two are *overload* and *short circuit.*

An overload is a condition where, for various reasons, the current level within an electric circuit exceeds its specified limits but continues along its designated path. An excellent example is a stalled motor. This situation, at least initially, exceeds the normally specified steady-state condition. Overload will occur at start-up or if the rotation of the motor is impeded in some way. This overload will generally be in the area of 300 to 700 percent of the rated current level of the device. Start conditions are allowed for within most circuit parameters. Other stalled conditions must be protected.

Short circuits, on the other hand, are conditions which occur when the current path to the load is bypassed with a very low or negligible resistance path.

In general, an overload is defined by a magnitude of 200 to 800 percent of normal rating, and a short circuit is anything greater than this. However, because of its limited impedance-resistance path, a short circuit is usually considerably higher than 1000 percent of normal rating.

A short circuit involves circuit damage. An overload may or may not involve circuit damage and many times is easily eliminated without repair, even though *sustained overloads result in short circuits over time if not protected.* (1)

PHYSICS OF CURRENT INTERRUPTION

Kirchhoff's Laws

As discussed in Chap. 1, but worth repeating here, the voltage and current in a complete electric circuit obey Kirchhoff's voltage and current laws. Simply stated, these laws are:

- The rises and drops in voltage around any closed circuit (a circuit loop) must sum to zero.
- The total current flow into any one junction (connection point) must also sum to zero.

If we wish to interrupt the current in a circuit, we must do so in accordance with these laws. A great deal of Chap. 1 is devoted to the anything but simple process of circuit interruption, including arc behavior and the detailed physics of liquid metal vapor plasma states at the contact interface. The reader is directed to that discussion to emphasize the difficulty of the circuit interruption task.

Overcurrent Clearing Times

Any discussion of the physics of current interruption must address the question of detection of an overcurrent state. Before the interruption process is initiated—that is, when the contacts start to open—the interrupting device must first make a trip–no-trip decision. The period of time between the initiation of an overcurrent condition within a circuit and the initiation of interruptive action by the circuit protection device is termed the *detection period.* The different types of protection devices detect overcurrents in different ways. Thus they can have different detection periods for the same overcurrent condition.

In a thermal breaker, dissimilar metals, bonded together along a single surface, expand differently under the direct or indirect resistive heating of the overcurrent. This forces a lateral mechanical movement, perpendicular to the bonded surface, which releases a latched contact separation mechanism. In some types of thermal breakers, the contact mechanism can be formed using the bimetal material itself. In these devices, the bimetal arms and contacts snap open when they absorb sufficient energy from the circuit overcurrent. Another form of thermal breaker utilizes the longitudinal expansion of a *hot wire,* which carries the overcurrent, to release a contact latch.

The detection portion of a magnetic breaker is comprised of an electromagnet driven by the circuit current. An overcurrent will develop, within the electromagnet, enough magnetic pull to trip a spring-restrained latch which, as in the thermal breaker, allows the contacts to separate.

A solid-state switch detects overcurrents electronically, in many cases by simply monitoring the voltage drop across a low-value resistance which carries the circuit current.

Obviously, the faster a protection device can detect an overcurrent, the shorter the detection period. But, in the majority of cases, the fastest possible detection speed is not desirable. The speed of detection must be controllable and inversely matched to the severity of the overcurrent.

As noted earlier in this chapter, series-connected protection devices must be coordinated. For a given level of overcurrent, the device nearest to, and upstream from, the cause of the overcurrent must have the fastest response. Devices which are farther upstream must have a delayed response, such that the minimum circuit removal principle is adhered to. When we speak of response, we are referring to the total response time, or *total clearing time,* of the interruption device, from the time of the overcurrent initiation to the final current-zero at which interruption is completed. Since it is far easier to engineer the extent of the detection period for a given level of overcurrent than it is to control the extent of the actual current-interruption process, the total response time of any protection device is, by design, determined principally by the size of, and the time required to detect, the overcurrent state.

The *interruption period* is defined as the length of time between the start of interruptive action—for example, when the contacts start to part—and the final current-zero. The sum of the detection period is then the total clearing time, or total trip time, of the protection device. These different time periods are shown in Fig. 15.5.

In contrast to the detection period, the interruption period cannot be engineered to decrease as the intensity of an overcurrent increases. The interruption period is, however, almost always designed to be as short as possible, since during this period the protection device is absorbing energy, owing to the overcurrent flowing through the voltage drop across the contacts (or terminals in the case of a solid-state device).[2] If protection devices other than fuses do not clear the overcurrents fast enough during this period, they can be destroyed by

---

[2]A potential concern in shortening the interruption period is the possibility of overvoltages in inductive circuits due to high rates of current reduction. These inductive overvoltages could cause component damage due to insulation flashover.

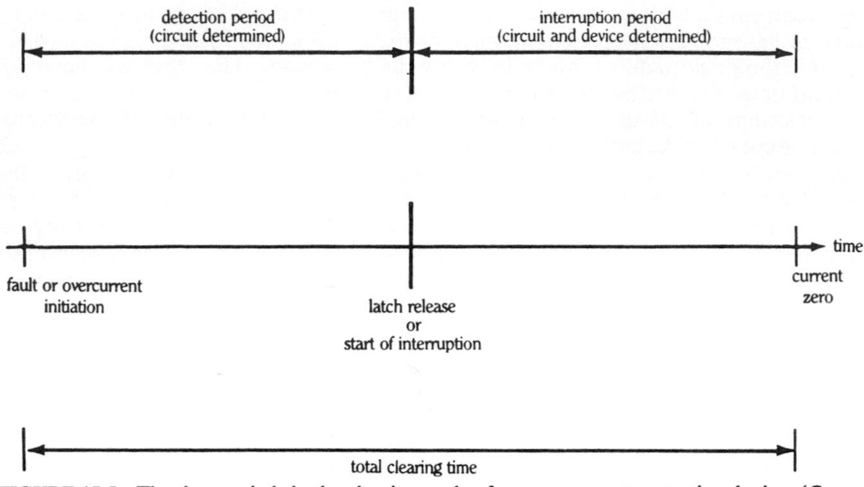

FIGURE 15.5 The time periods in the clearing cycle of an overcurrent protection device. (*Courtesy of Mechanical Products, Inc., Jackson, Mich.*)

their own power dissipation (of course fuses, by design, are always destroyed when they interrupt a circuit).

In ac circuits, the interruption period will last until either the first forced current-zero or the first natural current-zero at which the switching medium (arc or solid-state material) can reach its nonconducting blocking state. In dc circuits, the current-zero state is always a result of a forcing action by the interrupting device.

There are additional time periods of interest during the current-interruption process, such as contact travel time, arc restrike voltage transient time, thermal recovery time, and charge storage time (for solid-state devices).

General Concepts of Overcurrent Detection and Circuit Interruption

All interruption devices absorb energy from the circuit in which they operate. Even a simple mechanical switch absorbs energy within its switching arc during the time period when the arc is present. How much energy the switching device can absorb and still function in additional operations is a measure of the device's interrupt rating. For example, the on/off switch in a laboratory power supply is rated to switch X number of amperes at a given level of input line voltage. Such a switch, however, is not designed to interrupt high levels of overcurrent and could fail (i.e., be destroyed) if it is used to do so. The switching contacts in a circuit breaker, however, *are* designed for overcurrent interruption. Therefore, a breaker in series with an on/off switch would have a higher interrupt rating than that of the switch.

A portion of the energy absorbed during the operation of an overcurrent protection device is used in the overcurrent detection process. Since detection is a binary trip–no trip decision process, there must exist a threshold portion of the total absorbed energy within the device which will trigger the device's interruption process.

In a thermal circuit breaker, the threshold is reached when a certain level of thermally induced expansion or deflection is attained within the thermal element. In a magnetic circuit breaker, the threshold is reached when a movable armature has been magnetically attracted to a certain position.

We term the *detection threshold* value of energy W_{dt}. The time period between the initiation of the circuit overcurrent and the time at which the absorbed energy within the detection mechanism of the protection device surpasses W_{dt} is termed the detection period t_d (see Fig. 15.5). If the overcurrent is initiated at $t = 0$, we then have

$$W_d = \int_0^{td} v_B i_B dt$$

where W_d is the total amount of energy absorbed within the protection device during the detection period, v_B is the voltage drop across the device terminals, and i_B is the device current. Note that in all cases

$$W_d > W_{dt}$$

but only slightly, owing to inefficiencies within the protection device internal circuitry. Such inefficiencies include contact resistance losses, wiring resistance losses, conduction heat transfer away from thermal elements, etc.

The interruption period of a protection device begins immediately after the detection energy threshold has been exceeded. Within this period, contacts separate and initiate an arc in a circuit breaker. The total energy absorbed by the device during this period W_i is given by

$$W_i = \int_{td}^{tc} v_B i_B dt$$

where t_c is the total clearing time of the interruption device (again referring to Fig. 15.5). Within the interruption period $t_i = t_c - t_d$ is the arc time t_a, where $t_a \leq t_i$. The energy absorbed within the arc W_a is given by

$$W_a = \int_{tc - ta}^{tc} v_a i_B dt$$

where v_a is the voltage drop across the device arc. In all practical devices, the arc voltage drop dominates the device voltage so that $v_a \cong v_B$, and the dominant portion of the total energy absorbed during the interruption period is consumed in the arc. Thus $W_a \cong W_i$.

The total clearing energy absorbed W_c by the protection device during the total clearing time t_c is then given by

$$W_c = W_d + W_i$$

It is this total absorbed energy W_c which doubly concerns the protection device designer. In many cases the designer would like the detection threshold energy W_{dt} to be small, such that for large overcurrents the detection period can be short and the total W_c small. But the designer would also like the device to be tough (i.e., be able to interrupt very large overcurrents, with their associated large W_c's) and to survive. Within these two conflicting goals lies a compromise design.

Nowhere is this compromise more evident than in the design of a thermal circuit breaker. It is obvious that the detection of overcurrents by thermal means can be fast. A high-speed semiconductor fuse is, in fact, the fastest electromechanical device available. But a fuse can afford to be fast, since it is designed to self-destruct when it operates. It is deliberately designed to have a low thermal mass. Thus it can reach its operating (melting and vaporization) temperatures in very short periods of time. A thermal circuit breaker must have sufficient thermal mass such that it will not self-destruct during operation. This additional mass slows the response time to such a degree that the action of a pure magnetic circuit breaker can be significantly faster. A thermal circuit breaker designer is forced, therefore, to trade off device speed for device survivability.

In many respects survivability is the essence of protection science. The user-engineer—the one who specifies the particular overcurrent protection device to be used in a particular circuit—is concerned with the survivability of the circuit components. It is this engineer's task to choose the protection device which will, under a set of known fault or overload conditions, limit the amount of destructive overcurrent energy that is absorbed by the circuit components.

A nearly universal measure of the potential for damage in an overcurrent situation is the total I^2t that a particular circuit component can absorb and still survive. It is only natural then that overcurrent protection devices would also be characterized by the amount of I^2t that they let through in a given overcurrent condition. Mathematically, the *let-through* I^2t of a protection device is given by

$$I^2t = \int_0^{t_c} i_B{}^2 dt$$

Note that we are concerned with the aggregate heating current, integrated over the total clearing period. To emphasize this, we define a mean square current $<i_B{}^2>$:

$$<i_B{}^2> = \frac{1}{t_c}\int_0^{t_c} i_B{}^2 dt$$

The protection device let-through I^2t is thus given by $<i_B{}^2>t_c$.

Note that the I^2t value is an average value. Many times, two different overcurrent waveforms can have the same I^2t value, yet one can be potentially more destructive than the other. To be safe, when device I^2t limits are specified, the actual current waveform should always be specified, i.e., dc, sinusoidal, offset sinusoidal, pulse, triangular, etc. [Adapted from (2).]

Benefits of Limiting the Let-through Energy

Protection against the harmful effects of overcurrent focuses mainly on avoiding excessive heat and—to a lesser degree—on avoiding mechanical damage. Both of these harmful effects are proportional to the square of the current. Thermal energy is proportional to the square of the rms value; magnetic forces are proportional to the square of the peak value. The most effective way of providing protection is therefore *to drastically limit the let-through energy,* which provides the following advantages:

- Far less damage at the location of the short circuit.
- Fast electrical separation of a faulty unit from a system. (This applies in particular to power supplies connected in parallel which would be automatically switched off if the voltage of the bus bar drops below a certain level.)
- Far less wear in the breaker itself.
- Better protection of all components in the short-circuit path.
- Far wider range of selective action in conjunction with an "upstream" fuse (i.e., no nuisance shutdown due to a feeder-line interruption which causes a blackout in all its connected branches) (3).

Joule Heating

Current in a metallic conductor results in *Joule heating* within the conducting medium. Joule heating is the transfer of electron kinetic energy to a surrounding lattice atomic structure via collisions between current flow electrons and the host lattice atoms. The more current, the faster the average drift velocity of the electrons, leading to more kinetic energy transfer, or heating, upon an electron–lattice atom collision.

Since the average electron velocity between lattice atom collisions is directly proportional to the total current, the average kinetic energy of the electrons $1/2mv^2$ is proportional to the square of the total current. And since this kinetic energy is transferred completely to the lattice structure, the Joule heating in a conductor is also proportional to the square of the total current. The energy transfer per unit time is the heating power or I^2R power loss within the conductor. (2)

THERMAL CIRCUIT BREAKERS

How Thermal Circuit Breakers Function

Thermal circuit breakers operate by sensing the status of current flow, through the calibration of a proportional correlation to resistance heating of an active element. Since the performance of various materials under resistance heating is highly predictable, thermal sensing circuit breakers can be calibrated to provide consistent, reliable protection in a wide variety of electronic and electrical equipment. (1)

Thermal breakers simulate the electrothermal behavior of the protected component (conductors in wiring, motors, transformers, etc.) by a simple, but effective, device: the bimetal. This simple mechanical element can simulate the power losses, integrate the heating effect of such losses (integral $I^2R\,dt$), transform electrical energy into motion (deflection), and trigger a mechanism to cause automatic interruption of the current which produced these effects.

Using the heat created by the current instead of the magnitude of the current itself offers an advantage. *Heat* determines the allowable stress of the insulation and the allowable duration of the various overload conditions encountered in practical applications.

Heat-activated devices therefore easily manage the surplus energy required for startup or high-torque operation of motors. They cope well with high inrush spikes which occur in switching power supplies, transformers, tungsten filament lamps, etc., and avoid nuisance tripping due to such transients.

Similar, in fact, to the load (e.g., motor or wire), they react to the electrical energy (and energy dissipation) during overload conditions and thus allow good use to be made of the withstand capacity of the load.

Bimetals can also handle frequencies in a fairly wide range, e.g., from dc to 400 Hz, without requiring any change in ratings or characteristics.

A number of attributes associated with thermal breakers are:

- Good simulation of the thermal behavior of the protected component
- Capability of coping with start-up and inrush currents
- Suitability for a wide range of frequencies
- Simplicity and reliability

Thermal breakers are temperature-sensitive. This can be an advantage in some applications because the withstand capacity of the component to be protected is almost always temperature-sensitive as well. The variation of the operating characteristics of thermal breakers with ambient temperatures is closely matched to the allowable thermal stress of PVC insulations. For other insulations, the matching is not as close but the similarity exists, at least in principle, in any application where the protective device and the component to be protected are operating in an environment of practically identical ambient air temperature.

In special applications, however, as for the protection of submerged pumps, the ambient temperatures differ. In this type of application, the dependence on ambient temperature places the thermal breaker at a disadvantage. Breakers with compensation against ambient change or hydraulic magnetic breakers may be an alternative to thermal breakers for such an application. Nevertheless, thermal breakers can, to a certain degree, be adjusted to special requirements in regard to the withstand capacity of the protected item. Their delay time can be influenced in several ways where special calibration would not be satisfactory. The task may be accomplished by using a different method of heating the bimetal.

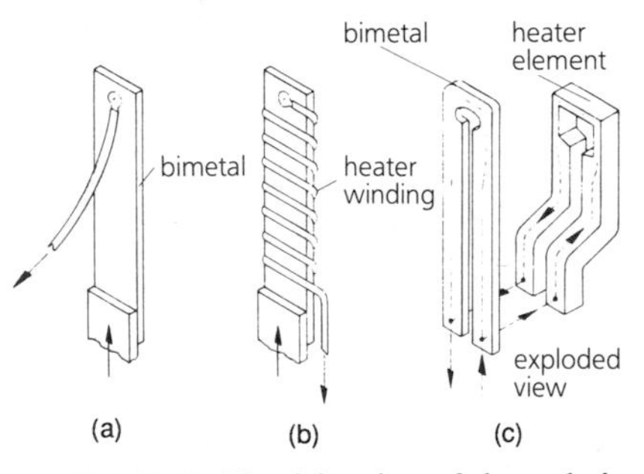

FIGURE 15.6 The delay time of thermal circuit breakers can be influenced by heating the bimetal. Three methods are shown: (*a*) directly heating the bimetal by the internal losses produced by the current passing through the bimetal, (*b*) indirectly heating the bimetal by wrapping a heater winding around the bimetal, and (*c*) combined heating by paralleling a U-shaped bimetal and a U-shaped heating element. (*Courtesy of Weber U.S.A., Inc., Malvern, Pa., and Weber, Ltd., Emmenbrücke, Switzerland.*)

The most widely used method is *direct heating* of a bimetal strip by the internal losses produced by the current passing through the bimetal (Fig. 15.6*a*). Where such losses are insufficient to produce enough heat and to cause sufficient deflection, a heater winding is wrapped around the bimetal strip to obtain the required heat (Fig. 15.6*b*). Since the heat has to pass through insulation before it reaches the bimetal, a time lag will occur and a delayed action will result.

The third example (Fig. 15.6*c*) shows combined heating by paralleling a U-shaped bimetal and a U-shaped heater element, as illustrated. This heating method simulates the behavior of electrical motors, where the iron losses contribute substantially to the heating process. The bimetal represents the copper windings, while the heater represents the iron. The delaying effect of the big mass of iron is simulated by shaping the heater in such a way that its heat is produced at a location remote to the bimetal. The heat has to first pass through a cooler section of substantially bigger cross-sectional area before it can reach the bimetal.

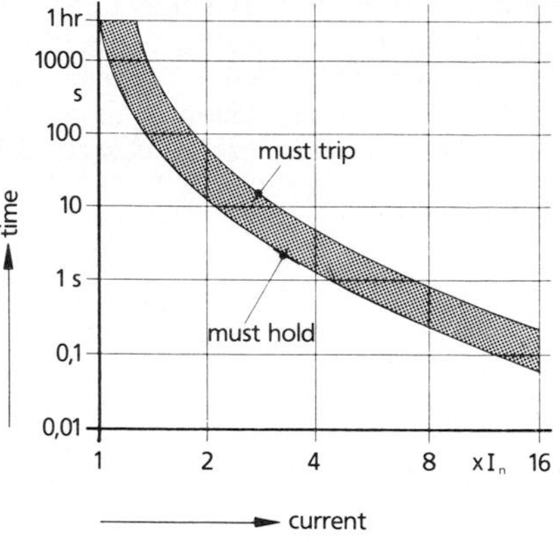

FIGURE 15.7 Typical trip curve of a thermal circuit breaker. (*Courtesy of Weber U.S.A. Inc., Malvern, Pa., and Weber, Ltd., Emmenbrücke, Switzerland.*)

The typical tripping characteristic of thermally operating breakers is shown in Fig. 15.7.

The interruption of overload currents is achieved by the opening of contacts and the setting up of an electric arc (see Chap. 1).

In case of ac interruption, the arc will continue to burn until the alternating current reaches its first natural passage through zero when reversing its direction. It will then usually extinguish and not reignite. Simple arc chambers can cope with the stress imposed on them by the comparatively low currents which have to be interrupted. By relying on the natural passage of the alternating current through zero, breakers do not have to set up a high arc voltage, nor do they have to be fast. The duration of the arc is, for thermal breakers, practically independent of the setup of the breaker. This is also true for some magnetic breakers.

In the case of dc interruptions, an arc voltage exceeding the supply voltage has to be set up to ensure interruption. This can be done by:

- Using a large contact opening
- Stretching the arc by electrodynamic blast
- Quenching the arc
- Subdividing the arc (3)

The irreversible action which triggers the interruption process in a thermal circuit breaker is the mechanical movement of the thermal element past a detection threshold position. If the thermal element movement at the threshold point is a smooth, gradual transition from its prethreshold to its postthreshold position, the device is termed a *creep-type* device. But, if the thermal element undergoes a sudden and relatively large buckling transition at the threshold point, the device is termed a *snap-type* device. Regardless of the manner of thermal element movement, creep or snap, the detection threshold is a thermal energy absorption process. In the process, a sufficiently heated bimetallic element physically moves, or a sufficiently heated wire physically elongates, a prescribed distance.

The thermal expansion of any material in any one direction, neglecting all other elastic effects, can be described by

$$x = x_0(1 + \alpha_t \Delta T) \tag{15.1}$$

where x is the length in question, x_0 is the length at the reference temperature, ΔT is the temperature rise above reference, and α_t is the coefficient of thermal expansion. For small temperature rises α_t is a constant. But for large differences between the material temperature and reference, α_t becomes a nonlinear function of the rise ΔT. Expansions of the form described in (15.1) are the basis for operation in *all* thermal circuit breakers.

Hot-wire Thermal Circuit Breakers. In a hot-wire circuit breaker a wire of length x_0 at reference temperature T_0 will expand by length $\Delta x = x_0 \alpha_t \Delta T$ for temperature rise ΔT. This temperature rise can be generated by the I^2R loss in the wire itself. If a given wire extension Δx_{trip} trips a latch mechanism (Fig. 15.8), the needed temperature rise within the wire at the detection threshold is then

$$\Delta T_{th} = \frac{\Delta x_{trip}}{x_0 \alpha_t} \tag{15.2}$$

We now have a situation exactly analogous to that of a fuse's detection mechanism. In a fuse the irreversible action triggered at the detection threshold level is the melting and vaporization of at least a portion of the fuse link material. The fuse link construction is designed such that energy absorption and subsequent

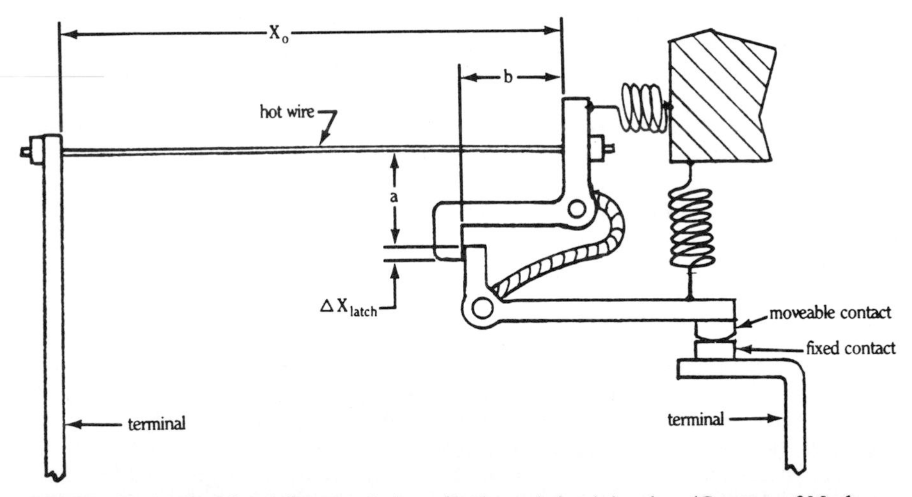

FIGURE 15.8 A functional diagram of a hot-wire thermal circuit breaker. (*Courtesy of Mechanical Products, Inc., Jackson, Mich.*)

melting is concentrated in particular weak link areas.[3] At high enough heating levels, portions of the link reach the melting temperature of the link material. Once a section of the link is molten, only a slight amount of additional heating is needed to vaporize a portion of the molten region. The exception in the hot-wire thermal circuit breaker is that we need only raise the element temperature to the value given by (15.2), not to the melting point.

Bimetal Creep Thermal Circuit Breakers. In the hot-wire thermal circuit breaker, the thermal expansion and the mechanical movement is in the longitudinal direction of the wire. In a bimetal thermal circuit breaker, the elements also expand in longitudinal directions, but the major or useful mechanical movement is in a lateral direction to the elements. It is interesting to note that thermal "creep" breakers must be derated at high ambient temperatures. However, by special design techniques, such as the addition of a second bimetal (complementary) mechanism, thermal breakers can be *ambient temperature compensated.* These compensated mechanisms show little variation over their specified operating temperature range.

Bimetal Snap Thermal Circuit Breakers. Bimetal snap thermal circuit breakers depend on the sudden lateral buckling of a bimetal element. The theory of snap-action bimetal elements is a special application of the more general theory of elastic stability of flexible structures. In one of the first mathematical discussions of the theory of elastic stability, Euler showed that column structures under axial compressive loading were subject to failure by lateral buckling at certain critical values of the axial load. In essence, this same theory applies to snap bimetal elements.

Consider the precurved bimetal blade structure shown on edge in Fig. 15.9. This precurved blade has thickness h, width b, and a slight reference temperature upward curvature, such that the midspan y-axis deflection at reference temperature is δ_0. The blade is held firm at its ends by rigid frame members A and B, restricting any x-axis elongation. If the higher-expansion side of the bimetal blade is on the inside of the precurve, the blade midspan deflection δ will tend to diminish as the blade temperature is raised owing to the bimetal bending moment. As the blade's curvature diminishes as the blade's temperature is raised, there will develop within the blade a longitudinal compressive force P, since the rigid frame will not allow any extension in its x-axis separation distance L_0. If the precurvature is large enough and the temperature is raised high enough, the blade compressive force P will reach the Euler critical value and the blade will suddenly buckle downward in an attempt to relieve the force and reach a new postbuckling locus, shown by the dotted line in Fig. 15.9. This sudden buckling is the snap we refer to when we describe a bimetal element as a snap element.

As the temperature is raised beyond the buckling temperature, the blade's downward curvature is simply increased. There is no further buckling action. If the temperature of the buckled blade is lowered, however, the downward deflection will diminish, and the compressive force P will start to rise again. As the temperature is lowered still further, the compressive force P will again reach the Euler critical value, and the blade will suddenly reverse buckle, this time, however, in the upward direction (i.e., the blade snaps back).

---

[3]Low-cost, low-current, noncritical performance fuses are fabricated without weak link regions and consist of simple uniform lengths of fusible wire.

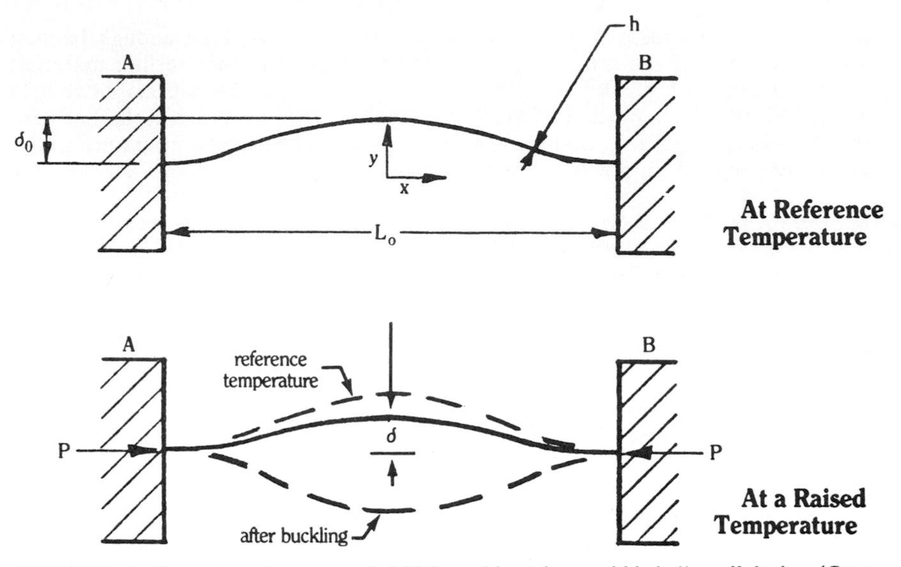

FIGURE 15.9 Geometry of a precurved rigid-frame bimetal curved-blade "snap" device. (*Courtesy of Mechanical Products, Inc., Jackson, Mich.*)

The structure shown in Fig. 15.9 buckles because it can develop enough longitudinal compressive force *P*. Are there other structures, or variations of this structure, which can also allow buildup of compressive force to the Euler critical value in a bimetal element? The answer, of course, is yes.

Consider the flexible frame structure shown in Fig. 15.10. This assembly differs from the fully rigid structure of Fig. 15.9 only in the degree of rigidity in which frame members *A* and *B* are held at separation distance L_0. In the flexible frame, any incremental increase in frame separation ΔL from the original length L_0 incurs a spring retention force $\gamma \Delta L$, where γ is the effective spring constant of the frame. This retention force is then the bimetal blade longitudinal compressive force.

There are several practical implementations of the bimetal snap structure. They all differ in the manner in which the blade longitudinal force *P* is formed. They all operate, however, by principles similar to those we have developed for the simple compressive frame structure of Fig. 15.10.

The circular disk structure in Fig. 15.11 was, perhaps, the first snap bimetal device. A snap disk supplies its own longitudinal compressive force *P* owing to its circular symmetric structure. Any lateral deflection in that disk face increases or decreases circumferential tension. This circumferential force results in a radially directed force, which can induce lateral buckling, just as in the simple blade structure. Theories concerning the stability of an isolated bimetal structure can only be approximately applied to practical disk structures, since real devices support the disk by an arm or frame, and this support destroys the circular symmetry on which these theories are established. This invalidation of theoretical solutions forces most practical disk devices to be designed using the empirical method, otherwise known as "cut, try, and remember." (2)

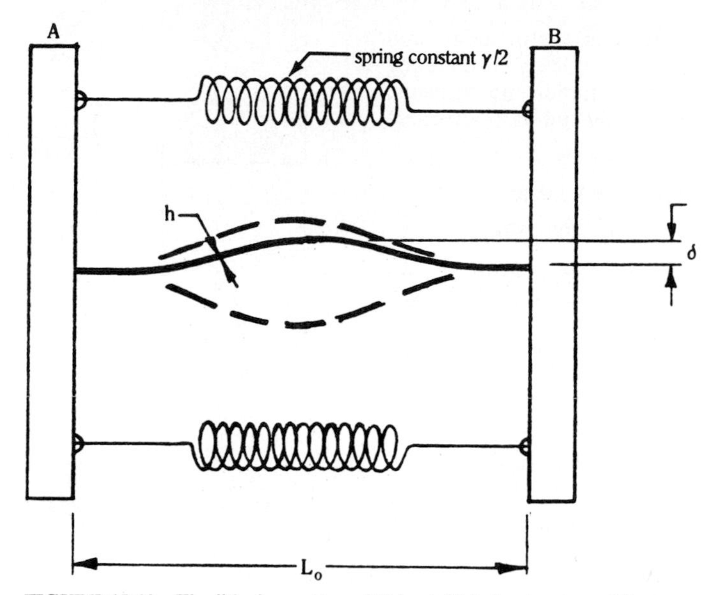

FIGURE 15.10 Flexible frame "snap" bimetal blade structure. (*Courtesy of Mechanical Products, Inc., Jackson, Mich.*)

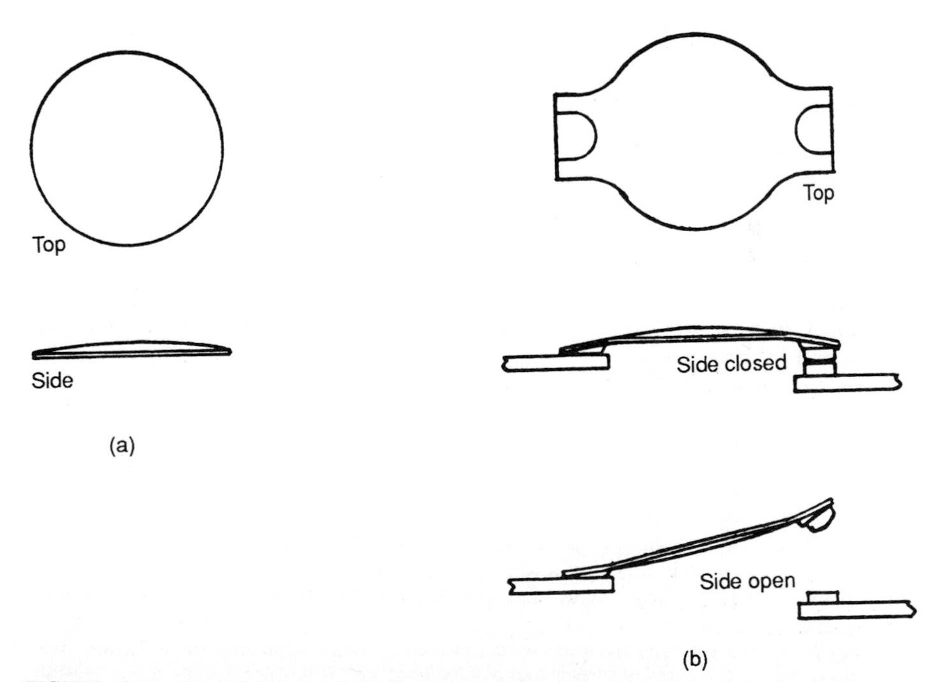

FIGURE 15.11 "Snap" bimetal disk structures: (*a*) free disk structure, and (*b*) practical disk structure. (*Courtesy of Mechanical Products, Inc., Jackson, Mich.*)

Thermal breakers are manufactured in many different frames and mounting styles. Depending on their mechanisms, they can be subdivided into breakers with:

1. A latch-type mechanism
2. A spring-type mechanism

The principal difference is illustrated by Fig. 15.12. In a latch-type mechanism, the contact force is fully retained until the deflection of the bimetal causes the unlatching of the mechanism and opening of the contact.

In a spring-type mechanism, the contact force decreases as the bimetal deflects. The initially already low contact force will be further reduced to extremely low values when the loading comes close to the "must trip" value of the breaker. Low contact force increases such risks as:

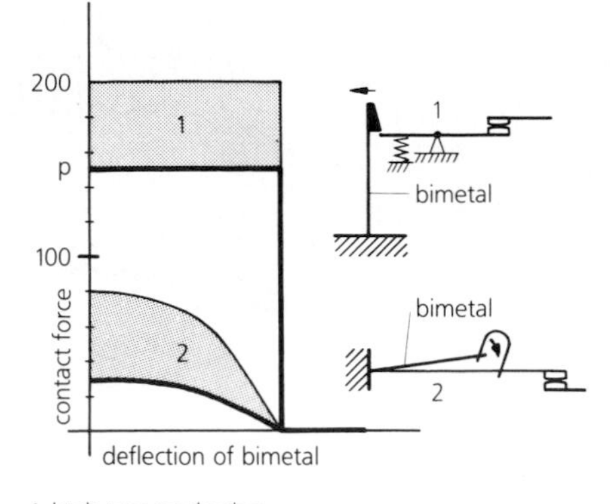

1 latch-type mechanism
2 spring-type mechanism
→ direction of deflection

FIGURE 15.12 Comparison of contact force vs. deflection in latch-type and spring-type thermal breakers. (*Courtesy of Weber U.S.A., Inc., Malvern, Pa., and Weber, Ltd., Emmenbrücke, Switzerland.*)

- Contact "noise" due to contact bounce.
- Undue wear on contacts.
- Welding of contacts if high fault currents flow through the breaker. These currents may be interrupted by a backup device, but will cause dynamic contact separation if the contact force is low. (3)

MAGNETIC CIRCUIT BREAKERS

How Magnetic Circuit Breakers Function

A magnetic circuit breaker is, in effect, an electromagnetic coil and armature device that opens a set of contacts quickly to protect the circuit whenever current exceeds a predetermined value. This occurs because the current in the coil generates sufficient magnetic flux to attract the armature. (4)

Magnetic Force. It is well known that electromagnets can exert a lifting or attractive force on *ferromagnetic* materials such as iron. The force mechanism is the same mechanism by which permanent magnets attract iron objects. Simply stated, near the surfaces of a magnetic material (a ferromagnetic material, one which has a low resistance to the flow of *magnetic flux*), the density of the energy stored in the magnetic field is much higher on the exterior than on the interior of the material. By the principle of virtual displacement,[4] there will be a mechanical

[4]All mechanical force expressions can be derived by the principle of an imagined or "virtual" displacement. A body is assumed to move in a given direction by an incremental distance Δx, and a change in total system energy ΔW, in this case, total system stored magnetic energy, is calculated for this movement. The mechanical force is then $F = \Delta W/\Delta x$. If the total system energy is lowered by the displace-

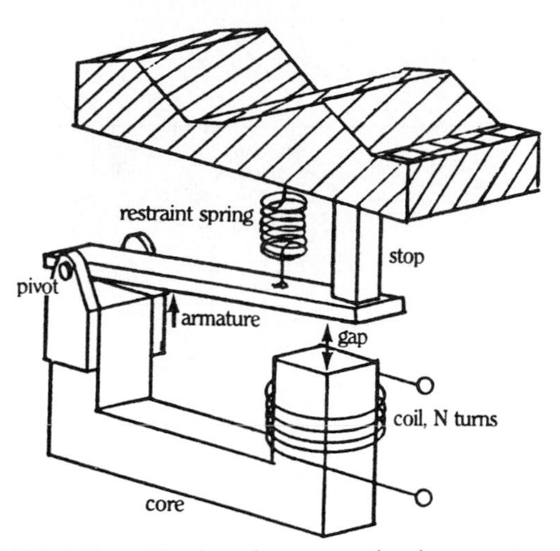

FIGURE 15.13 An electromagnetic-relay structure with a spring-restrained armature. (*Courtesy of Mechanical Products, Inc., Jackson, Mich.*)

pressure in the direction of the outward normal at the surface of the magnetic material. Since there will be more magnetic field flux at the surfaces of the material that are closest to a nearby magnet or electromagnet, the total net force on the magnetic material body will be an attractive force, toward the magnet or electromagnet.

Consider the electromagnet structure shown in Fig. 15.13. In it a coil on N turns of wire is wrapped around one leg of a ferromagnetic structure. A movable ferromagnetic armature is hinged to another leg of the core structure. At one end position of swing the armature closes the core structure and completes a closed path of ferromagnetic matter through which magnetic flux can flow. The armature is held away from the core closing position by a spring mechanism, creating a classic "relay" structure.

Coil current will induce magnetic flux within the core material, in the armature material, and in the gap between the armature and coil leg of the core. At a sufficient level of coil current the magnetic attractive force on the armature will exceed the retention force of the spring and the armature will move to its core closed position. If, by its movement, the armature can trip a latch mechanism—releasing a spring-driven contact-opening mechanism—then based on the level of coil current, we have a trip–no trip decision mechanism (i.e., we have a magnetic circuit breaker).

A simplified armature-latch release mechanism is shown in Fig. 15.14. Note that the armature's path is composed of two sections, a free-movement (spring constraint only) portion, and a latch-release (spring constraint and latch restraint force) portion. It is similar to the deflection path of the bimetallic element in a creep-type thermal circuit breaker.

ment Δx, the force is attractive, but if the total system energy is raised by the Δx movement, the force is repulsive.

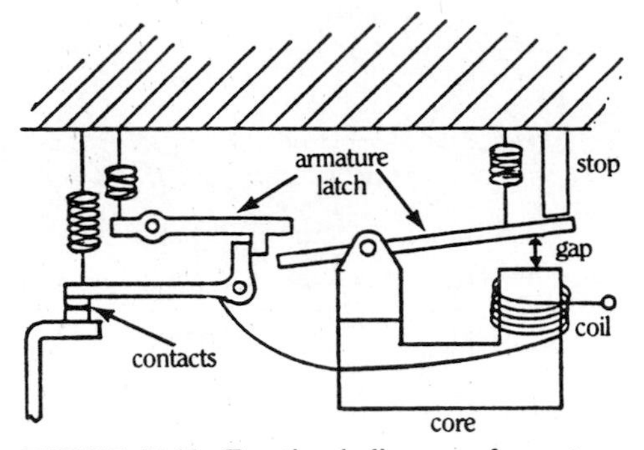

FIGURE 15.14 Functional diagram of armature-latch release and movable contact mechanisms in a magnetic circuit breaker. (*Courtesy of Mechanical Products, Inc., Jackson, Mich.*)

The Magnetic Circuit. The *magnetic flux* ϕ_g that flows through the air gap in the structure of Fig. 15.14 is a portion of the total flux generated by the current flowing in the coil which surrounds the core material. This total flux ϕ, which flows through the coil-enclosed cross-sectional area, is produced by the coil current and is proportional to both the magnitude of the coil current i and the total number of turns N of wire that makes up the coil. The product of N and i is referred to as the magnetomotive force, or *mmf,* of the coil. The proportionality factor in the relationship between the total flux produced by the coil current and the coil mmf has units of flux (measured in *webers*) per unit mmf (measured in *ampere-turns*).

Since the total flux ϕ is proportional to a force term—the mmf—a simple analogy can be made between a magnetic circuit and an electric circuit. In a dc electric circuit the current i flows because of an electromotive force (emf) E.

Ohm's law for an electric circuit states that

$$i = \text{current} = \frac{\text{emf}}{\text{resistance to current flow}} = \frac{E}{R}$$

or

$$E = iR$$

where R is the resistance of the circuit to the flow of current. A *magnetic Ohm's law* is then

$$\phi = \text{flux} = \frac{\text{mmf}}{\text{resistance to flux flow}} = \frac{Ni}{R}$$

or

$$Ni = \phi R$$

where R is the resistance of the magnetic circuit to the flow of flux. This resistance to flux flow R has been assigned a special name to differentiate it from resistance to current flow. We refer to it as the *reluctance* of the magnetic circuit.

The reluctance of a magnetic circuit will be approximately constant as long as the flux density in any one portion of the circuit is below the saturation flux density for that portion of the circuit. Ferromagnetic materials become saturated with magnetic flux at density levels of approximately 1 to 2 teslas = 1 to 2 webers/(meter)^2. At density levels near this saturation value, the effective reluctance of the material rises rapidly. At density levels below the saturation value, the reluctance of ferromagnetic elements is far below that of comparably sized elements constructed of nonmagnetic materials.

To construct a representation of a simple lumped magnetic circuit for the magnetic circuit breaker structure of Fig. 15.14, we must remember that the total flux created by the coil is made up of two components: a gap component ϕ_g which flows through the coil, the core structure, the armature, and the gap, and a leakage component ϕ_l which flows through the coil, a portion of the core structure, and a leakage air path. Figure 15.15*a* illustrates the two components flowing in their physical paths. Figure 15.15*b* illustrates an electrical equivalent "magnetic circuit" for the device. Here, the coil mmf is shown as a dc voltage source of magnitude *Ni*, and the magnetic reluctances of the different flux paths are shown as equivalent resistances.

The reluctance portion of the core which carries both the leakage flux and the gap flux is labeled R_c, the reluctance of the air path portion of the leakage flux path is labeled R_l, the reluctance of the core and armature portion of the gap flux path is labeled R_{ca}, and the gap reluctance is termed R_g. The actual values of the reluctances R_c, R_l, R_{ca}, and R_g are determined by the effective cross-sectional areas of the respective flux paths, the effective lengths of the respective flux paths, and the magnetic permeabilities of the respective flux paths. If the path

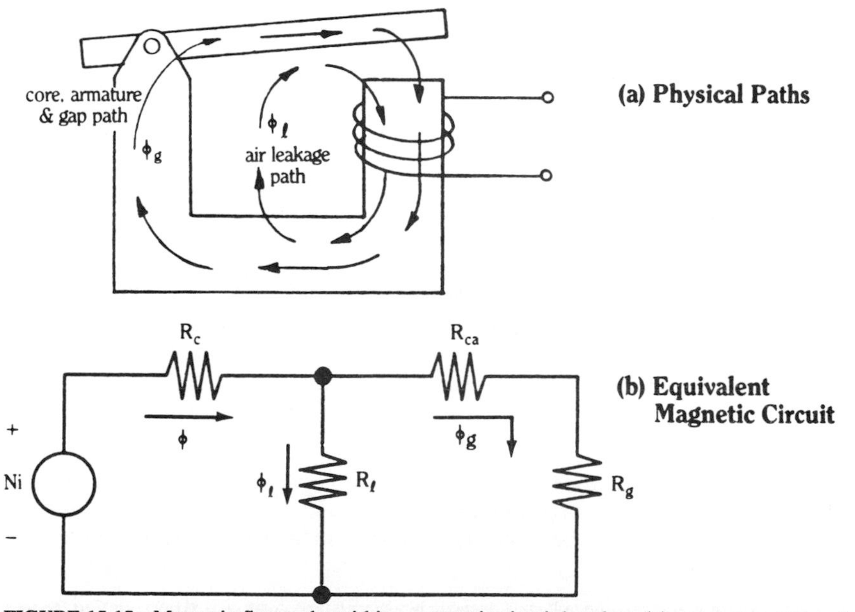

FIGURE 15.15 Magnetic flux paths within a magnetic circuit breaker: (*a*) gap component and leakage component flowing in their physical paths, and (*b*) an electrically equivalent "magnetic circuit" for the same device. (*Courtesy of Mechanical Products, Inc., Jackson, Mich.*)

material is a ferromagnetic material, such as iron, the path reluctance will also be a function of the level of flux density within the path, if the density level is near or above the saturation level.

Detection Threshold Current for Magnetic Circuit Breakers. Recall earlier in this chapter we derived expressions for what we termed the long-time dc operating current or dc threshold current. Operation of these devices at their respective levels of threshold current would ensure that eventually the devices would trip. Magnetic circuit breakers also have dc threshold currents. When magnetic devices are operated at their threshold levels, the trip or detection times are also "long" times. Action at the onset of mechanical movement in a magnetic circuit breaker, however, is more abrupt than in a thermal circuit breaker.

Mechanical movement of the thermal element is evident up to the detection threshold value in a thermal breaker. In a magnetic breaker, there is the possibility of no mechanical action (i.e., no movement whatsoever) until the threshold current is exceeded.

Time Response of Magnetic Circuit Breakers. Analysis of magnetic circuit breaker operation must include (as opposed to analysis of thermal circuit breaker operation) the mechanical dynamics of the armature. Complete analysis requires a simultaneous solution of the electric circuit which contains the breaker, the magnetic circuit of the breaker, and the mechanical equation of motion of the armature.

Eddy Currents. In our discussion of magnetic circuits to this point, we have assumed that the production of magnetic flux is instantaneously proportional to the coil current i. Strictly speaking, this is true only for magnetic circuits which do not contain any electrically conductive magnetic flux paths. When the level of magnetic flux changes (i.e., is raised or lowered) in a path or medium which can also conduct electric current, such as iron, there will arise, by Faraday's law,[5] circulating eddy currents. These eddy currents will, in turn, produce reaction magnetic flux which will tend to cancel out a portion of, or all of, the original excitation flux. The eddy currents will flow in closed paths, spatially perpendicular to the excitation flux.

As a result of eddy currents, flux cannot be instantaneously produced in conducting mediums. There will always be a time "lag" between the flux flow and the exciting mmf. A simple magnetic equivalent circuit representative of this time-delay effect can be made by including "inductive" elements in series with reluctance elements which represent electrically conductive paths. For example, the structure shown in Fig. 15.15 has two magnetic paths which are made up of ferromagnetic material: the core path and the armature path. We include eddy current effects in these two paths by placing inductive elements in series with the reluctive elements which represent these paths. This modified magnetic circuit is shown in Fig. 15.16.

Delayed Response in Magnetic Circuit Breakers. In comparison with the detection time response of thermal circuit breakers, we can classify the detection time response of magnetic circuit breakers as "fast." In many cases, magnetic breakers

[5]Faraday's law states that the voltage induced in any closed path in space is proportional to the time rate of change of the net magnetic flux which flows through the closed path cross-sectional area. If the closed path is in an electrically conducting medium, there will then be a circulating current around that closed path proportional to the generated voltage and the electrical conductivity of the medium.

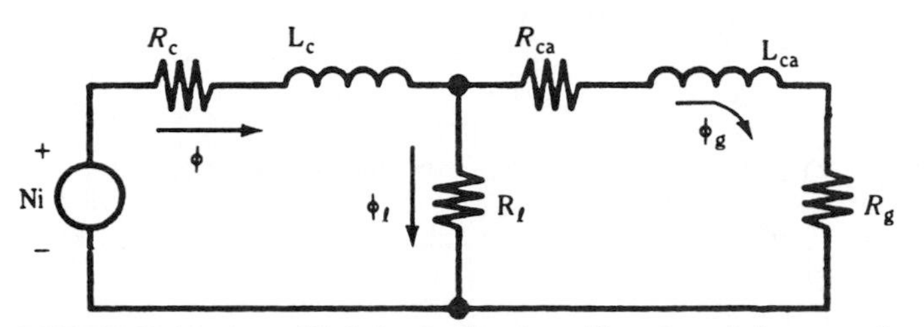

FIGURE 15.16 A modified circuit, showing eddy current inductances for the magnetic circuit breaker structure of Fig. 15.15*a* and *b*. (*Courtesy of Mechanical Products, Inc., Jackson, Mich.*)

are, in fact, *too* fast, and are subject to *nuisance trips* due to transient inrush currents. Whereas thermal breakers can "ride through" transient inrush currents by means of their relatively long thermal time constants, magnetic breakers tend to respond to the instantaneous magnitudes of inrush currents because of their relatively low trip energy requirements once the threshold current has been exceeded. Over the years magnetic circuit breaker designers have developed several schemes to delay the detection response to transient inrush currents. In essence, the goal is to mimic the dual-element "slow blow" fuse structure. The ideal "dual-element" magnetic breaker would have two detection mechanisms in series: one slow low threshold current mechanism to ride through inrush currents; and one fast high threshold current mechanism to respond quickly to true high-level fault currents.

A true, almost ideal dual-element characteristic can be achieved in a magnetic circuit breaker through use of an inertial core-delay tube. This clever device is shown in Fig. 15.17*a*. The drive coil core is a hollow tube which contains a movable but spring-restrained core ferromagnetic slug. At low coil currents the slug is spring-restrained in a recessed position, $x = 0$.

At coil currents above a certain operating current threshold, the attractive magnetic force (solenoid effect) of the coil is enough to overcome the spring force and to initiate movement of the slug. The slug moves toward the coil center and begins to fill the hollow coil core with ferromagnetic material. As the coil core fills with magnetic material the core section reluctance falls to lower values, enabling the total magnetic flux produced by the coil to increase. When the core slug reaches a certain point in the core section, dependent on the level of drive coil current, the core reluctance is decreased to a value low enough that the magnitude of the armature gap flux is sufficient to cause the armature to break away from its stopped position. The breaker then trips as described previously.

The dynamics of the slug movement can be even further slowed by the addition of a viscous fluid within the hollow core section. If the initial coil current is high enough, the gap flux will have enough strength to trip the armature without the need of flux enhancement by slug movement through the core section. The dynamics of this high-level trip are, then, the fast dynamics of a pure, simple magnetic breaker, unaffected by the flow dynamics of the core slug mechanism.

In terms of a magnetic equivalent circuit, the movable-core dual-element magnetic circuit breaker can be described as shown in Fig. 15.17*b*. The core path reluctance R_c is now a function of the slug displacement from its restrained $x = 0$ position. It is a maximum when the slug is at $x = 0$, and a minimum when the slug has advanced to its core-filled position at $x = x_{max}$. Note that the "slow" behavior of the complete system is determined by the slug movement, the "fast" be-

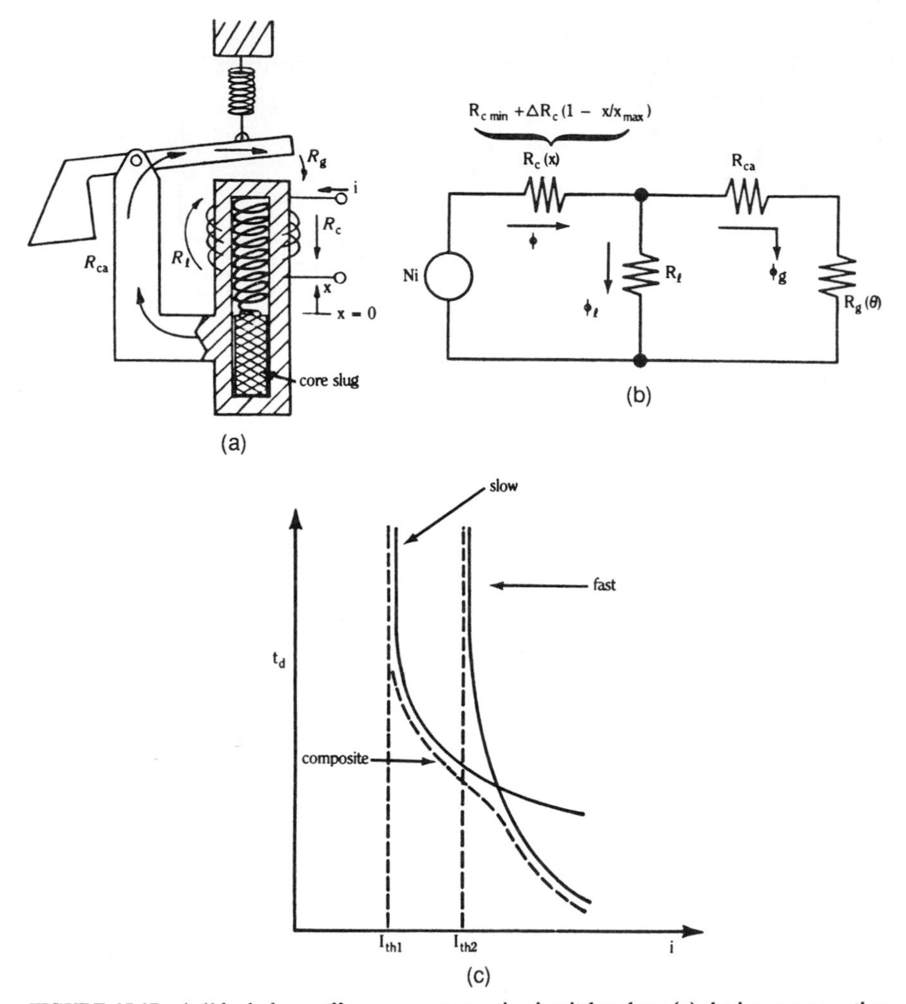

FIGURE 15.17 A "dual-element" response magnetic circuit breaker: (*a*) device cross section showing movable core slug, (*b*) equivalent magnetic circuit, and (*c*) time-current response. (*Courtesy of Mechanical Products, Inc., Jackson, Mich.*)

havior of the system is determined by the armature movement, and coupling between the two is determined by the magnetic circuit. The resultant combined detection time–device current characteristics are shown in Fig. 15.17*c*.

Other methods of desensitizing the response of magnetic breakers to inrush currents include the tailoring of the core flux reluctance path as a function of core slug position and the addition of an inertial device, similar to a flywheel, to the armature structure.

Flux can be "bled off" from the core-gap-armature path through use of flux shunts (sometimes referred to as flux busters) or through the use of an elongated core path which is not covered by the drive coil. In either situation a major portion of the core flux produced by the drive coil tends not to flow through the core-armature gap until a movable internal core slug (similar to the one in Fig. 15.17*a*) has reached its fully advanced position. Rather, this flux "bleeds" into leakage paths, producing no useful armature torque. However, when the core slug is at its most advanced position these leakage paths are effectively "shorted" and the major portion of the core flux paths have an enhanced, true, dual-element response characteristic. Extra inertial mass, when added to the armature mechanism, increases the total effective armature moment of inertia or, equivalently, the total effective armature characteristic time. This addition does not change the sensitivity of the detection mechanism; it only slows its response. It slows it, however, across the board. It does not create the desired dual-element response; rather it simply burdens the armature with additional inertia, enabling it to ride through transient inrush currents by means of sheer sluggishness. (2)

THERMAL MAGNETIC CIRCUIT BREAKERS

Thermal magnetic breakers have two means to achieve automatic interruption of an overcurrent:

1. A thermo bimetal (for the overload currents)
2. An electromagnet (for the short-circuit currents)

Consequently, the operating characteristic is essentially composed of two zones (see Fig. 15.18), connected by another zone (denoted 3 in the figure), where either one or the other mode of actuation will be effective.

The electromagnet in the breaker should be specified such that it will not trip during transients likely to occur in the intended application. This determines the level of the current below which instantaneous tripping should not occur.

The upper level, indicating the current above which instantaneous tripping must occur, is of interest in factors concerning the selective action of two protective devices.

In the short-circuit range of overcurrents (greater than 8 to 12 times the rated current), the faster interruption obtainable with the magnetic actuator is an advantage, and can help save the heater windings of indirectly heated bimetals from overheating while improving the breaking capacity of the circuit breaker.

Depending on the design of the magnetic release and the mechanism and arc chamber of the breaker, the achievable performances can vary over wide limits. Typical examples are shown in Fig. 15.19*a, b,* and *c.*

The breakers primarily intended for overload protection are usually capable of interrupting, without backup assistance, currents up to 100 to 500 A and are capable of further use after such an interruption. Performance at higher fault levels usually relies on backup assistance by fuses or breakers. (3)

SOLID-STATE OVERCURRENT PROTECTION DEVICES

Ever since the introduction of the germanium power bipolar transistor in the early 1950s, electronic circuit designers have attempted to replace mechanical

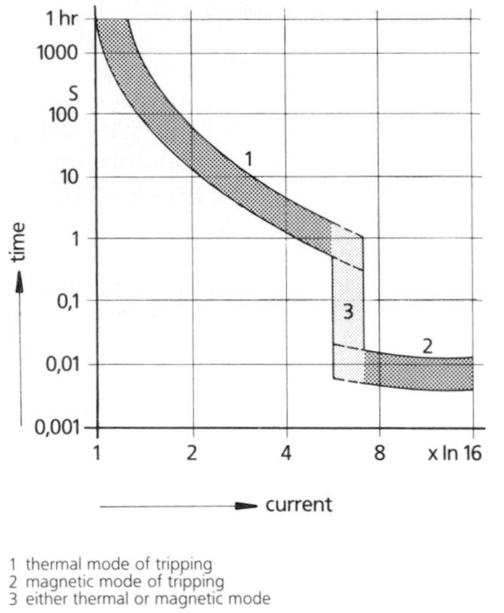

1 thermal mode of tripping
2 magnetic mode of tripping
3 either thermal or magnetic mode

FIGURE 15.18 Automatic interruption of an overcurrent in a thermal magnetic circuit breaker is accomplished by (1) a thermo bimetal (for the overload currents) and (2) an electromagnet (for the short-circuit currents). Accordingly, the operating characteristic is essentially composed of two zones, linked by another zone, (3), where either one or the other mode of actuation will be effective. (*Courtesy of Weber U.S.A., Inc., Malvern, Pa., and Weber, Ltd., Emmenbrücke, Switzerland.*)

switches in power circuits with solid-state counterparts. In certain instances, in low-voltage dc applications, or in applications which required very high speed operation or electronic remote control, they have succeeded. However, even today, for the majority of overcurrent protection applications, the cost of solid-state power switching devices exceeds the value of the functions they provide. There is little doubt, however, in the minds of solid-state device engineers, that this situation will change soon.

The technology which enables integration of solid-state power switching devices and associated heat sinking structures into and onto the same silicon substrates that contain "smart" control electronics is advancing rapidly. And, as the levels of device-control integration rise, the prices of switching modules fall.

The basic configuration of a solid-state overcurrent protection system, for ac or dc circuits, is shown in Fig. 15.20. In comparison with a simple circuit breaker, the complexity of the solid-state protection system may at first seem overwhelming, but it must be remembered that the majority (and soon perhaps all) of the functional blocks shown in Fig. 15.20 can be placed on a single piece of silicon. And thus the trade-off between the simplicity of a fuse, thermal, or magnetic circuit breaker and the flexibility, controllability, and reliability of a solid-state system must be based on a performance-cost measure for each system, not on a parts count or a configuration layout. Table 15.1 compares the major attributes of mechanical and solid-state switches as applied to overcurrent protection devices.

Power Semiconductor Switching Devices

Power semiconductor devices are used as pure on-off switching devices in overcurrent protection applications. They are not operated in their linear or amplifying mode. Therefore, the devices are rated by their on-state current-carrying capabilities and their off-state voltage-blocking capabilities. The product of the device on-state current capability and the off-state voltage-blocking capability is referred to as the device switching volt-amp (VA) capability and is an overall measure of the usefulness of the device as a power switch.

Devices are also classified by the speed at which they switch from the off- to the on-state, and vice versa. The switching speed requirements are seldom below tenths of milliseconds in protection applications. Since all of the devices dis-

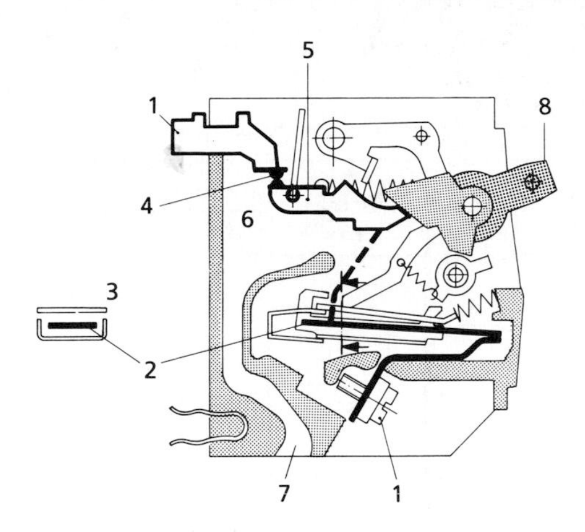

1 terminals
2 bimetal
3 magnetic release
4 contact
5 contact arm
6 arc chamber
7 arc vent
8 toggle

(a)

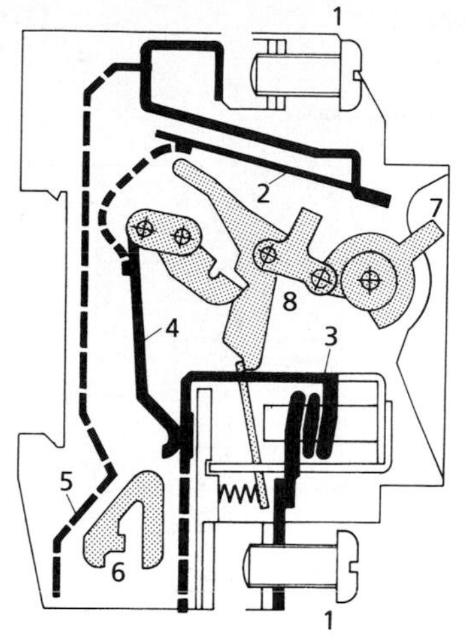

1 terminals
2 bimetal
3 magnet
4 contact arm
5 arc runner
6 arc blade
7 toggle
8 latch

(b)

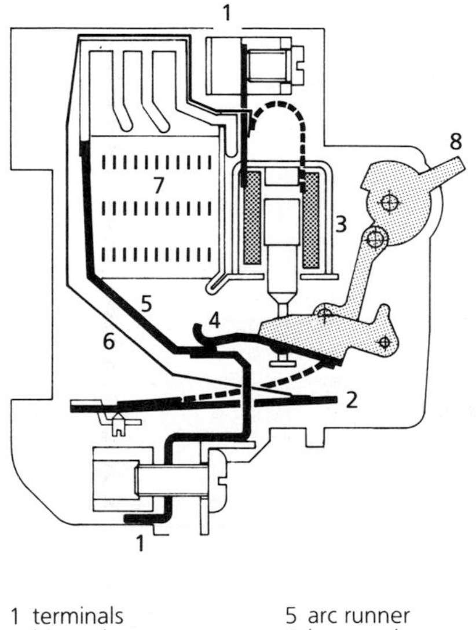

1 terminals
2 bimetal
3 magnetic release
4 contact arm
5 arc runner
6 booster-loop
7 arc chute
8 toggle

(c)

FIGURE 15.19 Cross sections of typical thermal magnetic breakers: (*a*) U.S.-style, residential version, (*b*) European-style, not energy limiting, and (*c*) energy limiting version. (*Courtesy of Weber U.S.A., Inc., Malvern, Pa., and Weber Ltd., Emmenbrücke, Switzerland.*)

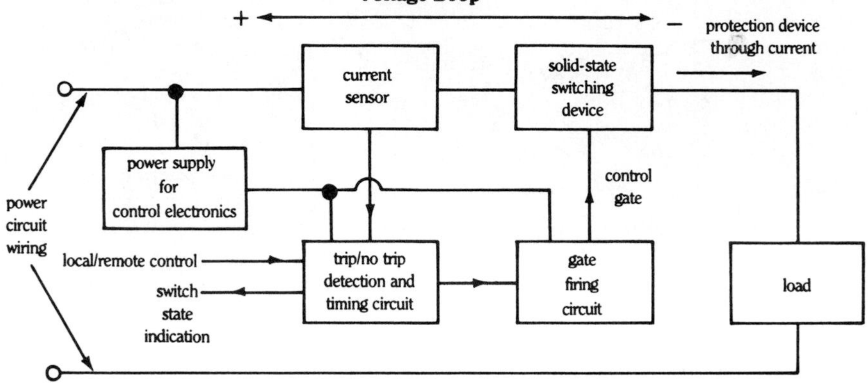

FIGURE 15.20 A solid-state overcurrent protection system. (*Courtesy of Mechanical Products, Inc., Jackson, Mich.*)

TABLE 15.1 Comparison of Mechanical and Solid-State Switches

Attribute	Electromechanical	Solid state
Reliability	Moderate (moving parts)	High (no moving parts)
Cost-switched amp	Low	Moderate–high (at present, will improve with time)
Lifetime for a given duty cycle	Moderate	Long
On-state losses	Low	Moderate
Switching losses	Moderate	Moderate
Switching speed	Slow (mechanically limited)	Fast
Ease of local control and ability to interface remote control	Moderate (requires mechanical or electromechanical interface)	Very easy (requires only electronic interface, will improve as integration levels grow)
Susceptibility to overvoltage transients	Low	High (must be protected with additional solid-state transient protection devices)
Ability to provide current-limiting function	Hard to do (can be done only with special arc control)	Easy to do (device speed enables fast switching at a given instantaneous current level)

Source: Mechanical Products, Inc., Jackson, Mich.

cussed in this section are capable of operation at these switching times and below, the choices among the various devices are those of switching volt-ampere capability and on-state loss vs. device-system cost.

Power semiconductor switching devices are classified into two broad categories: continuous drive devices and latching devices.

Continuous drive devices require a continuous control signal, either a gate terminal voltage or a gate terminal current, to maintain forward or on-state conduction. Power devices which require continuous control are the power bipolar junction transistor (BJT), the power Darlington BJT, the power MOS (metal-oxide-semiconductor) field effect transistor (MOSFET), and the insulated gate transistor (IGT).

Latching power devices need only a gating pulse signal to switch states. These thyristor devices include the semiconductor controlled rectifier (SCR), the TRIAC, the gate turn-off thyristor (GTO), and the MOS controlled thyristor (MCT).

We can further subdivide the latching power devices into a group containing those which can be turned on, but not off, under gate control; and another group containing those which can be both turned on and off with gate signals. The SCR and the TRIAC belong in the first subgroup. They can be turned off only by temporary reversal of device current. External force commutation circuits must be used if these devices are to be applied as current interrupters. Without forced commutation, these devices cannot be used in dc circuits and must be sized to survive the initial overcurrent surge, before the first current-zero, in ac circuits.

The GTO and MCT are members of the second subgroup. Current in these devices can be commutated by simple gate control.

All the devices mentioned above, both continual control and latching, are normally off devices. That is, they remain in the voltage-blocking or off state until a suitable gate signal is applied.

Electronic Detection of Overcurrents

Solid-state overcurrent protection devices usually rely on electronic means to detect overcurrent levels. Electronic detection is more flexible. That is, detection levels and tripping decision delays can be easily changed by choosing different element values, adjusting variable element values, or changing resident software. Changes can even be made from a remote location via electronic means. Electronic detection can easily mimic dual-element fuse behavior such that normal inrush currents will be masked yet large fault currents will be rapidly tripped. Electronic detection is even fast enough to provide control for fault current limiting.

Almost all electronic detection systems utilize the same basic circuitry. Current flow through a low value sampling or shunt resistor produces a voltage which is proportional to the instantaneous value of the current. This voltage signal is then "processed" and ultimately compared with a reference level or set of reference levels, to determine the state (trip, no trip, alert, etc.) of the protection system. (2)

SELECTION OF OVERCURRENT PROTECTION DEVICES

The most important aspect of circuit protection selection is a complete understanding of the system to be protected. Because the protection design is usually the last consideration, and time is always at a premium, this aspect of design is usually rushed. However, if time is devoted to this important aspect of design, and the following methodology is utilized, the final system design should be safe, economical, and properly specified.

McCleer (2) suggests that the seven-step procedure given below be followed whenever selecting an overcurrent protection device. The first six steps in this procedure define the problem in a detailed engineering sense such that the seventh step, the actual choice of a particular protection device, can be made on a logical and relevant knowledge base.

The seven steps in the selection process are listed in a condensed form, followed by a discussion of each step in detail:

1. Determine *what* is to be protected and *why:* device(s), component(s), or circuit(s).
2. Determine *how* damaging overcurrents and natural inrush currents and surges can flow through the devices listed in step 1.
3. Determine *where* a current interruption device should be placed to check the currents of step 2 and to coordinate with other protection devices in the system.
4. *Calculate* the magnitude and duration of the potential fault currents of step 2 with respect to damage thresholds of the device(s), component(s), or circuit(s) of step 1.
5. List any *supplementary requirements* for the protective device such as an auxiliary switch for an alarm circuit, lighted actuation, environmental considerations, electrical configurations (shunt trip, relay trip, etc.).
6. Determine which (if any) *regulatory requirements* (UL, CSA, VDE, MIL-SPEC, etc.) apply to the prospective protection device applications.
7. Choose a reliable protection device which:
   a. Meets the performance requirements determined in step 4.
   b. Allows natural transient inrush currents, considered in step 2, to pass unhindered.
   c. Functions at the required rated current and system voltage level and coordinated with other protection devices within the system as considered in step 3.
   d. Meets the ancillary requirements determined in step 5.
   e. Meets the regulatory considerations determined in step 6.

Discussion of Selection Steps

1. *What and Why.* Although seemingly trivial, this step is the most important of all. It is here that the application and the objective of the protection scheme are clearly defined.

   Often, these considerations are not obvious. For example, consider the fuse which is in series with the 120-V input line in almost all electronic equipment (electronic test equipment, stereo amplifiers, etc.). What is it protecting and why is it there? Is it protecting all the interior electronic components and wiring from overcurrent damage? Emphatically not! From its exterior position this would be, in general, impossible. From the line side of the unit an input fuse can protect only major interior components—components which, by design, pass major portions of the normal current draw of the device. Such components are input power transformers, power supply rectifiers, or output power transistors, etc.

   Components within low power draw subcircuits, such as preamplifiers, timing circuits, or logic circuits, could easily be damaged by overcurrent flow caused by failure of another component, such as a bypass electrolytic capacitor. Yet these damaging overcurrents will be of limited magnitude and, when added to the total current draw of the unit, will be masked since the total unit current draw will still be within the margin of the steady-state value specified for the line side fuse.

Each component or set of components that should be protected from overcurrent damage must be identified at the start of any protection system design. Of near equal importance is to note if a component or set of components:

- Is not deemed critical
- Is not economical to protect
- Is not considered a fire hazard, if subject to overcurrent flow

2. *How.* There are several different situations in which the selected component or set of components from step 1 can be damaged by overcurrent flow. These should be identified. The source(s), path(s), and cause(s) of overcurrents should be listed in order of the magnitude of potential damage to the component(s) to be protected. Also noted should be the natural inrush currents (if any) which flow through the component(s) of interest.
3. *Where.* The path(s) of overcurrent flow identified in step 2 must be opened by one or more protection devices. The location of a suitable protection device should be chosen carefully. It should protect the maximum amount of circuitry without sacrificing protection of downstream components (recall the example of the line side fuse and its inability to protect all the interior components). The placement of the protection device must also coordinate with other protection devices within the system. For example, it would be redundant to place a 20-A rated protection device in the input line of a piece of equipment plugged into a branch circuit which is already protected by a 20-A breaker at the branch source.
4. *Calculate.* This step is the most technical of all seven steps. In it we quantify the detailed timing requirements which must be satisfied if protection devices placed at the locations specified in step 3 are to minimize damage due to the overcurrents identified in step 2.

   All available overcurrent damage information concerning the component(s) identified in step 1 should be collected. For example, conductors can carry overload currents for short periods of time without undergoing permanent damage. The curve showing the "no permanent damage" limiting amount of overcurrent vs. the "time the overcurrent is present" is referred to as the wire's *smoke curve.* An example set of smoke curves for thermoplastic insulated copper conductors is given in Fig. 15.21*a* and *b.* Often, though, we cannot obtain information as detailed as that given in a smoke curve. Often only a single value of permanent damage I^2t is given at a single value of device exterior reference temperature. If no information is available, then at a minimum, a damage I^2t must be estimated.

   Next, the overcurrent scenarios identified in step 2 must be quantified. Analytic methods, graphical aids supplied by various protection device manufacturers, or computer-aided engineering circuit analysis programs can be utilized to determine worst case overcurrent time waveforms.

   Data for the overcurrent calculations must be obtained for each component within the system under consideration. The required data are available from component data sheets supplied by the component manufacturers. Many times, however, simple worst case estimates are entirely adequate. For example, consider the case of choosing a circuit breaker for a 120-V, 60-Hz appliance. The worst case fault current for the breaker would be a solid short on the appliance side of the protection device. If we cannot be sure of the exact details of the 120-V system which powers the appliance we must assume the "worst"; that is, the appliance is connected directly to a 120-V distribution

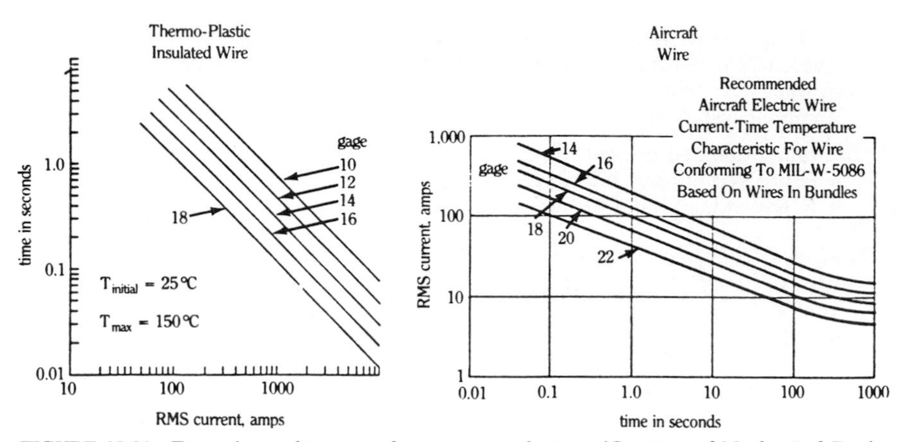

FIGURE 15.21 Example smoke curves for copper conductors. (*Courtesy of Mechanical Products, Inc., Jackson, Mich.*)

box or panel board. The short-circuit rating for most 120-V distribution boxes is 5000 A(rms). This amount of short-circuit current is not necessarily available at the outlet of each distribution box that is rated for 5000 A(rms), but since we cannot be sure of the supply system "behind" the box, we must assume this maximum value.

The magnitude Z for the Thevenin equivalent system impedance is then

$$Z = \frac{120 \text{ V}}{5000 \text{ A}} = 0.024\ \Omega$$

To find the worst case fault current at the breaker location we then add the line cord impedance and any breaker internal resistance to the Thevenin system impedance. The fault current is then 120 V divided by the total impedance. For convenience, typical values of resistances[6] of various size copper conductors are given in Table 15.2. To complete our example, assume the line cord is 6 ft (1,83 m) of paired No. 16 conductor. To be conservative we neglect any breaker impedance. We then have for the total system impedance, assuming (as a worst case) an inductive system impedance

$$Z_{\text{total}} = \sqrt{\left(2 \times 6 \times \left(\frac{4.10}{1000}\right)^2 + (0.024)^2\right.} = 0.055\ \Omega$$

The potential worst case, steady-state fault current i_f, at the breaker is then

$$i_f = \frac{120 \text{ V}}{0.055\ \Omega} = 2180 \text{ A rms}$$

If the actual feed line (from the distribution box) is known or if any interior (within the appliance) wiring can be specified, the added contribution to the system impedance Z would lower this calculated value of the worst fault current.

[6]For wire sizes No. 10 and above, the 60-Hz impedance is almost completely resistive.

TABLE 15.2 DC and 60-Hz Resistances of Copper Conductors at 77°F (25°C)

Conductor size, AWG	Resistance per 1000 ft, Ω
18	6.51
16	4.10
14	2.57
12	1.62
10	1.02

Source: Mechanical Products, Inc., Jackson, Mich.

Figure 15.22 graphically shows the objective of this step. A time trajectory of the rms overcurrent is drawn on the same plot as the component damage curve. This *component damage curve* is referred to as the device *"must protect"* curve. In Fig. 15.22 we show two sample rms current trajectories: a nondamaging transient inrush current and a fault current. We see that, in the case of the fault current, we must alter the device current trajectory (i.e., interrupt the current) before time t_x, the time at which the fault trajectory crosses the "must protect" curve. This time determines the required speed of a potential interruptive device.

5. *Supplemental Requirements.* In this step we simply list all the desired noncritical (as related to the actual interruption process) attributes we wish the protection device to have. These attributes can be chosen from the list given in Table 15.3.
6. *Regulatory Requirements.* If regulatory agency approvals are required for a protection device in the application under consideration, they should be listed in this step.

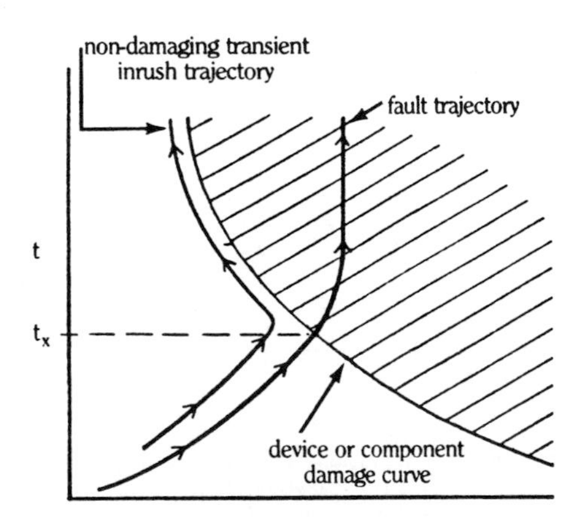

FIGURE 15.22 A component "must protect" curve plotted with fault and transient inrush trajectories. (*Courtesy of Mechanical Products, Inc., Jackson, Mich.*)

TABLE 15.3 Supplemental Requirements of Overcurrent Protection Devices

Performance parameter	Application requirement
Ratings:	
Current	
Tolerance	
% must not trip	
% ultimate trip	
Operating temperature	
Rupture capacity	
Vibration range	
Shock method	
Voltage:	
dc	
ac	
Number ϕ	
Voltage line to line	
Dielectric test:	
Sea level	
Altitude	
Altitude	
Seal:	
Sealed	
Hermetic	
Indication	
Ability to remote control	
Type reset	
Auxiliary contacts number and type	
Insulation	
Endurance	
Overload cycling	
Reclosing	
Acceleration	
Sand and dust	
Corrosion	
Humidity	
Explosionproof	
Applicable documents	
Watts loss	
Voltage drop	
Temp. rise	
Temp. cycling	
Cascade application	
Compatible operation	
Configuration	
Other	

Source: Mechanical Products, Inc., Jackson, Mich.

The Protection Device in Question

	Can satisfy interruption requirements as determined in step 4	Can be supplied with supplemental functions as listed in step 5	Can meet agency approval considerations as detailed in step 6	Other considerations, cost, etc.
Fuse				
Thermal CB				
Thermal-Mag CB				
Magnetic CB				
Solid-State Switch				

FIGURE 15.23 Overcurrent protection device comparison matrix for a particular application. (*Courtesy of Mechanical Products, Inc., Jackson, Mich.*)

7. *Choose.* If we have completed the detailed work required in steps 1 through 6, this last step should be nearly automatic. We can, if documentation is required, form a "comparison matrix," which shows at a glance the capabilities of the various interruption device technologies as related to our specific application. A sample comparison matrix is shown in Fig. 15.23. We can answer a simple yes or no to each entry, or qualify an answer with an "almost" or "needs study." The last column is a catch-all category wherein any special or overriding considerations can be noted.

Overcurrent Protection Device Selection Example

It is best to explain McCleer's seven-step procedure for choosing a protection device for a particular application by going through a detailed example. Our example will involve the protection of printed circuit traces:

1. *What and Why.* We wish to determine if traces on expensive printed circuit fiberglass boards, which feed linear voltage regulators, can be protected from permanent damage. A schematic of the regulator feeds is shown in Fig. 15.24. Unregulated, capacitor filtered, full wave rectified voltage, at a nominal 8 V dc, is routed to the linear 5.0-V regulators using 0.1-in-wide, 1-oz copper traces,[7] which carry a rated steady-state direct current of 4.0 A. From standard design curves for etched copper conductors (see Fig. 15.25), we determine that the steady-state temperature rise of the traces will be approximately 50°F (10°C) at rated current. We stipulate that permanent damage to the traces will take place if the momentary temperature rise of a trace exceeds 212°F (100°C) over ambient.
2. *How.* Linear regulator chips draw negligible inrush overcurrents. Thus we need only be concerned with fault currents. Worst case faults would be direct shorts to ground along a trace, such as at position FI in Fig. 15.24. We consider these in-line faults to be highly improbable and worry only about direct shorts at the input terminals to the regulator chips, such as at position F2. These terminal faults could be caused by inadvertent placement of probe tips during troubleshooting of the board while it is powered.
3. *Where.* We can place interrupting devices at the current summing point *A* (see Fig. 15.24) or in individual leads, such as at positions *B, C, D,* and *E.* A single device at position *A* has the obvious advantage of reduced parts count and reduced cost.

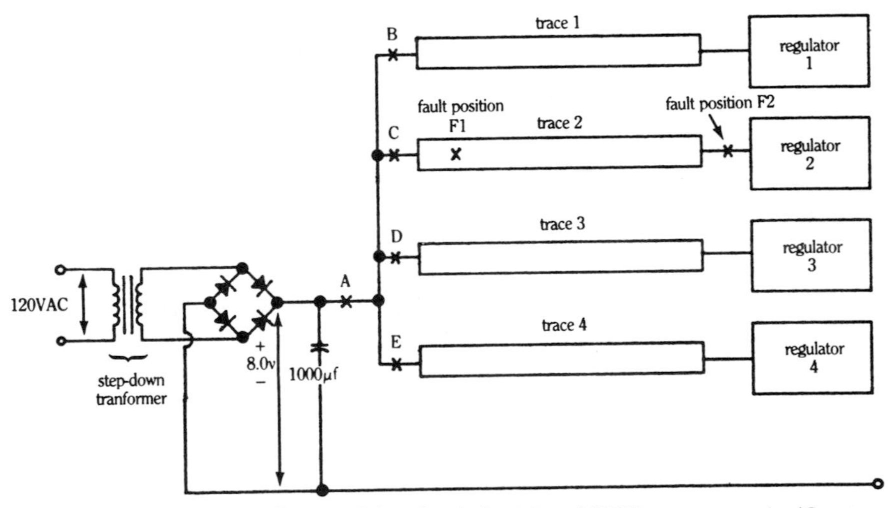

FIGURE 15.24 Schematic diagram of the printed circuit board (PCB) traces example. (*Courtesy of Mechanical Products, Inc., Jackson, Mich.*)

[7]One ounce of copper per square foot of trace surface area.

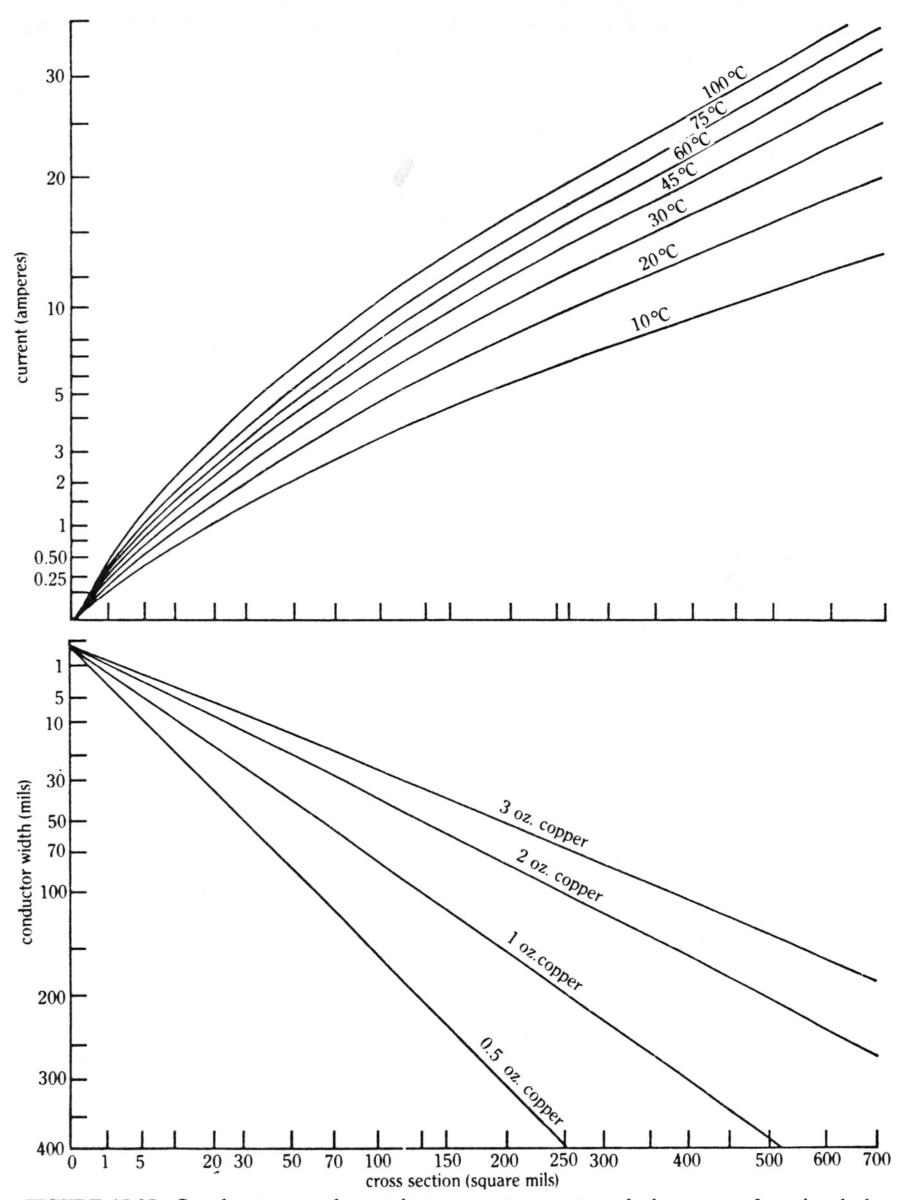

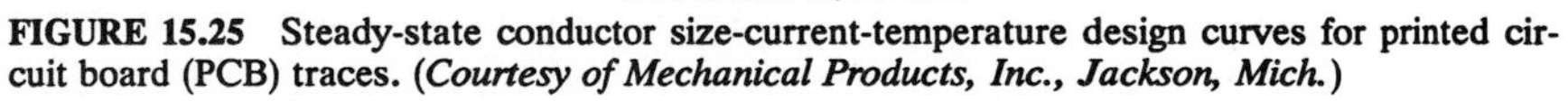

FIGURE 15.25 Steady-state conductor size-current-temperature design curves for printed circuit board (PCB) traces. (*Courtesy of Mechanical Products, Inc., Jackson, Mich.*)

4. *Calculate.* A direct short at the input to the regulator, connected to the shortest trace, length equal to 6 in (152,4 mm), presents a resistance[8] of

$$R_{\text{trace}} = \frac{\rho_r \times \text{length}}{\text{cross-sectional area}}$$

$$= \frac{0.679 \times 10^{-6}\ \Omega\ \text{in} \times 6\ \text{in}}{130 \times (10^{-3}\ \text{in})^2}$$

$$= 31.3\ \text{m}\Omega$$

at the input to the trace path. We assume that the board is connected to the current summing point by 2 ft (609,6 mm) of No. 18 copper wire, which has a total resistance of approximately 12.8 mΩ. We further approximate the total effective connector resistance as an additional 10 mΩ. The total fault path resistance seen across the filter capacitor terminals is then approximately 54 mΩ. The internal effective series resistance ESR of the 1000 μF electrolytic capacitor (from its data sheet) is approximately 50 mΩ. Thus the total resistance seen by the filter capacitor, during its discharge into the fault, is 104 mΩ.

We can quickly calculate two limiting cases for fault current flow. One limiting case would be to assume that the voltage across the filter capacitor is a "stiff" 8.0 V dc, held firm by the rectifier supply. In this case the fault current would be a steady

$$i = \frac{8.0\ \text{V}}{0.054\ \Omega} = 148\ \text{A}$$

The second limiting case is to assume that the rectifier supply has such a high internal impedance that, essentially, all the fault current comes from the discharge of the 1000-μF filter capacitor into the fault. For this case, the current falls with a simple RC time constant τ_e. We have

$$i = I_0 e^{-t/\tau_e}$$

where $I_0 = \dfrac{8.0\text{V}}{0.104\ \Omega} = 77\text{A}$

$$\tau_e = 0.104 \times 1000 \times 10^{-6} = 0.104\ \text{ms}$$

The rms heating current as a function of time for this second case is

$$\sqrt{<i^2>} = \sqrt{\frac{l}{t}\int_0^t (I_0 e^{-t/\tau_e})^2 dt}$$

$$= I_0 \sqrt{\frac{1 - e^{-\phi}}{\phi}}$$

where

$$\phi = \frac{2t}{\tau_e}$$

[8]To be conservative, we neglect any increase in the copper trace resistance due to the temperature rise.

To calculate the trace damage or "must protect" curve, assume that from experimental results we have determined that the thermal time constant of the 100-mil-wide trace $\tau = R_t C_t$ is 55 s (where R_t is the thermal resistance in temperature rise per power (watts) dissipated per unit length and C_t is the thermal capacitance of the trace and board in energy stored (joules) per unit length per degree temperature rise). We then solve the thermal transient equation for the trace.

The base temperature rise in the per unit notation is taken as the damage temperature rise and the base current is taken as the steady-state current, from Fig. 15.25, which will cause the damage temperature rise. In this case, the damage rise is 212°F (100°C) and, from Fig. 15.25, the base current which will give this rise for a 100-mil-wide trace is 12.5 A. For a given rms current above 12.5 A we can solve the thermal transient equation for the time at which the trace temperature rise equals 212°F (100°C). A plot of corresponding overcurrents and 212°F (100°C) temperature rise times is then the trace "must protect" curve.

The "capacitor only" discharge current trajectory, the stiff 8.0-V supply fault trajectory, and the trace "must protect" curve are all given in Fig. 15.26.

5. *Supplemental Requirements.* We require only that the protection device give a visual indication if it has been tripped.
6. *Regulatory Requirements.* We require a UL recognized component as this is not a branch rated circuit breaker.
7. *Choose.* First, we note that no single protection device at position *A* is acceptable, because the steady-state damage current for a single trace is 12.5 A, and a single device at *A* must have a rating of at least 16 A.

   At positions *B, C, D,* and *E* it should be obvious that almost any interruption device with a rating of approximately 6.0 A[9] will work. Since there is no inrush to delay for, there is no "too fast" restriction on a potential protection device. The trip curve for an inexpensive 6.0-A thermal snap device is depicted in Fig. 15.26. Note that the breaker trip response curve tracks the trace "must protect" curve in its constant I^2t region quite well. This is typical for thermal breaker protection of "wire"-type devices. (2)

When considering what is to be protected and why, keep in mind the *dynamics* of circuit protection—in order to avoid nuisance trips attributable to start-up inrush and harmless surges within power systems, it is necessary to provide a margin of tolerance between the steady-state current of the circuit and the rating of the protector. In general, for circuit protectors it is 15 to 20 percent. In addition, there is a trip window or tolerance on the calibration of the protection device. For precise circuit breakers this tolerance is between 25 and 35 percent. This means that a circuit breaker will hold 100 percent and will trip between 100 percent and 125 to 135 percent within an hour. Based upon this common industry specification as an example, a 10.0-A rated protector can be expected to hold 10.0 A, or 100 percent. It can also be expected to trip at 12.5 or 13.5 A within an hour. The expected trip point is governed by the maximum ultimate trip (MUT) specification. In this example, the MUT is 125 or 135 percent depending on the circuit breaker's specifications.

[9]We must allow some margin at the steady-state rating point. Most manufacturers recommend at least a 15 percent "safety factor." In any event, we must choose from a limited set of devices with a limited set of available ratings supplied by manufacturers.

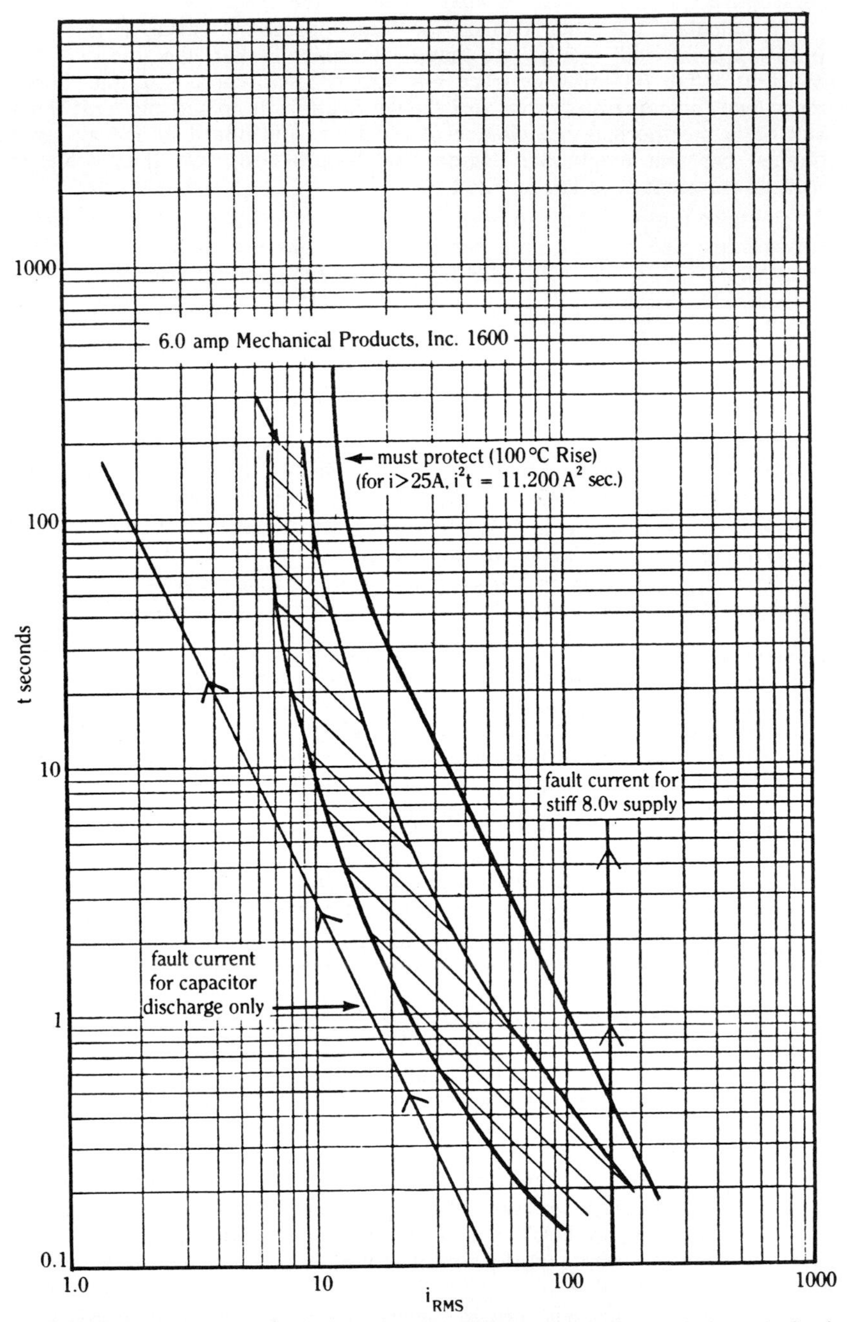

FIGURE 15.26 "Must protect" curve, fault current trajectories, and thermal circuit breaker trip curve for the printed circuit board (PCB) traces example. (*Courtesy of Mechanical Products, Inc., Jackson, Mich.*)

The most important considerations when matching potential fault protection to a circuit protector are the I^2t factor and the fault trajectories. The I^2t factor projects the potential for damage in a component—i.e., wire, motor, power rectifiers, transformers, etc. Generally, this factor is a measure of what a device can absorb and still survive. The measure is a function of current in amperes vs. time in seconds.

The fault trajectory is simply a graphic representation of a fault and, if it is specified as a current in amperes and time in seconds, then both the I^2t and fault trajectory can be put on the same graph. The point at which the two lines cross on the graph represents the condition where circuit or component damage can be expected. For further clarification, the trip curve of the specified circuit breaker can be superimposed upon this graph, giving a visual indication of the level of protection offered by the circuit breaker and its suitability for the application.

It is necessary to consider all aspects of the circuit protector which affect its published operating characteristics when considering its trip curve graph. Specifically, variations in performance can be encountered when factoring (1) position differences in magnetic circuit breakers, (2) ambient temperature changes in thermal circuit breakers and magnetic circuit breakers, (3) the potential for overreaction to inrush currents often encountered in magnetic breakers, and (4) the effect of voltage on the published trip characteristics of magnetic circuit breakers.

Having considered all these variables, the circuit protection device user must also weigh the ancillary product considerations of circuit protection—i.e., allowances between product cost and performance, supplementary requirements, and regulatory approvals. (1)

Table 15.4 summarizes the physical characteristics of the overcurrent protection devices discussed in this chapter.

For a comprehensive discussion on the physics of circuit protection and selection methodology, the reader is directed to "The Theory and Practice of Overcurrent Protection" by Patrick J. McCleer (2), available from Mechanical Products, Inc., Jackson, Mich.

GLOSSARY

Ambient temperature. Refers to the temperature of the air immediately surrounding the circuit breaker and protector.

Ambient temperature compensation. Any technique used to mitigate the effects of off-nominal operating temperatures on the response time and trip level of an overcurrent protection device.

Automatic reset. Device which will automatically open an overloaded circuit. It then will automatically close or complete the circuit after a period of time. If the overload is still present, the device will continue to cycle until either the power or the overload is removed.

Backup. A term given to a second overcurrent protection device "upstream" to a fault or overload which will interrupt the overcurrent flow if a primary "downstream" device fails to do so. The response characteristics of the two protection devices should be chosen such that the two trip response times "cascade" for any given overcurrent value, with the device closest to the fault or overload (the primary or downstream device) responding first.

Bimetal. A layered arrangement of two dissimilar metal sheets bonded together such that thermal expansion of the combination results in a curvature along the bonded surfaces.

Branch circuit. Portion of a wiring system extending beyond final overcurrent device protecting circuit.

TABLE 15.4 Typical Values and Ranges for Protection Devices Rated Below 30 A

Parameters sensed	Fuse current, temp.	Thermal current, temp.	Magnetic current, temp.	Thermal-magnetic current, temp.	Solid-state current, voltage, temp.
Manufacturer's listing of useful ambient temperature range, C°	—	−40 to 70 −55 to 121*	−40 to 85	−10 to 60	−55 to 85
Ultimate trip derating with temperature	Yes	Yes No*	No	Yes	No
Overload trip derating with temperature	Yes	Yes No*	Yes	Yes	No
Minimum current rating, A	0.002	0.05	0.02	0.5	0.001
Interrupting capacity, fail safe, A	1500–10,000	1000–6000	1000–5000	1500–10,000	†
Power loss	Low	Low	Low	Low	Med.
Voltage drop at rated current:					
At 2 A	0.16	0.75	0.50	0.60‡	0.20§
At 10 A	0.12	0.03	0.13	0.13‡	0.30§
Cycle life at rated current	¶	5000	10,000	10,000	1,000,000
Switchable	No	Yes	Yes	Yes	Yes
Remotely controllable	No	No	No	No	Yes
Position sensitive	No	No	Yes	No	No
Vibration and shock tolerance	High	High	Med.	Med.	Very high
Cost	Low	Low	Med.	Med.	High
Maintenance	High	Low	Low	Low	Low

Source: Mechanical Products, Inc., Jackson, Mich.
*Temperature-compensated thermal circuit breaker.
†Solid-state circuit protectors are not current withstand devices, they are current-limiting.
‡Applies to ac only.
§Applies to dc only.
¶Not a switchable device.

Cascading operation. The delayed time operation of upstream overcurrent protection devices which are backup overcurrent protection devices to those closer to the cause of the overcurrent.

Circuit breaker and protector. An automatic switching device that opens the circuit and interrupts the flow of current when an overload condition occurs.

Common trip. Multipole circuit breaker in which overload on any pole will cause all poles to open simultaneously.

Component damage curve. See *Must protect curve* or *Smoke curve,* below.

Coordination. A term used to describe the backup action of secondary, upstream overcurrent protection devices (see also *Backup* and *Coordination delay*).

Coordination delay. The time period required to ensure correct backup action between a primary overcurrent protection device and a secondary backup protection device.

Creep thermal circuit breaker. A thermal circuit breaker which monitors the gradual movement (due to *Joule heating*) of a bimetal blade to detect the presence of an overcurrent.

Current rating. Designation or rating given in amperes at which the device will not trip. A specific temperature is usually assigned.

Derating factor. Factor by which the ampere rating of a circuit protector must be altered to allow for the effects of ambient temperature on the protector.

Detection period. That period of time, starting at the initiation of an overcurrent, during which an overcurrent protection device makes a "trip" decision.

Detection threshold. The amount of overcurrent energy that an overcurrent protection device absorbs in order to make a "trip" decision.

Electromechanical circuit breaker. An overcurrent protection device which utilizes an arc between separated contacts to interrupt a current.

Electromotive force (emf). A voltage due to an energy conversion process such as the internal voltage of a battery due to an electrochemical reaction, or the internal winding voltage of a motor due to electromechanical action.

Ferromagnetic. A metallic material in which it is relatively easy to "line up" the magnetic axes of groups of atoms (referred to as magnetic domains) by applying an exciting magnetic field. For a given value of exciting magnetic field the magnetic flux density in a ferromagnetic material is much higher (up to a saturation limit) than that in a nonferromagnetic material.

Hot wire (thermal circuit breaker). A type of electromechanical circuit breaker which monitors the elongation of a (*Joule*) heated wire to determine the presence of an overcurrent.

I^2t. The time integral of the square of the current through a conductor or device. An I^2t value is a measure of the heating energy that a conductor or device has absorbed.

Inrush. A common name for the natural, transient, startup overcurrent of a device.

Instantaneous trip (opening). "Instantaneous" indicates delay is not introduced purposely into action of the device.

Interruption period. The time period, measured from the time at which a "trip" decision is made, required by an overcurrent protection device to completely interrupt the flow of an overcurrent.

Interruptive capacity. The highest level of fault current that a circuit protective device is intended to interrupt. Devices qualified to UL1077 may be inoperable following the fault but must have failed-safe (no loss of case integrity, no emissions capable of igniting external materials, or no internal arc-over to grounded parts). Devices qualified to UL489 must be operable following the fault interruption and capable of clearing a 200 percent overload.

Inverse-time. See *Time-inverse,* below.

Joule heating (i^2R). The process of transferring electrical kinetic energy in conduction electrons to the host lattice atoms in a conducting material via collisions between the electrons and the lattice atoms. In the process the temperature of the lattice atoms is raised.

Let-through I^2t. The total I^2t that an overcurrent protection device passes during its total clearing time.

Magnetic circuit breaker. An electromechanical overcurrent protection device that can be reset, which detects the presence of an overcurrent by monitoring the mechanical movement of an "arm" which is attracted by the magnetic pull of an electromagnet carrying the overcurrent.

Magnetic flux. The integral of the normal component of the magnetic flux density through a surface. The magnetic flux density is a vector function which defines the magnetic (vector) force on a differential volume element carrying unit current density.

Magnetic Ohm's law. The ratio between the magnetic flux density and the exciting magnetic field intensity at a particular point in space. In free space the magnetic permeability is equal to $4\pi \times 10^{-7}$ H/m. In a ferromagnetic material (for magnetic flux density values below the saturation limit) the magnetic permeability is several thousand times that of the free space value.

Magnetomotive force (mmf). The excitation or drive force which creates a magnetic field. Measured in ampere-turns.

Manual reset. Refers to those breakers in which the electrical contacts remain open after a trip until someone physically closes or completes the circuit by either pushing a reset button or throwing a switch.

Maximum ultimate trip. Current rating at which a circuit protection device will trip within a certain period of time at a specified temperature.

Minimum ultimate trip. Current rating for which a circuit protection device will not trip for an extended period of time at a specified temperature.

Must protect curve. A curve in a current-time diagram which indicates the maximum amount of time a device can pass a particular rms level of current without permanent damage.

Nuisance trips. Those trips caused by a response to nondamaging inrush or startup current surges, as opposed to an actual overcurrent trip.

Ohm's law. "The electromotive force (emf) in a static electric circuit is equal to the electric current times the total electric resistance of the circuit."

Overcurrent. A current with rms value exceeding the rated current of a circuit or a device.

Overcurrent protection. Protection achieved by limiting the duration and magnitude of exposure to an overcurrent.

Overcurrent protection device. Any electromechanical or electronic device which detects and limits the flow of an overcurrent in an electric circuit.

Overload. An electrical load or current flow greater than that which a circuit is designed to handle.

Overload capacity. The highest level of overload current that a device will interrupt and remain in operable condition capable of clearing additional overloads.

Rated current (for an overcurrent protection device). The maximum nominal value of steady-state rms current at which the device will not respond (trip) at a specified ambient temperature.

Reluctance. A measure of the difficulty with which magnetic flux flows through a medium. The lower the magnetic reluctance of a magnetic path, the higher the flow of magnetic flux for a given level of excitation.

RMS (root mean square). The square root of the time average of the square of a quantity. The rms value of a current is the equivalent "heating" value of the current, that is, equal to the steady-state direct current which gives the same amount of *Joule* heating in a device or conductor.

Rupture capacity. A term applied to reusable protective devices. See *Interruptive capacity.*

Safety factor. The allowance added to the steady-state application current to ensure that the protective device selected will be more than sufficient to handle the application without nuisance trips.

Selective system. See *Selectivity.*

Selectivity. The property of an overcurrent protection system by which only the minimum amount of circuit necessary is isolated by overcurrent protection devices during an overcurrent situation. See *Selective system.*

Sensitivity. A measure of the ability of an overcurrent protection device to detect overcurrents above the rated value of the device.

Short circuit. A condition which occurs when a current path to a load is bypassed with a very low or negligible resistance path. Under this condition, excessively high current flows, which represents a significant hazard to both the equipment and the attendant personnel.

Smoke curve. A plot of the maximum rms current that can flow through a component vs. the time it can flow before the onset of permanent damage. See *Must protect curve.*

Snap thermal circuit breaker. A thermal circuit breaker which monitors the abrupt or buckling movement (due to *Joule* heating) of a bimetal element to detect the presence of an overcurrent.

Solid-state power switch. A switch which uses a solid-state power device rather than mechanically separated contacts to open an electric circuit.

Thermal circuit breaker. An electromechanical overcurrent protection device that can be reset, which detects the presence of an overcurrent by monitoring the mechanical movement (due to *Joule* heating) of a metallic element.

Thermal-magnetic circuit breaker. An electromechanical overcurrent protection device that can be reset, which detects the presence of an overcurrent by both thermal and magnetic means. The thermal detection mechanism is usually set to trip at a lower overcurrent value than the magnetic mechanism.

Time-delay. Qualifying term indicating that purposely delayed action is introduced.

Time-inverse. Time-current relationship where protective device opening time decreases as current increases.

Total clearing time. The total time period, measured from the initiation of an overcurrent, required by an overcurrent protection device to detect and completely interrupt an overcurrent. It is the sum of the device detection period and interruption period.

Trip curve. The graphical presentation of overcurrent protection device response characteristics. The curve consists of overcurrent magnitudes plotted against ranges of corresponding total clearing times. A trip curve is sometimes referred to as a time-current curve.

Trip free. A device's ability to automatically open a circuit in a fault situation even though its actuator is physically held in the "on" position.

Ultimate trip current. The rms magnitude of the minimum current at which an overcurrent protection device will operate (i.e., trip) at a specified ambient temperature. If the ultimate trip current is given as a range of currents, the lower value of the range is referred to as the minimum ultimate trip current, and the upper value of the range is referred to as the maximum ultimate trip current.

UL489. UL Standard (requirements and specifications) for "Circuit Breakers and Circuit Breaker Enclosures."

UL1077. UL Standard (requirements and specifications) for "Supplementary Protectors for Use in Electrical Equipment."

Voltage drop. The voltage decrease across the protector and breaker due to the internal resistance of the device.

Webers. The unit of measure of magnetic flux.

REFERENCES

1. Olson, E. David, and David Hunt, "A Comprehensive Guide to Circuit Protection Systems," Mechanical Products, Inc., Jackson, Mich., 1991.
2. McCleer, Patrick J., "The Theory and Practice of Overcurrent Protection," Mechanical Products, Inc., Jackson, Mich., 1987.
3. Kirchdorfer, Josef, "Circuit Protection in Electrical Equipment," 2d ed., Weber, Ltd., Electrotechnical Apparatus and Systems, Emmenbrücke, Switzerland, 1987.
4. "Airpax Magnetic Circuit Breakers," Philips Technologies, Airpax Protector Group, Cambridge, Md., 1990.

APPENDIX 1

INTERNATIONAL STANDARDS & TESTING AGENCIES

COUNTRY & MARK	STANDARD AGENCY SYMBOL	STANDARDS AGENCY	TEST AGENCY SYMBOL	TESTING AND/OR CERTIFICATION AGENCY
ARGENTINA	IRAM	IRAM Instituto Argentino de Racionalization de Materiales Chile 1192, C. Postal 1098 Buenos Aires, ARGENTINA Telephone: (54) 1 37 37 51		INTI Instituto Nacional de Tecnologia Industrial LN Alem No. 1067 1001 Buenas Aires ARGENTINA Telex: 21859 INTI AR
AUSTRALIA		SAA — Standards Association of Australia 80 Arthur Street North Sydney, NSW 2060, AUSTRALIA Telephone: (61) 2 963 4111 FAX: (61) 2 959 3896 Telex: 99-26514		SECV — State Electrical Commis. of Victoria 15 William Street Melbourne, Victoria 3000, AUSTRALIA Telephone: (61) 3 392 2253 Telex: 31153 Sydney County Council Test. Labs 14 Nelson Street Chatswood, New South Wales, AUSTRALIA Telephone: (61) 02 410511 Telex: 22810 EANSW — Energy Authority of New South Wales 1 Castlereagh Street Box 485, GPO Sydney, NSW 2001, AUSTRALIA Telephone: 61 2 234 4444 FAX: 61 2 221 6229 Telex: NSWEA AA170320 ETSA — Electricity Trust of South Australia 26-56 Burbridge Road Mile End, SA 5031, AUSTRALIA Telephone: (61) 08 352 0719 Telex: 88655 SECQ — State Electricity Commission of Queensland GPO Box 10 Brisbane, QLD 4001, AUSTRALIA SECWA — State Energy Commission of Western Australia GPO Box L921 Perth, WA 6001, AUSTRALIA Telephone: (61) 09 277 2488 Telex: 92674 HEC — Hydro-Electricity Commision of Tasmania
AUSTRIA	ÖVE	ÖVE — Österreichischer Verband Für Elektrotechnik Eschenbachgasse 9 A-1010 Wien, AUSTRIA Telephone: (43) 222 587 6373 Telex: 32 22 603 OEVE		
BELGIUM		CEB — Comité Electrotechnique Belge B Brussels BELGIUM Telephone: (32) 2 512 0028 FAX: (32) 2 511 2938	CEBEC	CEBEC — Comité Electrotechnique Belge Service Rodestraat 125 B-1630 Linkebeek Bruxelles, BELGIUM Telephone: (32) 2 380 8520 FAX: (32) 2 380 6133 Telex: 62834 CEBEC B
BRAZIL	ABNT	ABNT — Associação Brasileira de Normas Tecnicas Rua Marquês de ITU 88-4° Andar 01223 São Paulo-SP, BRAZIL Telephone: (55) 11 35 94 33 FAX: None Telex: 112 1452 CELB BR		SINMETRO — National System of Metrology Standardization and Industrial Quality
BULGARIA		State Comm./Science & Tech. Prog. Standards Office 21, 6th September St. Sofia 1000 BULGARIA Telephone: 85 91 Telex: 22570 DKS BG		
CANADA	SA	CSA — Canadian Standards Association 178 Rexdale Blvd. Rexdale, Ontario M9W 1R3, CANADA Telephone: (416) 747 4000 FAX: (416) 747 2475 Telex: 06-989344		

Source: **Panel Components Corp., Santa Rosa, Calif.**

COUNTRY	STANDARD AGENCY SYMBOL	STANDARDS AGENCY	TEST AGENCY SYMBOL	TESTING AND/OR CERTIFICATION AGENCY
CHINA, PEOPLE'S REPUBLIC OF		CSBS China State Bureau for Standardization P. O. Box 820 Beijing PEOPLE'S REPUBLIC OF CHINA Telegrams: 0621 Beijing		
CZECHOSLOVAKIA		Urad pro Normalizacia Mereni (Office for Standards and Measurements) Václavské nam.19 113 47 Praha 1 CZECHOSLOVAKIA Telephone: (42) 2 26 22 51 FAX: (42) 2 265 795 Telex: 12948 UNM	ESC	Elektrotechnicky zkusebni ustav Post Office 71 CS-171 02 Praha 8-Troja CZECHOSLOVAKIA Telegrams: ELEZUM Praha Telephone: (42) 2 840 641-9 Telex: 122880 EZU C
DENMARK		DEK Dansk Elektroteknisk Komite (Danish Electrotechnical Committee) Strandgade 36 DK-1401 Copenhagen K DENMARK Attn: DANELKOMITE Telephone: (01) 57 50 50 Telex: FOTEX DK 16600 Fax: 011-45-1576350	D	DEMKO Postbox 514 DK-2730 Herlev DENMARK Telephone: (45) 42 94 72 66 FAX: (45) 42 94 7261 Telex: 35125 DEMKO DK
EGYPT	EOS	EOS — Egyptian Organization for Standardization & Quality Control 2 Latin America St. Garden City, Cairo EGYPT Telephone: (20) 29720 Telex: 93296 EOS UN		
FINLAND	FI	SETI — Electrical Inspectorate Sarkiniementie 3 P.O. Box 21 SF-00210 Helsinki 21 FINLAND Telephone: (358) 0-69631 FAX: (358) 0-69254 74 Telex: 122877 SETI SF		
FRANCE		UTE — Union Technique de l'Électricité Cedex 64 F-92052 Paris la Defense FRANCE Telephone: (33) 1 4723 7257 FAX: (33) 1 4723 6860 Telex: CEFUTE 620816 F		Laboratorie Central des industries électriques (LCIE) 33, Avenue du Général Leclerc B.P. 8 F-92260 Fontenay aux-Roses FRANCE Telephone: (33) 1 40 95 60 60 Telex: LABEL EC 250080 F FAX: (33) 1 40 95 60 95
GERMANY		DKE Deutsche Electrotechnische Kommission im DIN und VDE (German Electrotechnical Commission of DIN and VDE) Stresemannallee 15 D-6000 Frankfurt 70 GERMANY Telephone: (49) 69 6308 0 FAX: (49) 69 6308 273 Telex: 4-12871	DVE	VDE VDE-Prüfstelle Merianstrasse 28 D-6050 Offenbach am Main FEDERAL REPUBLIC OF GERMANY Telephone: (49) 69 8306 1 FAX: (49) 69 8306 555 Telex: 4152796 VDEP D TUV — Technischer Uberwachungs-Verein Albionstrasse 56 D-1000 Berlin FEDERAL REPUBLIC OF GERMANY Telephone: (49) 30 75 3021 Telex: 1 84 517 TUV Rheinland 111 Deerwood Place Suite 160 San Ramon, CA 94583, U.S.A. Telephone: (415) 820-8444 FAX: (415) 820-8467

Source: Panel Components Corp., Santa Rosa, Calif.

COUNTRY	STANDARD AGENCY SYMBOL	STANDARDS AGENCY	TEST AGENCY SYMBOL	TESTING AND/OR CERTIFICATION AGENCY
				TUV Rheinland of North America 108 Mill Plain Road Danbury, CT 06811, U.S.A. Telephone: (203) 798-0811 FAX: (203) 798-0694 Telex: 4990573 TUV America, Inc. 1416 NW 9th Street Corvallis, OR 97330 Telephone; (503) 753-4438 & (503) 753-4439 FAX: (503) 753-4510
GREECE		ELOT — The Hellenic Organiz. of Standardization Didotou 15 GR-106 80 Athens, GREECE Telephone: (30) 1 360 9517 FAX: (30) 1 364 4569 Telegrams: ELOTYP Telex: 219621 ELOT GR		
HONG KONG		Hong Kong Standards & Testing Centre 10 Dai Wang St. Taipo Industrial Estate Taipo, N.T., HONG KONG Telephone: (852) 0-6530021 FAcsimile: (852) 0-6534353 Telex: 84652 HKIND HX		
HUNGARY		MSZH Magyar Szabvanyugi Hivatal (Hungarian Office for Standardization) Pf.24 H-1450 Budapest 9 HUNGARY Telephone: (36) 183-011 Telex: 225723 NORM H		MEEI Magyar Elektrotechnikai Ellenorzo Intezet (Hungarian Institute for Testing Electrical Equipment) Váci út 48/a-b Pf. 441 H-1395 Budapest XIII, HUNGARY Telephone: (36) 1 495-561 Telegrams: TESTHUNGARIA Telex: 224931 MEEI H
ICELAND		Rafmagnesftirlit Riskisins (Electrical Inspection Agency) P. O. Box 8240 IS-128 Reykjavik ICELAND Telephone: (354) 1-84133 FAX: (354) 1-689256		
INDIA	ISI	BIS-Bureau of Indian Standards "Manak Bhavan" 9 Bahadur Shah Zafar Mrg. New Delhi 110002 INDIA Telephone: (91) 11 266021/270131 Telex: 031-3970 ISI/ND		
INDONESIA	SII	Badan Kerjasama Standardisasi LIPI-YDNI (LIPI-YDNI Joint Standardization Committee) Jalan Teuku Chiek Ditiro 43 P. O. Box 250, Jakarta INDONESIA Telephone: (62) 351658		
IRAN		Institute of Standards and Industrial Research of Iran P. O. Box 2937, Tehran IRAN Telephone: (98) 2221 6031-8		
IRELAND		NSAI — Nat. Standards Authority of Ireland Glasnevin Ballymun Road IRL-Dublin 9, IRELAND Telephone: (353) 1 370 101 FAX: (353) 1 369 821 Telex: 32502 OLASEI		NETH—National Electrical Test House Ballymun Road IRL–Dublin 9, IRELAND Telephone: (353) 1 370 101 FAX: (353) 1 379 620 Telex: 32502 OLASEI

Source: Panel Components Corp., Santa Rosa, Calif.

COUNTRY	STANDARD AGENCY SYMBOL	STANDARDS AGENCY	TEST AGENCY SYMBOL	TESTING AND/OR CERTIFICATION AGENCY
		(Ireland, continued)		NASI of America 5 Medallion Ctr. (Greeley St.) Merrimack, NH 03054 Telephone: (603) 424-7070 FAX: (603) 429-1427
ISRAEL		SII — Standards Institution of Israel 42, University Street IL-Tel Aviv 69977, ISRAEL Telephone: (972) 3 42 28 11 FAX: (972) 3 41 96 83 Telex: 35508 SIIT IL		
ITALY		CEI Comitato Elettrotecnico Italiano (Italian Electrotechnical Comm.) Viale Monza 259 I-20126 Milano, ITALY Telephone: (39) 2 25 50 641 Telex: 312207 CEITAL 1		IMQ Instituto Italiano del Marchio di Qualita Via Quintiliano, 43 I-20138 Milano, ITALY Telephone: (39) 2 5073216 FAX: (39) 2 5073271 Telex: 310494 IMQ I
JAPAN		All certification and test agencies are part of "MITI"—The Ministry of International Trade and Industry JIS (apply for Dentori approvals here) Japanese Industrial Standards Office Agency of Industrial Science & Technology and Technology 1-3-1 Kasumigaseki 1-Chome Chiyoda-ku, Tokyo 100 JAPAN Telephone: (81) 3 501-9296 FAX: (81) 3 580-1418 *UL has information on MITI requirements. Contact:* UL Overseas Inspection Services 1285 Walt Whitman Road Melville, New York 11747 Phone: (516) 271-6200 • FAX: (516) 271-8250	(Dentori mark)	JET — Japan Electrical Testing Laboratory 5-14-12 Yoyogi Shibuya-ku, Tokyo 151 JAPAN Telephone: (81) 3 466-5121 • FAX: (81) 3 468-9090 JMI Institute (formerly Japan Machinery & Metals Inspec. Instit.) 21-25 Kinuta 1-Chome Setagaya-ku, Tokyo, JAPAN Telephone: (81) 3 416-0192 FAX: (81) 3 416-9691 Telex: 2423301 JMI J Electrical Appliance Safety Office (same address as JIS) Phone: (81) 3 501-1511 • FAX: (81) 3 501-1836
KOREA, DEM. PEOPLE'S REP. OF (NORTH)		Committee for Standardization Sosong guyok Ryonmod dong Pyongyang DEMOCRATIC PEOPLE'S REPUBLIC OF KOREA Telephone: 3327 Telegrams: STANDARD		
KOREA, REP. OF (SOUTH)		Industrial Advance. Admin. 94-267 Yongdeungpo-Dong Yongdeungpo-Ku Seoul, REPUBLIC OF KOREA Telephone: (82) 2 633-8815		Korea Institute of Mach. & Metals 222-13, Guru-Dong, Guru-ku Seoul 140, REPUBLIC OF KOREA Telephone: (82) 2 855-0611-5 Telex: 28456
LUXEMBOURG		ITM 26, Zithe Bte Postale 26 L-2010 Luxembourg Telephone: (352) 49 921 2106 FAX: (352) 49 14 47 Telex: 2985 MINIES		
MEXICO	DGN	DGN — Dirección General de Normas Secretaria de Patrimonio y Fomento Industrial Puente de Tecamachalco No. 6, Lomas de Tecamachalco Seccion Fuentes Naucalpan de Juarez, ESTADO DE MEXICO, C.P. 53950 Telephone: (520) 395-36-43 Telex: 1775690		
NETHERLANDS		NEC — Nederlands Elektrotechnisch Comite Kalfieslaan 2 Postbus 5059 2600 GB Delft, THE NETHERLANDS Telephone: (31) 15 69 03 90 FAX: (31) 15 69 01 90 Telex: 38144 NNI NL	KEMA EUR	N.V. KEMA — NV tot Keuring van Elektrotechnische Materialen Utrechseweg 310 Postbus 9035 NL-6800 ET Arnhem, THE NETHERLANDS Telephone: (31) 85 56 28 53 FAX: (31) 51 56 06 Telex: 75132 KLTI NL

Source: Panel Components Corp., Santa Rosa, Calif.

COUNTRY	STANDARD AGENCY SYMBOL	STANDARDS AGENCY	TEST AGENCY SYMBOL	TESTING AND/OR CERTIFICATION AGENCY
NEW ZEALAND	SANZ	SANZ — Standards Association of New Zealand Private Bag Wellington, NEW ZEALAND Telephone: (64) 4 842 108 FAX: (64) 4 843 938 Telex: 3850 SANZ NZ		
NORWAY		NEK Norsk Elektroteknisk Komite (Norwegian Electrotechnical Committee) Oscarsgate 20 P. O. Box 7099 Homansbyen Oslo 3, NORWAY Telephone: (47) 2 606 697 Telex: 17206 NENEKN	N	NEMKO Norges Elektriske Materiellkontroll Gaustadallen 30 Postboks 73 Blindern N-0314 Oslo 3 NORWAY Telephone: (47) 2 691950 FAX: (47) 2 698636 Telex: 77260 NEMKO N
PAKISTAN		Pakistan Standards Institution 39 Garden Road Saddar, Karachi 3 PAKISTAN Telephone: (92) 73088		
POLAND	B	Polski Komitet Normalizacji Miar i Jakosci Polish (Committee for Standardization, Measures and Quality Control) ul. Elektoralna 2 00-139 Warsaw POLAND Telephone: 200241 Telex: 813642		Electrical Laboratory of Biuro Znaku Jakosci ul. Swietojerska 17 Warsaw, POLAND Telephone: 31 63 67 Association of Polish Electrical Engineers Quality Testing Office ul. Wsopolna 32/46 Warsaw, POLAND Telephone: 21 90 38 Telegrams: BEBETOT-Warszawa
PORTUGAL		CEP — Comissão Electrotécnica Portuguesa Portuguesa Rua Infantaria 16, n°41-2° 1200 Lisbon PORTUGAL Telephone: (351) 681048-681049		PQ — Instituto Portugués da Qualidade Rua José Estevão, 83A P-1199-Lisboa Codex, PORTUGAL Telephone: (351) 1 539891 FAX: (351) 1 530033 Telex: 13042 QUALIT P
ROMANIA	STAS	Institutul Roman de Standardizare (Romanian Standards Institute) Bucaresti, Sect. 1 str. Roma. 32-34 R71219-RS, ROMANIA Telephone: (40) 337660 Telex: 011312		
SAUDIA ARABIA		SASO — Saudia Arabian Standards Organization Sitteen Street P.O. Box 3437 Riyadh 11471 Kingdom of Saudia Arabia Telephone: (966) 1 479 0406 FAX: (966) 1 479 3063 Telex: 401610 SASO SJ		
SINGAPORE		Singapore Institute of Standards and Industrial Research 1 Science Park Drive Singapore Science Park Singapore 0511, REPUBLIC OF SINGAPORE Telephone: (65) 778 7777 Telex: RS 28499 SISIR FAX: (65) 778 0086		Postal address: P. O. Box 1128 Singapore 9111
SOUTH AFRICA	SABS	SABS — South African Bureau of Standards Private Bag X191 Pretoria 0001 REPUBLIC OF SOUTH AFRICA Telephone: (012) 428-7911 Telex: 3626 SA		

Source: Panel Components Corp., Santa Rosa, Calif.

COUNTRY	STANDARD AGENCY SYMBOL	STANDARDS AGENCY	TEST AGENCY SYMBOL	TESTING AND/OR CERTIFICATION AGENCY
SPAIN		IRANOR Instituto Nacional de Racionalización y Normalización Zurbano 46 Madrid 10 SPAIN Telephone: (34) 410 46 76 Telex: 46545 UNOR E		AEE Asociación Electrotécnica y Electronica Española Avda. de Brasil, 7 E-28020 Madrid SPAIN Telephone: (34) 1 456 7664 FAX: (34) 1 270 4972 Telex: 27626 UNESA E
SWEDEN		SEK Svenska Elektriska Kommissionen (Swedish Electrotechnical Commission) Box 5177 S-10244 Stockholm, SWEDEN Telephone: (46) 8 23 3195 Telex: 17109 ELNORM S		SEMKO — Svenska Elektriska Materielkontrollanstalten AB (Swedish Institute for Testing & Approval of Electrical Equip.) Box 1103 S-16422 Krista-Stockholm, SWEDEN Telephone: (46) 8 750 0000 FAX: (46) 8 750 6030 Telex: 8126010 SEMKO S
SWITZERLAND		SEV — Schweizerischer Elektrotechnischer Verein Postfach CH-8034 Zürich, SWITZERLAND (for parcels: Seefeldstrasse 301, CH-8008 Zürich) Telephone: (41) 1 384 9111 FAX: (41) 1 551426 Telex: 56047 SEVCH		
TAIWAN		BCIQ — Bureau of Commodity Inspection and Quarantine Ministry of Economic Affairs, Republic of China 4, Section 1, Chinan Road Taipei, TAIWAN ROC Telephone: (886) 2 351-2141 FAX: (886) 2 393-2324 Telex: 27247 BCIQ		
UNITED KINGDOM		BSI British Standards Institution 2 Park Street London W1A 2BS, ENGLAND Telephone: (44) 01 629 9000 Telex: 266933 (BSILON G) FAX: 0908 320856		BSI — British Standards Institution, Cert. & Assessment Dept. P.O. Box 375 Milton Keynes MK14 6LL, ENGLAND Telephone: (44) 908 315555 FAX: (44) 908 320856 Telex: 82424 BSI HHC G
			BEAB	BEAB — British Electrotechnical Approvals Board Mark House/The Green 9/11 Queen's Road, Hersham, Walton-on-Thames GB-Surrey KT12 5NA, ENGLAND Telephone: (44) 932 24 44 01 FAX: (44) 932 22 66 03 Telex: 8812027 BEAB G
				BASEC — British Approvals Service for Electric Cables P.O. Box 390 Milton Keynes GM-MK14 6LN, ENGLAND Telephone: (44) 908 31 55 55 Telex: 82682 BSIQAS G FAX: (44) 908 32 08 56
				BASEEFA — British Approval Services./Electronic Equipment for Flammable Atmosphere Health & Safety Executive Harpur Hill, Buxton, Derbyshire, ENG. Telephone: 0298 (Std. Code: 6211)
			ASA	ASTA — Assoc. of Short Circuit Testing Authorities, Inc. 23/24 Market Place Rugby CV21 3DU ENGLAND
UNTED STATES OF AMERICA		ANSI — American National Standards Institute 1430 Broadway New York, NY 10017 Telephone: (212) 354-3300		ARL — Applied Res. Labs of Florida 5371 N.W. 161 Street Miami, FL 33014 Telephone: (305) 624-4800 FAX: (305) 624-3652
		NEMA — National Electrical Manufacturers Association 2101 L. Street, N.W., Suite 300 Washington D.C. 20037 Telephone: (202) 457-8400 FAX: (202) 457-8468 Telex: 904077		DS&G — Dash, Straus & Goodhue 593 Massachusetts Ave. Boxborough, MA. 01719 Telephone: (508) 263-2662 FAX: (508) 263-7086 Telex: 317-632-DASH

Source: Panel Components Corp., Santa Rosa, Calif.

COUNTRY	STANDARD AGENCY SYMBOL	STANDARDS AGENCY	TEST AGENCY SYMBOL	TESTING AND/OR CERTIFICATION AGENCY
		NFPA National Fire Protection Agency 1 Battery March Park Quincy, MA. 02269		ETL — ETL Testing Labs, Inc. Industrial Park Cortland, NY 13045 Telephone: (607) 753-6711 — ALSO (800) 354-3851 FAX: (607) 756-9891 Other ETL offices: 660 Forbes Blvd. So. San Francisco, CA 94084 Telephone: (415) 871-1414 FAX: (415) 873-7357 5855-P Oakbrook Parkway Norcross, GA 30093 Telephone: (404) 446-7294 FAX: (404) 446-7025 Factory Mutual Insurance Co. 1151 Boston-Providence Hwy. Norwood, MA. 02062 Telephone: (617)769-7900 Telex: 92-4415 MET — MET Electrical Testing Co. 916 W. Patapsco Avenue Baltimore, MD 21230 Telephone: (301) 354-2200 FAX: (301) 354 1624
			UL	UL — Underwriters Laboratories, Inc. 1285 Walt Whitman Road Melville, L.I., NY 11747 Telephone: (516) 271-6200 FAX: (516) 271-8259 Telex: 6852015 Other UL offices: 1655 Scott Blvd. Santa Clara, CA 95050 Telephone: (408) 985-2400 FAX: (408) 296-3256 Telex: 470607 333 Pfingsten Road Northbrook, IL 60062 Telephone: (708) 272-8800 FAX: (708) 272-8129 Telex: 6502543343 12 Laboratory Drive P. O. Box 13995 Research Triangle Park, NC 27709 Telephone: (919) 549-1400 FAX: (919) 549-1842 Telex: 4937928
U.S.S.R.	CCCP	GOSSTANDART — USSR State Committee for Standards Leninsky Prospekt 9 117049 Moscow M-49, USSR Telephone: (7) 95 236 40 44 Telex: 411378 GOST SU		
VENEZUELA	NV	CODELECTRA en Conjunto con COVENIN Avda. Ppal. Las Mercedes-Edf. Centro Victorial-Piso 1 Caracas 1060 VENEZUELA Telephone: (58) 91 99 06		COVENIN - Ppal. Comis. Venezolana de Normas Industriales Av. Boyaca (COTA MIL) Edf. Fundacion La Salle, 5° Piso Caracas 105, VENEZUELA
YUGOSLAVIA		Federal Instit. for Standardization Slobodana Penezica Krcuna 35, 11000 Beograd, YUGOSLAVIA Telephone: (38) 11 644-066 Telex: 12089 YU JUS		

Source: **Panel Components Corp., Santa Rosa, Calif.**

APPENDIX 2
IP CODES (INGRESS PROTECTION)

IEC (International Electrotechnical Commission) **is a significant worldwide organization with specifications for electrical and electronic components.**

IEC 529 outlines an international classification system for the sealing effectiveness of enclosures of electrical equipment against the intrusion into the equipment of foreign bodies (i.e., tools, dust, fingers) and moisture. This classification system utilizes the letters "IP" ("Ingress Protection") followed by two digits.

Degrees of Protection - First Digit
The first digit of the IP code indicates the degree that persons are protected against contact with moving parts (other than smooth rotating shafts, etc.) and the degree that equipment is protected against solid foreign bodies intruding into an enclosure.

0 No special protection
1 Protection from a large part of the body such as a hand (but no protection from deliberate access); from solid objects greater than 50 mm in diameter
2 Protection against fingers or other objects not greater than 80 mm in length and 12 mm in diameter
3 Protection from entry by tools, wires, etc., with a diameter or thickness greater than 2.5 mm
4 Protection from entry by solid objects with a diameter or thickness greater than 1.0 mm
5 Protection from the amount of dust that would interfere with the operation of the equipment
6 Dust-tight

Degrees of Protection - Second Digit
Second digit indicates the degree of protection of the equipment inside the enclosure against the harmful entry of various forms of moisture (e.g. dripping, spraying, submersion, etc.).

0 No special protection
1 Protection from dripping water
2 Protection from vertically dripping water
3 Protection from sprayed water
4 Protection from splashed water
5 Protection from water projected from a nozzle
6 Protection against heavy seas, or powerful jets of water
7 Protection against immersion
8 Protection against complete, continuous submersion in water

The IP Code Symbols:
The chart at right illustrates the use of special symbols in the IP classification system. In the "1st digit" columns, note the grid-like symbols next to numbers 5 and 6. In the "2nd digit" columns numbers 3-8 are symbolized by teardrop shaped symbols, sometimes enclosed in a box or a triangle, sometimes unenclosed (#7-8). These symbols can be placed on equipment to illustrate the IP protection provided.

IP 54 = IP 5 4
IP letter code → IP
1st digit → 5
2nd digit → 4

1st digit		2nd digit	
0	Non protected	0	Non protected
1	Protected against solid objects greater than 50mm	1	Protected against dripping water
2	Protected against solid objects greater than 12mm	2	Protected against dripping water when tilted up to 15°
3	Protected against solid objects greater than 2.5mm	3	Protected against spraying water
4	Protected against solid objects greater than 1.0mm	4	Protected against splashing water
5	Dust protected	5	Protected against water jets
6	Dust tight	6	Protected against heavy seas
		7	Protected against the effects of immersion
		8	Protected against submersion

Note: IEC 529 does not specify sealing effectiveness against the following: mechanical damage of the equipment; the risk of explosions; certain types of moisture conditions, e.g., those that are produced by condensation; corrosive vapors; fungus; vermin.

Source: **Panel Components Corp., Santa Rosa, Calif.**

APPENDIX 3

APPLICABLE TESTS FOR NEMA HAZARDOUS AND NONHAZARDOUS LOCATIONS

A3.2

NEMA (National Electrical Manufacturer's Association) prepares standards which define a product, process or procedure with reference to one or more of the following: nomenclature, composition, construction, dimensions, tolerances, safety, operating characteristics, performance, quality, electrical rating, testing and the service for which designed. The reference standards herein reflect the latest data in the NEMA Standards Publication #250 - 1985 Revision 2 on Enclosures for Industrial Controls and Systems.

NEMA REFERENCE NONHAZARDOUS LOCATIONS	APPLICABLE TESTS
TYPE 1 enclosures are intended for indoor use primarily to provide a degree of protection against contact with the enclosed equipment in locations where unusual service conditions do not exist.	***Rod entry*** ***Rust resistance***
TYPE 2 enclosures are intended for indoor use primarily to provide a degree of protection against limited amounts of falling water and dirt. They are not intended to provide protection against conditions such as dust or internal condensation.	***Rod entry*** ***Drip*** ***Rust resistance***
TYPE 3 enclosures are intended for outdoor use primarily to provide a degree of protection against windblown dust, rain, and sleet; and to be undamaged by the formation of ice on the enclosure. They are not intended to provide protection against conditions such as internal condensation or internal icing.	***Rain*** ***Dust*** ***External icing*** ***Rust resistance***
TYPE 3R enclosures are intended for outdoor use primarily to provide a degree of protection against falling rain; and to be undamaged by the formation of ice on the enclosure. They are not intended to provide protection against conditions such as dust, internal condensation, or internal icing.	***Rod entry*** ***Rain*** ***External icing*** ***Rust resistance***
TYPE 3S enclosures are intended for outdoor use primarily to provide a degree of protection against windblown dust, rain, and sleet, and to provide for operation of external mechanisms when ice laden. The are not intended to provide protection against conditions such as internal condensation or internal icing.	***Rain*** ***Dust*** ***External icing*** ***Rust resistance***
TYPE 4 enclosures are intended for indoor or outdoor use primarily to provide a degree of protection against windblown dust and rain, splashing water, and hose directed water; and to be undamaged by the formation of ice on the enclosure. They are not intended to provide protection against conditions such as internal condensation or internal icing.	***Hosedown*** ***External icing*** ***Rust resistance***
TYPE 4X enclosures are intended for indoor or outdoor use primarily to provide a degree of protection against corrosion, windblown dust and rain, splashing water, and hose-directed water; and to be undamaged by the formation of ice on the enclosure. They are not intended to provide protection against conditions such as internal condensation or internal icing.	***Hosedown*** ***External icing*** ***Corrosion resistance***
TYPE 6 enclosures are intended for indoor or outdoor use primarily to provide a degree of protection against the entry of water during temporary submersion at a limited depth; and to be undamaged by the formation of ice on the enclosure. They are not intended to provide protection against conditions such as internal condensation, internal icing, or corrosive environments.	***Submersion*** ***External icing*** ***Rust resistance***
TYPE 6P enclosures are intended for indoor or outdoor use primarily to provide a degree of protection against the entry of water during prolonged submersion at a limited depth; and to be undamaged by the formation of ice on the enclosure. They are not intended to provide protection against conditions such as internal condensation or internal icing.	***Air pressure*** ***External icing*** ***Corrosion resistance***
Type 12 enclosures are intended for indoor use primarily to provide a degree of protection against dust, falling dirt, and dripping noncorrosive liquids. They are not intended to provide protection against conditions such as internal condensation.	***Drip*** ***Dust*** ***Rust resistance***
Type 13 enclosures are intended for indoor use primarily to provide a degree of protection against dust, spraying water, oil, and noncorrosive coolant. They are not intended to provide protection against conditions such as internal condensation.	***Oil exclusion*** ***Rust resistance***

Source: Micro Switch, Freeport, Ill.

HAZARDOUS LOCATIONS

Type 7 enclosures are intended for indoor use in locations classified as Class I, Groups A, B, C, or D, as defined in the *National Electrical Code*. Type 7 enclosures shall be capable of withstanding the pressures resulting from an internal explosion of specified gases, and contain such an explosion sufficiently that an explosive gas-air mixture existing in the atmosphere surrounding the enclosure will not be ignited. Enclosed heat generating devices shall not cause external surfaces to reach temperatures capable of igniting explosive gas-air mixtures in the surrounding atmosphere. **Group A – Acetyline.** **Group B – Atmospheres containing hydrogen manufactured gas.** **Group C – Atmospheres containing diethyl ether, ethylene, or cyclopropane.** **Group D – Atmospheres containing gasoline, hexane, butane, naptha, propane, acetone, toluene, or isoprene.**	***Explosion*** ***Hydrostatic*** ***Temperature***
Type 9 enclosures are intended for indoor use in locations classified as Class II, Groups E, F or G, as defined in the *National Electrical Code*. Type 9 enclosures shall be capable of preventing the entrance of dust. Enclosed heat generating devices shall not cause external surfaces to reach temperatures capable of igniting or discoloring dust on the enclosure or igniting dust-air mixtures in the surrounding atmosphere. **Group E – Atmosphere containing metal dust.** **Group F – Atmospheres containing carbon black, coal dust or coke dust.** **Group G – Atmospheres containing flour, starch, or grain dust.**	***Dust penetration*** ***Temperature***

Source: Micro Switch, Freeport, Ill.

BIBLIOGRAPHY

Charkey, Edw.: *Electromechanical System Components,* Wiley-Interscience, New York, 1972.

Dreyfuss, Henry: *Symbol Sourcebook: An Authoritative Guide to International Graphic Symbols,* Van Nostrand Reinhold, New York, 1984.

Handbook of Photoelectric Sensing, Banner Engineering Co., Minneapolis, Minn., 1991.

Harper, Charles: *Handbook of Electronic Systems Design,* McGraw-Hill, New York, 1979.

Harper, Charles: *Electronic Packaging and Interconnection Handbook,* McGraw-Hill, New York, 1991.

Holm, R.: *Electric Contacts, Theory and Application,* 4th ed, Springer-Verlag, New York, 1967.

Hughes, Frederick W.: *Illustrated Guidebook to Electronic Devices and Circuits,* Prentice-Hall, Englewood Cliffs, N.J., 1983.

Lee, T.H.: *Physics and Engineering of High Power Switching Devices,* MIT Press, Cambridge, Mass., 1975.

Lockwood, J. P.: *Applying Precision Switches, A Practical Guide,* Micro Switch, Freeport, Ill., 1989.

McCleer, Patrick J.: *The Theory and Practice of Overcurrent Protection,* Mechanical Products Inc., Jackson, Mich., 1987.

Sanders, M. S., and E.J. McCormick: *Human Factors in Engineering and Design,* 6th ed, McGraw-Hill, New York, 1986.

Woodson, Wesley E.: *Handbook of Human Factors Engineering Data: Information and Guidelines for the Design of Systems, Facilities, Equipment, and Products for Human Use,* McGraw-Hill, New York, 1981.

Woodson, Wesley E., B. Tillman, and Peggy Tillman: *Human Factors Design Handbook,* 2d ed, McGraw-Hill, New York, 1991.

U.S. Government Publications

MIL-STD-1472D: *Human Engineering Design Criteria for Military Systems, Equipment and Facilities,* U.S. Department of Defense, Washington, D.C., 1989.

INDEX